Fundamental Problems of Gauge Field Theory

NATO ASI Series

Advanced Science Institutes Series

A series presenting the results of activities sponsored by the NATO Science Committee, which aims at the dissemination of advanced scientific and technological knowledge, with a view to strengthening links between scientific communities.

The series is published by an international board of publishers in conjunction with the NATO Scientific Affairs Division

A	**Life Sciences**	Plenum Publishing Corporation
B	**Physics**	New York and London
C	**Mathematical and Physical Sciences**	D. Reidel Publishing Company Dordrecht, Boston, and Lancaster
D	**Behavioral and Social Sciences**	Martinus Nijhoff Publishers
E	**Engineering and Materials Sciences**	The Hague, Boston, and Lancaster
F	**Computer and Systems Sciences**	Springer-Verlag
G	**Ecological Sciences**	Berlin, Heidelberg, New York, and Tokyo

Recent Volumes in this Series

Volume 134—Fundamental Processes in Atomic Collision Physics
edited by H. Kleinpoppen, J. S. Briggs, and H. O. Lutz

Volume 135—Frontiers of Nonequilibrium Statistical Physics
edited by Garald T. Moore and Marlan O. Scully

Volume 136—Hydrogen in Disordered and Amorphous Solids
edited by Gust Bambakidis and Robert C. Bowman, Jr.

Volume 137—Examining the Submicron World
edited by Ralph Feder, J. Wm. McGowan, and Douglas M. Shinozaki

Volume 138—Topological Properties and Global Structure of Space-Time
edited by Peter G. Bergmann and Venzo De Sabbata

Volume 139—New Vistas in Nuclear Dynamics
edited by P. J. Brussaard and J. H. Koch

Volume 140—Lattice Gauge Theory: A Challenge in Large-Scale Computing
edited by B. Bunk, K. H. Mütter, and K. Schilling

Volume 141—Fundamental Problems of Gauge Field Theory
edited by G. Velo and A. S. Wightman

Series B: Physics

Fundamental Problems of Gauge Field Theory

Edited by

G. Velo

University of Bologna
Bologna, Italy

and

A. S. Wightman

Joseph Henry Laboratories
Princeton, New Jersey

Springer Science+Business Media, LLC

Proceedings of the sixth course of the International School of Mathematical Physics on Fundamental Problems of Gauge Field Theory, held July 1-14, 1985, in Erice, Sicily, Italy

Library of Congress Cataloging in Publication Data

International School of Mathematical Physics (6th: 1985: Ettore Majorana International Centre for Scientific Culture)
Fundamental problems of gauge field theory.

(NATO ASI series. Series B, Physics; v. 141)
"Proceedings of the Sixth International School of Mathematical Physics and NATO Advanced Study Institute on Fundamental Problems of Gauge Field Theory, held July 1-14, 1985, at the Ettore Majorana Center for Scientific Culture, Erice, Sicily, Italy."
"Published in cooperation with NATO Scientific Affairs Division."
Bibliography: p.
Includes index.
1. Gauge fields (Physics)—Congresses. I. Velo, G. II. Wightman, A. S. III. NATO Advanced Study Institute on Fundamental Problems of Gauge Field Theory (1985: Ettore Majorana International Centre for Scientific Culture) IV. Title. V. Series.
QC793.3.F5I576 1985 530.1′43 86-25331

ISBN 978-1-4757-0365-8 ISBN 978-1-4757-0363-4 (eBook)
DOI 10.1007/978-1-4757-0363-4

Originally published by Plenum Press, New York in 1986
Softcover reprint of the hardcover 1st edition 1986

PREFACE

The sixth Ettore Majorana International School of Mathematical Physics was held at the Centro della Cultura Scientifica Erice, Sicily, 1-14 July 1985. The present volume collects lecture notes on the session which was devoted to Fundamental Problems of Gauge Field Theory. The School was a NATO Advanced Study Institute sponsored by the Italian Ministry of Public Education, the Italian Ministry of Scientific and Technological Research and the Regional Sicilian Government.

As a result of the experimental and theoretical developments of the last two decades, gauge field theory, in one form or another, now provides the standard language for the description of Nature; QCD and the standard model of the electroweak interactions illustrate this point. It is a basic task of mathematical physics to provide a solid foundation for these developments by putting the theory in a physically transparent and mathematically rigorous form.

The lectures and seminars of the school concentrated on the many unsolved problems which arise here, and on the general ideas and methods which have been proposed for their solution. In particular, we mention the use of rigorous renormalization group methods to obtain control over the continuum limit of lattice gauge field theories, the exploration of the extraordinary enigmatic connections between Kac-Moody-Virasoro algebras and string theory, and the systematic use of the theory of local algebras and indefinite metric spaces to classify the charged C* states in gauge field theories.

In addition, there was the stochastic realization of quantum field theory which provides not only an alternative framework but a possible practical numerical approach to the solution of gauge theories. Finally, we had the remarkable application of topological and differential geometric tools to the analysis of anomalies.

We hope that the lecture notes will provide both a summary of current research in this field and an introduction for those who are just entering it.

G. Velo
A. S. Wightman

CONTENTS

FUNDAMENTAL PROBLEMS OF GAUGE FIELD THEORY:

INTRODUCTION TO THE PROBLEMS

Arthur S. Wightman

Physics Department
Princeton University
Princeton, NJ 08540

Preliminary Remarks

We have now had nearly a decade and a half of successful applications of gauge field theory to electroweak and strong interactions. The pioneering experiments, and the more systematic experiments that have followed, have given rough agreement for QCD and rather precise agreement for the standard model of the electroweak interactions. There is a general consensus that the new theories represent important progress. However, the calculations to be compared with experiment are at the moment still rather tentative in places, and there are many questions of principle the answers to which have not yet been given in an unambiguous way.

While all this was going on, constructive quantum field theory had its first real victories, the construction of the solutions of the ϕ_3^4 theory in the mid 1970's being the most striking. Clearly an attack on four-dimensional gauge theories was in order, in the hope that eventually a systematic constructive development of the theory would provide a solid basis for QCD, the theory of electroweak interactions and grand unified theories. Progress did not come immediately; the methods used to make constructions of such theories as ϕ_3^4 seemed to have reached their limits. Some new ideas had to be added, and - in fact - in the last few years they have been. We are now in the middle of another period of explosive development of constructive quantum field theory, and in this school you will be immersed in a bath of rigorous renormalization groups, strings, geometric anomalies, etc., ideas the promise of which has given rise to an atmosphere of optimism.

The purpose of the present introduction is to describe some of the main questions which it is hoped that the theory will eventually answer, if it has not already, thereby to provide some perspective on the whole enterprise. Of course, for reasons of space, I have had to make a selection from a rich collection of questions deserving discussion.

The Importance of the Higgs and Confinement Mechanism

The standard model of the electroweak interactions represents enormous progress over the preceding [QCD + Fermi Interaction] theory in at least two respects. First, it unifies the theories of the electromagnetic and weak interactions. Secondly, it provides a whole new range of predictions since the theory is renormalizable. The renormalizability implies that the radiative corrections are finite, and determined in terms of a finite number of coupling constants.

The observation of W^+, the W^-, and the Z° with masses in good agreement with predictions is accounted for by a theory in which these particles are described by gauge field A, and a gauge invariant dynamics. Thus, the field action contains no mass term m^2A^2; such a term would not be gauge invariant. From the point of view of the 1950's, such a field A would be regarded as massless. It would have been, and - in fact - was, regarded as a problem how to make the intermediate bosons A described into massive particles. I will outline the complicated trajectory that led to our present understanding shortly. The point I want to make here is that whatever may happen to electroweak theory in the future, it must incorporate the two fundamental features of the standard model: it must yield finite radiative corrections and a mass for the intermediate vector bosons. These are indispensible attributes of the so-called Higgs mechanism.

For the strong interactions and QCD, the experimental situation is almost equally clear (provided we ignore the Stanford quarks). The quarks, and gluons, must be there but must be confined. That raises the second big problem: how to describe confined particles and to show that confinement actually happens with an appropriate Yang-Mills theory.

For the grand unified theories of strong and electromagnetic interactions the same questions arise. Although there is at the moment no standard model GUT, if such a theory is found it must predict massive intermediate vector bosons and confined quarks and gluons.

If constructive field theory is going to be of any significance in providing a foundation for particle physics it must provide mathematically rigorous and physically satisfying solutions to the problem of the Higgs mechanism and to the problem of confinement of gluons and quarks.

The Tangled History of the Higgs Mechanism

At the end of the 1950's there were many attempts to use gauge fields to describe various aspects of strong and weak interactions. For example, Lee and Yang [1] discussed the possibility of describing baryon conservation by analogy with electric charge conservation. The idea was that there should be a field analogous to the electromagnetic field for which the baryon current would be the source. Unfortunately, this model gave rise to long-range forces between baryons, analogous to Coulomb forces between electric charges, but not seen in experiments. On the other hand, Sakurai [2] proposed to use non-Abelian gauge fields to describe mesons giving rise to the strong interaction. At first, such theories also seemed problematical, because the short range of nuclear forces appeared to require a massive Yang-Mills field, and the theory of such a field seemed non-renormalizable.

An important clarifying remark was made by Schwinger [3]. He pointed out that the theory of a gauge field might yield massive vector mesons for dynamical reasons even if the inertial Lagrangian contains no mass term. He also displayed a simple model (massless fermions coupled to a massless vector field in two-dimensional space-time, now known as the

Schwinger model) in which this very phenomenon occurs. Then Anderson pointed out that analogous phenomena occur in condensed matter physics: a plasmon is an excitation of a gas of charged particles which behaves like a particle of mass greater than zero; when combined with transverse modes, it is analogous to the massive vector meson [4]. Finally, Higgs [5], Guralnik, Hagen and Kibble [6], and Brout and Englert [7] showed that in what is now known as the Higgs model (a charged scalar field, ϕ , coupled to the electromagnetic field but with the usual term $m^2/2\, \phi^*\phi$ in the Lagrangian replaced by $-m^2/2\, \phi^*\phi$); the assumption that the vacuum expectation value, $\langle\phi\rangle_0$, of ϕ is non-vanishing implies, in a semi-classical approximation, that the vector field indeed develops a mass. Furthermore, the two degrees of freedom associated with the charged field reduce to one (uncharged) degree of freedom, the extra degree of freedom providing the longitudinal polarization of the vector meson. Of course, when $\langle\phi\rangle_0 \neq 0$, the $U(1)$ global gauge invariance under $\phi \mapsto e^{i\theta}\phi$ is broken and the local gauge invariance $\phi(x) \mapsto e^{i\theta(x)}\phi(x)$, $A_\mu(x) \mapsto A_\mu(x) - \partial_\mu\theta$ as well.

This acount of the behavior of the Higgs model and its generalizations to analogous situations in which a massless gauge field coupled to some matter field acquires a mass has become conventional wisdom in gauge field theory [8]. It is this which is meant by the phrase "the Higgs mechanism".

The discovery of the Higgs mechanism opened the way for the general application of Yang-Mills theory to problems of the strong and weak interactions. An indispensible technical development was the proof that the solutions for which gauge symmetry is broken are renormalizable [9].

Meanwhile, there were developments in statistical mechanics which indicated that the simple picture of the Higgs mechanicsm just described is not correct, or at least requires some alteration. Here, the paper of F. Wegner Duality in Generalized Ising Models and Phase Transitions Without Order Parameter, Jour. Math. Phys. 12 (1971) 2259-2272, deserves mention. Wegner started from the fact, established by Merwin and Wagner [10], that the Heisenberg model of interacting spins in two space dimensions has vanishing spontaneous magnetization: $\langle\vec{\sigma}_j\rangle = 0$. (Here $\vec{\sigma}_j$ is the spin at the lattice site labeled by j. The expectation value $\langle\vec{\sigma}_j\rangle$ is analogous to $\langle\phi(x)\rangle_0$ in the Higgs model). For the Heisenberg model in three space dimensions $\langle\vec{\sigma}_j\rangle$ is non-vanishing in the ferromagnetic phases for temperatures below the Curie point and vanishing above, i.e. $\langle\vec{\sigma}_j\rangle$ is an order parameter for the phase transition. It is called a local order parameter because it only depends on spins at a finite number of sites. Wegner made a systematic investigation of Ising models (in which the spins take the values ± 1) and showed that there are models in two and three dimensions with a phase transition but no local order parameter. The models are what nowadays would be called $\mathbf{Z}_2$ gauge theories. Incidentally, the emphasis in the paper is on establshing that the models have a phase transition; the absence of a local order parameter is proven but seems to be regarded as rather clear.

It may have been clear to the statistical mechanics, but it was not to the particle theory people, seduced as they were by the conventional wisdom on the Higgs mechanism. It was only in 1975 tht Elitzur examined a lattice gauge field model which is a simplified lattice version of the Higgs model and showed that a natural looking order parameter vanishes [11]. More specifically, Elitzur assumed a theory in which there are variables ϕ_j, $0 \leq \phi_j < 2\pi$ with j running over the sites of a lattice, and variables A_b, $0 \leq A_b < 2\pi$ with b running over the bonds of the lattice. The action was assumed to be, for a finite chunk Λ of the lattice,

$$S_\Lambda = \kappa \sum_{\substack{j,k \\ b=\{jk\} \\ b\in\Lambda}} \cos(\phi_j - \phi_k - A_b) + \frac{1}{g^2} \sum_{P\subseteq\Lambda} \cos\Big(\sum_{b\in\partial P} A_b\Big)$$

where P runs over all plaquettes which touch the chunk Λ of the lattice and ∂P is the oriented boundary of the plaquette P. Then in the standard manner of lattice gauge theory, following Wilson, one makes a probability measure

$$d\mu_\Lambda = \frac{1}{Z_\Lambda} e^{-S_\Lambda} \prod_{i\in\Lambda} d\phi_i \prod_{b\in\Lambda} dA_b$$

$$Z_\Lambda = \int e^{-S_\Lambda} \prod_{i\in\Lambda} d\phi_i \prod_{b\in\Lambda} dA_b$$

If j stays away from the boundary of Λ the measure $d\mu_\Lambda$ is invariant under the local gauge transformation

$$\phi_j \mapsto \phi_j + c_j$$

$$A_b \mapsto A_b + c_j - c_k \quad \text{where} \quad b = \{jk\}$$

Elitzur proved $\langle \cos A_b \rangle = 0$ in the thermodynamic limit in which $\Lambda \to$ lattice.

Elitzur left open the question of the effect of boundary conditions on his argument. He did, however, prove explicitly that with the inclusion of a coupling to an external magnetic field, B, the local order parameters are continuous in B at B = 0. The effects of boundary conditions were investigated by de Angelis, de Falco, and Guerra [12] as part of a program in which they carefully constructed the thermodynamic limits of lattice versions of the X-Y model, scalar electrodynamics and the Abelian Higgs model. The conclusion was the same: no local order parameter. In fact, they showed that all the traditional gauge dependent Green's functions, for example $\langle \phi(x)\phi^*(y)\rangle$, vanish identically for $x \neq y$.

Looked at globally, these results are a simple consequence of the local gauge invariance of the measure in the thermodynamic limit. Explicitly, if g is a local gauge transformation and $d\mu_\infty$ is the measure on the configurations, ϕ, of the system assuming some fixed boundary conditions, then what is proven is

$$d\mu_\infty^g(\phi) = d\mu_\infty(g\phi) = d\mu_\infty(\phi)$$

This can be used to establish, for example,

$$\langle \phi(x)\rangle_0 = \langle e^{i\theta(x)}\phi(x)\rangle_0$$

Then averaging over $\theta(x)$ we get $\langle\phi(x)\rangle_0 = 0$. One can also see this from the DLR equations which are left invariant by local gauge transformations. The crucial fact is that, in the thermodynamic limit, boundary conditions cannot favor one choice of local gauge over another [14].

There was no great fuss made in particle physics over the results. Local order parameters simply stopped appearing in leading articles on the subject, although justifiably, Higgs phases, in which vector mesons acquire mass, did not [13].

However, there was one last point not covered by the above described discussion: the influence of local gauge fixing terms. Such terms make the measure, $d\mu$, non-invariant under local gauge transformation by definition and the question arises whether they could alter the above conclusions about local order parameters (that was the conjecture of Elitzur [11]). Fröhlich, Morchio, and Strocchi examined this question and were able to show that the lattice version of

$$\langle \phi(x) \exp i \int_x^y A_\mu(\xi) d\xi^\mu \phi^*(y) \rangle_0 \longrightarrow \text{ as } |x-y| \to \infty, \quad (1)$$

with exponentially fast approach to the limit [14]. They then noted that this implies, in the temporal gauge,

$$\langle \phi(x) \rangle_0 = 0$$

Thus, at least in this gauge, the local order parameter proposed by the conventional wisdom does not work.

What is then to be done? Fröhlich, Morchio and Strocchi and, independently, Banks and Rabinovici [13] and 't Hooft [15] proposed that all the physics of these models be described in terms of gauge independent fields. Fröhlich, Morchio and Strocchi proved that a complete set of such gauge invariant fields exists and outlined how to express interesting quantities such as the mass of the vector meson in terms of them. Thus, it is not necessary to use gauge dependent local order parameters in order to describe the physical predictions for a state which is a Higgs phase.

Of course, all this leaves quite open whether the Higgs model even in its lattice version actually has a Higgs phase in which there is a massive vector meson and a massive spin zero meson, both of charge zero. Here there has been some recent progress.

1. C. King has shown the existence of the continuum limit for the solutions of the Higgs model in three space-time dimensions [16]. The construction of the continuum limit of the Higgs model in two space-time dimensions was already completed by Brydges, Fröhlich and Seiler [17]. Relatively little information is yet available on the properties of these continuum limit solutions. However, for the lattice Higgs models, a much more detailed description exists.

2. Balaban, Brydges, Imbrie and Jaffe have shown that for a suitable range of parameters in the lattice models, there is a Higgs phase in which the vector meson is massive [19]. That there is a phase transition in this model from this Higgs phase in which fractionally charged infinitely massive quarks are not confined to a QED phase in which they are confined and the vector mesons are massless (= photons) was already stated in [20]; the proof is given in Seiler's book [18] (see also [21]). The latter proof has the advantage that it shows the existence of the QED phase both for bosonic matter (Higgs model) and for fermionic matter.

3. T. Kennedy and C. King [22] have introduced an order parameter which is the lower bound of the Euclidean lattice version of

$$\langle \phi(x) \exp i \int_x^y \partial_\mu \left(\frac{1}{\Delta}\right) \partial^\nu A_\nu(\xi) d\xi^\mu \phi^*(y) \rangle_0 \tag{2}$$

(this replaces the exponential factor in (1) by the exponential of the integral of the longitudinal part of A_μ. Notice that this is a non-local affair because of the inverse Laplacean). It should be emphasized that the Kennedy-King order parameter is defined only for Abelian gauge fields and for the non-compact form of the field action. Kennedy and King show that for a suitable range of parameters this quantity has a strictly positive lower bound, while for another range of parameters it is zero. Thus, the vector meson mass and the Kennedy-King lower bound give two different order parameters for the Higgs to QED phase transition. It would be nice to show that they agree and to have a detailed phase diagram on the lattice as well as in the continuum limit.

The description of the solutions of the Euclidean Higgs model in terms of gauge invariant quantities is of primary physical importance but it leaves the status of the conventional description of the theory in terms of local fields in Minkowski space up in the air.

There are numerous questions to be answered here before we can be sure we have really understood the Higgs mechanism.

a) Is there a neat description of the different phases in terms of the Minkowski space fields ϕ, ϕ^* and A_μ?

b) What is the status of the charge current operator

$$j^\mu(x) = ei[\phi^*(x)(\partial^\mu + ei A^\mu)\phi(x) - (\partial^\mu - ei A^\mu)\phi^*(x)\phi(x)]$$

and the charge

$$Q = \int j^0(x) d^3x$$

in the different phases. In a confined phase does j^μ exist as an operator valued distribution, but with $Q = 0$? Or does Q define a quadratic form which is not an operator having the vacuum in its domain?

You will hear an interesting point of view on these questions in Strocchi's lectures based on work by Morchi and Strocchi.

c) The argument of Fröhlich, Morchio and Strocchi shows that

$$\langle \phi(x) \phi^*(y) \rangle_0 \to 0 \qquad \text{as} \qquad |x-y| \to \infty$$

in the temporal gauge while that of Kennedy and King shows

$$\langle \phi(x) \phi^*(y) \rangle_0 \not\to 0 \qquad \text{as} \qquad |x-y| \to \infty$$

in the Landau gauge. Could it be that, in the temporal gauge, the constructed solution is not a pure phase, but that the pure phases into which it decomposes have local order parameters? Does the Landau gauge solution belong to a one parameter family of solutions which are carried into one another by $\mathcal{U}(1)$ (global) gauge transformations?

d) Is the Higgs mechanism always accompanied by the absence of charged particles: This question has been answered with 'Yes' by Buchholz and Fredenhagen for any theory in which all particles are massive [28], but the general question appears to be open.

Some Questions about Confinement

I will not attempt to trace the history of ideas about confinement. It does not involve the kind of blatant paradoxes that I have just described for the Higgs mechanism. Furthermore, it has been discussed many times before; see, for example, Seiler's book [18]. Instead, I will make some miscellaneous remarks and raise some questions.

From the outset of modern gauge field theory there have been two general points of view on the desirable way to develop the subject. One puts the emphasis on gauge invariant expression (Wilson loops, etc.) and tries to put the theory in a form in which everything significant is expressed in an explicitly gauge invariant fashion; see, for example, pp. 163-181 of Seiler's book [18], and also [21]. The other works in terms of the full panoply of gauge dependent notions (gauge potentials, Faddeev-Popov ghosts, indefinite metric, etc.). The goal in this latter approach is to display the theory as a special kind of (indefinite metric) local field theory. A priori, these two points of view might well be consistent; they might simply lead to different descriptions of the same theory. It seems reasonable to assume such consistency until it is proven otherwise. In practice, the two points of view complement each other. The former makes important contact with statistical mechanics while renormalization theory is more easily worked out in the latter. Although there is considerable difference of opinion as to the fruitfulness of the two approaches, I hope we can avoid an analogue of the ludicrous spectacle of the 1960's in which the S-matrix was looked at from two points of view, analytic S-matrix theory and field theory, and it was argued by some analytic S-matrix theorists that to be ignorant of field theory is a virtue.

There are, of course, fundamental problems common to both approaches. In each one can attempt to construct theories as limits of lattice approximations with various lattice actions. Do these all produce the same continuum limit? The answer is not clear, but there are clear indications that this is not a trivial problem. For example, in QED_3 there are arguments that the use of the compact (Wilson) form of the lattice action yields a continuum limit theory in which there are magnetic monopoles, and confinement holds for all non-zero values of the electric charge [25]. This is in contrast to the continuum limit which is the free quantized electromagnetic field [24]. As of the moment, we are lucky to have proofs that a continuum limit exists with any choice of lattice action, not to speak of a proof of independence of the form of the action. In fact, the only relativistic theories I know about, for which existence has been established (apart from the super-renormalizable models considered a decade ago), are the free electromagnetic field in three dimensions [26], $Higgs_{2,3}$ [17], [16] and $G(ross)\text{-}N(eveu)_2$ [27], [28], and the last was treated directly without a passage via the lattice.

Now let me turn to confinement. How does it appear in the two approaches? In the former, one looks for gauge invariant order parameters. Of the many which have been proposed, let me mention two: the Wilson loop and the Fredenhagen-Marcu parameter. The Wilson loop $\langle W(C)\rangle$ is the expectation of a path-ordered product around a closed curve C and Wilson's original criterion for confinement was a bound

$$\langle W(C)\rangle \leq \text{const} \exp -\alpha A(C)$$

where α is some positive constant and A(C) is the area of any surface enclosed by C. For non-confinement on the contrary one only expects

$$W(C) \geqslant \text{const } \exp -\alpha^1 P(C)$$

where P(C) is the length (= perimeter) of C [18]. This criterion works well for lattice gauge fields not coupled to matter and there is a large body of rigorous results based on it; this is the confinement of infinitely heavy quarks. It was shown to hold generally for strong coupling by Osterwalder and Seiler [29]. On the other hand, in the presence of matter, pair production alters the situation and another criterion has to be sought. Fredenhagen and Marcu apply the idea that one can make a charged state by making a pair of oppositely charged particles and then letting one run away to infinity [30]. This is somewhat delicate since one must not let the energy of the state run away to infinity. Furthermore, one has to be sure that the resulting state is really something new; it should be orthogonal to the vacuum. Fredenhagen and Marcu propose the criterion for the existence of charged states which in picturesque terms is

$$\rho = \lim_{|x-y| \to \infty} \frac{\left| \sum_i \langle \, {}_{x,i} \boxed{} {}_{y,i} \rangle \right|^2}{\langle \, {}_x \boxed{} {}_y \, \rangle}$$

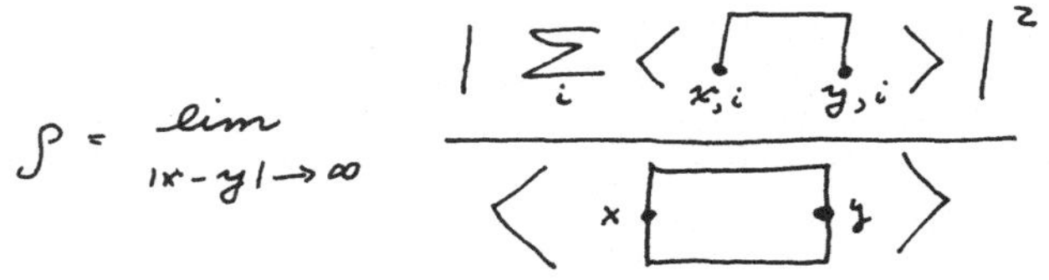

(The main trick is to make the vertical dimension in the paths grow proportional to the horizontal).

$\rho = 0$ means charged states = no confinement

$\rho \neq 0$ means no charged states = confinement.

Fredenhagen and Marcu have shown that this criterion when combined with convergent expansions really works to distinguish two phases in the Z_2 lattice gauge theory which Wegner studied in 1971 [28].

I have cited these results to give the flavor of the subject. Of course, there is much more to the story, e.g. Mack and Meyer [32], Bricmont and Fröhlich [33], and we surely have not seen the last of the order parameters relevant to this kind of analysis. There is plenty of opportunity here for further physical cleverness and mathematical analysis.

I now turn to attempts to treat the same phenomena within the Gupta-Bleuler formalism. Here in the absence of concrete examples of solutions of quantized gauge field theories, opportunities for irrelevant confusions abound. It is not really clear a priori what one should assume about the non-physical states of a theory in order that the physical states it predicts satisfy the standard conditions. F. Strocchi has probably carried the program furthest [31]. See also [35].

To be more specific, let me recall that in field theory in indefinite metric spaces, one has a triplet of vector spaces $\mathcal{H}, \mathcal{H}', \mathcal{H}''$ with $\mathcal{H} \supseteq \mathcal{H}' \supseteq \mathcal{H}''$, and a form $\langle \cdot , \cdot \rangle$ on $\mathcal{H}$, which is positive on $\mathcal{H}'$:

$$\langle \Phi, \Phi \rangle \geqslant 0 \quad \text{for} \quad \Phi \in \mathcal{H}'$$

$\mathcal{H}''$ is defined as the subspace of $\mathcal{H}'$ consisting of all vectors Φ for which $\langle\Phi,\Phi\rangle = 0$. The physical subspace is defined as the closure $\mathcal{H}_{PHYS} = \overline{\mathcal{H}'/\mathcal{H}''}$ of the quotient space $\mathcal{H}'/\mathcal{H}''$, closure being understood in the metric defined on $\mathcal{H}'/\mathcal{H}''$ by $\|\Phi\| = \sqrt{\langle\Phi,\Phi\rangle}$. All this can be done without defining a topology in $\mathcal{H}$, but, for the later development, it is essential to have some notion of continuity in $\mathcal{H}$ consistent with its linear structure and such that $\langle\cdot,\cdot\rangle$ is continuous. Usually, it is assumed that the topology arises from a Hilbert space structure (under some fairly general hypotheses that is no loss of generality). However, the scalar product $(\cdot,\cdot)$ in $\mathcal{H}$ that defines the auxiliary Hilbert space structure has no physical meaning apart from its rôle in the definition of continuity. All this comes to a head when one undertakes to prove a reconstruction theorem for a field theory in indefinite metric. Then one is given the expectation values

$$\langle \Psi_0, \prod_i \phi_i(x_i)\Psi_0\rangle$$

in the $\langle\cdot,\cdot\rangle$ inner product of products of the fields, and wants to construct a triplet $\mathcal{H}, \mathcal{H}', \mathcal{H}''$ a vacuum state, Ψ_0, and field operators ϕ_i with these expectation values and <u>some</u> definition of continuity. One can certainly not expect $(\cdot,\cdot)$ to be determined uniquely up to unitary equivalence. That is not a serious physical objection, but it does give rise to technical problems.

The space $\mathcal{H}$ is introduced to give elbow room for gauge potentials and other non-gauge invariant objects which it is impossible to squeeze into $\mathcal{H}_{PHYS}$, but we have the problem of how the basic properties are to be extended. For example, one usually assumes that relativistic invariance is guaranteed by the existence of a continuous unitary representation of the universal covering group of the Poincaré group $\{a,A\} \mapsto U(a,A)$, where a is a space-time translation and $A \in SL(2,C)$. The group representation property is

$$U(a,A)\,U(b,B) = U(a+\Lambda(A)b, AB)$$

where $\Lambda(A)$ is the element of the restricted Lorentz group determined by A. Here it is natural to assume instead of unitarity

$$\langle U(a,A)\Phi, U(a,A)\Psi\rangle = \langle\Phi,\Psi\rangle$$

This does not imply that the $U(a,A)$ are unitary in terms of $(\cdot,\cdot)$ or even bounded. In fact, in the representation appearing in the Gupta-Bleuler gauge for the free electromagnetic field $U(0,A)$ is unbounded when $\Lambda(A)$ contains a boost. When it comes to the spectral condition for $U(a,A)$ we are really in the dark. Strocchi has proposed that for a large family of Φ,Ψ in $\mathcal{H}$, $\langle\Phi, U(a,1)\Psi\rangle$ have a Fourier transform in the variable, a, satisfying the spectral condition [31]. This hypothesis is supported by perturbation theory calculations of concrete gauge theory models, but one should not lose sight of the fact that perturbation theory can be misleading when it comes to support properties. For example, in the calculation of the retarded fundamental solution of a relativistic wave equation of spin 3/2 in an external field, perturbation theory gives a (divergent) sum of terms each with support in or on the light cone, even though the exact solution has support which extends beyond the light cone. However, Strocchi's spectral hypothesis works well in the reconstruction theorem, which is a non-perturbation affair. Nevertheless, it would be very reassuring to see how this all turns out in some non-trivial examples, especially in confined phases. $Higgs_2$ and $Higgs_3$ come to mind as good candidates.

Morchio and Strocchi have introduced a further distinction concerning the matrix elements $\langle\Phi, U(a,1)\Psi\rangle$. Suppose Φ and Ψ are of the form $\mathcal{P}(\phi)\Psi_0$ where $\mathcal{P}$ is a polynomial in the fields smeared with test functions of compact support. Then a theory is defined to have non-confining infra-red singularities if the Fourier transform of all $\langle\Phi, U(a,1)\Psi\rangle$ are measures in some neighborhood of the mass shell $p^2 = 0$; the theory has confining infra-red singularities if in some neighborhood of $p^2 = 0$, the Fourier transform of some matrix element is a distribution of higher order than a measure. The reason for this terminology can be understood from the behavior of the two-point function of fields in the two cases. It is a general fact that a two-point function in a positive metric theory must fall off at least as fast as a Coulomb field in space-like directions, because such a theory always has non-confining infra-red singularities [32]. On the other hand, if a two-point function has appropriate confining infra-red singularities, it will grow linearly in space-like directions thus verifying one of the primitive conceptions about a confining potential. Just as in the case of the Higgs mechanism it would be very reassuring to see these ideas on infra-red singularities verified in models.

When it comes to confinement in non-Abelian gauge theories, specific proposals for the characterization of the physical subspace in the Gupta-Bleuler formalism have been put forward by Curci and Ferrari [36] and by Kugo and Ojima [37]. They argue that, in non-Abelian gauge theory in local gauges, one has to introduce Faddeev-Popov ghost fields and the invariance group of the theory is promoted: from gauge transformations one passes to BRS transformations. There is associated with the BRS transformations a BRS charge, Q_{BRS}. They propose to characterize vectors, Φ, of the subspace, $\mathcal{H}'$, by

$$Q_{BRS} \quad = 0.$$

This beautifully simple characterization needs further study, especially in view of the fact that the general arguments of Kugo and Ojima seem to depend on the use of the asymptotic condition in $\mathcal{H}$. The validity of the asymptotic condition in $\mathcal{H}$, especially in a confining phase, is open to question.

If, as everyone hopes and expects, (and as we will hear from Jaffe and Balaban), the several attempts to construct gauge field theory via rigorous renormalization group arguments are about to succeed, we may have reached a point where the kind of questions raised above are answerable on non-trivial examples.

THE CLASSIFICATION OF STATES IN QUANTUM THEORY USING THE THEORY OF LOCAL ALGEBRAS

The origins of the theory of local algebras of observables are in the work of Haag and Araki at the end of the 1950's and Haag and Kastler in the early 1960's [38], [39]. The theory reflects the deep physical intuition that a large fraction of what can be measured in the laboratory can be described in purely geometrical terms: an event either did or did not take place in a given region of space and in a given time interval. The first great success of the approach was a theory of super-selection rules, but it was little noticed by practising particle theory people, who rediscovered bits of it a decade later. However, the conceptual machinery has been gradually seeping into statistical mechanics and elementary particle theory.

Progress in the theory of local algebras has been slow for at least two reasons:

1) The questions to which the theory naturally leads are quite general and require a powerful general analysis.

2) The language in which the theory is developed, the theory of C^* algebras and von Neumann algebras, is very technical.

There is an irony in 2): the theory has developed a justified reputation for mathematical density but, in fact, the assumptions are very physical and it is only the hard work necessary to reach physical conclusions that is very mathematical.

I want to give an example to show the kind of sweeping generality and power that the theory offers. It is chosen from the work of D. Buchholz [40].

Here is the physical intuition. The super-selection rules of a theory ought to be labeled, in part, by charges e.g. electric charge, baryonic charge. Such charges should be carried by massive particles whose orbits eventually enter any future light cone. Thus, measurements in any future light cone should determine all the charges of a theory.

Here is how the theory describes this situation. Everything is expressed in terms of C^* algebras $\mathfrak{A}(\mathcal{O})$ associated with bounded open sets of space-time. They are all regarded as subalgebras of the <u>quasi-local</u> algebra $\mathfrak{A}$ which is the closure in the operator norm of $\cup_{\mathcal{O}} \mathfrak{A}(\mathcal{O})$. A <u>state</u> ρ on $\mathfrak{A}$ is a positive linear functional normalized to be 1 on the unit element of $\mathfrak{A}$. Such a state generates a cyclic representation, π_ρ, of $\mathfrak{A}$ in a Hilbert space $\mathcal{H}_\rho$. It is supposed that the space-time translation is represented in $\mathcal{H}_\rho$ by a continuous unitary representation satisfying the spectral condition: $a \mapsto U(a)$.

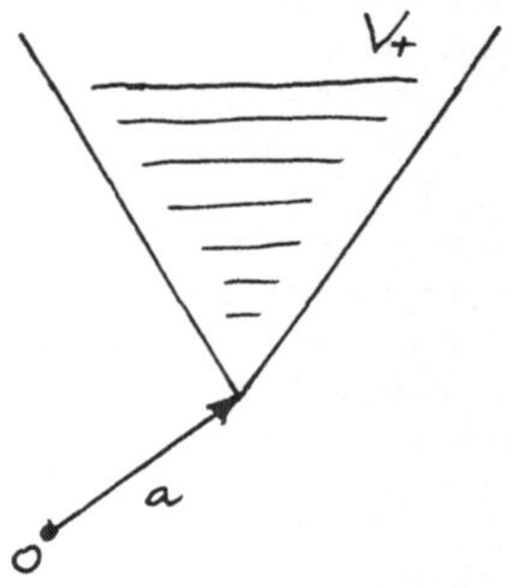

The future cone $V_+ + a$

Consider the region $a + V_+$, i.e. the future light cone with vertex at a. Then $\pi_\rho(\mathfrak{A}(a+V_+))$ is the algebra generated by the observables which can be measured in $a + V_+$ in the representation π_ρ. The center of the quasi-local algebra is defined

$$\mathfrak{Z} = \pi_\rho(\mathfrak{A})'' \cap \pi_\rho(\mathfrak{A})'$$

where the prime stands for the commutant. The center of $\pi_\rho(\mathfrak{A}(a+V_+))$ is

$$\mathfrak{Z}_+ = \pi_\rho(\mathfrak{A}(a+V_+))'' \cap \pi_\rho(\mathfrak{A}(a+V_+))'$$

The operators in $\mathfrak{Z}$ commute with all observables in $\pi_\rho(\mathfrak{A})$ and therefore define super-selection rules. The operators in $\mathfrak{Z}_+$ only have to commute with all the observables which can be measured in $a + V_+$. However, Buchholz proves the following lemma.

<u>Lemma</u>

The elements of $\mathfrak{Z}_+$ commute with translations

$$U(a)\, Z\, U(a)^{-1} = Z \quad \text{for} \quad Z \in \mathfrak{Z}_+$$

and

$$\mathfrak{Z}_+ \subseteq \mathfrak{Z} \, .$$

Thus, the commutative algebra $\mathfrak{Z}_+$ is independent of the choice of a and its elements define super-selection rules. We can diagonalize all the operators of $\mathfrak{Z}_+$ simultaneously and consider the subrepresentation of π_ρ which has all the corresponding charge quantum numbers fixed. This Lemma gives a precise general geometric realization of the above-described intuition.

As you will hear from Fredenhagen, the next stage in theory is to classify the states in the theory using as criterion of equivalence unitary equivalence to the vacuum representation on the space-like complement of space-like cones (these space-like cones can be regarded as fattened strings running to infinity, so strings work their way into the theory of local algebras also). The hope is to refine the theory sufficiently that its classification of charges makes contact with the classification arising from gauge field theory itself.

As a last remark, on local algebras I note that the theory of local algebras has led, in the context of ordinary electrodynamics, to an improved understanding of the scattering theory of charged particles. I want to call your attention to the fact, if you do not know it already, that this scattering theory is very different from that for particles coupled to massive fields. In particular, Lorentz invariance is broken in the charged vectors; the boosts are not unitarily implemented [14]. The text books on the scattering theory of infra-particles are going to have to be rewritten and the theory of local algebras is in the process of explaining exactly why.

CONCLUSION

Now I find that in this somewhat lengthy introduction, I have managed to say nothing about the two subjects most talked about by particle physicists during this last year: strings and anomalies. I also have managed to say nothing about new applications of the conformal group, which seem to put the subject of exactly soluble 2-dimensional models in a whole new light, nor about stochastic quantization which is not only interesting but possibly practically important in calculations.

So be it. Take it as an indication that our subject so flourishes, that it is impossible to comment on all the relevant interesting developments in an introductory lecture. In any case, you will hear about these things from Neveu, Alvarez-Gaumé, Olive and Zwanziger.

REFERENCES

[1] T.D. Lee and C.N. Yang: Conservation of Heavy Particles and Generalized Gauge Transformations. Phys. Rev. 98 (1955) 1501

[2] J.Sakurai: Theory of Strong Interactions. Ann. Phys. (N.Y.) 11 (1961) 1-48

[3] J. Schwinger: Gauge Invariance and Mass. Phys. Rev. 125 (1962) 397-398;
Gauge Theories of Vector Particles. pp. 89-134 in Theoretical Physics, 1962 Int. Atomic Energy Agency

[4] P.W. Anderson: Plasmons, Gauge Invariance and Mass. Phys. Rev. 130 (1963) 439-442

[5] P. Higgs: Broken Symmetries, Massless Particles, and Gauge Fields. Phys. Lett. 12 (1964) 132-133;
Spontaneous Symmetry Breaking Without Massless Particles. Phys. Rev. 145 (1966) 1156-1163

[6] G. Guralnik, C. Hagen, T. Kibble: Global Conservation Laws and Massless Particles. Phys. Rev. Lett. 13 (1964) 585-587

[7] R. Brout and F. Englert: Broken Symmetry and the Mass of Gauge Vector Bosons. Phys. Rev. Lett. 13 (1964) 321-323

[8] S. Weinberg: Recent Progress in Gauge Theories of the Weak, Electromagnetic and Strong Interactions. Rev. Mod. Phys. 46 (1974) 255-277

[9] B.W. Lee: Renormalization of the σ-Model. Nucl. Phys. B9 (1969) 649-672;
G. 't Hooft: Renormalizable Lagrangians for Massive Yang-Mills Fields. Nucl. Phys. B35 (1971) 167-188

[10] N.D. Mermin and H. Wagner: Absence of Ferromagnetism or Antiferromagnetism in One or Two Dimensional Isotropic Heisenberg Models. Phys. Rev. Lett. 17 (1966) 1133-1136

[11] S. Elitzur: Impossibility of Spontaneously Breaking Local Symmetries. Phys. Rev. D12 (1975) 3978-3982

[12] G. De Angelis, D. de Falco, and F. Guerra: Note on the Abelian Higgs Model on a Lattice: Absence of Spontaneous Magnetization. Phys. Rev. D17 (1978) 1624-1628

[13] T. Banks and E. Rabinovici: Finite Temperature Behavior of the Lattice Abelian Higgs Model. Nucl. Phys. B160 (1979) 349-379

[14] J.Fröhlich, G. Morchio and F. Strocchi: Higgs Phenomenon Without Symmetry Breaking Order Parameter. Nucl. Phys. B190 (1981) 553-582

[15] G. 't Hooft: Naturalness, Chiral Symmetry Breaking and Spontaneous Chiral Symmetry Breaking. pp. 135-157 in Recent Developments in Gauge Theories. Cargèse Lectures 1974, ed. G. 't Hooft et al.

[16] C. King: The $\mathcal{U}(1)$ Higgs Model. I. The Continuum Limit. II. The Infinite Volume Limit. To appear in Comm. Math. Phys.

[17] D. Brydges, J. Fröhlich, E. Seiler: Construction of Quantized Gauge Fields.
I. General Results. Ann. of Phys. 121 (1979) 227-284,
II. Convergence of the Lattice Approximation. Comm. Math. Phys. 71 (1980) 159-205,
III. The Two Dimensional Abelian Higgs Model Without Cutoffs. Comm. Math. Phys. 79 (1981) 353-399

[18] E. Seiler: Gauge Theories as a Problem of Constructive Field Theoery and Statistical Mechanics. In Lecture Notes in Physics 159, Springer Verlag 1982; see also: The Confinement Problem, MPI-PAE/PTh 47/85, July 1985

[19] T. Balaban, D. Brydges, J. Imbrie and A. Jaffe: The Mass Gap for Higgs Models on a Unit Lattice. Ann. of Phys. (N.Y.) 158 (1984) 281-319

[20] D. Brydges, J. Fröhlich and E. Seiler: Diamagnetic and Critical Properties of Higgs Lattice Gauge Theories. Nucl. Phys. B152 (1979) 521-532

[21] D. Brydges and E. Seiler: Absence of Screening in Certain Lattice Gauge and Plasma Models. To appear

[22] T. Kennedy and C. King: Symmetry Breaking in the Lattice Abelian Higgs Model. Phys. Rev. Lett. 55 (1985) 776-778

[23] D. Buchholz and K. Fredenhagen: Locality and the Structure of Particle States. Comm. Math. Phys. 84 (1982) 1-54

[24] J. Fröhlich, K. Osterwalder and E. Seiler: On Virtual Representations of Symmetric Spaces and their Analytic Continuation. Ann. of Math. 118 (1983) 461-489

[25] M. Göpfert and G. Mack: Proof of Confinement of Static Quarks in 3-Dimensional Lattice Gauge Theory for All Values of the Coupling Constant. Comm. Math. Phys. 82 (1982) 545-606

[26] L. Gross: Convergence of $\mathcal{U}(1)_3$ Lattice Gauge Theory to its Continuum Limit. Comm. Math. Phys. 92 (1983) 137-162

[27] K. Gawedski, A. Kupiainen: Exact Renormalization for the Gross-Neveu Model of Quantum Fields. Phys. Rev. Lett. 54 (1985) 2191-2194

[28] J. Feldman, J. Magnen, V. Rivasseau and R. Sénéor: A Renormalizable Field Theory: The Massive Gross-Neveu Model in Two Dimensions. École Polytechnique preprint 1985

[29] K. Osterwalder and E. Seiler: Gauge Field Theories on a Lattice. Ann. of Phys. (N.Y.) 110 (1978) 440-471

[30] K. Fredenhagen and M. Marcu: A Confinement Criterion for QCD with Dynamical Quarks. DESY preprint 85-008, January 1985

[31] K. Fredenhagen and M. Marcu: Charged States in $\mathbb{Z}_2$ Gauge Theories. Comm. Math. Phys. 92 (1983) 81-119

[32] G. Mack and H. Meyer: A Disorder Parameter that Tests for Confinement in Gauge Theories with Quark Fields. Nucl. Phys. 200B (1982) 249

[33] J. Bricmont and J. Fröhlich: An Order Parameter Distinguishing Between Different Phases of Lattice Gauge Theories with Matter Fields. Phys. Lett. 122B (1983) 73-77

[34] F. Strocchi: Spontaneous Symmetry Breaking in Local Gauge Quantum Field Theory; The Higgs Mechanism. Comm. Math. Phys. 56 (1977) 57-78;
Local and Covariant Gauge Quantum Field Theories. Cluster Property, Super-selection Rules and the Infra-Red Problem. Phys. Rev. D17 (1978) 2010-2021

[35] G. Morchio and F. Strocchi: Infra-Red Singularities, Vacuum Structure and Pure Phases in Local Quantum Field Theory. Annales de l'Institut Henri Poincaré XXXIII (1980) 251-282

[36] G. Curci and R. Ferrari: An Alternative Approach to the Proof of Unitarity for Gauge Theories. Nuovo Cim. 35A (1976) 273-279

[37] T. Kugo and I. Ojima: Local Covariant Operator Formalism of Non-Abelian Gauge Theories and Quark Confinement. Suppl. Prog. Theor. Phys. 66 (1979) 1-130

[38] S. Doplicher: Local Aspects of Superselection Rules. Comm. Math. Phys. 85 (1982) 73-86;
the references and introduction of this and the following paper provide general background for the remarks of this section.

[39] J. Roberts: Localization in Algebraic Field Theory. Comm. Math. Phys. 85 (1982) 87-98

[40] D. Buchholz: The Physical State Space of Quantum Electrodynamics. Comm. Math. Phys. 85 (1982) 49-71

TOPICS IN STRING THEORY

A. Neveu

CERN
Geneva
Switzerland

1. INTRODUCTION

String models were initially invented around 1970 to describe the phenomenology of soft hadronic interactions. They were motivated by the following phenomenological observation, called duality[1]: in a first approximation, inelastic amplitudes, e.g. $\pi^- p \to \pi^0 n$, can be described by either a sum of direct channel resonances (Δ, N^*, ...) or a sum of crossed-channel Regge poles (ρ, ρ', ...). This approximation, called the narrow resonance approximation, neglects unitarity corrections. Its validity for inelastic amplitudes is of the order of 10%. It is much better than adding the separate contributions of Regge poles and resonances, as naïve field theory would suggest should be done, which would come out wrong by a factor of 2.

Duality can be described by the pictorial equations shown in Fig. 1.

The first closed formula exhibiting such a behaviour was given by Veneziano[1] for a simplified model involving only one species of mesons: the four-point Veneziano amplitude is

$$F(s,t) = g^2 \frac{\Gamma[-\alpha(s)]\Gamma[-\alpha(t)]}{\Gamma[-\alpha(s) - \alpha(t)]} , \tag{1.1}$$

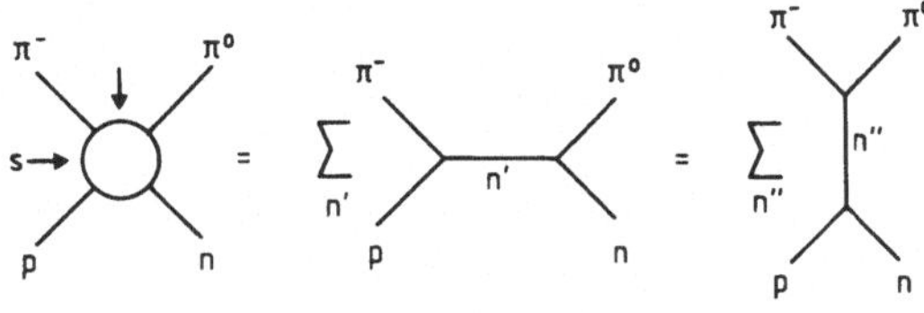

Fig. 1. Equation (1.1), the fundamental duality equation

where $\alpha(s) = \alpha_0 + \alpha' s$; α_0 is the intercept, α' the slope of the leading Regge trajectory, and g^2 is a coupling constant. This crossing symmetric function can indeed be written either as a sum of Regge poles, using the asymptotic behaviour of the Γ function, or, using its poles, as a sum of resonances, with masses m_n^2 such that $\alpha(m_n^2) = n$, and spin $\leq n$. It also contains the important experimental fact that hadrons made out of the light (u, d, s) quarks lie on linear trajectories, to an excellent approximation. We note that the full four-point amplitude in this model is actually

$$F(s,t) + F(t,u) + F(u,s) ,$$

each term being associated with a different non-cyclic permutation of the four external lines of Fig. 1.

Duality and the Veneziano formula (1.1) were generalized from four- to N-body processes[1,2]. The N-body generalization was obtained by requiring that the amplitude can be written as a sum of Regge poles *or* resonances in *any planar* subchannel (Fig. 2). Just as for the four-point case, the full amplitude is obtained by summing the contributions of the various non-cyclic permutations of (1, 2, ..., N). Modified by hand to account for trajectories with positive intercept (ρ, ω), SU(3) quantum numbers by Chan-Paton[3] factors, and baryons, it was used to fit, with very few parameters, a large class of processes, with reasonable success.

After finding the spectrum of those resonances[1,2,4,5], it was realized that they correspond to all the excitation levels of a relativistic string[6], and that the scattering of those string states can be understood in a purely geometrical fashion: the two initial strings come together to form a single one (Fig. 3), which then decays into the two final ones; summation over all possible intermediate string states corresponds to the series of Eq. (1.1).

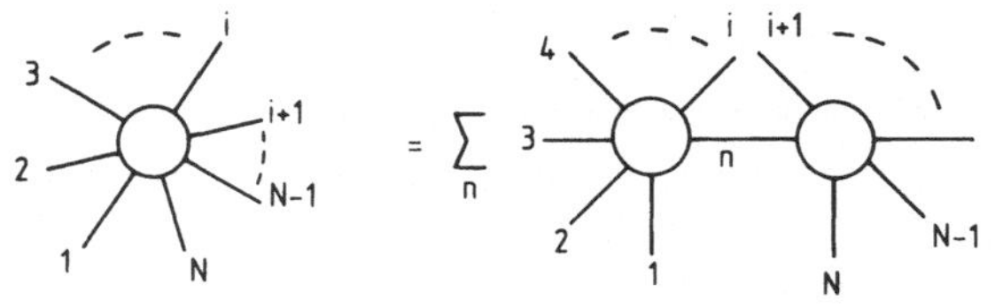

Fig. 2. Factorization of an N-point function as a sum of resonances

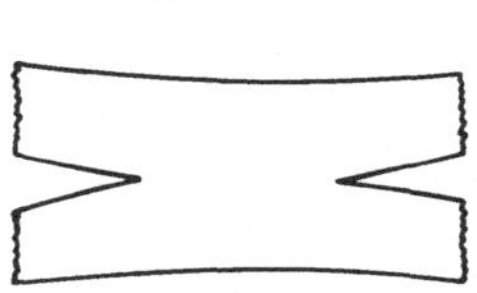

Fig. 3. Interacting string picture of a tree amplitude

Fig. 4. The four-string interaction

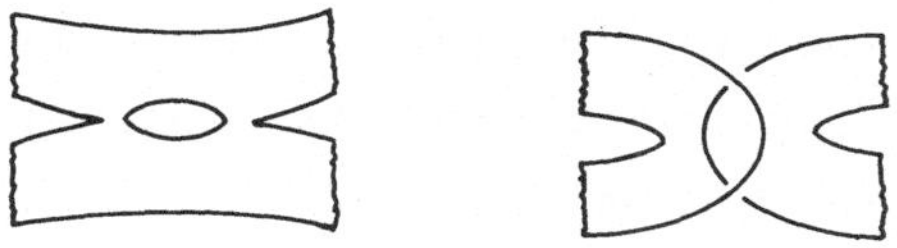

a) b)

Fig. 5. Unitarity corrections, a) planar, b) non-planar

The string-joining vertex is just the probability amplitude that the two initial strings have a common end and coincide with the intermediate string. The string-splitting vertex is its Hermitian conjugate. It was found that a second type of interaction, or four-string vertex (Fig. 4) is needed to recover the Veneziano amplitude and its generalizations[7]. This four-string vertex, where two strings touch and exchange strands, is the only one present when only closed strings and their interactions are considered. In this interacting string picture, unitarity corrections correspond to intermediate states with more than one string, and non-simply connected surfaces (Fig. 5), with various topologies, or to intermediate states with closed strings (Fig. 6). Unitarity corrections were indeed computed by rather primitive methods very soon in the history of this subject[5], and it seems that one can thus perturbatively build a consistent theory in the critical dimension D of space-time: D = 26 for the original purely bosonic string; D = 10 for superstrings. At least in lowest order of perturbation theory (one loop), superstrings have the remarkable property of being finite, and it is strongly believed by the aficionados that this finiteness is true to all orders. The D = 26 bosonic string is infinite in one loop, but this infinity is infrared in nature, rather than ultraviolet; this is due to the fact that the massless scalar state, and the two scalar tachyons present in this theory can disappear into the vacuum. This divergence might be absent in an appropriately infrared regulated version of the theory.

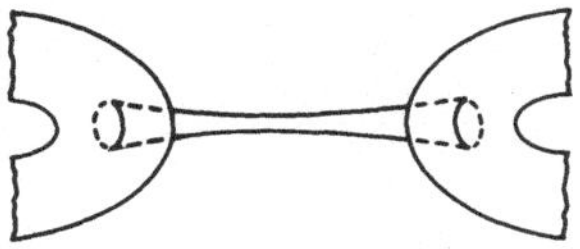

Fig. 6. The surface of Fig. 5b can be stretched into a tube exhibiting the closed-string intermediate state

As alluded to above, there exist several string models. In the first lecture, we shall describe in some detail the simplest one, the open bosonic string, which turns out to live most naturally in 26 dimensions.

In the second lecture, we review the closed bosonic strings, and the open and closed 10-dimensional strings[8] (superstrings). In the third lecture, we deal with various compactification schemes which have been proposed to deal with the extra space dimensions, from 4 to 10 or 26; in particular, we describe the Frenkel-Kac construction which builds non-Abelian internal symmetry groups out of the compactified dimensions, and the resulting heterotic string. Finally, in the fourth lecture, we address the important problem of the second quantization of string theories, and of the underlying gauge invariance which is responsible for the possibility of dealing, in a consistent fashion, with interacting high-spin states without negative metric.

2. FREE BOSONIC OPEN STRING

As it moves in time, the one-dimensional string spans a two-dimensional Minkowskian manifold, described by D functions $X^\mu(\sigma,\tau)$, $\mu = 0, \ldots, D-1$, of two parameters σ and τ. Our space-time metric is $-{+}{+}\ldots{+}$. We use the notation

$$X'^\mu = \frac{\partial X^\mu}{\partial\sigma}\ ,\ \dot{X}^\mu = \frac{\partial X^\mu}{\partial\tau}\ . \tag{2.1}$$

In the conventional treatment of this geometrical system, it is assumed that one of the directions, $\dot{X}^\mu$, on the surface is time-like and the other, X'^μ, is space-like. the classical dimensionless Nambu-Goto[6] action is just the area of the surface:

$$S = -\frac{1}{2\pi\alpha'}\int d\sigma\, d\tau\, [(\dot{X}^\mu X'_\mu)^2 - \dot{X}_\mu{}^2 X'_\nu{}^2]^{1/2}\ . \tag{2.2}$$

This is the most natural generalization of the action of a point particle, which is just the length of its world path. The dimensional parameter α' will ultimately be the slope of the Regge trajectories. The area of the surface is purely geometrical, and S is invariant under reparametrizations $\sigma \to \tilde{\sigma}(\sigma,\tau)$, $\tau \to \tilde{\tau}(\sigma,\tau)$. This local classical reparametrization invariance eliminates degrees of freedom. That all D components of X^μ are not independent degrees of freedom can be noticed by first computing the canonical momentum densities:

$$P_\mu = \frac{\delta S}{\delta\dot{X}^\mu} = -\frac{(-g)^{1/2}}{2\pi\alpha'}\,[\dot{X}^\nu X'_\nu)X'_\mu - X'_\nu{}^2\dot{X}_\mu]\ , \tag{2.3}$$

where

$$-g = (\dot{X}^{\mu}X'_{\mu})^2 - \dot{X}_{\mu}^{2}X'_{\nu}^{2} , \tag{2.4}$$

and finding that

$$P_{\mu}^{2} + \frac{X'^{2}_{\mu}}{(2\pi\alpha')^2} \equiv 0 , \tag{2.5}$$

$$X'_{\mu}P^{\mu} \equiv 0 , \tag{2.6}$$

so that one cannot solve uniquely for $\dot{X}^{\mu}$ in terms of P^{μ} and X^{μ}, and the canonical Hamiltonian is non-unique. There are several ways of dealing with this situation. We shall first choose a parametrization such that only independent degrees of freedom appear, eliminating by hand the dependent ones, i.e. solving analytically the constraints (2.5) and (2.6). This is the original light-cone treatment of Ref. 6. It uses light-cone coordinates defined by

$$X^{\pm} = X_{\mp} = \frac{1}{\sqrt{2}} (X^{1} \pm X^{0}) , \tag{2.7}$$

with the scalar product of two vectors U and V given by

$$U^{\mu}V_{\mu} = U^{+}V^{-} + U^{-}V^{+} + U_{i}V_{i} , \quad i = 2, \ldots, D-1 . \tag{2.8}$$

It is a geometrical fact that one can partly fix the parametrization of a time-like surface by choosing orthogonal coordinates by the conditions

$$(\dot{X}_{\mu} \pm X'_{\mu})^2 = 0 \leftrightarrow \begin{cases} \dot{X}_{\mu}^{2} + X'^{2}_{\mu} = 0 . \\ \dot{X}^{\mu}X'_{\mu} = 0 . \end{cases} \tag{2.9}$$

One has then simply

$$P_{\mu} = \frac{1}{2\pi\alpha'} \dot{X}_{\mu} , \quad \Pi_{\mu} \equiv \frac{\delta S}{\delta X'_{\mu}} = - \frac{1}{2\pi\alpha'} X'_{\mu} , \tag{2.10}$$

and the equations of motion

$$\frac{\partial}{\partial\tau} P_{\mu} + \frac{\partial}{\partial\sigma} \Pi_{\mu} = 0 \tag{2.11}$$

simply become those of D free, massless, two-dimensional fields:

$$\ddot{X}_\mu - X''_\mu = 0 . \tag{2.12}$$

The orthogonality constraints (2.9) still allow for conformal reparametrizations, defined by

$$\tilde{\sigma} + \tilde{\tau} = f(\sigma + \tau) , \quad \tilde{\sigma} - \tilde{\tau} = g(\sigma - \tau) , \tag{2.13}$$

where f and g are arbitrary functions. From Eq. (2.13) it follows that $\tilde{\tau}$ itself satisfies the equation of motion (2.12), so that one can choose

$$X^+ = c\tilde{\tau} , \tag{2.14}$$

c being a constant to be determined, and one can suppress the $\sim$ above the characters.

To determine c, let us first say a few words about the boundary conditions which come with the equations of motion (2.11). These boundary conditions are obtained by writing that boundary terms coming from partial integrations in the variational problem vanish. These boundary terms are easily found to be

$$- \frac{1}{2\pi\alpha'} \int d\tau \left[\Pi_\mu + \mathcal{P}_\mu \frac{d\sigma}{d\tau} \text{(boundary)}\right] \delta X^\mu . \tag{2.15}$$

Hence, for a string with free ends, we require

$$\Pi_\mu + \mathcal{P}_\mu \frac{d\sigma}{d\tau} \text{(boundary)} = 0. \tag{2.16}$$

Looking at the + component of this equation, one obtains, in the light-cone orthogonal parametrization chosen above

$$\frac{d\sigma}{d\tau} \text{(boundary)} = 0 . \tag{2.17}$$

In our parametrization, the σ density $\mathcal{P}^+$ of p^+ momentum, obtained by Noether's theorem, is a constant:

$$\mathcal{P}^+ = \frac{c}{2\pi\alpha'} . \tag{2.18}$$

This equation gives a simple interpretation of the parameter σ, as the integrated p^+ momentum, a physical quantity. The total centre-of-mass momentum

$$p_\mu = \int d\sigma\, \mathcal{P}^\mu \tag{2.19}$$

is conserved, and, from Eq. (2.17), one can fix the range of σ to be between 0 and π, obtaining

$$c = 2\alpha' p^+ \ . \tag{2.20}$$

Alternatively, one could find it more convenient to set $c = 1$, and let the range of σ depend on the total p^+ momentum. This can be convenient when a string splits into two, or when the ends of the string are not free, and p^+ momentum can flow in and out of the ends, according to the applied external forces.

The next step is to use the constraints (2.5) and (2.6) to compute the - components $\mathcal{P}^-$ and X^- in terms of the transverse components $\mathcal{P}^i$, X^i. From Eq. (2.5), one obtains

$$\frac{2p^+}{\pi}\mathcal{P}^- + \mathcal{P}^{i2} + \frac{1}{(2\pi\alpha')^2} X'^{\,2}_i = 0 \ , \tag{2.21}$$

which determines $\mathcal{P}^-$, and forms Eq. (2.6),

$$X^-(\sigma) = q^- + \frac{\pi}{p^+}\int_0^\pi d\sigma' \left[\theta(\sigma-\sigma') - \frac{\sigma'}{\pi}\right] X'_i\, \mathcal{P}^i(\sigma') \ , \tag{2.22}$$

where q^- is the centre-of-mass position in the - direction:

$$q^- = \frac{1}{\pi}\int_0^\pi d\sigma\, X^-(\sigma) \ . \tag{2.23}$$

The transverse coordinates, together with p^+ and q^-, are thus the independent degrees of freedom of the string. From the boundary conditions (2.16), one can write the following Fourier decomposition:

$$X^\mu(\sigma,\tau) = q^\mu(\tau) + \sum_{\substack{n=-\infty \\ n\neq 0}}^{+\infty} (2\alpha')^{1/2}\, \frac{i}{n}\, \alpha_n{}^\mu(\tau) \cos n\sigma \ , \tag{2.24}$$

$$\mathcal{P}^\mu(\sigma,\tau) = \frac{p^\mu}{\pi} + \frac{1}{2\pi}\left(\frac{2}{\alpha'}\right)^{1/2} \sum_{n\neq 0} \alpha_n{}^\mu(\tau) \cos n\sigma \ . \tag{2.25}$$

The light cone Hamiltonian H, which generates translations in the + direction, is proportional to the p^- momentum. More precisely,

$$H = -cp^- = -2\alpha' p^- p^+ \ . \tag{2.26}$$

Integrating Eq. (2.21) over σ, one finds

$$H = 2\pi\alpha' \int_0^\pi \frac{1}{2}\left[P_i^{\,2} + \frac{1}{(2\pi\alpha')^2} X'^{\,2}_i\right] d\sigma \ . \tag{2.27}$$

The equations of motion are then obtained by using H as a Hamiltonian, and the Poisson brackets:

$$\begin{aligned} &\{X_i, X_j\} = \{P_i, P_j\} = 0 \\ &\{X_i(\sigma), P_j(\sigma')\} = \delta(\sigma-\sigma')\delta_{ij} \\ &\{q^-, p^+\} = 1 \ . \end{aligned} \tag{2.28}$$

In terms of α_n,

$$\{\alpha_n^{\ i}, \alpha_m^{\ j}\} = -in\ \delta_{n,-m}\delta_{ij} \qquad (a_{-n} = a_n^{\ *}) \tag{2.29}$$

and

$$H = \alpha' p_i^{\,2} + \frac{1}{2}\sum_{p\neq 0} \alpha_{-p}^{\ \ i}\alpha_p^{\ i} \ . \tag{2.30}$$

From Eqs. (2.29) and (2.30), one sees that one is dealing with a system of decoupled harmonic oscillators. Using Eqs. (2.26) and (2.30), one finds the mass spectrum

$$-\alpha'(2p^+p^- + p_i^{\,2}) \equiv \alpha' M^2 = \frac{1}{2}\sum_{p\neq 0} \alpha_{-p}^{\ \ i}\alpha_p^{\ i} \ . \tag{2.31}$$

Hence, classical string excitations may have any positive M^2, the ground state being massless.

To get a feeling for this system, one can easily describe the simplest possible motion of the string, which corresponds to exciting only the α_1

mode. We take as solutions of the equations of motion, boundary conditions, and constraints, the following expressions:

$$X_2 = \frac{1}{2} \cos \tau \cos \sigma$$

$$X_3 = \frac{1}{2} \sin \tau \cos \sigma \tag{2.32}$$

$$X_4 \ldots X_{D-1} = 0$$

$$X^+ = 2\pi\alpha' \, p^+ \tau$$
$$X^- = 2\pi\alpha' \, p^- \tau \tag{2.33}$$

$$M^2 = 2p^+ p^- = \frac{\pi}{8} \frac{l^2}{(2\pi\alpha')^2} \tag{2.34}$$

where l is an arbitrary constant. We thus obtain a motion which is a rigid rotation at uniform angular velocity, determined as a function of l, by Eqs. (2.32) and (2.33), and such that the ends of the string move at the velocity of light, according to the constraints (2.9) and the boundary conditions (2.16). This motion describes the leading trajectory, which has maximal angular momentum at a given mass.

It is interesting to work out $\alpha^-{}_n$ from Eqs. (2.21), (2.22). Taking Fourier components, one finds

$$\alpha^-{}_n = \frac{(2\alpha')^{-1/2}}{p^+} L_n \, , \tag{2.35}$$

where L_n is defined by

$$L_n = \frac{1}{2} \sum_{p=-\infty}^{+\infty} \alpha_{n+p}{}^i \alpha_{-p}{}^i \, . \tag{2.36}$$

In Eq. (2.35) and the following, we define $\alpha_0{}^\mu$ by

$$\alpha_0{}^\mu = p^\mu \sqrt{2\alpha'} \, . \tag{2.37}$$

The L_n's generate the Virasoro algebra

$$\{L_n, L_m\} = -i(n-m) L_{n+m} \, , \tag{2.38}$$

and one has

$$L_0 = H \ . \tag{2.39}$$

According to Noether's theorem, the Lorentz generators are given by

$$M_{\mu\nu} = \int_0^\pi d\sigma\, (X_\mu P_\nu - X_\nu P_\mu) \ . \tag{2.40}$$

The M_{ij} operators are bilinear in the modes $\alpha_n{}^i$. the M^{i-} boost operators are more interesting, because they are trilinear in $\alpha_n{}^i$. Because boosts in the i direction mix + and i components, they do not naïvely preserve condition (2.14), which must be restored by an appropriate conformal reparametrization. This is the intuitive explanation for the appearance via $\alpha^-{}_n$ of the conformal generators L_n in M^{i-}, which perform this reparametrization. Checking that the Poisson brackets of $M_{\mu\nu}$ close to the Lorentz algebra is an exercise left for the reader.

In this light-cone treatment, quantization is done though the correspondance principle, by the rule

$$[\ ,\ ] = i\{\ ,\ \} \tag{2.41}$$

for the independent degrees of freedom. The question of operator ordering shows up in L_0, and in M^{i-}. We define L_0 as the normal-ordered form of the classical expression

$$L_0 = \alpha' p_i{}^2 + \sum_{n=1}^{\infty} \alpha_{-n}{}^i \alpha_n{}^i \ , \tag{2.42}$$

and in M^{i-}, products of $\alpha_n{}^i$ and L_m are written with the positive index to the right.

The first difference between classical Poisson brackets and commutators occurs in the Virasoro algebra, for $[L_n, L_n]$. This commutator does not immediately give L_0 in the normal-ordered form of Eq. (2.42). Normal reordering leaves a c-number term, which is found to be

$$[L_n, L_{-n}] = 2nL_0 + \frac{D-2}{12}(n^3 - n) \ . \tag{2.43}$$

The rest of the Virasoro algebra is unchanged.

Lorentz invariance of the quantum theory is not *a priori* guaranteed, and must be checked by explicit calculation. The only commutator which requires reordering is $[M^{-i},M^{-j}]$. It comes out to be

$$[M^{-i},M^{-j}] = -2i\left(L_0 + 2\alpha' p^+ p^- - \frac{D-2}{24}\right) M^{ij}$$

$$-i\left(\frac{D-2}{24} - 1\right) \sum_{n\neq 0} n\alpha_{-n}{}^{i}\alpha_{n}{}^{j} \ . \tag{2.44}$$

The term $L_0 + 2\alpha p^+ p^-$ is just the classical result [which vanishes classically according to Eqs. (2.26), (2.30), and (2.39)]. However, in the course of the calculation, it does not come out in this form, but rather with L_0 sandwiched between the two $a_n{}^i$ operators of M^{ij}. Pushing L_0 to the left generates the last term: $i\ \Sigma\ n\alpha_{-n}{}^{i}\alpha_{n}{}^{j}$. The terms (D-2)/24 just come from Eq. (2.43). In Eq. (2.44), we thus see that the Lorentz algebra closes if and only if

$$D = 26 \tag{2.45}$$

$$\alpha' M^2 = -1 + \sum_{n>0} \alpha_{-n}{}^{i}\alpha_{n}{}^{i} \ . \tag{2.46}$$

The second condition means that the lowest-lying state is a tachyon, and that the first excited state is a massless vector particle: the intercept α_0 of the leading trajectory is equal to 1. That this should be so could be seen from the beginning: there are only D-2 transverse polarizations available for the first excited level, which must thus be massless, by Lorentz invariance. At the second excited level, one may have 1/2(D-1)(D-2) states of the form $\alpha_{-1}{}^{i}\alpha_{-1}{}^{j}|0\rangle$ and (D-2) states of the form $\alpha_{-2}{}^{i}|0\rangle$, hence a total of

$$\frac{D^2 - D - 2}{2} = \frac{D(D-1)}{2} - 1 \ , \tag{2.47}$$

which is exactly the number of components of a 'pure massive spin-2' for any D [a traceless symmetric second-rank tensor, an irreducible representation of SO(D-1)].

At the third excited level, one has 1/6 D(D-1)(D-2) states $\alpha_{-1}{}^{i}\alpha_{-1}{}^{j}\alpha_{-1}{}^{k}|0\rangle$, $(D-2)^2$ states $\alpha_{-1}{}^{i}\alpha_{-2}{}^{j}|0\rangle$, and D-2 states $\alpha_{-3}{}^{i}|0\rangle$. Under the little group SO(D-1), they rearrange into a symmetric third-rank tensor with traces removed, and an antisymmetric second-rank tensor:

$$\frac{1}{6} D(D-1)(D-2) + (D-2)^2 + D - 2 = \left[\frac{1}{6} D(D-1)(D+1) - (D-1)\right] + \frac{1}{2}(D-1)(D-2) \ . \tag{2.48}$$

Although this counting of states and rearrangement is correct for any D, and also for higher excited states[9], only in 26 dimensions can the Lorentz group be realized with the simple choice of M^{-i} given here.

The existence of the tachyon is a problem, but not a deadly disease. First, in superstrings (described in the next lecture), there are no tachyons. Secondly, once interactions are introduced, one should in principle proceed as in the sigma model: shift the fields by some vacuum expectation value to a stable minimum of the potential, if it exists. However, to examine this problem, one will have to wait for a satisfactory covariant second quantized Lagrangian formalism for the interacting string theory, constructed as outlined in the last lecture.

The covariant quantization of the string assumes the commutation relations

$$[X^{\mu}(\sigma),P^{\nu}(\sigma')] = i\ \delta^{\mu\nu}\delta(\sigma-\sigma') \tag{2.49}$$

and imposes the constraints (2.5) and (2.6) as selecting the physical subspace, in order to eliminate the negative metric states (ghosts) implied by the indefinite metric of Eq. (2.49). As in the Gupta-Bleuler formalism for QED, one only imposes that matrix elements of the constraints between physical states vanish. Defining L_n as the covariant extension of $\mathcal{L}_n$, one has

$$L_n = \frac{1}{2}\pi\alpha' \int_0^{\pi} d\sigma\ e^{in\sigma}\left(P_{\mu} + \frac{1}{2\pi\alpha'}X'_{\mu}\right)^2$$

$$+ \frac{1}{2}\pi\alpha' \int_0^{\pi} d\sigma\ e^{-in\sigma}\left(P_{\mu} - \frac{1}{2\pi\alpha'}X'_{\mu}\right)^2 , \tag{2.50}$$

so that the matrix elements of the constraints between physical states vanish for

$$L_n|\text{physical state}\rangle = 0\ , \qquad n > 0 \tag{2.51}$$

$$(L_0-\alpha_0)|\text{physical state}\rangle = 0\ . \tag{2.52}$$

In Eq. (2.52), $\mathcal{L}_0$ has been normal ordered, allowing for an arbitrary intercept α_0. The advantage of the covariant quantization is precisely its covariance. However, the norms of the states selected by Eqs. (2.51) and (2.52) must be computed. For $\alpha_0 = 1$, the solutions of Eqs. (2.51) and (2.52) contain no negative norm state if and only if $D \leq 26$ [10,11]. The positive norm solutions are in one to one correspondence with the states of the light-cone quantization only if $D = 26$. For $D < 26$, there are extra, positive norm, longitudinal states[11]. For $\alpha_0 < 1$, the solutions of Eq. (2.51) and (2.52) contain no negative norm state if and only if $D \leq 25$ [9], but no interacting theory is known, even at tree level, for this case.

Twelve years ago, a functional integral approach was developed in Refs. 12 and 23. In this approach, the unmanageable non-linear action (2.2) is first linearized by choosing a parametrization such that

$$(\dot{X}_\mu - X'_\mu)^2 = 0 \qquad (2.53)$$

and using the identity

$$(\dot{X}^\mu X'_\mu)^2 - \dot{X}_\mu^{\,2} X'^{\,2}_\mu = \frac{1}{4}(\dot{X}_\mu^{\,2} - X'^{\,2}_\mu)^2 - \frac{1}{4}(\dot{X}_\mu - X'_\mu)^2(\dot{X}_\mu + X'_\mu)^2 \ . \qquad (2.54)$$

In this way, one obtains an action bilinear in X^μ, which is just the action of D free massless fields in two dimensions. Sources are added by constraining the surface to go through a set of n fixed points $X_1^{\,\mu}, X_2^{\,\mu}, \ldots, X_n^{\,\mu}$ in the form of a product of δ functions:

$$\delta[X^\mu(\sigma_1,\tau_1) - X_1^{\,\mu}]\ \delta[X^\mu(\sigma_2,\tau_2) - X_2^{\,\mu}]\ \ldots\ \delta[X^\mu(\sigma_n,\tau_n) - X_n^{\,\mu}] \ , \qquad (2.55)$$

or rather, since one is interested in scattering amplitudes with well-defined external momenta p_i $(i = 1, \ldots, n)$, in the Fourier transformed form:

$$\exp i \sum_i p_i^{\,\mu} X^\mu(\sigma_i,\tau_i) \ . \qquad (2.56)$$

This source is linear in the X^μ fields, so that a simple shift of integration variables in the X^μ functional integral leads to the integrand of the Koba-Nielsen formula:

$$\exp \sum_{i \neq j} p_i p_j \ln |z_i - z_j| \ , \qquad (2.57)$$

where $z_j = \sigma_j + i\tau_j$ (a Wick rotation has been performed on the τ variable); $\ln |z_i - z_j|$ is of course just the Euclidean propagator of a free, massless, two-dimensional field between z_i and z_j. We refer the reader to Refs. 12

and 13 for details about the functional integration measure and the Faddeev-Popov ghost fields. This quantization of Eq. (2.2) by functional integrals was Lorentz invariant only in D = 26 and gave the same answers as the more straightforward method using the operator formalism.

It had been known for some time[14] that there exists another Lagrangian density classically equivalent to the Nambu-Goto Lagrangian,

$$L_{NG} = [(\dot{X}^\mu X'_\mu)^2 - \dot{X}_\mu^2 X'^2_\mu]^{1/2} \quad . \tag{2.58}$$

It is

$$L_{BDVH} = -\frac{1}{2}\sqrt{-g}\; g^{ab}\partial_a X_\mu \partial_b X^\mu \ , \tag{2.59}$$

where $(a,b) = (\sigma,\tau)$, and one has introduced the two-dimensional metric tensor $g^{ab}(\sigma,\tau)$ on the surface and its inverse g_{ab}, with $g = \det g_{ab}$. The Lagrangian L_{BDVH} is the familiar Lagrangian density of massless fields in general relativity, and naturally gives a reparametrization invariant action. Although it looks as though g^{ab} introduces new degrees of freedom, it is not so classically. Variation of Eq. (2.59) with respect to g^{ab} gives

$$-\frac{1}{2} g_{ab}\sqrt{-g}\; g^{a'b'}\partial_{a'}X^\mu\partial_{b'}X_\mu + \sqrt{-g}\;\partial_a X^\mu \partial_b X_\mu = 0 \ , \tag{2.60}$$

which is solved by

$$g_{ab} = f(\sigma,\tau)\;\partial_a X^\mu \partial_b X_\mu \ , \tag{2.61}$$

f being an arbitrary function. Putting Eq. (2.61) into Eq. (2.59) gives back Eq. (2.58) for any f, thus proving the classical equivalence of L_{NG} and L_{BDVH}. The arbitrariness of the function f reflects the invariance of L_{BDVH} under the change

$$g_{ab} \to e^{\varphi(\sigma,\tau)}\; g^{ab} \tag{2.62}$$

for arbitrary φ, called Weyl invariance, and valid only in two space-time dimensions for scalar fields.

For quantization, L_{BDVH} has the advantage of being linear in X^μ, contrary to L_{NG}, so that standard regularization methods are more easily applicable. Formally, one would start from the functional integral

$$\int Dg^{ab} DX^{\mu} \exp i \int d\sigma\, d\tau\, (L_{BDVH} - \mu_0{}^2 \sqrt{-g}) \,. \tag{2.63}$$

The term $\mu_0{}^2 \sqrt{-g}$ has been added to make the integration over the absolute magnitude of g^{ab} convergent, but the final answer does not depend on $\mu_0{}^2$, as long as it is non-zero. In Eq. (2.63), the exponent is still reparametrization-invariant, so that appropriate terms must be added which fix the parametrization, together with the corresponding Faddeev-Popov ghosts. For example, one can choose the parametrization by imposing conditions on X^{μ}, such as those of Refs. 12 and 13, and do the g^{ab} integration for each value of σ and τ. This is an elementary exercise, and one recovers all the details of the approach of Refs. 12 and 13, in particular their strange integration measure. This procedure is valid provided that the regularization of the other, X^{μ}, integrations does not introduce any extra dependence on g^{ab}.

It turns out that this is the case in general, as pointed out by Polyakov[15]. In these papers, Polyakov takes another approach imposing gauge conditions on g^{ab} from the geometrical fact, already mentioned for Eq. (2.9), that in two dimensions one may always choose the parametrization such that the manifold is conformally flat. This means that g^{ab} is of the form

$$g_{ab} = e^{\varphi}\, \delta_{ab} \,. \tag{2.64}$$

In this gauge, the metric disappears from L_{BDVH}, and one again has D two-dimensional massless scalar fields for the orbital X_{μ} degrees of freedom of the string. One would thus think that there is not much more to say and that L_{NG} and L_{BDVH} are indeed equivalent. However, this is not so: the X^{μ} integration must be regularized, and the regularization breaks the φ independence of L_{BDVH}, whether one uses Pauli-Villars regulators or higher derivative kinetic terms. For example, Pauli-Villars regulator fields $X^{\mu}{}_R$ have large mass terms of the type $M_R{}^2 \sqrt{-g\, X_R{}^{\mu^2}}$, M_R being the cut-off. Since one insists on reparametrization invariance (to be dealing with a geometrical object), one must choose $M_R{}^2 \sqrt{-g\, X_R{}^{\mu^2}}$, rather than $M_R{}^2 X_R{}^{\mu^2}$. This breaking of the independence of L_{BDVH} on the choice of φ is quite analogous to the breaking of chiral invariance by regulators in massless fermion QED or QCD, which gives rise to the Adler anomaly. Hence, via regularization, a φ dependence creeps in, which was first computed in Ref. 15. Both the X^{μ} integration and the integration over Faddeev-Popov ghosts of the choice (2.64) contribute. The contribution of the former is manifestly proportional to D, whilst the latter is D independent, and one finds that the two contributions cancel when D is precisely equal to 26. When D is different from 26, one obtains an effective action for the field φ:

$$S_{eff}(\varphi) = \frac{26 - D}{48\pi} \int d\sigma\, d\tau \left[\frac{1}{2} (\partial_\tau\varphi)^2 - \frac{1}{2} (\partial_\sigma\varphi)^2 - e^\varphi\right] . \quad (2.65)$$

Hence, an extra dynamical degree of freedom, φ, appears for $D \neq 26$. It can be interpreted as a longitudinal excitation of the string. Quantization of the action (2.65) with appropriate boundary conditions has been performed for the free string[16], and one finds that the Regge trajectories remain linear and ghost-free, with shifted intercepts. Unfortunately, the corresponding interacting string theory has not yet been worked out.

3. CLOSED STRINGS AND SUPERSTRINGS

As mentioned in the first lecture, closed strings appear naturally when one computes unitarity corrections to open-string tree scattering amplitudes. However, they can also be treated by themselves following the same outline as described in the previous section. One starts from the same geometrical action, Eq. (2.2), but the boundary conditions disappear and are replaced by simple periodicity conditions on $X^\mu(\sigma,\tau)$. The Fourier mode expansion is thus

$$X^\mu(\sigma,\tau) = q^\mu(\tau) + \sum_{n\neq 0} \frac{i}{n}[\alpha_n{}^\mu(\tau)\, e^{in\sigma} + \bar{\alpha}_n{}^\mu(\tau)\, e^{-in\sigma}] . \quad (3.1)$$

The right-moving modes $\alpha_n{}^\mu(\tau)$ are now independent of the left-moving modes $\bar{\alpha}_n{}^\mu(\tau)$, whilst for the open string they are related by the boundary conditions. Both sets of modes are harmonic oscillators with the commutation relations (2.29), and they mutually commute:

$$[\alpha_n,\bar{\alpha}_m] = 0 \quad \text{for all } n,\, m . \quad (3.2)$$

Hence, except for the zero mode (centre-of-mass position) $q^\mu(\tau)$, one is dealing with a system which looks very much like the direct product of two open strings. The light-cone and the covariant treatments of Section 2 extend trivially, and they are again equivalent in 26 space-time dimensions. The only new feature is that an extra condition must be imposed on the physical states $|\psi\rangle$, which is

$$(L_0 - \bar{L}_0)|\psi\rangle = 0 . \quad (3.3)$$

This condition must be imposed both in light-cone and Lorentz-covariant cases. Its interpretation comes from Eq. (2.22). This equation, which reconstructs X^- from the transverse modes, selects a particular point (here zero) as the end-point of the integral. However, for the closed string, there is nothing special about that point, and one must thus impose that the generator of global σ translations be zero when applied to any physical state. This is the meaning of condition (3.3). Note that in this condition the zero mode cancels between L_0 and $\bar{L}_0$, and Eq. (3.3) simply imposes that there be as much excitation in the left-moving as in the right-moving modes; on the basis where the occupation numbers of the various modes are diagonal, it is thus a trivial condition to enforce.

As well as forcing the space-time dimension to be 26, closure of the Lorentz algebra shows that the leading intercept is two, and thus there is a massless spin-2 particle, the graviton. Also, one finds that the slope of the leading trajectory is one half that of the open string. In the covariant treatment, the set of equations (2.51) and (2.52) is now

$$(L_0 + \bar{L}_0 - 2)|\psi\rangle = 0$$

$$L_n|\psi\rangle = \bar{L}_n|\psi\rangle = 0 \qquad n \geq 1 \tag{3.4}$$

$$(L_0 - \bar{L}_0)|\psi\rangle = 0 .$$

When open and closed strings are coupled together, there is the possibility that some of the two-dimensional manifolds mentioned in the Introduction do not occur[8]. Namely, one can consistently ignore all non-orientable manifolds if the strings have an orientation which implies a corresponding orientation for its world sheet. For example, when U(N) quark and antiquark labels are attached to the end of our open string, it becomes oriented, since the end which carries the quark is different from the one which carries the antiquark. In such a model, the Moebius strip, for example, is manifestly excluded, since it corresponds to quark-quark (or quark-antiquark) states. However, for O(N) or Sp(2N) quark indices at the ends, the two ends are indistinguishable because the N (2N) representation of O(N) [Sp(2N)] is real (pseudoreal); in these cases, the strings are not oriented and non-orientable manifolds occur. Oriented closed strings have more states than the non-oriented ones, which are symmetric in right- and left-moving modes. For example, oriented closed strings have a massless antisymmetric tensor field of rank two in their spectrum. Both also have a massless scalar, commonly called the dilaton. In the zero slope limit, the graviton couples to itself and to the other states as described by Einstein's action. This is why the string has been proposed[17] as a finite unitary quantum mechanical theory of gravity, the slope α' playing the role of a unitary cut-off.

Fermionic degrees of freedom were added to the bosonic string as early as 1971 in Refs. 18 and 19. In the initial Lorentz covariant formulation of these references, on top of the D-orbital free, massless, two-dimensional fields $X^{\mu}(\sigma,\tau)$ of the previous section, a set of D two-dimensional Majorana free fermions $\lambda^{\mu}(\sigma,\tau)$ were introduced, with the action

$$S_f = -\frac{1}{4\pi\alpha'}\int d\sigma\, d\tau\; i\, \bar{\lambda}^{\mu}\rho^{\alpha}\partial_{\alpha}\lambda^{\mu} , \tag{3.5}$$

ρ^{α} being the 2 x 2 Dirac matrices. The boundary conditions on λ at $\sigma = 0$ and $\sigma = \pi$ for an open string are again obtained by looking at the boundary terms of the variational treatment of Eq. (3.5), and one finds that there are essentially two possibilities. One can always define λ such that

$$\lambda^{1\mu}(0,\tau) = \lambda^{2\mu}(0,\tau) , \tag{3.6}$$

where the indices 1 and 2 refer to the two components of the two-dimensional spinor field. At the other boundary, however, one may choose either

$$\lambda^{1\mu}(\pi,\tau) = \lambda^{2\mu}(\pi,\tau) , \tag{3.7}$$

or

$$\lambda^{1\mu}(\pi,\tau) = -\lambda^{2\mu}(\pi,\tau) . \tag{3.8}$$

Hence, the Fourier-mode expansions of the free fields λ^1 and λ^2 are

$$\lambda^{1\mu} = \sum_{n=-\infty}^{+\infty} d_n{}^{\mu}\, e^{-in(\tau-\sigma)} , \qquad \lambda^{2\mu} = \sum_{n=-\infty}^{+\infty} d_n{}^{\mu}\, e^{-in(\tau+\sigma)} , \tag{3.9}$$

for the boundary condition (3.7), and

$$\lambda^{1\mu} = \sum_{r=-\infty}^{+\infty} b_r{}^{i}\, e^{-in(\tau-\sigma)} , \qquad \lambda^{2\mu} = \sum_{r=-\infty}^{+\infty} b_r{}^{i}\, e^{-in(\tau+\sigma)} , \tag{3.10}$$

where the index r in Eq. (3.10) runs over half-integers, whilst the index n of Eq. (3.9) runs over integers. For both cases, one has the anticommutation rules:

$$\{d_m{}^{\mu}, d_n{}^{\nu}\} = \delta_{m,-n}\delta^{\mu\nu} , \tag{3.11}$$

$$\{b_r{}^{\mu}, b_s{}^{\nu}\} = \delta_{r,-s}\delta^{\mu\nu} . \tag{3.12}$$

Note that the $m = n = 0$ case of Eq. (3.11) is nothing but the definition of the Clifford γ-matrix algebra. Hence, the boundary conditions (3.7) describe space-time fermions (commonly referred to as the Ramond sector), whilst the boundary conditions (3.8) describe space-time bosons (Neveu-Schwarz sector).

Because the Ramond sector contains space-time fermions, it is natural to replace the Klein-Gordon-like equation (2.52) by a Dirac equation:

$$(F_0 - im)\psi = 0 , \tag{3.13}$$

where

$$F_0 = \sum_{n=-\infty}^{+\infty} \alpha_{-n} d_n . \tag{3.14}$$

In Eq. (3.14), one sees that the zero mode piece of F_0 is just the Dirac operator $\not\partial$. Upon multiplying (3.13) by $F_0 + im$, one finds

$$(L_0 + m^2)|\psi\rangle = 0 , \tag{3.15}$$

where L_0 is the Virasoro generator of the fermionic string, containing a piece coming from the d modes, namely

$$L_0 = \frac{1}{2} \sum_{n=-\infty}^{+\infty} (:\alpha_{-n}\alpha_n: + n:d_{-n}d_n:) . \tag{3.16}$$

Hence one also obtains in the Ramond sector a linear spectrum of equally spaced Regge trajectories, with half-integral spin states.

It is clear that by introducing new oscillators d^{μ} whose index can be time-like, one introduces new ghosts via the anticommutation relation of their time component in Eq. (3.11). Thus new gauge conditions, more powerful than those of Eq. (2.51), must be imposed. In Ref. 18, these were taken to be

$$F_n|\psi\rangle = 0 , \tag{3.17}$$

together with the original Virasoro conditions

$$L_n|\psi\rangle = 0 . \tag{3.18}$$

In terms of Fourier modes, one has

$$F_n = \sum_{p=-\infty}^{+\infty} \alpha_{n-p} d_p \ , \tag{3.19}$$

$$L_n = \frac{1}{2} \sum_{p=-\infty}^{+\infty} \left[\alpha_{n-p}\alpha_p + \left(p - \frac{n}{2}\right) d_{n-p} d_p\right] . \tag{3.20}$$

It was later shown that conditions (3.17) and (3.18) do indeed remove all negative metric states for space-time dimensions $D \leq 10$, and that for a ground-state mass $m = 0$, one is left with exactly D-2 transverse positive norm modes, much as for the bosonic string. The algebra of the F_n and L_n operators

$$\{F_n, F_m\} = 2L_{n+m} + \frac{D}{2} n^2 \delta_{n,-m} \ , \tag{3.21}$$

$$[L_n, F_m] = \left(\frac{n}{2} - m\right) F_{n+m} \ , \tag{3.22}$$

$$[L_n, L_m] = (n-m)L_{n+m} + \frac{D}{8} n^3 \delta_{n,-m} \ , \tag{3.23}$$

was the first example of a supersymmetry algebra, soon recast into ordinary field theory language[20]. In the present case, it is referred to as the two-dimensional superconformal algebra.

In the Neveu-Schwarz sector, one is dealing with space-time bosonic states, and the equations of motion and ghost-removing conditions were formed[21] by looking at what states factorize the tree amplitudes of Ref. 19. They are

$$\left(L_0 - \frac{1}{2}\right)|\psi_{NS}\rangle = 0 \ , \tag{3.24}$$

$$G_r|\psi_{NS}\rangle = 0 \ , \qquad r = \frac{1}{2}, \frac{3}{2}, \frac{5}{2}, \ldots \ , \tag{3.25}$$

$$L_n|\psi_{NS}\rangle = 0 \ , \qquad n = 1, 2, \ldots \ ,$$

where the L_n Virasoro operators are defined by the same formula (3.20) with b replacing d, and the G_r operators are the analogues of the F_n's:

$$G_r = \sum_{p=-\infty}^{+\infty} b_{r-p}\alpha_p \ . \tag{3.26}$$

As for the Ramond sector, the Neveu-Schwarz sector contains only D-2 transverse positive-norm modes for D = 10. The algebra of the G_r and L_n operators is quite similar to that of the F_m and L_n of the Ramond sector.

Whilst the ground state of the Ramond sector is a massless spin 1/2, the ground state of the Neveu-Schwarz sector is still a tachyon with mass $\alpha' M^2 = -1/2$. Note, however, that the tachyon state of the leading, intercept-one, trajectory has been removed. It was also quickly noticed that one would easily consistently remove this last tachyon by restricting oneself to the 'positive G parity' sector. This is the sector of solutions of Eqs. (3.24) and (3.25) which have an odd number of b excitations. In this sector, the ground state of the Neveu-Schwarz model is a massless vector.

After the discovery of space-time supersymmetry and supergravity, it was noticed[22] that in D = 10, the spectrum of states of the NS-R model has the right counting of modes, at all levels, to be supersymmetric in space-time, together with being supersymmetric on the two-dimensional world sheet. This is obtained only if a further chiral truncation is made on the Ramond sector, keeping only the solutions of Eqs. (3.13), (3.17), and (3.18) which have only one sign of the operator,

$$\Gamma_{11} = \gamma^0 \gamma^1 \ \ldots \ \gamma^9 (-1)^{\Sigma_{n=1}^{\infty} d_{-n} d_n} \quad . \tag{3.27}$$

It is surely somewhat strange to describe space-time fermions with the oscillators d_n^{μ} which have a Lorentz vector index, although no contradiction arises between spin and statistics. This prompted Green and Schwarz[23] to reformulate the NS-R model in terms of two-dimensional oscillators S_n, instead of d_n and b_r, which carry a space-time spinor index. So far, this has been achieved in simple form in the light-cone frame; Green and Schwarz thus have a set of transverse S_n^a operators, a = 1, ..., 8, with the same free field anticommutation properties as the d_n operators, but which transform as the eight-dimensional spinorial representation of the little group SO(8). In this new formulation, space-time supersymmetry is manifest, instead of being realized through the complicated fermion-emission vertex of the original formulation. That the two formulations are equivalent is proved by explicitly constructing the S_n^a modes in terms of the b_r^i and d_n^i modes. A Lorentz covariant two-dimensional action was later constructed[24] with space-time spinors, but so far remains intractable because the corresponding two-dimensional fields are not free (except in the light-cone frame, of course), but interact through Thirring-like couplings. It might well be that mapping these fields onto free fields would actually mean going back to the old formalism with d_n^{μ} and b_r^{μ}.

Like the bosonic string, superstrings can be open or closed, oriented or not oriented, as reviewed in Ref. 8. Type I superstrings are unoriented and can be either open or closed. They have N = 1 supersymmetry in 10 dimensions. In the zero slope limit, type I superstrings reduce to N = 1 Yang-Mills (from open strings) coupled to N = 1 supergravity (from closed strings). Type II superstrings have N = 2 supersymmetry in 10 dimensions, and are closed strings only. Type II superstrings further fall into two classes: type IIa, where the two space-time supersymmetries (one for the right-moving modes, one for the left-moving modes on the closed string) have the opposite chirality; and type IIb, where they have the same chirality. In the zero slope limit, type IIa reduces to the previously known non-chiral supergravity in 10 dimensions[25], while type IIb yields a new, chiral, previously unknown supergravity, which does not seem to admit a simple covariant formulation.

Like their field theoretic counterparts, superstrings enjoy remarkable properties, when compared with the bosonic string. First, they are free of tachyons. Secondly, the type II superstrings are one-loop finite, the massless scalar 'dilaton' being forbidden by supersymmetry from going into the vacuum, as mentioned in Section 1. More remarkably, although they are chiral, type IIb superstrings have no one-loop anomalies[26] (actually, this comes together with the fact that they are finite). Even more remarkably, it was shown last year[27] that type I superstrings are also one-loop anomaly free for the open string Yang-Mills gauge group SO(32). In the zero slope field theory limit[28], it was soon realized that the anomaly cancellation mechanism would also work for the Yang-Mills gauge group $E_8 \times E_8$ large enough to contain all known grand unification schemes. This sparked an intense renewed activity and interest in the string model. The heterotic string, which can have $E_8 \times E_8$ [as well as SO(32)] as the Yang-Mills gauge group, will be constructed in the next lecture.

4. NON-ABELIAN COMPACTIFICATION, AND THE HETEROTIC STRING

In order to deal with the embarassingly large number of space-time dimensions (10 or 26), it was proposed by Scherk and Schwarz[29] that the unwanted ones (6 or 22) be compactified. This is how the old Kaluza-Klein scheme came back into particle physics. Initially, the compactification was made on a simple hypercube, with periodic boundary conditions. This proceeds in a straightforward fashion. For the open bosonic string, the analysis of Section 2 is basically unchanged; the only modification is that the wave function of the centre-of-mass be periodic in the compactified directions. Hence, momentum is quantized by integers in those directions. The mass formula (2.31) then becomes

$$\alpha' M^2 = \alpha' p_\gamma^{\ 2} + \sum_{n=1}^{\infty} \alpha_{-n}\alpha_n - 1 , \tag{4.1}$$

p_γ being the momentum in the compactified directions. For closed strings, the mode decomposition of Eq. (3.1) must be modified to allow for the possibility of the string winding around the compactified dimensions:

$$X^i(\sigma,\tau) = x^i + p^i\tau + 2L^i\sigma + \sum_{n=-\infty}^{+\infty} (\alpha_n e^{-in\sigma} + \bar{\alpha}_n e^{in\sigma}) . \tag{4.2}$$

In this formula, L^i is a vector of the lattice of periodicities of the compactified directions. The corresponding mass formula now reads,

$$\alpha' M^2 = \alpha' p^2 + 2L^2 + \sum_{n=1}^{+\infty} (\alpha_{-n}\alpha_n + \bar{\alpha}_{-n}\bar{\alpha}_n) . \tag{4.3}$$

This winding of the closed string, corresponding to non-zero values of L, was actually discovered[30] by computing the non-planar graph of Fig. 5b and factorizing it in the closed-string channel.

Although this compactification of extra space-time dimensions is formally consistent, it has, like almost all Kaluza-Klein schemes, the unappealing feature of butchering the original SO(25,1) Lorentz symmetry to SO(25-d,1) × $U^d(1)$, if d directions are compactified; $U^d(1)$ is the group of translations in the compactified directions. It is a unique feature of string theories that this compactification can be improved from a $U^d(1)$ symmetry to any non-Abelian semi-simple, simply laced group of rank d. This is the so-called Frenkel-Kac construction[31,32]. We shall explain how it works for the simplest case of two compactified dimensions and the group SU(3). The two compactified directions are assumed to have the shape of the rhombus of Fig. 7, with sides $\vec{L}_1$ and $\vec{L}_2$ of equal length. On this space, the momentum of the centre-of-mass is quantized to be on the hexagonal lattice spanned by the two vectors P_1 and P_2 of Fig. 7, defined by

$$P_1L_1 = P_2L_2 = 1 , \qquad P_1L_2 = P_2L_1 = 0 , \tag{4.4}$$

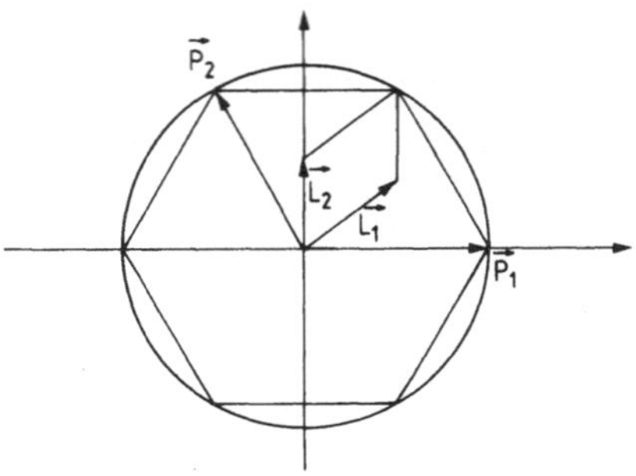

Fig. 7. The fundamental cell of the hexagonal lattice is the root diagram of SU(3)

Now, let us suppose that $\alpha' = 1/2$ and that L_1 and L_2 are such that $P_1^2 = P_2^2 = 2$. In formula (4.1), we then see that we can have several kinds of states at mass zero. Six of them are obtained by taking all harmonic oscillators in their ground state, and p_γ on the fundamental cell of the hexagonal lattice of Fig. 7. Two more are obtained by taking $p_\gamma = 0$, and exciting to its first level the harmonic oscillators α_{-1} in one of the two compactified directions. These eight massless states are scalars under the remaining 26 - 2 = 24 dimensional Lorentz group. They actually form the adjoint representation of SU(3). To prove this, Frenkel and Kac have actually constructed the corresponding generators out of the harmonic oscillators α_n of the string, and shown that they obey SU(3) commutation relations. These generators are called vertex operators because they are essentially identical to the operators used in factorizing the free amplitudes for external ground-state tachyons of the string model. The Lie algebra comes naturally in the Cartan basis: the vertex operators corresponding to the harmonic excitations $\alpha_{-1}|0\rangle$ in the compactified directions span the Cartan subalgebra; those corresponding to a p_γ at one of the vertices of the fundamental cell of the hexagonal lattice are the remaining ones, with p_γ as a root.

Since these group generators commute with the Virasoro operators, they map physical states into physical states. Hence not only the massless states but the whole theory has SU(3) symmetry. The tachyonic ground state is an SU(3) scalar, whilst higher and higher dimensional representations of SU(3) occur in the higher excited states. We note here the difference when using the Chan-Paton way of introducing quantum numbers at the ends of the string, where only few representations (the adjoint and the second-rank symmetric tensor) ever occur. This difference is even more pronounced when one considers closed strings, which in the Chan-Paton procedure have no internal quantum numbers. The Frenkel-Kac construction applies also to closed strings. Here, non-zero windings around the torus in the compactified directions play a crucial role, which is best seen when constructing the closed string from open ones (by the process of Fig. 6). (See Ref. 33 for details.) The result is that at zero mass the spectrum of closed-string states displays Lorentz vectors which are in the adjoint representation of SU(3). Hence this symmetry becomes a local space-time gauge symmetry.

We now turn to the construction of the heterotic string; it starts from the mode expansion of the closed string, Eq. (3.1). In this equation, as in most other aspects of closed strings, left-moving and right-moving modes are essentially independent. The basic idea of Ref. 34 is to take the right-moving modes to be those of 10-dimensional superstrings, and the left-moving modes those of the 26-dimensional bosonic strings, for which 16 of the 26 dimensions have been compactified using the Frenkel-Kac construction. Using the Green-Schwarz modes S_n^a, the mode expansion of the right-moving modes is thus in the light-cone frame:

$$X_R^i(\sigma,\tau) = \frac{1}{2} p^i(\tau-\sigma) + \sum_n \alpha_n^i e^{-in(\tau-\sigma)} , \qquad i = 1, \ldots, 8 , \tag{4.5}$$

for the orbital coordinates, and

$$S^a(\sigma,\tau) = \sum_n S_n^a e^{-in(\tau-\sigma)} , \qquad a = 1, \ldots, 8 , \tag{4.6}$$

for the fermionic coordinates. For the left-moving modes we have

$$X_L^i(\sigma,\tau) = \frac{1}{2} p^i(\tau+\sigma) + \sum_n \bar{\alpha}_n^i e^{-in(\tau+\sigma)} \tag{4.7}$$

for the space-time transverse index $i = 1, \ldots, 8$, and

$$X_L^I(\sigma,\tau) = \frac{1}{2} p^I(\tau+\sigma) + \sum_n \alpha_n^I e^{-in(\tau+\sigma)} \tag{4.8}$$

for $I = 9, \ldots, 24$ running over the 16 compactified directions. Following Refs. 31 and 32, p^I is on the lattice whose fundamental cell is the root lattice of a semi-simple group. When one goes to the one-loop level, this group is restricted to SO(32) and $E_8 \times E_8$ (see below).

Closed superstrings have N = 2 supersymmetry because there is a supersymmetry acting separately on the right-moving mode and one on the left-moving one. Thus, the heterotic string still has N = 1 supersymmetry. As for other closed strings, there is a restriction on physical states coming from the absence of a distinguished value of σ. This restriction, Eq. (3.3) for strings which treat symmetrically right- and left-movers, is now[34]

$$(L_0 - \bar{L}_0 + 1)|\text{physical state}\rangle = 0 \,. \tag{4.9}$$

The ground state of the heterotic string is thus obtained from $L_0 = 0$ and $\bar{L}_0 = 1$. We have

$$L_0 = \frac{1}{2} \sum_{p=-\infty}^{+\infty} :[\alpha_p^{\ i} \alpha_p^{\ i} + p\, S_{-p}^{\ a} S_p^{\ a}]: \,, \tag{4.10}$$

$$\bar{L}_0 = \frac{1}{2} \sum_{p=-\infty}^{+\infty} :[\bar{\alpha}_p^{\ i} \bar{\alpha}_p^{\ i} + \bar{\alpha}_p^{\ I} \bar{\alpha}_p^{\ I}]: \,. \tag{4.11}$$

The ground state of the right-moving modes is either in the vector or spinor representation of the SO(8) transverse rotation group. For the left-moving modes, in order to have $\bar{L}_0 = 1$ one may have an orbital excited state $\bar{\alpha}_{-1}^{\ i}|0\rangle$. This, when combined with the vector right-moving vacuum, gives a second-rank symmetric tensor (the transverse polarization states of the graviton) or a second-rank antisymmetric tensor, and when combined with the spinor right-moving vacuum it gives the transverse polarization states of the gravitino. All these states form the massless multiplet of supergravity. There are other ways of obtaining $\bar{L}_0 = 1$, by having a non-trivial momentum or harmonic excitation in the 16 compactified directions. These are Lorentz scalars, and give the massless multiplet of SO(32) or $E_8 \times E_8$ supersymmetric 10-dimensional Yang-Mills, when combined with the vacuum of the right movers.

We shall now attempt to describe briefly in words how the restriction to SO(32) and $E_8 \times E_8$ comes from the one-loop diagrams. As explained in Section 2, one must solve the free wave equation $\ddot{X} - X''$ with appropriate boundary conditions, and source terms corresponding to the external on-shell particles. For a one-loop diagram of closed strings, the topology of the world sheet is a torus (Fig. 8) of width $\sigma = 2\pi$ and length τ: the width 2π,

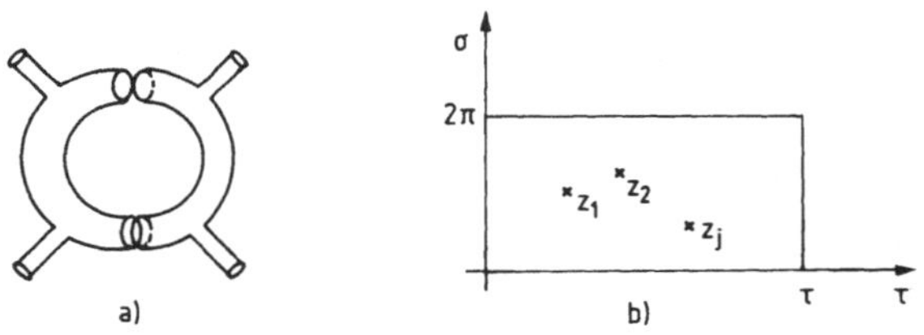

Fig. 8. a) The world sheet of a one-loop diagram of closed strings. b) The corresponding region in parameter space

with periodic boundary conditions in σ, corresponds to the parametrization of the world sheet described in Section 2. The length τ would be infinite for a tree diagram. Here, taking it finite with periodic boundary conditions in τ, corresponds to taking the trace and closing the loop. Taking the trace is summing over all possible string configurations that can propagate around the loop. This trace has three parts, one coming from the trace over the harmonic oscillator excited states, another from the continuous 10-dimensional integration over the centre-of-mass momentum, and finally one from the discrete sum over the root lattice of the group for the 16 compactified directions. The length τ as well as the positions $z_j = \tau_j + i\sigma_j$ of the external sources are integrated over. The integrand is singular when τ goes to zero, and all sources coincide with their electrostatic images with respect to the lines 0 and τ. This singularity occurs in all closed-string theories, and its interpretation and removal have been given in Ref. 35: at least for the ordinary bosonic strings and for closed superstrings, the integrand is invariant under the change $\tau \to 4\pi^2/\tau$, and the singularity at $\tau = 0$ is just duplicating the singularity at $\tau = \infty$, which is a unitarity cut. Hence, the integration region over τ must be restricted to $\tau > 2\pi$. For the heterotic string, the symmetry of the integrand under $\tau \to 4\pi^2/\tau$ occurs only if the lattice of discrete momenta in the 16 compactified directions is self-dual. Hence the restriction to $E_8 \times E_8$ and SO(32) [technically, spin $(32)/Z_2$].

The discovery of the heterotic string has generated much excitement because of its possible use as a grand unified finite anomaly-free theory incorporating gravity, chiral fermions, and a large enough Yang-Mills gauge group. At present, however, there does not yet exist any elegant compelling way of compactifying down to four dimensions.

5. SECOND QUANTIZED STRING FIELD THEORY

In the previous lectures, we have described strings mostly in the language of first quantization, summing over world histories of the string. In principle, this is sufficient to answer questions that arise in perturbation theory. However, for non-perturbative phenomena -- such as spontaneous symmetry breaking, classical solutions with non-trivial space-time dependence, renormalization theory to all orders, etc. -- a second quantized formalism is clearly preferable. This should involve an infinite component field, one component to describe each of the excitation levels of the string. Such an infinite component field theory had already been worked out more than 10 years ago[36] in the light-cone frame of Section 2. In this frame, the open bosonic string is described by the functional field $\varphi[x^+, p^+, x^i(\sigma)]$; this can be interpreted as the probability amplitude that the string at light-cone time x^+, with total + momentum p^+, has in the transverse coordinates the shape described by the function

$$x^i(\sigma) = \sum_{n=0}^{\infty} x_n{}^i \cos n\sigma , \qquad \sigma \in [0,\pi] . \tag{5.1}$$

According to Section 2, this is the Fourier decomposition appropriate to open strings; $x_0{}^i$ is just the centre-of-mass position in the transverse D-2 coordinates; $x_n{}^i$ are the positions of the harmonic oscillators of Section 2. This string field φ satisfies the following Schrödinger equation:

$$\left[2ip^+ \frac{\partial}{\partial x^+} + H_T\right]\varphi = 0 , \tag{5.2}$$

with

$$H_T = - \frac{\partial^2}{\partial x_0{}^{i2}} + \sum_{n=1}^{\infty} \alpha_{-n}{}^i \alpha_n{}^i . \tag{5.3}$$

In Eq. (5.3), $\alpha_{-n}{}^i$ and $\alpha_n{}^i$ are the creation and annihilation operators of the nth harmonic mode $x_n{}^i$. Rather than deal with a formulation where the position $x_n{}^i$ is diagonal, it may be more convenient to diagonalize the Hamiltonians of these harmonic oscillators. Hence, we introduce a ket$|\psi\rangle$ in the Hilbert space of all these harmonic oscillators, and by definition we have

$$\varphi[x^+, p^+, x^i(\sigma)] = \langle x^i(\sigma)|\psi\rangle . \tag{5.4}$$

We can then expand $|\psi\rangle$ on the occupation number basis:

$$|\psi\rangle = \left[s(x^\mu) + \alpha_{-1}{}^i A^1{}_i(x^\mu) + \alpha_{-1}{}^i \alpha_{-1}{}^j h_{ij}(x^\mu) + \alpha_{-2}{}^i A^2{}_i(x^\mu) + \ldots\right]|0\rangle . \tag{5.5}$$

In this expansion, $s(x^\mu)$ is an ordinary scalar field which describes the ground-state tachyon, $A^1{}_i$ is a transverse Maxwell field for the massless vector particle, etc. The Schrödinger equation (5.2) then reduces to Klein-Gordon equations for these component fields, with a mass operator M which is just

$$\alpha' M^2 = \sum_{n=1}^{\infty} \alpha_{-n}\alpha_n . \tag{5.6}$$

Ten years ago, the cubic and quartic interaction terms which ultimately appear in the equation of motion (5.2) had also been worked out[36]; this resulted in a classical string field theory in the light-cone frame, which, upon quantization, reproduced the S matrix computed earlier by first quantization (very primitive) methods.

In order to use general non-perturbative methods, it is clearly desirable to have a Lorentz covariant formulation. In this formulation, the expansion of the string field $|\psi\rangle$ is much the same as that of Eq. (5.5), except that the oscillators carry a Lorentz vector index μ instead of the transverse index i. Correspondingly, the fields A_μ^{1}, A_μ^{2}, $h_{\mu\nu}$, also carry Lorentz indices. According to Section 2, the string field $|\psi\rangle$ then satisfies the Klein-Gordon equation

$$(L_0 - 1)|\psi\rangle = 0 \tag{5.7}$$

together with the constraints

$$L_n|\psi\rangle = 0 \ , \qquad n = 1, 2, \ldots \ . \tag{5.8}$$

In terms of component fields, Eq. (5.7) is just the on-mass-shell condition. It is interesting to work out the meaning of Eq. (5.8) for the first few fields A_μ^{1}, A_μ^{2}, $h_{\mu\nu}$. Using the explicit expression of L_1 and L_2 given in Section 2, one finds

$$-i\partial_\mu A_\mu^{1} = 0 \ , \tag{5.9}$$

$$-i\partial_\mu h_{\mu\nu} + A_\mu^{2} = 0 \ , \tag{5.10}$$

$$-i\partial_\mu A_\mu^{2} + h_{\mu\mu} = 0 \ . \tag{5.11}$$

Equations (5.9) and (5.10) are given by $L_1|\psi\rangle = 0$, Eq. (5.11) by $L_2|\psi\rangle = 0$. We see that they are clearly gauge conditions on the component fields, and, considering the form of Eq. (5.9), we shall refer to Eq. (5.8) as fixing the Landau gauge. From Section 2 we know that it removes all negative norm states from the theory.

It is then a natural question to ask for more general gauge-invariant equations of motion, from which Eqs. (5.7) and (5.8) would arise in a particular gauge. We expect an infinite number of gauge invariances, one for every one of the field components A_μ^{1}, $h_{\mu\nu}$, A_μ^{2}, etc. This can be achieved mass level by mass level, by successively releasing the constraints on $|\psi\rangle$.

At the first level, we release the constraint $L_1|\psi\rangle = 0$, but $|\psi\rangle$ is still subject to

$$L_2\psi = L_1{}^2\psi = L_3\psi = \ldots = 0 \ . \tag{5.12}$$

Consider now the gauge transformation of $|\psi\rangle$,

$$\delta|\psi\rangle = L_{-1}|\Lambda_1\rangle \ , \tag{5.13}$$

where $|\Lambda_1\rangle$ is a string field of the same type as $|\psi\rangle$,

$$|\Lambda_1\rangle = \left[\lambda_1(x^\mu) + \lambda_1\alpha_{-1}{}^\nu(x^\mu) + \ldots\right]|0\rangle \ , \tag{5.14}$$

and is still subject to $L_1\Lambda_1 = L_2\Lambda_1 = 0$. Using the form of L_{-1} given in Section 2, we find that the transformation (5.13) contains the transformation

$$\delta A_\mu{}^1 = \partial_\mu\lambda_1(x^\mu) \ , \tag{5.15}$$

which is precisely the Abelian transformation expected for a linearized Yang-Mills theory.

An action invariant under $\delta\psi = L_1\Lambda_1$ is given by

$$\frac{1}{2}\langle\psi|\left(L_0 - 1 - \frac{1}{2}L_{-1}L_1\right)|\psi\rangle \ . \tag{5.16}$$

The equation of motion is given by

$$\left(L_0 - 1 - \frac{1}{2}L_{-1}L_1\right)|\psi\rangle = 0 \ . \tag{5.17}$$

Explicitly testing this, we find

$$\left(L_0 - 1 - \frac{1}{2}L_{-1}L_1\right) L_{-1}|\Lambda_1\rangle = (L_0L_{-1} - L_{-1} - L_{-1}L_0)|\Lambda_1\rangle = 0 \tag{5.18}$$

since $L_1|\Lambda_1\rangle = 0$ at this level.

Equation (5.18) was probably known to a few people in the old heyday of string theory but has been rediscovered more recently[37]. The projector P of a string field onto the physical states of Eq. (5.8) has been given elsewhere[38]. For the lowest level, it is given by

$$P = 1 - \frac{1}{2}L_{-1}\frac{1}{L_0}L_1 \ . \tag{5.19}$$

We note that *at this level* the equation of motion is given by

$$(L_0 - 1)\, P\psi = 0 \,. \tag{5.20}$$

It has been proposed more recently[37] that Eq. (5.20) is the correct equation of motion of all levels of the string, and this speculation has been encouraged by the above coincidence for the spin-1 at the first level. In Refs. 37 and 38, P has been formally computed for all levels, and Eq. (5.20) is explicitly non-local at the second level, and more and more so for higher levels. One's suspicions are further aroused by the fact that P can be constructed in an arbitrary space-time dimension, and that D = 26 is not particularly favoured. The clearest way to show that Eq. (5.20) is not the right generalization is to consider the first level of the closed non-oriented bosonic string. At this level, the covariant degrees of freedom are described by a single symmetric field $h_{\mu\nu}$. The analogue for the closed string of the Landau gauge-fixing condition (5.8) is that it satisfies

$$\partial^2 h_{\mu\nu} = 0 \,, \qquad \partial_\mu h_{\mu\nu} = 0 \,. \tag{5.21}$$

These equations tell us that at this level the closed string contains only a spin-2 and a spin-0. The generalization of Eq. (5.20) for this level is

$$\partial^2 R_\mu{}^\varrho R_\nu{}^\lambda h_{\varrho\lambda} = 0 \,, \tag{5.22}$$

where

$$R_\mu{}^\varrho = \left(\delta_\mu{}^\varrho - \frac{\partial_\mu \partial^\varrho}{\partial^2}\right) . \tag{5.23}$$

We immediately recognize that Eq. (5.22) is a non-local equation, which does not admit a Hamiltonian formulation. Making it local by multiplication of ∂^2 leads to additional states. In fact, it is known[39] that there is no Lorentz-invariant, gauge-invariant, local action constructed from $h_{\mu\nu}$ alone, which describes both spin-2 and spin-0. The only way to achieve this is to introduce another field (a scalar) to describe the spin-0, and have the well-known Einstein + massless scalar action. This illustrates the general pattern: the original string field $|\psi\rangle$ does not provide enough degrees of freedom to describe its gauge-covariant propagation. Supplementary string fields are needed. However, it should be possible to set them equal to zero in the Landau gauge [Eqs. (5.7) and (5.8)]. In a recent paper[40] the local gauge-covariant formulation of all known string theories for all levels was given. It requires only a finite number of supplementary string fields, which can all be gauged away to zero. All open strings require only 6 supplementary fields. Closed strings require 20 supplementary fields, except the Ramond-Ramond sector of closed superstrings, which requires 24. It is remarkable that, when made explicit in terms of component fields, this also provides a complete and surprisingly simple solution to the problem of a causal, covariant, and unitary description of massive fields of arbitrarily high 'spin', at least at the non-interacting level. Here we shall only describe the open bosonic string.

We first remark that the gauge invariances of the non-local equation (5.20) can be taken to be

$$\delta|\psi\rangle = L_{-1}|\Lambda_1\rangle , \tag{5.24}$$

$$\delta'|\psi\rangle = L_{-2}|\Lambda_2\rangle , \tag{5.25}$$

and that the Landau gauge of Eq. (5.8) is just

$$L_1|\psi\rangle = L_2|\psi\rangle = 0 , \tag{5.26}$$

all higher L_{-n} gauge invariances and L_n Landau gauges being a consequence of the first two, combined with the Virasoro algebra, Eq. (2.43).

Since supplementary fields are introduced, they will have their own equations of motion, and will also appear in the equation of motion of $|\psi\rangle$. Hence it is natural to introduce two supplementary string fields, $|\varphi_1\rangle$ and $|\varphi_2\rangle$, such that Eqs. (5.26) are their equations of motion in the gauge where they are zero. Thus, the equation of motion of $|\psi\rangle$ becomes

$$(L_0 - 1)|\psi\rangle + L_{-1}|\varphi_1\rangle + L_{-2}|\varphi_2\rangle = 0 . \tag{5.27}$$

This is invariant under the gauge transformations (5.24) and (5.25), provided that

$$\begin{aligned} \delta|\varphi_1\rangle &= -L_0|\Lambda_1\rangle , \qquad \delta|\varphi_2\rangle = 0 \\ \delta'|\varphi_1\rangle &= 0 , \qquad\qquad\;\; \delta'|\varphi_2\rangle = -(L_0 + 1)|\Lambda_1\rangle . \end{aligned} \tag{5.28}$$

This cannot yet be the whole story since the conditions (5.26), now interpreted as a piece of the equations of motion of $|\varphi_1\rangle$ and $|\varphi_2\rangle$, are not gauge invariant by themselves. Gauge invariance can be restored rather simply by adding four more supplementary string fields: $|\zeta_2\rangle$, $|\zeta_3\rangle$, $|\zeta'_3\rangle$, and $|\zeta_4\rangle$, with gauge transformations

$$\delta|\zeta_2\rangle = -L_1|\Lambda_1\rangle\ , \quad \delta|\zeta_3\rangle = 0\ , \quad \delta|\zeta'_3\rangle = -L_2|\Lambda_1\rangle\ , \quad \delta|\zeta_4\rangle = 0\ ,$$

$$\delta'|\zeta_2\rangle = -3|\Lambda'\rangle\ , \quad \delta'|\zeta_3\rangle = -L_1|\Lambda'\rangle\ , \quad \delta'|\zeta'_3\rangle = 0\ , \quad \delta'|\zeta_4\rangle = -L_2|\Lambda'\rangle\ , \tag{5.29}$$

and the equations of motion of $|\varphi_1\rangle$ and $|\varphi_2\rangle$ become

$$L_1\psi + 2|\varphi_1\rangle + L_{-1}|\zeta_2\rangle + L_{-2}|\zeta_3\rangle = 0 \tag{5.30}$$

$$L_2\psi + L_1|\varphi_2\rangle + L_{-1}|\zeta'_3\rangle + L_{-2}|\zeta_4\rangle + 3|\varphi_2\rangle = 0\ . \tag{5.31}$$

In checking the gauge invariance of these equations, the special space-time dimension 26 plays a crucial role, appearing in the commutator of L_2 and L_{-2}.

One might think that this procedure of introducing supplementary fields does not terminate. Remarkably enough, however, no further fields are required to make the equations of motion of $|\zeta_2\rangle$, $|\zeta_3\rangle$, $|\zeta'_3\rangle$, and $|\zeta_4\rangle$, gauge invariant. They are

$$\begin{aligned} &L_1|\varphi_1\rangle - (L_0 + 1)|\zeta_2\rangle + 3|\varphi_2\rangle = 0\ , \\ &L_2|\varphi_1\rangle - (L_0 + 2)|\zeta'_3\rangle = 0\ , \\ &L_1|\varphi_2\rangle - (L_0 + 2)|\zeta_3\rangle = 0\ , \\ &L_2|\varphi_2\rangle - (L_0 + 3)|\zeta_4\rangle = 0\ . \end{aligned} \tag{5.32}$$

The gauge-invariant action from which the above equations follow is

$$\begin{aligned} S = \tfrac{1}{2}\langle\psi|(L_0 - 1)|\psi\rangle &+ \langle\psi|L_{-1}|\varphi_1\rangle + \langle\psi|L_{-2}|\varphi_2\rangle + \langle\varphi_1|\varphi_1\rangle \\ &+ \langle\varphi_1|L_{-1}|\zeta_2\rangle + \langle\varphi_1|L_{-2}|\zeta_3\rangle + 2\langle\varphi_2|\varphi_2\rangle - \tfrac{1}{2}\langle\zeta_2|(L_0 + 1)|\zeta_2\rangle \\ &- \langle\zeta_3|(L_0 + 2)|\zeta'_3\rangle - \tfrac{1}{2}\langle\zeta_4|(L_0 + 3)|\zeta_4\rangle\ . \end{aligned} \tag{5.33}$$

To show that the above answer is correct, we must be able to recover the on-shell gauge conditions (5.7) and (5.8) and set all supplementary fields equal to zero. For this, we start with some particular solution of the gauge-invariant equations of motion (5.27), (5.30), (5.31), and (5.32), and perform on all fields the gauge transformation defined by

$$|\Lambda_1\rangle = \frac{1}{L_0}|\varphi_1\rangle\ , \tag{5.34}$$

$$|\Lambda_2\rangle = \frac{1}{L_0 + 1}|\varphi_2\rangle\ . \tag{5.35}$$

It is a trivial matter to check that one thus obtains

$$(L_0 - 1)(|\psi\rangle + \delta|\psi\rangle + \delta'|\psi\rangle) = 0 \ ,$$
$$L_1(|\psi\rangle + \delta|\psi\rangle + \delta'|\psi\rangle) = L_2(|\psi\rangle + \delta|\psi\rangle + \delta'|\psi\rangle) = 0 \ , \tag{5.36}$$
$$|\varphi_1\rangle + \delta|\varphi_1\rangle + \delta'|\varphi_1\rangle = |\varphi_2\rangle + \delta|\varphi_2\rangle + \delta'|\varphi_2\rangle$$
$$= |\zeta_2\rangle + \delta|\zeta_2\rangle + \delta'|\zeta_2\rangle = \ldots = 0 \ ,$$

so that indeed the action (5.33) describes the same dynamical system as do the original equations (5.7) and (5.8).

We have thus constructed a gauge-invariant free string field theory. The general method for the introduction of interactions has been described in Ref. 41. It proceeds by analogy with the construction of the Yang-Mills theory starting from the free Lagrangian using the Noether method. The interacting gauge transformation of $|\psi\rangle$ contains, besides the inhomogeneous term of Eq. (5.13), a term of first order in the coupling constant g. By analogy with the Yang-Mills gauge transformation, this then must be separately linear in $|\Lambda\rangle$ and in $|\psi\rangle$. The analogue of the Yang-Mills structure constants which appear in this bilinear term thus combines two string fields to give a third one. In the old string theory, there appear such an object, which is nothing else than the three-string vertex. Since this vertex is known to have nice properties with respect to the Virasoro operators L_n, it is a natural building block in the construction of an interacting gauge transformation, and the corresponding invariant action. This construction has been carried out explicitly for the first levels in Ref. 41, to which we refer the reader for the details.

REFERENCES

1. G. Veneziano, Phys. Rep. 9C:199 (1973).
2. J.H. Schwarz, Phys. Rep. 8C:269 (1973).
3. J. Paton and Chan Hong-Mo, Nucl. Phys. B10:519 (1969).
4. J. Scherk, Rev. Mod. Phys. 47:123 (1975).
5. V. Alessandrini, D. Amati, M. Le Bellac and D. Olive, Phys. Rep. 1C:269 (1971).
6. P. Goddard, J. Goldstone, C. Rebbi and C.B. Thorn, Nucl. Phys. B56:109 (1973).
   C. Rebbi, Phys. Rep. 12C:1 (1974).
7. S. Mandelstam, Phys. Rep. 13C:259 (1974).
8. J.H. Schwarz, Phys. Rep. 89C:223 (1982).
9. J. Goldstone and C.B. Thorn, private communication.

10. P. Goddard and C.B. Thorn, Phys. Lett. 40B:235 (1972).
P. Goddard, C. Rebbi and C.B. Thorn, Nuovo Cim. 12A:425 (1972).
11. R.C. Brower, Phys. Rev. D6:1655 (1972).
12. J.-L. Gervais and B. Sakita, Phys. Rev. Lett. 30:719 (1973).
13. S. Mandelstam, Nucl. Phys. B64:205 (1973) and B69:77 (1974).
14. L. Brink, P. Di Vecchia and P. Howe, Phys. Lett. 65B:471 (1976).
15. A.M. Polyakov, Phys. Lett. 103B:207 and 211 (1981).
16. J.-L. Gervais and A. Neveu, Nucl. Phys. B199:59 (1982), B209:125 (1982), B238:125 and 396 (1984).
17. J. Scherk and J.H. Schwarz, Nucl. Phys. B81:118 (1974).
18. P. Ramond, Phys. Rev. D3:2415 (1971).
19. A. Neveu and J.H. Schwarz, Nucl. Phys. B31:86 (1971); Phys. Rev. D4:1109 (1971).
20. J.-L. Gervais and B. Sakita, Nucl. Phys. B34:477 (1971).
21. A. Neveu, J.H. Schwarz and C.B. Thorn, Phys. Lett. 35B:441 (1971).
22. F. Gliozzi, D. Olive and J. Scherk, Phys. Lett. 65B:282 (1976); Nucl. Phys. B122:253 (1977).
23. M.B. Green and J.H. Schwarz, Nucl. Phys. B181:502 (1981), B198:252 and 441 (1982); Phys. Lett. 109B:444 (1982).
24. M.B. Green and J.H. Schwarz, Phys. Lett. 136B:367 (1984).
25. E. Cremmer and B. Julia, Nucl. Phys. B103:399 (1976).
26. L. Alvarez-Gaumé and E. Witten, Nucl. Phys. B234:269 (1983).
27. M.B. Green and J.H. Schwarz, CalTech preprint CALT-68-1224 (1984).
28. M.B. Green and J.H. Schwarz, Phys. Lett. 149B:117 (1984).
J. Thierry-Mieg, Phys. Lett. 156B:199 (1985).
29. J. Scherk and J.H. Schwarz, Phys. Lett. 57B:463 (1975).
30. E. Cremmer and J. Scherk, Nucl. Phys. B103:399 (1976).
31. I.B. Frenkel and V.G. Kac, Inv. Math 62:23 (1980).
32. P. Goddard and D. Olive, Univ. Cambridge preprint DAMTP83/22 (1983), to be published in Proc. Mathematical Sciences Research Institute Workshop on Vertex Operators, Berkeley, 1984, ed. J. Lepowsky (Springer-Verlag, New York, 1984).
33. F. Englert and A. Neveu, preprint CERN TH.4168 (1985).
34. D.J. Gross, F. Harvey, E. Martinec and R. Rohm, Phys. Rev. Lett. 54:502 (1985), Nucl. Phys. B256:253 (1985).
35. J.A. Shapiro, Phys. Rev. D5:1945 (1972).
36. E. Cremmer and J.-L. Gervais, Nucl. Phys. B76:209 (1974) and B90:410 (1975).
M. Kaku and K. Kikkawa, Phys. Rev. D10:1110 and 1823 (1974).
37. T. Banks and M. Peskin, Gauge invariance of string fields, to appear in Proc. Symposium on Anomalies, Geometry and Topology, Argonne, 1985.
M. Kaku and J. Lykken, Supergauge field theory of superstrings, ibid.
D. Friedan, Univ. Chicago preprint EFI85-27 (1985).
38. R.C. Brower and C.B. Thorn, Nucl. Phys B31:163 (1971).
39. P. van Nieuwenhuizen, Nucl. Phys. B60:478 (1973).
40. A. Neveu, H. Nicolai and P.C. West, preprint CERN TH.4233 (1985).
41. A. Neveu and P.C. West, preprint CERN TH.4200 (1985).

KAC-MOODY AND VIRASORO ALGEBRAS IN LOCAL QUANTUM PHYSICS

D.I. Olive

The Blackett Laboratory
Imperial College
London SW7 2BZ

ABSTRACT

These notes are based on lectures given at the Erice International School in Mathematical Physics in July 1985, and treat the Sugawara construction of the Virasoro algebra in terms of a given Kac-Moody algebra, thereby complementing previous lecture notes. Deductions are made and applied to the string theory of unified particle interactions, conformally invariant two dimensional quantum field theories and the phase transitions of two dimensional lattice systems. The 'quark model' representation of Kac-Moody algebras is also treated. A symmetric space criterion is derived for the quantum equivalence of a chiral model with Wess-Zumino term to a free fermion theory.

CHAPTER 1: INTRODUCTION AND BACKGROUND

(1.1) Introduction

Since physics is the study of symmetry in nature, and group theory the mathematical analysis of symmetry it follows that group theory naturally provides appropriate mathematical tools for theoretical physics. Indeed the use of finite groups and infinite groups which are finite dimensional (e.g. the rotation group) is well established. An important feature of nature is locality (or causality) and it seems that the theory of affine Kac-Moody algebras (and their associated Virasoro algebras), provide an extremely powerful yet natural framework for unifying the concepts of symmetry and locality. These algebras are infinite dimensional and so would exponentiate into infinite dimensional groups but at present it is easier to study them as algebras, thereby regarding them as the second stage generalisation of angular momentum theory beyond ordinary finite dimensional Lie algebra theory. They nevertheless constitute the class of infinite dimensional algebra in which the dimension diverges to infinity in the most controlled way.

These lectures are intended to complement other notes on the subject, already produced by myself (OLIVE) and my collaborator, Peter GODDARD. These latter notes will therefore constitute background and parallel reading for these lectures which deal with two related topics much studied in the physics literature in the last year or so, namely the construction of Virasoro generators from Kac-Moody generators by Sugawara's construction (Chapter 2) and the use of the "quark model" to construct Kac-Moody, and hence Virasoro generators (Chapter 3). The treatment will elaborate and synthesise the results of a series of papers written by P. Goddard, A. Kent, W. Nahm and myself.

We shall consider the affine, untwisted Kac-Moody algebra $\hat{g}$ which has commutation relations :

$$[T^i{}_m, T^j{}_n] = if^{ij\ell}T^\ell{}_{m+n} + k\delta^{ij}m\delta_{m+n,0}\,, \qquad (1.1.1)$$

where the central (c-number) term k commutes with all $T^i{}_m$. The suffixes m and n take integer values and $f^{ij\ell}$ are the totally antisymmetric structure constants of the compact Lie algebra g, an orthonormal basis for whose generators is given by $T^i{}_o$($i = 1, \ldots.$ dim g) More discussion of (1.1.1.), the structure of the root system of $\hat{g}$ and its representation theory can be found in my Srní notes (OLIVE) as well as in the mathematical literature (see KAC 1983).

Associated with $\hat{g}$ is a VIRASORO algebra with generators L_m satisfying :

$$[L_m, T^j{}_n] = -nT^j{}_{m+n}\,, \qquad (1.1.2)$$

$$[L_m, L_n] = (m-n)L_{m+n} + (c/12)m(m^2-1)\delta_{m+n,0}\,, \qquad (1.1.3)$$

where c is another central (c-number) term.

Equations (1.1.1.) - (1.1.3.) constitute the system to be studied in these lectures. They constitute the semi-direct product of the Kac-Moody algebra $\hat{g}$ with the Virasoro algebra. One of the key questions of physical interest will concern the possible allowed values of the c-numbers k and c and their interrelation, and this will depend on the physical applications we have in mind as now reviewed.

(1.2) Survey of Physical Applications

There are at least three disparate physical applications of the algebraic structure above, two quite old, namely the current algebra theory of the 1960's and the string theory of particle interactions of the early 1970's. The third is recent and a surprise, the theory of the behaviour of spin systems on two dimensional lattices at their critical temperature at which phase transitions occur. The feature common to all these is the connection with conformally invariant quantum field theories in two dimensions. The fact that we talk about two rather than four dimensions is the price paid for an algebraic structure which is tractable according to present knowledge. Nevertheless it is not such a limitation from the physical point of view since the string theories are currently thought to be the most realistic theory of unified particle interactions and since experimentalists can readily make and study two dimensional substances with lattice structure.

To see the connection with current algebras let us first define the Kac-Moody "field":-

$$T^j(z) = \sum_{m\in Z} z^{-m} T^j_m \tag{1.2.1}$$

where z is a complex variable usually considered in the vicinity of the unit circle. Now define the "current":-

$$J^j(\xi) = (\hbar/2\pi R)\, T^j(e^{i\xi/R}) \tag{1.2.2}$$

In terms of this, the Kac-Moody algebra (1.1.1) reads :

$$[J^i(\xi),J^j(\xi)] = i\hbar f^{ij\ell} J^\ell(\xi)\delta(\xi-\eta) + \frac{i\hbar^2 k\delta^{ij}}{2\pi} \frac{\partial}{\partial\xi}\delta(\xi-\eta). \tag{1.2.3}$$

This is recognizable as a "current algebra" relation (see ADLER AND DASHEN) with the c-number term appearing as the derivative of a delta-function and hence as a SCHWINGER (1959) term. The current J has period $2\pi R$ in its argument yet R does not appear explicitly in (1.2.3.) and so can be as large as we choose. The appearance of Planck's constant, $\hbar$, squared in the Schwinger term emphasises that that term is a second order quantum effect and in fact comes from a loop Feynman diagram as will be verified in section (3.2). The integer suffix in (1.1.1) is seen to be $R/\hbar$ times the momentum conjugate to ξ, quantised because of the period in ξ.

Equations (1.2.3) can arise in at least two apparently different two dimensional quantum field theories. In the theory of free massless "quarks" (or fermions) the commutation relations of currents $(j^{t;i}(t,x), j^{x;i}(t,x))$, $(i=1,2,..\ \dim g)$, bilinear in the quarks are calculated in the usual way. The Schwinger term occurs in the commutator of $j^{t;i}$ with $j^{x;j}$ only. When $j^{t;i}$ and $j^{x;i}$ are added (or subtracted) to obtain the light cone components $j^{+;i}(j^{-;i})$ it is found that $j^{+;i}$, commutes with $j^{-;j}$ and that each individually satisfies (1.2.3). Since the quarks are free and massless their equations of motion imply that $j^{+;i}$ depends on t and x in the combination t - x only. Likewise $j^{-;i}$ depends on t + x only. Hence the current algebra constitutes two commuting copies of (1.2.3) satisfied by $j^{+;i}(t-x)$ and $j^{-;i}(t+x)$.

The energy momentum tensor $\theta^{\mu\nu}$ is the standard one bilinear in the free massless fermions and hence traceless as well as symmetric. Thus there are only two independent components which can be taken to be θ^{++} and θ^{--} depending on t-x and t+x respectively by virtue of the conservation law. The DIRAC-SCHWINGER commutation relations take the form of two commuting Virasoro algebras with θ^{++} related to L_m (and θ^{--} to $\bar{L}_m$) by an analogue of equations (1.2.1) and (1.2.2). Altogether we have two commuting copies of the semidirect product of the Virasoro algebra with $\hat{g}$.

When we try to construct the current algebra (1.2.3) with bosonic fields instead of fermionic ones we find that the most promising possibility occurs when we consider dim g fields constrained to lie on the manifold of the Lie group G obtained by exponentiating g. Unfortunately the commutator of $j^{x;i}$ with $j^{x;j}$ vanishes instead of yielding a term proportional to $f^{ij\ell} j^{t;\ell}$ as in the fermionic case and as a consequence it is impossible to obtain the $\hat{g}$ Kac-Moody algebra unless we heed the recent observation of WITTEN and add to the usual G-invariant kinetic energy a new term called the Wess - Zumino term. When this is added as prescribed by WITTEN yet again we obtain two commuting copies of the semidirect product of a Virasoro algebra with $\hat{g}$, and hence a conformally invariant theory.

When z is related to space and time as explained above, i.e. identified with exp $(i(t+x)/R)$, it follows that the Virasoro generators L_o, $\bar{L}_o$ are proportional to $(H\pm P)$, the translation generators on the light cone.

$$L_o \sim (H+P)/2 \qquad (1.2.4)$$

and hence, in a quantum field theory, intrinsically positive according to the usual desiderata.

In string theory (see the review edited by JACOB) z is related to the parameters τ and σ on the string world sheet in a similar way so that $2\pi R$ is now simply the length of the string measured in σ. A similar sort of algebraic structure to eqns.(1.1.1) - (1.1.3) arises with the Kac-Moody generators now corresponding to the vertex operators for particle emission. Again L_o is intrinsically positive. It is well known from the constructions in string theory that the value of the c-number c occurring in the Virasoro algebra (1.1.3) is given by:

$$c = \text{number of bosons} + \tfrac{1}{2}\,\text{number of fermions}, \qquad (1.2.5)$$

and thus quantised. In quantum field theory these constructions correspond to the energy-momentum tensor for free massless, real bosons and fermions. Thus the spectrum (1.2.5) is a necessary condition for a conformal quantum field theory to be "free". Presumably other values of c indicate "interaction".

Lattice systems automatically possess a fundamental scale, the lattice spacing, but this effectively diverges at the critical temperature leaving a theory which is scale invariant, and local (because of the nearest neighbour interactions) and hence conformally invariant as all conformal currents are conserved by virtue of the tracelessness of the energy-monentum tensor.

In two dimensions the algebra of conformal transformations is infinite dimensional (the sum of two commuting Virasoro algebras with generators L_n and $\bar{L}_n$) as can be seen by considering the infinitesimal conformal transformation:-

$$\delta_n z = \varepsilon_n z^{n+1} \quad (\varepsilon_n \text{ small})$$

generated by L_n. If we identify the Cartesian co-ordinates x, y

$$z = x + iy \tag{1.2.6}$$

instead of exp $(i(t+x)/R)$ as before, we see, putting n=o that, if ε_o is real, δ_o is a scale transformation and that if ε_o is imaginary, δ_o is a rotation. We conclude that $L_o + \bar{L}_o$ is the scale or dilatation generator, D, and $L_o - \bar{L}_o$ the rotation generator, S. Hence, instead of (1.2.4), we have

$$L_o \sim (D+S)/2 \tag{1.2.7}$$

This identification has sometimes been made in quantum field theory too. In either case, (1.2.4) or (1.2.7), L_o has the fundamental property that its spectrum has to be positive.

The realisation that the Virasoro algebra, originally found in string theory is also relevant to conformally invariant quantum field theory has a long history. See GERVAIS AND SAKITA; DEL GIUDICE, DI VECCHIA AND FUBINI; FERRARA, GATTO AND GRILLO; FUBINI, HANSON AND JACKIW; FRIEDAN; BELAVIN, POLYAKOV AND ZAMOLODCHIKOV.

Finally let us mention that there are other applications of Kac-Moody algebras, in which k vanishes, yielding what is called a loop algebra (for example in soliton theory). They are fundamentally different from the applications above with $L_o \geqslant 0$, as we shall now see, and will not be further discussed in these lectures, despite their great interest.

(1.3) The Permitted values of k, c and h

We have explained that in the physical applications considered L_o is fundamentally positive:

$$L_o \geqslant 0 \tag{1.3.1}$$

In addition we usually require that we have a state space with positive definite scalar product and hermiticity conditions:

$$T^{i\,\dagger}_n = T^i_{-n} \quad ; \quad L_n^{\dagger} = L_{-n} \tag{1.3.2}$$

Such representations are called unitary and only these will be considered henceforth.

According to the commutation relations (1.1.2) and (1.1.3) the action of T^i_n or L_n ($n \geq 1$) on an eigenstate of L_o, lowers its eigenvalue by n. In view of (1.3.1) this procedure must stop eventually and hence there must be states $|\Psi\rangle$ satisfying:

$$L_n|\Psi\rangle = 0 \quad , \quad n\geq 1 \qquad (1.3.3a)$$

$$T^i_n|\Psi\rangle = 0 \quad , \quad n\geq 1 \qquad (1.3.3b)$$

These states must form a representation of $g + L_o$ and this representation is irreducible precisely when that of the complete algebra $\hat{g} + L_o$ is. Thus, given irreducibility, all the states $|\Psi\rangle$ have the same L_o eigenvalue, h, say, which indeed has to be positive as

$$0\leq \left|\left|L_{-1}|\Psi\rangle\right|\right| = \langle\Psi|L_1L_{-1}|\Psi\rangle = \langle\Psi|[L_1,L_{-1}]|\Psi\rangle = 2\langle\Psi|L_o|\Psi\rangle = 2h\left|\left||\Psi\rangle\right|\right|. \quad (1.3.4)$$

The physical requirement that L_o is positive, (1.3.1), has implied that we have what is perversely called a "highest weight" representation of $\hat{g}$, and the unitary such possibilities have been classified by mathematicians in a way very close to that of finite dimensional Lie algebra theory. See KAC 1983 and the Srní lectures for a sketch (GODDARD;OLIVE).

In such positive L_o, unitary representations it is possible to establish a remarkable theorem concerning the permitted values of the c-numbers k and c appearing in (1.1.1) and (1.1.3).

<u>Theorem on permitted c-numbers</u>

(i) $2k/\psi^2 = 0,1,2,3,\ldots$ (and is called the level, x) (1.3.4a)

(ii) either $c\geq 1$ or $c=1-6/(m+2)(m+3)$; $m = 0,1,2,3\ldots$ (1.3.4b)

Further if m=0 (so c=0) only the trivial representation with T^i_m and L_n vanishing exists. Likewise if x=0, T^i_m vanishes.

Part (i) of this theorem is well known and proofs can be found in KAC 1983; OLIVE and GODDARD. The result means that k is quantised in units determined by the size of the structure constant of g, conveniently taken to be $\psi^2/2$ where ψ is the highest root of g. We could simplify matters by choosing $\psi^2=2$ but as explained later this condition cannot be maintained simultaneously for a algebra and a subalgebra in general so we prefer to allow ourselves flexibility.

The Virasoro generators L_n are (unlike the T^i_n) intrinsically normalised by the fact that their structure constants are integers and thus c is likewise intrinsically normalised. Part (ii) of the theorem concerning the allowed values of c, is a recent, surprising discovery due to FRIEDAN, QIU AND SHENKER. It states that the spectrum of c resembles that of a scattering problem with a threshold at c=1, with a discrete sequence of bound states accumulating there.

The result part(ii) was established in an indirect way, by tracing zeros of the KAC(1978) determinant (of matrices of scalar products of states $L_{-1}|\Psi\rangle$, $L_{-2}|\Psi\rangle$, $(L_{-1})^2|\Psi\rangle$, etc where $|\Psi\rangle$ satisfies (1.3.3a)). This argument was incomplete in that it failed to establish the existence of unitary representations of (1.1.3) for all the values of c in (1.3.4b). This will be shown by different methods in section (2.6).

That the vanishing of c implies the vanishing of all L_n follows from a related result of FQS (eqn. (1.3.5) below) namely that, if c=0, h=0. Given this, eqn. (1.3.4) and the positive definite nature of the scalar product shows that L_{-1} (and likewise all L_{-n} $n \geq 1$) annihilate $|\Psi\rangle$.

The possible eigenvalues of L_o exceed by a positive integer (≥ 0) the eigenvalues h for a state satisfying (1.3.3a). FRIEDAN, QIU AND SHENKER also showed that there were only a finite number of possibles h values for each of the discrete sequence of c values less than unity, and that these were rational numbers:

$$h = h_{p,q} \equiv \frac{((m+3)p - (m+2)q)^2 - 1}{4(m+2)(m+3)} ; \qquad 1 \leq p \leq m-1 \; ; \; 1 \leq q \leq p. \quad (1.3.5)$$

For example

if c = 0; h = 0

c = 1/2; h = 0, 1/16, 1/2

c = 7/10; h = 0, 3/80, 1/10, 7/16, 3/5, 3/2.

Physically these results are of great importance since the eigenvalues of $L_o + \bar{L}_o$ yield, by (1.2.7) the dimensions of possible fields and hence critical exponents. These are measurable and found to be rational. This is now explained by (1.3.5) which is potentially a consequence of the representation theory of the Virasoro algebra. FRIEDAN, QIU AND SHENKER compared (1.3.5) with the exponents of known lattice models and concluded that c=1/2 for the Ising model, c=7/10 for the tri-critical Ising model and so on.

The only non-trivial value of c common to the sequence (1.3.4b) and equation (1.2.5) is c=1/2. This tells us that the Virasoro algebra for the Ising model can be realised by the energy-momentum tensor for a single, real, free, massless fermion. Thus a new light is shed on a result of ONSAGER, originally found forty years ago. One question to be dealt with now concerns the corresponding construction for higher terms in the sequence (1.3.4b) of discrete c values (see section 2.6).

The purpose of this introductory section has been to set the background and motivate the physical importance of the representation theory of the Virasoro algebra. The reader is urged to consult the references cited above for more information.

CHAPTER 2: SUGAWARA'S CONSTRUCTION OF THE VIRASORO ALGEBRA

(2.1) Sugawara's Construction for g simple

We have written down, in eqns. (1.1.1) to (1.1.3), the semidirect product of the Kac-Moody and Virasoro algebras as if the generators were totally independent, but in fact there is a construction of the Virasoro algebra in terms of bilinears in the Kac-Moody generators which it will be the purpose of these notes to explain and develop. The idea arose naturally in the theory of current algebras (GELL-MANN; GELL-MANN AND NE'EMAN; ADLER AND DASHEN) where it was argued that the full dynamics of the theory should be formulated in terms of currents. This means that the energy momentum tensor should be expressed in terms of currents and it was realised in 1968 by SUGAWARA that this was possible if this tensor was bilinear in currents and if the SCHWINGER (1959) term (often ignored hitherto) was taken into account. This idea was originally applied in four space time dimensions but it was soon realised that it worked most neatly in two dimensions. Since the currents correspond to the T^i_m, and the energy momentum tensor to L_m, Sugawara's construction translated into the present language reads:-

$$\mathcal{L}(z) = \sum_{n\varepsilon Z} z^{-n}\mathcal{L}_n = \frac{1}{2k + Q_\psi} {}^{x}_{x}\sum_{i=1}^{\dim g} T^i(z)\, T^i(z){}^{x}_{x} \tag{2.1.1}$$

or, equivalently:

$$\mathcal{L}_n = \frac{1}{2k + Q_\psi} \sum_{m\varepsilon Z} {}^{x}_{x}\sum_{i=1}^{\dim g} T^i_{m+n}\, T^i_{-m}{}^{x}_{x} \tag{2.1.2}$$

The first statement emphasises that two quantum field operators $T^i(z)$ (see (1.2.1)) are multiplied together at the same point. In order to avoid a singularity and obtain a quantity with finite matrix elements in a highest weight representation it is necessary to introduce a normal ordering denoted by the double crosses whereby the T^i_m with positive suffices are moved to the right of those with negative suffices (since, by (1.3.2), T^i_n ,n≥1, are "like" destruction operators). The singularity is exhibited by the Wick contraction following from this definition and (1.1.1).

$$T^i(z)T^i(\zeta) = {}^{x}_{x}T^i(z)T^i(\zeta){}^{x}_{x} + \frac{kz\zeta}{(z-\zeta)^2} \quad ,|z|>|\zeta| . \tag{2.1.3}$$

Notice that Sugawara's construction is like a Kac-Moody generalisation of the quadratic Casimir operator for g. Notice also that the prefactor is not 1/(2k) (as Sugawara thought), but is subtly "renormalised" to be $(2k + Q_\psi)^{-1}$ where Q_ψ is the quadratic Casimir in the adjoint representation of g (so that ψ denotes its highest weight). The necessity for this prefactor is seen by calculating the L_m, T^i_n commutator, paying attention to the normal ordering. Given the bilinear form of $\mathcal{L}$ in terms of T^i_m a simple determination of the coefficient was given by KNIZHNIK AND ZAMOLODCHIKOV. Let the highest weight states (1.3.3) belong to a representation of g with matrix representation t^i. Then

$$T^i_o\,|\Psi\rangle = |\Psi\rangle\, t^i \tag{2.1.4}$$

If $1/\beta$ is the prefactor, we have by (1.3.3)

$$\mathcal{L}_{-1}|\Psi\rangle = \sum_{i=1}^{\dim g} \beta^{-1} T^i_{-1}\,|\Psi\rangle t^i$$

Acting on this equation with T^j_1 and using its commutation relations (1.1.1) and (1.1.2) with T^i_{-1} and $\mathcal{L}_{-1}$ yields $\beta = 2k+Q_\psi$.

Finally it has to be checked that $\mathcal{L}$ indeed satisfies (1.1.3) and that the c number is actually

$$c_g = \frac{2k \dim g}{2k + Q_\psi} = \frac{x \dim g}{x + \tilde{h}(g)} \quad , \qquad (2.1.5)$$

where x is the level $2k/\psi^2$ mentioned in section (1.3) and $\tilde{h}(g) = Q_\psi/\psi^2$ is called the dual Coxeter number. We shall prove it is an integer and list its values later.

The check is done by using (1.1.2) in the explicit form (2.1.2) for $\mathcal{L}_n$. The c number arises in restoring the normal ordering after the commutation, using (1.1.1).

Notice that Sugawara's construction is automatically unitary in the sense (1.3.2) if the Kac-Moody generators are. Further $\mathcal{L}_o$ is positive. In fact the states of an irreducible representation of $\hat{g}$, annihilated by $T^i{}_n$, n>0, (and hence by L_n, n>0,) form an irreducible representation of g whose quadratic Casimir operator is given by $Q_t = \Sigma\, (t^i)^2$, (see (2.1.4)). Then the eigenvalues of $\mathcal{L}_o$ on these states is :

$$\mathcal{L}_o \longrightarrow Q_t\,/(2k+Q_\psi) > 0 \quad . \qquad (2.1.6)$$

The results described above form the culmination of a long series of papers in the physics literature : CALLAN, DASHEN AND SHARP; SUGAWARA; SOMMERFIELD; COLEMAN, GROSS AND JACKIW; BARDAKCI AND HALPERN; DELL' ANTONIO, FRISHMAN AND ZWANZIGER, and DASHEN AND FRISHMAN, ending with the correct prefactor for level 1 representations of SU(N). More recently the correct general formula (2.1.1) has been given by KNIZHNIK AND ZAMOLODCHIKOV; GODDARD AND OLIVE 1985; and TODOROV. Applications in string theory have been discussed by NEMESCHANSKY AND YANKIELOWICZ and JAIN, SHANKAR AND WADIA. The Sugawara formula has also appeared in the mathematical literature : SEGAL; FRENKEL; KAC 1983 (p.161), and GOODMAN AND WALLACH.

(2.2) Sugawara's construction for g not simple

If the Lie algebra g is semisimple i.e. $g = g_1+g_2+\ldots$ then the relevant construction is simply

$$\mathcal{L}^{g} = \mathcal{L}^{g1} + \mathcal{L}^{g2} + \qquad (2.2.1)$$

where $\mathcal{L}^{g_i}$ denotes (2.1.1) for g_i. The c-number for $\mathcal{L}^g$ is also obtained additively:

$$c_g = c_{g_1} + c_{g_2} + .. \qquad (2.2.2)$$

The result is valid even if g is not semisimple. Thus if g is abelian its structure constants vanish, so $Q_\psi = 0$ and we have simply

$$c_g = \dim g = \text{rank } g \qquad (2.2.3)$$

A physical example of this case is VIRASORO's original construction in string theory. The quantities p^i_m of string theory generate the Kac Moody algebra:

$$[p^i_m, p^i_n] = mk\delta^{ij}\delta_{m+n,o} \qquad (2.2.4)$$

(usually with k=1 chosen) and Sugawara's construction (2.1.1) reduces to Virasoro's construction

$$L(z) = \frac{1}{2k} {}^{x}_{x} \sum_{i=1}^{\text{rankg}} p^i(z)^2 {}^{x}_{x} \qquad (2.2.5)$$

Here the group G is just real space $\mathbb{R}^d$ with rank and dimension d. That the c number is d was originally noted by WEIS. This result also applies if space is compactified to a torus $\mathbb{R}^d$/lattice since it is still an abelian group.

(2.3) Properties of c_g for g simple

We shall show that c_g is a rational number lying between dim g and rank g, attaining its lower bound, rank g, if and only if g is simply laced (i.e. has roots all of the same length) and a level 1 representation of $\hat{g}$ is considered. To do this we shall find an expression for Q_ψ in terms of the root system of g. Q_ψ was the quadratic Casimir in the adjoint representation of g and hence is defined in terms of structure constants of g:

$$Q_\phi \, \delta^{ij} = \sum_{m,n=1}^{\dim g} f^{imn} \, f^{jmn} \tag{2.3.1}$$

It will be useful for later work to consider a more general real representation of g than the adjoint. Let it have antihermitian real generators M^i satisfying

$$[M^i, M^j] = f^{ijk} M^k, \qquad M^{i*} = M^i \tag{2.3.2}$$

Since g is simple (and compact)

$$\mathrm{Tr}(M^i M^j) = - y_M \, \delta^{ij} = -x_M \, \phi^2 \, \delta^{ij} \tag{2.3.3}$$

for some real, positive y_M. Putting i = j and summing from 1 to dim g yields

$$Q_M \dim M = y_M \dim g \tag{2.3.4}$$

where Q_M is the quadratic Casimir $-\sum_1^{\dim g} M^i M^i$. Of course in the adjoint representation Q_M equals Q_ϕ. By summing instead over the rank g generators of the Cartan subalgebra we find

$$y_M = \sum \mu^2 / (\mathrm{rank}\ g) \tag{2.3.5}$$

where the sum is over the weights μ of the representation M (with multiplicities included). The weights of the adjoint representation are the roots and when g is simple they have at most two distinct lengths. Let there be n_L long roots and n_S short roots respectively. Then

$$\dim g = n_L + n_S + \mathrm{rank}\ g \tag{2.3.6}$$

and

$$\left(\frac{\text{long root}}{\text{short root}}\right)^2 \equiv (L/S)^2 = 1, 2 \text{ or } 3$$

By (2.3.5)

$$\tilde{h} \equiv Q_\phi / \phi^2 = (n_L + (L/S)^{-2} n_S)/\mathrm{rank}\ g \tag{2.3.7}$$

so that the dual coxeter number $\tilde{h}$ is certainly rational and hence so is c_g, (2.1.5).

Obviously c_g is less than dim g and we now show what is less obvious, that it is greater than or equal to rank g. By (2.1.5), (2.3.6) and (2.3.7).

$$c_g - \text{rank } g = \frac{n_L(x-1) + n_S(x-(L/S)^{-2})}{x + \tilde{h}} > 0 \qquad (2.3.8)$$

as we consider level $x > 1$ and $(L/S)^{-2} < 1$ by definition.

What is more important is that c_g equals rank g if, and only if, $(L/S)^2 = 1$, i.e. g is simply laced, and x=1, i.e. the $\hat{g}$ representation has level 1 . This conclusion remains true if g is semisimple in the sense that each component g_i is to be simply laced. These conditions are precisely those for the validity of the "vertex operator" representation of $\hat{g}$ and we shall show subsequently that this is no coincidence; there must be such a construction when c_g and rank g coincide.

Finally, introducing more Lie algebra theory (HUMPHREYS), we shall show that the dual Coxeter number $\tilde{h}$ is an integer. If M generates an irreducible representation of g it has a unique highest weight λ, and then

$$Q_M = \lambda(\lambda + 2\rho)$$

where ρ is half the sum of positive roots of g. Hence

$$\tilde{h} = Q_\psi/\psi^2 = 1 + 2\rho.\psi/\psi^2$$

as ψ is the highest weight of the adjoint representation. Now ψ/ψ^2 is a coroot of g and can be expanded as an integer linear combination of the simple coroots of g:

$$\psi/\psi^2 = \sum_{i=1}^{\text{rank } g} m_i \; \alpha_i \; / \; (\alpha_i)^2$$

Then as $2\rho.\alpha_i/(\alpha_i)^2 = 1$,

$$\tilde{h} = 1 + \sum_{1}^{\text{rank } g} m_i \qquad (2.3.9)$$

and is clearly an integer. Its value can easily be calculated for simple groups: A_n: n+1, C_n: n+1, E_6:12, E_7:18, E_8:30, F_4:9, G_2:4 and SO(n): n-2,(n> 5).

(2.4) The situation where g has a subalgebra $h \subset g$

In the first instance let us suppose that g and h are both simple. We can choose an orthonormal basis for g which includes as a subset an orthonormal basis for h, let us say i = 1,2,.. dim h. We automatically obtain a Kac-Moody algebra $\hat{h}$ inheriting the same central term k as $\hat{g}$ But the $\hat{h}$ level may differ from that of $\hat{g}$ because the highest roots of g and h, ψ and ϕ, say, may have unequal lengths.

We may suppose $\hat{g}$ has level $2k/\psi^2$ equal to 1. Then the level of $\hat{h}$, $2k/\phi^2$, must equal 1, 2 or 3 etc. Hence ψ^2/ϕ^2 equals an integer greater than or equal to unity. In general the $\hat{h}$ level must be greater than or equal to the $\hat{g}$ level.

Sugawara's construction can be applied to both $\hat{g}$ and $\hat{h}$ to obtain Virasoro generators $\mathcal{L}^g$ and $\mathcal{L}^h$ respectively. Of course they have different prefactors and different c-numbers in general. We have, by (1.1.2),

$$[\mathcal{L}^g_m, T^j_n] = -n\, T^j_{m+n} \quad , \quad j=1..\dim g,$$

$$[\mathcal{L}^h_m, T^j_n] = -n\, T^j_{m+n} \quad , \quad j=1..\dim h.$$

Hence, subtracting

$$[\mathcal{L}^g_m - \mathcal{L}^h_m, T^j_n] = 0 \quad , \quad j=1..\dim h\ , \qquad (2.4.1)$$

and so, by (2.1.2),

$$[\mathcal{L}^g_m - \mathcal{L}^h_m, \mathcal{L}^h_n] = 0\ . \qquad (2.4.2)$$

Thus $\mathcal{L}^g_m$ has been broken into two mutually commuting pieces.

$$\mathcal{L}^g_m = \mathcal{L}^h_m + K_m \qquad (2.4.3)$$

where K_m commutes with the $\hat{h}$ Kac Moody algebra (2.4.1) and can be thought of as relating to the coset G/H. Further, since by (2.4.2),

$$[\mathcal{L}^g_m, \mathcal{L}^g_n] = [\mathcal{L}^h_m, \mathcal{L}^h_n] + [K_m, K_n] ,$$

we deduce that K_m, like $\mathcal{L}^g_m$ and $\mathcal{L}^h_m$, satisfies a Virasoro algebra and that the c-number is by virtue of (2.1.5)

$$c_K = c_g - c_h = \frac{2k \dim g}{2k + Q_\psi} - \frac{2k \dim h}{2k + Q_\phi} \quad . \qquad (2.4.4)$$

Since the eigenvalues of $\mathcal{L}^g_o$ are bounded below (eqn. (2.1.6)), so are those of K_o. Therefore the highest weight representation of $\hat{g}$ must decompose into highest weight representations of the K_m Virasoro algebra so that in particular we must have c_K positive i.e.

$$c_K \geqslant 0 \qquad (2.4.5)$$

with c_K vanishing if, and only if, K_m vanishes, according to the theorem of section (1.3). Of course if c_K is less than unity it must take one of the values in the discrete series (1.3.4), and in section (2.6) we shall make choices of $h \subset g$ so as to obtain the complete series (1.3.4) with explicitly unitary representations. Thus by subtracting two Virasoro algebras whose c numbers exceed unity (section 2.3), we can obtain a Virasoro algebra with $0 <$ $c < 1$.

These results were due to GODDARD AND OLIVE 1985 and GODDARD, KENT AND OLIVE 1985a. Particular examples of related algebraic structures occur in the earlier work of BARDAKCI AND HALPERN and MANDELSTAM 1973.

Generalisation to the cases when g and h are not simple is easily made using the results of section (2.2).

(2.5) A Quantum Equivalence Theorem and the Vertex Operator Construction

A second use of the preceding argument concerns the deduction that the single numerical condition $c_g = c_h$ implies that

$$\mathcal{L}^g = \mathcal{L}^h \qquad (2.5.1)$$

since if c_K vanishes then so does K in a highest weight representation by the theorem of section (1.3). We call (2.5.1) a quantum equivalence theorem since it establishes the equality of two apparently different operators which, in a conformally invariant quantum field theory, correspond to components of the energy- momentum tensors of two apparently different theories. Two theories with the same energy-momentum tensor have the same Hamiltonian and so are indeed quantum equivalent.

This is a remarkably powerful result which we shall return to in Chapter 3 when we study fermions. As an immediate application consider the Cartan subalgebra t of g which exponentiates to a maximal torus T subgroup of G. T is an abelian group, isomorphic to $\mathbb{R}^{\text{rank } g}/\Lambda_R(g)$ where $\Lambda_R(g)$ is the root lattice of g. We saw in section (2.3) that $c_K = c_g - c_t = c_g - \text{rank } g$ vanishes if and only if g is simply laced, and a level 1 representation of $\hat{g}$ is considered. Thus, when these two conditions are satisfied, the quantum equivalence theorem (2.5.1) states that, choosing $\psi^2 = 2$;

$$\frac{\text{rank } g}{2 \dim g} \sum_{i=1}^{\dim g} {}^x_x T^i(z)T^i(z)^x_x = \tfrac{1}{2} \sum_{i=1}^{\text{rank } g} {}^x_x T^i(z)T^i(z)\,^x_x \,. \qquad (2.5.2)$$

Now the $T^i(z)$, i=1,2, ..rank g, appearing on the right hand side of (2.5.2) and corresponding to the Cartan subalgebra of g satisfy (2.2.4) (with k=1), according to (1.1.1). Thus the equivalence theorem (2.5.2) states that when g is simply laced and the level x=1, the Sugawara construction (2.1.2) equals Virasoro's construction (2.2.5) for a string moving on the maximal torus T of G. This suggests that it must be possible to construct the Kac Moody generators E^α_m corresponding to the step operators E^α of g from the T^i_m (i=1, ..rank g) and indeed this is precisely what the vertex operator construction achieves. One obtains the FUBINI-VENEZIANO vector $Q^i(z)$ by integrating

$$iz \frac{dQ^i}{dz} = T^i(z) \qquad i=1,\ldots\text{rank } g$$

Then the E^α_m are the Laurent coefficients in the expansion of the vertex operator

$$\sum_{m \in Z} z^{-m} \varepsilon(\alpha, T_0^i) E^{\alpha}_{m} = z : e^{i\alpha.Q(z)} : \quad , \tag{2.5.3}$$

where the normal ordering is with respect to the bosonic oscillators which are the Laurent coefficients of $T^i(z)$, $i=1, ..\text{rank } g$, and $\varepsilon(\alpha, T^i_o)$ is a Klein transformation needed to correct some signs in commutation relations. This construction is due to FRENKEL AND KAC and SEGAL. See GODDARD AND OLIVE 1984 for more information, generalisations and the observation that the construction relates to the motion of a string on the maximal torus of the group G. The explicit verification of (2.5.2) is due to FRENKEL.

Two commuting copies of the above result (2.5.2) can be interpreted in terms of two dimensional conformally invariant quantum field theory. In section (1.2) we saw that an interesting example was the "Wess-Zumino" theory, with a field confined to the manifold of a Lie group G, and described by an action consisting of the usual kinetic term plus a "Wess-Zumino" term, each normalised so that the energy momentum tensor was indeed bilinear (à la Sugawara) in conserved currents satisfying the $\hat{g}$ Kac-moody algebra. WITTEN showed that the level of $\hat{g}$ was equal to the only free parameter left in the action, the overall coefficient which had to be an integer to ensure the single valuedness of exp $\{i \text{ Action} / \hbar\}$.

Equation (2.5.2) shows that if g is simply laced and the level is unity then the Wess-Zumino model is quantum equivalent to a free scalar field theory on T, the maximal torus of G. At first sight this is highly surprising because the two equivalent theories possess different numbers of independent fields. Yet the total number of degrees of freedom is infinite in each case. The scalar field on the torus T "feels" the non-abelian structure of G through the periodicity structure of the torus which is specified by the root lattice of g ($T = \mathbb{R}^d / \Lambda_R(g)$).

The NAMBU action for the string can be regarded as a chiral model on the group $\mathbb{R}^d$ defined by flat space. The ghost free nature of the theory depends on the conformal symmetry of this action but then holds only in 26 dimensions (10 for the fermionic string). Actually it is the c-number of the Virasoro algebra which is critical and in flat space this equals the dimension, (2.2.3). If we try to follow the Kaluza-Klein philosophy and let the string move on the manifold of $G \times \mathbb{R}^d$ we must add the Wess-Zumino term to the action in order to retain conformal symmetry and hence the no ghost theorem. Since we must choose g simply laced and level 1 because of the string vertex operators, we have, by the above results, that the critical c number (26) equals d plus the rank of g rather than its dimension. Insofar as supergravity is a special limit of the superstring theory this shows that the recent attempts to use Kaluza Klein theory in supergravity were wrong, because the seven extra dimensions were apparently used in the wrong way.

We believe that the above comments on the relation between string theories and chiral models were due to Witten, as reported by FREUND, OH AND WHEELER. Later NEMESCHANSKY AND YANKIELOWICZ elucidated the point further.

(2.6) The discrete sequence of Virasoro c-numbers less than unity

We saw (equ. 2.3.8) that Sugawara's construction always yielded a Virasoro c-number c_g exceeding the rank of g and hence never less than unity. Nevertheless c less than unity could result from a judicious choice of $h \subset g$ (and level) in the construction $K = \mathcal{L}^g - \mathcal{L}^h$ of section (2.4). We now see that the complete sequence $c = 1 - 6/(m+2)(m+3)$ results from the choice G/H = Sp(m+1)/Sp(m) xSp(1) -with level 1. That Sp(m) x Sp(1) is a subgroup of Sp(m+1) and inherits the same level can be seen from the extended Dynkin diagram for the symplectic algebras;-

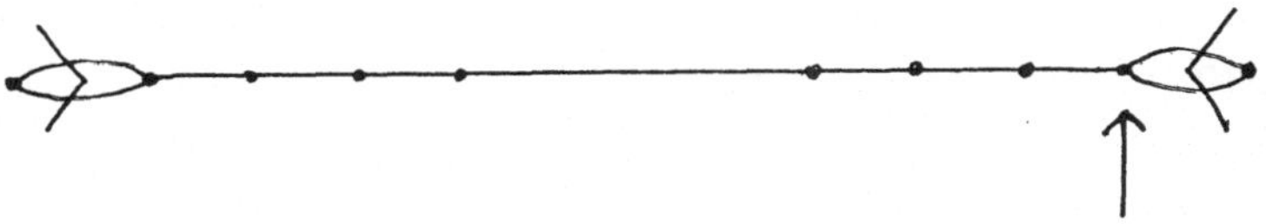

Deletion of the arrowed point leaves the Dynkin diagram for the desired subgroup. Since the deleted point corresponds to a short root it is evident that the highest roots of each factor of h have the same lengths as that of g. Hence as explained in section (2.4) the levels are all the same and hence all of level 1 if we choose that in the first place. Then by (2.1.5)

$$c_{Sp(m)} = \frac{\dim\ Sp(m)}{1 + \tilde{h}(Sp(m))} . \qquad (2.6.2)$$

Sp(m) has long roots $\pm 2e_i$, i=1,2, ..m, short roots $\pm e_i \pm e_j$ $(i \neq j)$, and rank m. Hence $n_L = 2m$, $n_S = 2m(m-1)$ and dim Sp(m) = (2m+1)m. Thus by (2.3.7) the dual Coxeter number $\tilde{h}$ equals

$$\tilde{h}(Sp(m)) = (n_L + n_S\ (L/S)^{-2})/m = (2m + m(m-1))/m = m+1,$$

as quoted at the end of section (2.3).

Hence inserting in (2.6.1)

$$c_{Sp(m)} = \frac{m(2m+1)}{m+2} = 2m - 3 + 6/(m+2),$$

and so, as desired (GODDARD, KENT, AND OLIVE 1985a):

$$c_K = c_{Sp(m+1)} - c_{Sp(m)} - c_{Sp(1)} = 1-6/(m+2)(m+3) \quad . \qquad (2.6.3)$$

We shall see in the next chapter that level 1 Sp(m) Kac-Moody algebra representations can be constructed in a unitary way by considering fermion fields in the defining representation of Sp(m). Partial results already exist on finding within the fermionic Fock space highest weight states of K_n corresponding to the critical exponents (1.3.5) (ALTSCHULER).

The relation between the above construction and the lattice models with the corresponding value of c is not yet clear but it is interesting that the defining representation of Sp(m) used is quaternionic since quaternions are essentially Pauli spin matrices. It is possible to reformulate the above construction in terms of SU(2) groups, using results of the next chapter and this may also aid the physical interpretation. It has led to a proof that all the possible h values (1.3.5) can occur in the above unitary representation (GODDARD,KENT AND OLIVE 1985a and b).

CHAPTER 3: THE QUARK MODEL

(3.1) Kac-Moody Generators Bilinear in Real Fermi Fields

So far we have been talking in general about unitary highest weight representations of (affine untwisted) Kac-Moody algebras without specifying any details of their construction. As was clear, the state space of these representations somewhat resembled a Fock space of a quantum field theory with the "highest weight state" corresponding to the vacuum. We saw in section (2.5) that the "vertex operator" representations (applicable at unit level when g is simply laced) acted in the Fock space of rank g real scalar fields. In this chapter we construct representations in the Fock space of real fermi fields, taking as our point of departure the "quark model" of the 1960's whereby currents are represented bilinearly in fermi fields:

$$T^i(z) = \sum_{n\varepsilon Z} z^{-n}\, T^i_{\ n} = \frac{i}{2} H^\alpha(z) M^i_{\alpha\beta} H^\beta(z) \quad . \qquad (3.1.1)$$

The real, antisymmetric matrices M^i satisfy the commutation relations (2.3.2). The fermi fields $H^\alpha(z)$ are correspondingly real in the sense explained below.

Thus we are using fermions to build a representation of $\hat{g}$ from a real representation of g. The extension to complex representations of g follows in section (3.3).

Because of the sum over integers in (3.1.1) the right hand side must be single valued when the complex variable z encircles the origin. Thus the dim M fermi fields $H^\alpha(z)$ must be either periodic or antiperiodic under this operation:-

$$H^\alpha(ze^{2i\pi}) = \begin{cases} H^\alpha(z) & : \text{periodic}, \quad (R) \qquad (3.1.2a) \\ -H^\alpha(z) & : \text{antiperiodic}, \quad (NS), \qquad (3.1.2b) \end{cases}$$

corresponding respectively to the RAMOND and NEVEU-SCHWARZ fermi fields of string theory.

As we shall see in section 3.6 it is essential to retain both possibilities since the representations (3.1.1) of $\hat{g}$ with different choices (3.1.2) are in general inequivalent. The fermi fields (3.1.2a or b) can be expanded in integral or half integral powers of z respectively:

$$H^\alpha(z) = \sum z^{-n} b^\alpha_n \quad ; \quad \begin{matrix} n \in Z \ (R) \\ n \in Z+\tfrac{1}{2} \ (NS) \end{matrix} \quad . \qquad (3.1.3)$$

The statement of reality for these fields is:

$$b^{\alpha\,\dagger}_n = b^\alpha_{-n} \quad . \qquad (3.1.4)$$

Their anticommutation relations are

$$\{b^\alpha_m, b^\beta_n\} = \delta^{\alpha\beta} \delta_{m+n,o} \quad . \qquad (3.1.5)$$

Thus, when m is distinct from zero, b^α_m and b^α_{-m} form a single fermionic harmonic oscillator. This applies to both Ramond and Neveu-Schwarz fields, but in the former case there is, in addition, a zero mode satisfying

$$\{b^\alpha_o, b^\beta_o\} = \delta^{\alpha\beta}, \qquad (3.1.6)$$

from which we deduce that

$$b^\alpha_o = \frac{1}{\sqrt{2}} \gamma^\alpha (-1)^{\sum_{\alpha=1}^{\dim M} \sum_{m=1}^{\infty} b^\alpha_{-m} b^\alpha_m} , \qquad (3.1.7)$$

where the γ^α are Dirac gamma matrices. The extra phase factor (Klein transformation) is needed to ensure anticommutation with the non-zero modes. We define a vacuum annihilated by the b^α_m with positive suffices m:

$$b^\alpha_m|0\rangle \quad = \quad 0 \qquad\qquad m > 0 \quad . \qquad (3.1.8)$$

This vacuum is unique in the Neveu-Schwarz case but has a $2^{[\dim M/2]}$-fold degeneracy in the Ramond case owing to a Dirac spinor component for the zero modes. ($[p]$ is the integral part of p).

Upon this vacuum we erect a Fock space, and associated with it we define a normal ordering operation denoted by open dots whereby destruction operators are moved to the right of creation operators (with the inclusion of appropriate signs for fermi statistics).

Then normal ordered products of Neveu-Schwarz fields will have finite matrix elements in the Fock space and will be totally antisymmetric in these fields. We require the same to be true for Ramond fields and this means that we extend the definition of normal ordering to the zero modes so as to guarantee antisymmetry:

$$^o_o\, b^\alpha_o\, b^\beta_o\, {}^o_o = (b^\alpha_o b^\beta_o - b^\beta_o b^\alpha_o)/2 = (\gamma_\alpha\gamma_\beta - \gamma_\beta\gamma_\alpha)/4.$$

Using these definitions we can define a Wick contraction function $\Delta(z,\zeta)$:

$$H^\alpha(z)H^\beta(\zeta) = {}^o_o H^\alpha(z)H^\beta(\zeta)^o_o + \delta^{\alpha\beta}\Delta(z,\zeta) \ ; \ |z| > |\zeta|, \qquad (3.1.9)$$

and evaluate it as

$$\Delta(z,\zeta) = \sum_{m=1/2}^{\infty} (\zeta/z)^m = \frac{\sqrt{z\zeta}}{z - \zeta} \qquad (NS), \qquad (3.1.10a)$$

$$= \sum_{m=1}^{\infty} \quad (\zeta/z)^m + 1/2 = \frac{(z+\zeta)/2}{z-\zeta} \qquad \text{(R)}, \qquad (3.1.10b)$$

in the Neveu-Schwarz and Ramond cases respectively. Note that equation (3.1.9) makes explicit the pole singularity in the operator product of the two fermi fields and that the residue is the same in each case. The convergence of the summations in (3.1.10) determines the inequality $|z|>|\zeta|$ in (3.1.9).

(3.2) The Kac-Moody Algebra Commutators

Now we wish to check the Kac-Moody commutation relations (1.1.1) paying particular attention to how the value of k (the central term) depends upon the choice of quark representation (with generators M^i). The product of two generators (3.1.1) is a product of four fermi fields and we shall use WICK's theorem to express this in normal ordered form (with respect to 3.1.9) with the singularities made manifest through the contraction function Δ. This is precisely DYSON's technique for deriving Feynman diagrams and was used extensively in string theory. Nowadays it is called an operator product expansion and it reads, providing $|z| > |\zeta|$, by virtue of (3.1.9):

$$\begin{aligned} {}^{o}_{o}H^{\lambda}(z)H^{\mu}(z)^{o}_{o}\;{}^{o}_{o}H^{\nu}(\zeta)H^{\rho}(\zeta)^{o}_{o} &= {}^{o}_{o}H^{\lambda}(z)H^{\mu}(z)H^{\nu}(\zeta)H^{\rho}(\zeta)^{o}_{o} \\ &+ \Delta(z,\zeta)^{o}_{o}\{H^{\lambda}(z)H^{\rho}(\zeta)\delta^{\mu\nu} + H^{\mu}(z)H^{\nu}(\zeta)\delta^{\lambda\rho} - H^{\lambda}(z)H^{\nu}(\zeta)\delta^{\mu\rho} \\ &\quad - H^{\mu}(z)H^{\rho}(\zeta)\delta^{\lambda\nu}\}^{o}_{o} + \Delta(z,\zeta)^{2}(\delta^{\mu\nu}\delta^{\lambda\rho} - \delta^{\mu\rho}\delta^{\lambda\nu}) \\ &\equiv h^{\lambda\mu\nu\rho}(z,\zeta) \; . \end{aligned} \qquad (3.2.1)$$

Because this has the symmetry property

$$h^{\lambda\mu\nu\rho}(z,\zeta) \quad = \quad h^{\nu\rho\lambda\mu}(\zeta,z)$$

both $T^i{}_m\, T^j{}_n$ and $T^j{}_n\, T^i{}_m$ can be written as double integrals of the same integrand:

$$(-1/4)\,\frac{d\zeta\zeta^n}{2\pi i\zeta}\int\frac{dzz^m}{2\pi iz}\,M^i_{\lambda\mu}M^j_{\nu\rho}\;h^{\lambda\mu\nu\rho}(z,\zeta)$$

differing only in the orientation of the contours of integration. In the first case $|z| > |\zeta|$ and in the second case this inequality is reversed. The only singularities of the integrand are at $z,\zeta = 0$ and ∞ and at $z = \zeta$. Thus the commutator $[T^i{}_m, T^j{}_n]$ is the same integrand integrated over the difference of the two contours which by Cauchy's theorem can be taken to be a z contour enclosing $z = \zeta$ once followed by a ζ integration contour enclosing the origin once:

$$[T^i{}_m, T^j{}_n] = -(1/4) \int_o \frac{d\zeta\zeta^n}{2\pi i\zeta} \int_\zeta \frac{dzz^m}{2\pi iz} M^i_{\lambda\mu} M^j_{\nu\rho} h^{\lambda\mu\nu\rho}(z,\zeta). \quad (3.2.2)$$

The term in (3.2.1) regular at $z = \zeta$ does not contribute to (3.2.2), the pole term (single contraction, Δ) yields $if^{ijk}T^k{}_{m+n}$ while the double pole (double contraction or loop diagram) yields the Schwinger term as follows. It is:

$$-\frac{Tr(M^iM^j)}{2} \int_o \frac{d\zeta\zeta^n}{2\pi i\zeta} \int_\zeta \frac{dzz^m}{2\pi iz} \Delta(z,\zeta)^2 . \quad (3.2.3)$$

By (3.1.10) we have the identity:

$$\Delta(z,\zeta)^2 = \frac{z\zeta}{(z-\zeta)^2} + 4\varepsilon \; ; \qquad \varepsilon = 0,(NS);\ 1/16,(R) . \quad (3.2.4)$$

The ε term drops out of (3.2.3) as it is regular at $z = \zeta$ (so that there is no distiction between the Neveu-Schwarz and Ramond cases as far as the commutation relations are concerned), leaving $-m\delta_{m+n,o}\, Tr(M^iM^j)/2$ for (3.2.3).

Since g is simple we have

$$-\,Tr\,(M^iM^j) = y_M\delta^{ij} = x_M\phi^2\delta^{ij}, \quad (3.2.5)$$

where x_M is positive as the M^i are antihermitian. Thus the central term k equals $y_M/2 = \phi^2 x_M/2$, and the level (1.3.3) is:

$$x = 2k/(\phi^2) = x_M . \quad (3.2.6)$$

Thus by the general theorem of section 1.3 whereby the levels of highest weight representations are quantised, x_M must be an integer and indeed it is, for representations M which are real as we have supposed. x_M is called the

DYNKIN index of the representation. M may describe a reducible representation which decomposes into two real components M_1 and M_2. Then

$$x_M = x_{M_1} + x_{M_2} \quad . \tag{3.2.7}$$

The Dynkin index x_M can be evaluated using the methods of section (2.3). Thus by (2.3.5) and (3.2.5), we have in terms of the weights μ of the representation M;

$$x_M = \sum \mu^2/(\psi^2 \text{ rank } g) \quad . \tag{3.2.8}$$

For example, the defining, n dimensional representation of SO(n) has weights $\pm e_i (i=1,2, \ldots [n/2])$ (and in addition 0 if n is odd) where the e_i are unit vectors and the highest root $\psi = e_1 + e_2$. Thus, since SO(n) has rank $[n/2]$, (3.2.8) shows that $x_M = 1$. Thus we can construct a level 1 representation of $\widehat{SO(n)}$ by choosing fermions in the defining representation of SO(n).

Taking fermions in the adjoint representation of g (assumed simple) yields representations of $\hat{g}$ of level $\tilde{h}(g)$, the dual Coxeter number of g discussed in section (2.3) and listed at the end of that section.

Further Dynkin indices can be calculated as above or read off tables (MACKAY AND PATERA or SLANSKY). It is clear that irreducible representations of g carried by the fermions cannot yield all the possible Kac-Moody levels, and even if reducible representations are allowed, the lowest levels may be missing. This is so for E_8. All representations are real, and the smallest Dynkin index is 30, corresponding to the adjoint representation. Therefore there is no way of obtaining Kac-Moody representations of $\hat{E}_8$ of level less than thirty by the construction (3.1.1). On the other hand level 1 can certainly be obtained by vertex operators since E_8 is simply laced.

(3.3) Quarks in Complex Representations of g

In the original quark model (GELL-MANN) the quarks were not in a real representation but in a complex one, the 3 of SU(3). We shall now treat such representations by building on our results of the previous two sections for real representations and see that new phenomena occur. Let N^i be a complex antihermitian matrix satisfying the Lie algebra g (eqn. 2.3.2) and split it into real and imaginary pieces A^i and B^i :

$$N^i = A^i + i B^i \quad . \tag{3.3.1}$$

A^i is real, antihermitian and hence antisymmetric. B^i is real, hermitian and hence symmetric. We shall construct a real representation of twice the dimension of N^i by replacing i by the 2dim N x 2dim N real matrix :

$$i \rightarrow J = \begin{pmatrix} 0 & I \\ -I & 0 \end{pmatrix} . \tag{3.3.2}$$

Obviously J is antisymmetric and

$$J^2 = - I_{(2 \dim N)} \quad . \tag{3.3.3}$$

Correspondingly

$$N^i \rightarrow M^i = \begin{pmatrix} A^i_i & B^i_i \\ -B^i & A^i \end{pmatrix} = I \otimes A + J \otimes B ,$$

if we think of the 2dim N space as a product of a 2 dimensional space times a dim N dimensional space. Remembering that Tr (A^iB^j) vanishes since A^i and B^j are respectively antisymmetric and symmetric we find :

$$\mathrm{Tr}\ (M^iM^j) = 2\ \mathrm{Tr}\ (A^iA^j - B^iB^j) = 2\ \mathrm{Tr}\ (N^iN^j) \ . \tag{3.3.4}$$

Now the real matrices M^i satisfy all the conditions of our construction (3.1.1) and we see from (3.3.4) that the resultant level is given by :

$$x_M = 2x_N \ . \tag{3.3.5}$$

Since the M representation decomposes into two irreducible components generated by N^i and N^{i*} this is in accord with (3.2.7) which holds even if M_1 and M_2 are complex.

If we denote the matrix $J = M^0$ we see that it fulfils the conditions of our Kac-Moody construction (3.1.1). The matrices M^i, i=0,1,.. dim g, generate the algebra $U_1 + g$ and hence we find the corresponding Kac-Moody algebra. This means that the generators T^0_m commute with the T^i_n ($i \neq 0$) and satisfy :

$$[T^0_m, T^0_n] = (\dim N)\ m\ \delta_{m+n,o} \tag{3.3.6}$$

by our previous result that $2k = -\mathrm{Tr}\ (M^{0^2}) = \mathrm{Tr}\ I = 2 \dim N$ by (3.3.3). Thus when our fermion representation is complex we automatically get an additional

U(1) commuting Kac-Moody algebra without adding any new fermions. When the U(1) generators are included in the Sugawara construction (2.1.1) of the Virasoro algebra the c-number is increased by unity since U(1) has unit rank (see 2.2.3) (GODDARD AND OLIVE 1985).

All we have said is illustrated by the defining, n dimensional representation of SU(n) which is complex, and which yields level one representations of $\widehat{SU(n)}$ as x_N equals 1/2 then so that x_M equals unity by (3.3.5).

When the complex representation is, in addition, pseudoreal, i.e. equivalent to its complex conjugate but not real, the above construction can be extended. It can be shown then that dim N is even and that the representation is actually quaternionic in that it can be written in terms of dimN/2 x dimN/2 matrices with real quaternionic entries. These quaternions can be represented by real 4 x 4 matrices, but commuting with these there exists a second set of 4 x 4 matrices representing quaternions (as SO(4) = SO(3) x SO(3)). Thus it is possible to construct a g + SU(2) algebra of real 2 dim N x 2 dim N matrices and hence the corresponding Kac-Moody algebra. Thus the g Kac-Moody algebra can automatically be extended by a $\widehat{SU(2)}$ Kac-Moody algebra commuting with it without adding any new fermions (GODDARD, KENT AND OLIVE 1985a). This phenomena is illustrated by the defining, 2m dimensional representations of the symplectic Sp(m) algebra of rank m, which is pseudoreal and indeed yields level one representations of $\widehat{Sp(m)}$. In the case of Sp(1) this is related to what used to be called the Pauli-Gursey symmetry in particle physics.

(3.4) Two Virasoro algebras

It is well known in fermionic string theory that it is possible to construct a Virasoro algebra bilinear in dim M fermions :

$$L(z) \;=\; \sum_{n\varepsilon Z} z^{-n} L_n \;=\; {}^{0}_{0}\, \tfrac{1}{2} \sum_{\alpha=1}^{\dim M} z \frac{dH^{\alpha}}{dz}(z)\, H^{\alpha}(z)\, {}^{0}_{0} + \varepsilon \dim M, \qquad (3.4.1)$$

where, as before, (3.2.4), ε equals 0 or 1/16 according as Neveu-Schwarz or Ramond fields are considered. The c-number for this L_n Virasoro algebra is:

$$c \;=\; \dim M/2\;. \qquad (3.4.2)$$

This could be proved directly using the Wick expansion (3.2.1) but we shall obtain an indirect proof by relating (3.4.1) to the Sugawara construction (3.1.1) and using the results of chapter 2.

In terms of conformally invariant quantum field theories in two dimensions (3.4.1) relates to the energy momentum tensor for dim M real free massless fermions of definite helicity.

Note that the addition of ε dim M affects L_0 only and implies that whereas L_0 vanishes on the Neveu-Schwarz vacuum it yields dimM/16 on the Ramond vacuum (which is degenerate). If dim M equals unity, c = 1/2, (3.4.2), and the weight 1/16 is as predicted by the FQS formula (1.3.5) and is relevant to the critical exponents of the Ising model.

When we constructed a $\hat{g}$ algebra out of dim M fermions by the quark model (3.1.1) Sugawara's construction (2.1.1) yields a second Virasoro algebra $\mathcal{L}_m$, apparently quadrilinear in fermions. When its c-number, (2.1.4), differs from dim M/2, (3.4.2), we can be sure that L_m, (2.4.1), and $\mathcal{L}_m$,(2.1.1), differ. Nevertheless we shall find a relation using the Wick identity (3.2.1) under the supposition that M now describes a real, irreducible representation of g. By the normal ordering of the T^i_m's denoted by crosses (see eqn. 2.1.3) :

$$\sum_{i=1}^{\dim g} T^i(z)T^i(\zeta) = {}^{\times}_{\times} \sum_{i=1}^{\dim g} T^i(z)T^i(\zeta) {}^{\times}_{\times} + \frac{k \dim g \, z\zeta}{(z-\zeta)^2} ,$$

where $|z|>|\zeta|$. But according to (3.2.1) involving the fermionic normal ordering denoted by open dots the same expression equals (for $|z|>|\zeta|$)

$$= {}^{\circ}_{\circ} \sum_{i=1}^{\dim g} T^i(z)\, T^i(\zeta) {}^{\circ}_{\circ} + Q_M {}^{\circ}_{\circ} \sum_{\alpha=1}^{\dim M} H^\alpha(z)\, H^\alpha(\zeta) {}^{\circ}_{\circ} \Delta(z,\zeta) +$$

$$(1/2)\dim g \; y_M \; \Delta(z,\zeta)^2,$$

where Q_M is the quadratic Casimir operator $(-\Sigma M^{i\,2})$ and is proportional to the unit matrix, using Schur's lemma and the irreducibility of M. The quantities which are normal ordered have finite limits as z and ξ coalesce. It follows that the singular quantities must cancel out of the above equality leaving something regular. The double pole indeed cancels if k equals $y_M/2$, thereby confirming our previous result (3.2.6). In fact, using (3.2.4) and (2.3.4) we have

$$(1/2)\dim g\; y_M\; \Delta(z,\zeta)^2 - \frac{kz\zeta \dim g}{(z-\zeta)^2} = 2\varepsilon \dim g\; y_M = 2\varepsilon\, Q_M \dim M\ .$$

The term with the single pole is regular as its residue, the normal ordered product of $H^\alpha(z)$ and $H^\alpha(\zeta)$ summed, vanishes at $z = \zeta$ by the antisymmetry property. By d'Hôpital's rule, the result is $Q_M\ {}^{0}_{0} \sum_{\alpha=1}^{\dim M} z \frac{dH^\alpha}{dz} H^\alpha\ {}^{0}_{0}$. Putting these results together, we have, in the limit $z = \zeta$, a relation between the two types of normal ordering :

$$
{}^{x}_{x} \sum_{i=1}^{\dim g} T^i(z)T^i(z)\ {}^{x}_{x} = {}^{0}_{0} \sum_{i=1}^{\dim g} T^i(z)T^i(z)\ {}^{0}_{0} + 2Q_M L(z) \qquad (3.4.3)
$$

with $L(z)$ given by (3.4.1). Notice that (3.4.3) holds for Neveu-Schwarz or Ramond fields.

According to the definition of fermionic normal ordering in section (3.1) the term with the open dots in (3.4.3) must be totally antisymmetric in the fermion fields, and hence can be written :

$$
1/3\ {}^{0}_{0}\, H^\alpha(z)H^\beta(z)H^\gamma(z)H^\delta(z)\, {}^{0}_{0} \left\{ \sum_{i=1}^{\dim g} (M^i_{\alpha\beta}M^i_{\gamma\delta} + M^i_{\beta\gamma}M^i_{\alpha\delta} + M^i_{\gamma\alpha}M^i_{\beta\delta}) \right\}.
$$

It follows that this term vanishes if, and only if :

$$
\sum_{i=1}^{\dim g} (M^i_{\alpha\beta}M^i_{\gamma\delta} + M^i_{\beta\gamma}M^i_{\alpha\delta} + M^i_{\gamma\alpha}M^i_{\gamma\delta}) = 0. \qquad (3.4.4)
$$

Eqn. (3.4.4) certainly holds for fermions in the adjoint representation of g since the M^i are the structure constants of g and eqn. (3.4.4) is simply the Jacobi identity for them. Then the identity (3.4.3) reads :

$$
\frac{1}{2Q_\phi}\ {}^{x}_{x} \sum_{i=1}^{\dim g} T^i(z)\, T^i(z)\ {}^{x}_{x} = L\,(z)\ .
$$

The left hand side is the Sugawara construction of the Virasoro algebra (2.1.1) for level $\tilde{h}(g)$ and is seen to equal the free fermion construction (3.4.1). By (2.1.4) the c number is $\dim g/2$. Thus we have established that the free fermion construction (3.4.1) does satisfy the Virasoro algebra with c number (3.4.2.) at least when $\dim M$ equals the dimension of a simple Lie algebra g, but it is not difficult to extend the result to $\dim M$ any integer.

Expression (3.4.4) also vanishes for fermions in the defining representation of g = SO(n). It is convenient to label the generators as an antisymmetric tensor so that :

$$(M^{ij})_{\alpha\beta} = \delta^{i}{}_{\alpha}\,\delta^{j}{}_{\beta} - \delta^{i}{}_{\beta}\delta^{j}{}_{\alpha}$$

and, by (3.1.1) :

$$T^{ij}(z) = i\ {}^{0}_{0}\ H^{i}(z)\ H^{j}(z)\ {}^{0}_{0}\ .$$

Evidently ${}^{0}_{0}\ T^{ij}(z)\ T^{ij}(z)\ {}^{0}_{0}$ vanishes by the antisymmetry property as $H^{i}(z)$ is repeated. It follows again that the free fermion construction (3.4.1) with n = dim M satisfies the Virasoro algebra with c number (3.4.2).

Whenever we have a real representation M of g with dimension dim M it follows automatically that :

$$g \subset SO(\dim M).$$

Therefore we have precisely the subalgebra situation discussed in section (2.4) with g replaced by SO(dim M) and h by g. Hence, by that analysis (eqn. 2.4.3) :

$$L(z) = \mathcal{L}^{g}(z) + K(z), \qquad (3.4.5)$$

where we use the fact, just shown, that the Sugarawa construction for SO(dim M) with dim M fermion equals the free fermion construction (3.4.1). Thus K_m is a Virasoro algebra commuting with the $\hat{g}$ Kac-Moody generators and hence with $\mathcal{L}^{g}$ (GODDARD AND OLIVE 85). By the 'quantum equivalence' theorem of section (2.5), K vanishes and $\mathcal{L}^{g}$ and L are equal if and only if c_K vanishes i.e.

$$c_g \equiv \frac{x_M \dim g}{x_M + \tilde{h}} = (\dim M)/2 \qquad (3.4.6)$$

By (2.3.3) this can be written

$$2Q_M = \psi^{2}(x_M + \tilde{h}) = 2k + Q_{\psi} \qquad (3.4.7)$$

which, by the identity (3.4.3), implies (3.4.4). Thus equations (3.4.6) and (3.4.4) are equivalent conditions for the equality of the Sugawara and free fermion constructions of the Virasoro algebra generators.

Let us illustrate condition (3.4.6) in the two cases in which we know K vanishes. If the fermions lie in the adjoint representation of g, x_M equals $\tilde{h}$, dim M = dim g and hence (3.4.6) is indeed satisfied. If the fermions lie in the defining representation of SO(n) (n>5) we saw that $x_M = 1$, dim M = n. As $\tilde{h}(SO(n)) = n-2$, dim SO(n) = n(n-1)/2, (3.4.6) is again satisfied.

Another interesting example concerns fermions in the n-dimensional representation of SU(n). As explained in the previous section we must consider 2n real fermions lying in a real, reducible representation with generators M. Then $x_M = 1$ and as $\tilde{h}(SU(n)) = n$, dim $(SU(n)) = n^2-1$, the left hand side of (3.4.6) equals $(n^2-1)/(1+n) = n-1$, one less than dim M/2 = 2n/2 = n, the right hand side of (3.4.6).

But we saw in the previous section that we can construct U(1) currents commuting with the SU(n) currents. Furthermore, as we saw in section (2.2) we can construct a Virasoro algebra $\mathcal{L}^{U(1)}$ with these U(1) currents. It commutes with $\mathcal{L}^{SU(n)}$ and has unit c-number. Hence, adding these two Virasoro generators $\mathcal{L}^{U(1)}$ and $\mathcal{L}^{SU(n)}$ we obtain a Virasoro algebra with c-number n now equal to precisely half the number of free real fermions. We conclude that the Sugawara construction for U(n) equals the free fermion construction (3.4.1). This example underlines the importance of recognising the U(1) currents as in the previous section. Actually this example has been known for a long time in the physics literature although it was derived quite differently (BARDAKCI AND HALPERN; DASHEN AND FRISHMAN).

Many more interesting solutions to (3.4.6) can be found by trial and error but we shall now see how to find the most general solution.

(3.5) Symmetric Spaces and a No-Interaction Theorem

We have seen that, at least if g is simple, the alternative conditions (3.4.4) or (3.4.6) are necessary and sufficient for the quantum equality of the two constuctions of the Virasoro algebra out of fermions, namely the Sugawara form (2.1.1) using fermion currents (3.1.1) and the free form (3.4.1). Since the Virasoro generators correspond to either the (++) or (--) components of the energy momentum tensor in two dimensional conformally invariant quantum field theories, we see that we have a numerical condition that an apparently interacting theory of fermions (with Sugawara energy momentum tensor) reduces to a free theory with no interactions. The mathematical significance of this result will be discussed in section (3.6).

We shall now find the complete solution to the alternative conditions (3.4.4) or (3.4.6) (and their generalisations when g is not simple). The key observation is that (3.4.4) constitutes the cyclic identity for the Riemann tensor of a symmetric space.

The theorem, due to GODDARD, OLIVE AND NAHM, is that the necessary and sufficient condition that the Sugawara $\mathcal{L}^g$ equals the free (3.4.1), is that there exists a group G' containing G, such that G'/G is a symmetric space whose tangent space generators transform under G in the same way as the fermions.

Thus, if we decompose g' into even and odd parts:

$$g' = g + p$$

The even generators, being g generators $T^i \equiv T^i_{o}$, satisfy

$$[T^i, T^j] = if^{ij\ell} T^\ell \tag{3.5.1}$$

while the odd generators p^α, being orthogonal to the even generators and transforming according to M, like the fermions, satisfy

$$[T^i, p^\alpha] = ip^\beta M^i_{\beta\alpha} \tag{3.5.2}$$

Finally because we have a symmetric space the odd generators must close on the even generators.

$$[p^\alpha, p^\beta] = X^{\alpha\beta}_{\ \ \ell}\, T^\ell$$

If g is simple we can choose

$$\mathrm{Tr}(T^i T^j) = y\delta^{ij} \quad ; \quad \mathrm{Tr}(p^\alpha p^\beta) = y\delta^{\alpha\beta}$$

Then we find

$$iyX^{\alpha\beta}_{\ell} = Tr(T^{\ell}[p^{\alpha},p^{\beta}]) = Tr([T^{\ell},p^{\alpha}]p^{\beta}) = iyM^{\ell}_{\alpha\beta}$$

Thus

$$[p^{\alpha},p^{\beta}] = i\, M^{\ell}_{\alpha\beta}\, T^{\ell}$$

so that by (3.5.2) and (3.5.3)

$$[[p^{\alpha},p^{\beta}],p^{\gamma}] = \sum_{i=1}^{\dim g} M^{i}_{\alpha\beta} M^{i}_{\gamma\delta} p^{\delta}$$

This means that $\sum_{i=1}^{\dim g} M^{i}_{\alpha\beta} M^{i}_{\gamma\delta}$ is the Riemann tensor of the symmetric space G'/G. The Jacobi identity for the p generators is the cyclic identity for the Riemann tensor and takes the form of our condition (3.4.4). Thus the existence of the symmetric space, as described, guarantees condition (3.4.4) and hence the equality of $\mathcal{L}^{g}$ and L as claimed. Conversely (3.4.4) guarantees that given (3.5.1) and (3.5.2) we can construct the symmetric space G'/G.

The virtue of the above result is that symmetric spaces have been classified by mathematicians, so that all possible cases of our "no interaction theorem" for fermions, $\mathcal{L}^{g} = L$, are listed in this classification, which can be found in HELGASON's book (p.518 for type I) or in the paper (GODDARD, OLIVE AND NAHM). In this list g need not be simple nor even semisimple, but the above arguments can easily be modified to include these possibilities. We have already mentioned the case g = U(n) when g is not semisimple. The corresponding symmetric space here is the complex projective space CP(n)=SU(n+1)/U(n). The example with n fermions transforming with respect to SO(n) corresponds to the sphere S^{n}=SO(n+1)/SO(n). These are the simplest examples of what are known as type I symmetric spaces. In our other example fermions transformed according to the adjoint representation of a simple group G. This corresponds to the type II symmetric space GxG/G. More details can be found in the original paper.

The above "no interaction theorem" was known fifteen years ago in the cases that we now see correspond to the sphere S^{n} and the complex projective space CP(n), but the proof was 'by hand'. Physicists seem to have concluded that all fermionic Sugawara models are therefore free and hence uninteresting but as we have seen, this is false. Even the result that certain fermionic Sugawara theories are free is now seen to be interesting in view of WITTEN's observation, mentioned earlier, that a totally different and apparently highly non linear theory, namely the chiral model with bosonic fields constrained to the manifold of a Lie group G, also exhibits the same Kac-Moody Sugawara structure if a Wess-Zumino term is added so as to yield equations of motion $\partial_+ j_- = 0$ and $\partial_- j_+ = 0$ where $j_- = g^{-1}\partial_- g$ and $j_+ = \partial_+ g\, g^{-1}$ are the "left" and "right" currents respectively. In order to ensure the single valuedness of exp(iAction/$\hbar$) the (suitably normalised) coefficient of the Wess-Zumino term has to be an integer x which turns out to be the level of the Kac-Moody $\hat{g}$ representation satisfied by the currents j_+ and j_- (which mutually commute). Further the energy-momentum tensor has the Sugawara form. As we have seen the same algebraic structure can be realized by the quark currents of free fermions but for dynamical consistency the Sugawara energy-momentum tensor must equal the free fermion energy-momentum tensor. The theorem of GODDARD, OLIVE AND NAHM above enumerates precisely the cases when this identity occurs in terms of the possible combinations of the choice of algebra g and level x. Thus we know precisely when the Wess-Zumino model can be quantum equivalent to a free fermion theory and have complemented the results of section (2.5) concerning the conditions for quantum equivalence to a free boson theory (on the maximal torus, in fact). The only overlap between the two results is for level 1 of $\widehat{SO(2n)}$ and this is a way of understanding the fermion-boson equivalence of SKYRME; COLEMAN; MANDELSTAM 1975 and others (see GODDARD AND OLIVE 1984).

(3.6) Finite Reducibility with an SO(n) Illustration

The representations of the Kac-Moody algebra $\hat{g}$ of physical interest, namely the highest weight, unitary ones (such that L_o is positive), have been classified and it is known that there exists a finite number of inequivalent irreducible representations with a given level x. This is because the states annihilated by the $T^i{}_m$, (m>1), form a representation of g which is irreducible under g if, and only if, the complete representation is irreducible under $\hat{g}$ since the two representations then share a common highest weight state $|\lambda, x\rangle$. In general we have

$$2\psi.\lambda/\psi^2 \leq x \qquad (3.6.1a)$$

or

$$\sum_{i=1}^{\text{rank } g} m_i x_i \leq x \qquad (3.6.1b)$$

in the notation of OLIVE. Clearly (3.6.1b) has a finite number of integer solutions x_i since the m_i are integers ≥ 1.

The quark model construction (3.1.1) yields a specific level, $x = x_M$ and it is natural to enquire how many irreducible components it contains (these being necessarily of the same level). The answer could be finite or infinite and unfortunately the latter case is the generic case. Thus the irreducible components usually occur with infinite multiplicity but there are special cases where this mathematical horror is evaded and these are furnished by the theorem of the previous section stating that $K = L-\mathcal{L}^g$ vanishes when there is a symmetric space G'/G.

To see this, first note that if K_m does not vanish it commutes with the $\hat{g}$ generators (see eq. (2.4.1)) so that the $\hat{g}$ highest weight states form a representation of the K_m Virasoro algebra. Since all non-trivial (i.e.$K \neq 0$) representations are infinite dimensional the infinite multiplicity follows (GODDARD OLIVE AND NAHM). Conversely (supposing g to be simple), if K_n vanishes, $\mathcal{L}_o$ and L_o are equal, in particular. Now from (2.1.1) we see that on a highest weight state of $\hat{g}$, $\mathcal{L}_o$ has the eigenvalue (by eqn.(2.1.6)):

$$\mathcal{L}_o \longrightarrow Q_\lambda/(2K+Q_\phi) \qquad (3.6.2)$$

which is bounded as λ varies over a finite range of possibilities given by (3.6.1b) as x is fixed. But $\mathcal{L}_o$ equals L_o, (3.4.1), whose eigenvalues are ϵdim M plus the total mode number (Σn_i) of the Fock space element considered, $(b^{\alpha_1}_{-n_1} b^{\alpha_2}_{-n_2} \ldots|0\rangle, n_i>0)$. Clearly there are only a finite number of states with given L_o eigenvalue and the finite multiplicity therefore follows. These statements can be extended to the case that g be semisimple but modification is needed otherwise. In the special case that g is simple, is of the same rank as g', and the Ramond fields only are used, the result has been already found by KAC AND PETERSON. They used the method of characters and also gained information about the decomposition into irreducible representations in their special situation.

Actually if K_m vanishes (by virtue of the symmetric space theorem) the expression (3.6.2) for the $\mathcal{L}_o$ eigenvalue simplifies (using 3.4.7):

$$\mathcal{L}_o \longrightarrow Q_\lambda/(2Q_M) \qquad (3.6.3)$$

But the L_o eigenvalue has to be p+(dim M)/16 or p/2 (where p is an integer

>0) according as Ramond or Neveu-Schwarz field are used. Hence

$$Q_\lambda/(2Q_M) = \begin{cases} p + (\dim M)/16 & (R) \quad (3.6.4a) \\ p/2 & (NS) \quad (3.6.4b) \end{cases}$$

This is highly surprising but we shall illustrate its validity by considering fermions in the defining representation of $g = SO(n)$, corresponding to the symmetric space $S^n = SO(n+1)/SO(n)$. As we have seen (after equation (3.2.8)) this yields level 1 and hence, by (3.6.1b) and the theory of representations of (SO(n) the possibilities $\lambda = 0$; $\lambda = \lambda_v \equiv e_1$ (the defining, n-dimensional representation of SO(n)) ; and $\lambda = \lambda_s$ (λ_t), the spinor representation of SO(n) which splits into two pieces when n is even. These weights are all fundamental weights and so correspond to specific points of the extended Dynkin diagrams in the usual way:

n even:

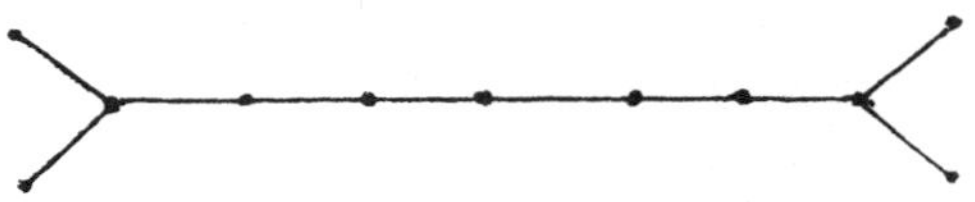

n odd:

If $\lambda = 0$, Q_λ vanishes, so by (3.6.3) $\mathcal{L}_0$ vanishes. There is only one state with zero L_0 eigenvalue and that is $|0_{NS}\rangle$, the vacuum in the Neveu-Schwarz sector. If $\lambda = \lambda_v$, we have, by (2.3.3) and (2.3.4) that Q_{λ_v} equals $\psi^2\dim SO(n)/n = (n-1)\psi^2/2$. Hence $\mathcal{L}_0$ simply has eigenvalue ½ by (3.6.3) and there are precisely n states with this L_0 eigenvalue, namely $b^\alpha_{-\frac{1}{2}}|0_{NS}\rangle$, $\alpha=1,2,\ldots n$. These indeed form an n dimensional representation of $g = SO(n)$.

For the spinor weight $\lambda = \lambda_s$ (or λ_t) the quadratic Casimir equals $Q_{\lambda_s} = y_{\lambda_s} \dim g/\dim M = (\Sigma\mu^2) \dim g/(\dim M \text{ rank } g)$ by equation (2.3.5). But any weight of the spinor representation has components $\pm\sqrt{(\psi^2/8)}$ so that its length squared is always ψ^2rank g/8 Thus

$$Q_{\lambda_s} = (\dim SO(n)\ \psi^2)/8 = n(n-1)\psi^2/16$$

Hence the $\mathcal{L}_0$ eigenvalue (3.6.3) equals

$$Q_{\lambda_s}/2Q_M = n/16 = (\dim M)/16$$

The only possible states are the degenerate Ramond vacua which indeed fall into the irreducible spinor representations of SO(n) with unit multiplicity (see the discussion after equation (3.1.8)).

We see that the minimal L_o eigenvalues indeed agree with (3.6.4), and that each possible level 1 irreducible representation occurs just once (with no degeneracy) and that both Ramond and Neveu-Schwarz fields are necessary to obtain this complete and neat picture.

These results were first obtained, differently, by FRENKEL. As we have mentioned before, when n is even, SO(n) is simply laced and its level one representations can therefore be obtained, equivalently, by the vertex operator construction. This equivalence is the basis of SKYRME's fermion-boson equivalence.

CHAPTER 4: REVIEW OF FURTHER DEVELOPMENTS

We close by mentioning some of the topics not treated in the preceding chapters but nevertheless closely related.

In section (2.5) we stated that, following FRENKEL AND KAC and SEGAL, vertex operators can be used to construct all the level 1 irreducible representations of the Kac-Moody algebras $\hat{g}$ when g is simply laced (i.e. of A, D or E type). The vertex operators can also be used (for g = D type) to construct fermion fields of either Ramond or Neveu-Schwarz type (FRENKEL), thereby making a link with the work of chapter 3. See GODDARD AND OLIVE 1984 for a review. These fermion fields must transform according to the defining (vector) representation of D_r, unless $g = D_4$,i.e. so(8), when spinorial fermion fields can also be constructed, owing to the triality property of so(8) (GODDARD, OLIVE AND SCHWIMMER). This is the mathematical fact which enabled GREEN AND SCHWARZ 1981 to formulate the superstring of GLIOZZI, OLIVE AND SCHERK in an explicitly supersymmetric way. This same fact can also be exploited to construct the unit level $\hat{E}_8$ representation out of fermions in an unconventional way thereby circumventing the limitations mentioned at the end of section (3.2) (GODDARD, OLIVE AND SCHWIMMER).

Yet more generalisations of the vertex operator construction are possible. One interesting possibility, developed by GODDARD AND OLIVE 1984 is to regard the FRENKEL-KAC construction as a "transverse" or "light cone gauge" construction in the sense of string theory and consider instead a "covariant" version with contour integrals of the vertex operator associated with length squared 2 points of any integral lattice, with Euclidean, singular or Lorentzian metric, thereby obtaining simply laced finite dimensional Lie algebras, affine untwisted Kac-Moody algebras or Lorentzian algebras respectively. Nestings of different sorts of algebra inside each other correspond to the nestings of the different types of lattice. Particularly interesting lattices to consider are the self-dual even ones because of their relation to electromagnetic duality conjectures (GODDARD, NUYTS AND OLIVE) and their associated modularity properties. The no-ghost theorem of string theory (BROWER; GODDARD AND THORN) implies that only three such Lorentzian lattices are of interest, those in 10, 18 or 26 dimensions. The first two contain the weight lattices of E_8, E_8+E_8 and Spin $(32)/Z_2$ in an interesting way as explained by GODDARD AND OLIVE 1984. After the dramatic discovery by GREEN AND SCHWARZ 1984 that the latter two Lie algebras constituted the unique anomaly free gauge algebras for supersymmetric gauge theories in ten dimensions, the above ideas were further developed by GROSS, HARVEY, MARTINEC AND ROHM.

Supersymmetric generalisations of the work of chapters 2 and 3 have been developed and the most striking feature is that supersymmetry can be realised with only fundamental fermion fields and no bosons (GODDARD AND OLIVE 1985; GODDARD, KENT AND OLIVE 1985a; FRIEDAN AND SHENKER; KAC AND TODOROV; DI VECCHIA, KNIZHNIK, PETERSEN AND ROSSI).

We end by remarking that although one of the remarkable and exciting features of the study of Kac-Moody and Virasoro algebras is the extent to which much of mathematics and physics becomes interrelated and unified, there is no doubt that much remains to be explored and discovered.

I wish to thank Peter Goddard, Adrian Kent, Werner Nahm and Adam Schwimmer for discussions.

REFERENCES

ADLER S. AND DASHEN R. : Current Algebras and applications to particle physics (Benjamin, New York 1968).

ALTSCHULER D. : Critical Exponents from Infinite - Dimensional Symplectic Algebras, University of Geneva preprint UGVA-DPT 1985/06-466.

BARDAKCI K. AND HALPERN M. : Phys. Rev. D3 (1971), 2493.

BELAVIN A.A., POLYAKOV A.M. AND ZAMOLODCHIKOV A.B. : Nucl. Phys. B241 (1984) 333.

BROWER R. : Phys. Rev. D6 (1972) 1655.

CALLAN C.G., DASHEN R.F. AND SHARP D.H. : Phys. Rev. 165 (1968) 1883

COLEMAN S. : Phys. Rev. D11 (1975) 2088.

COLEMAN S., GROSS D. AND JACKIW R. : Phys Rev. 180 (1969) 1359.

DASHEN R. AND FRISHMAN Y. : Phys. Rev. D11 (1975) 278.

DEL GIUDICE E., DI VECCHIA P. AND FUBINI S. : Annals of Phys. 70 (1972) 378.

DELL'ANTONIO G.F., FRISHMAN Y. AND ZWANZIGER D. : Phys. Rev. D6 (1972) 988.

DI VECCHIA P., KNIZHNIK V.G., PETERSEN J.L. AND ROSSI P. : Nucl. Phys. B253 (1985) 701.

DIRAC P.A.M. : Nuovo Cimento (10) 1 (1955) 16; Rev. Mod. Physics 34 (1962) 592.

DYNKIN E.B. : Transl. Amer. Math. Soc. (1) 9 (1962) 328 and (2) 6 (1957) 111.

DYSON F.J. : Phys. Rev. 75 (1949) 486.

FERRARA S., GRILLO A.F. AND GATTO R. : Nuovo Cimento 12A (1972) 959.

FRENKEL I.B. : Proc. Natl. Acad. Sci. U.S.A. 77 (1980) 6303; J. Funct. Anal. 44 (1981) 259.

FRENKEL I.B. AND KAC V.G. : Inv. Math. 62 (1980) 23.

FREUND P.G., OH P. AND WHEELER J.T. : Nucl. Phys. B246 (1984) 371.

FRIEDAN D. : 1982 Les Houches Summer School; ed. J.B. Zuber and R. Stora : Les Houches Session XXXIX, Recent Advances in Field Theory and Statistical Mechanics (North Holland 1984).

FRIEDAN D., QIU Z. AND SHENKER S. : Phys. Rev. Lett. 52 (1984) 1575; Vertex Operators in Mathematics and Physics; MSRI publication No. 3 (Springer 1984), p491.

FRIEDAN D. AND SHENKER S. : Proceedings of the Santa Fe meeting, edited T. Goldman and M.N. Nieto (World Scientific 1985), p437.

FUBINI S., HANSON A.J. AND JACKIW R. : Phys. Rev. D7 (1973) 1732.

FUBINI S. AND VENEZIANO G. : Nuovo Cimento 67A (1970) 29.

GELL-MANN M. : Phys. Rev. 125 (1962) 1067; Phys. Letters 8 (1964) 214.

GELL-MANN M. AND NE'EMAN Y. : The Eightfold Way (Benjamin, New York 1964).

GERVAIS J-L. AND SAKITA B. : Nucl. Phys. B34 (1971) 477.

GLIOZZI F., OLIVE D. AND SCHERK J. : Nuclear Phys. B122 (1977) 253.

GODDARD P. : Kac-Moody and Virasoro Algebras: Representations and Applications, DAMTP 85/7.

GODDARD P., KENT A. AND OLIVE D. 1985a : Phys. Lett. 152B (1985) 88.

GODDARD P., KENT A. AND OLIVE D. 1985b : Unitary Representations of the Virasoro and Super Virasoro Algebras, DAMPT preprint 85-21.

GODDARD P., NUYTS J. AND OLIVE D. : Nucl. Phys. B125 (1977) 1.

GODDARD P., NAHM W. AND OLIVE D. : Phys. Lett. 160B (1985) 111.

GODDARD P. AND OLIVE D. 1984 : Vertex Operators in Mathematics and Physics, MSRI Publication No. 3 (Springer 1984) p51.

GODDARD P. AND OLIVE D. 1985 : Nuclear Phys. B257 [FS14] (1985) 226.

GODDARD P., OLIVE D. AND SCHWIMMER A. : Phys. Lett. 157B (1985) 393.

GODDARD P. AND THORN C. : Phys. Letters 40B (1972) 235.

GOODMAN R. AND WALLACH N.R. : J. Reine. Angew. Math. 347 (1984) 69.

GREEN M. AND SCHWARZ J. 1981 : Nuclear Phys. B181 (1981) 502.

GREEN M. AND SCHWARZ J. 1984 : Physics Lett. 149B (1984) 117.

GROSS D., HARVEY J.A., MARTINEC E. AND ROHM R. : Phys. Rev. Lett. 54 (1985) 502; Nucl. Phys. B256 (1985) 253.

HELGASON S. : Differential Geometry, Lie Groups, and Symmetric Spaces, (Academic Press 1978).

HUMPHREYS J.E. : Introduction to Lie Algebras and Representation Theory, (Springer 1972).

JACOB M. (Editor) : Dual Theory, (North Holland 1974).

JAIN S., SHANKAR P. AND WADIA S. : Conformal Invariance and String Theory in Compact Space : Bosons, Tata Institute preprint TIFR/PH/85-3.

KAC V.G. 1978 : Proceedings of the International Congress of Mathematicians, Helsinki 1978 and Lecture Notes in Physics 94 (1979) 441. For the proof see B.L. Feigin and D.B. Fuchs: Functs. Anal. Prilozhen. 16 (1982) 47 [Funct. Anal. and App. 16 (1982) 114].

KAC V.G. 1983 : Infinite-dimensional Lie Algebras - An Introduction (Birkhauser 1983).

KAC V.G. AND PETERSON D.H. : Proc. Math. Acad. Sci. U.S.A. 78 (1981) 3308.

KAC V.G. AND TODOROV I.T. : Superconformal current algebras and their unitary representations, MIT preprint.

KNIZHNIK V.G. AND ZAMOLODCHIKOV A.B. : Nucl. Phys. B247 (1984) 83.

MACKAY W. AND PATERA J. : Tables of dimensions, indices and branching ratios for representations of simple groups, (Dekker 1981).

MANDELSTAM S. 1973 : Phys. Rev. D7 (1973) 3763 and 3777.

MANDELSTAM S. 1975 : Phys. Rev. D11 (1975) 3026.

NAMBU Y. : Lectures for the Copenhagen Symposium 1970 (Unpublished).

NEMESCHANSKY D. AND YANKIELOWICZ S. : Phys. Rev. Lett. 54 (1985) 620.

NEVEU A. AND SCHWARZ J.H. : Nucl. Phys. B31 (1971) 86; Phys. Rev. D4 (1971) 1109.

OLIVE D.I. : Kac-Moody Algebras: An Introduction for Physicists, Imperial/TP/84-85/14.

ONSAGER L. : Phys. Rev. 65 (1944) 117.

RAMOND P. : Phys. Rev. D3 (1971) 2415.

SCHWINGER J. 1959 : Phys. Rev. Lett. 3 (1959) 296 and ADLER AND DASHEN p235.

SCHWINGER J. 1962 : Phys. Rev. 127 (1962) 324, 130 (1963) 406, 130 (1963) 800.

SEGAL G. : Comm. Math. Phys. 81 (1981) 301.

SKYRME T.H.R. : Proc. Roy. Soc. A262 (1961) 237.

SLANSKY R. : Phys. Reports 79 (1981) 1.

SOMMERFIELD C. : Phys. Rev. 176 (1968) 2019.

SUGAWARA H. : Phys. Rev. 170 (1968) 1659.

TODOROV I.T. : Phys. Letters 153B (1985) 77.

VIRASORO M. : Phys Rev. D1 (1969) 2933.

WEIS J. : Unpublished.

WICK G.C. : Phys. Rev. 80 (1950) 268.

WITTEN E. : Commun. Math. Phys. 92 (1984) 455.

AN INTRODUCTION TO ANOMALIES

Luis Alvarez-Gaumé *

Lyman Laboratory of Physics
Harvard University
Cambridge, Massachusetts 02138, U.S.A.

I. INTRODUCTION

These lectures are dedicated to the study of the recent progress and implications of anomalies in quantum field theory. In this introduction we would like to recapitulate some of the highlights in the history of the subject. In its original form,[1] one considers a triangle diagram with two vector currents and an axial vector current. Requiring Bose symmetry and vector current conservation in the vector channels, one finds that the axial vector current is not conserved, leading to the breakdown of chiral symmetry in the presence of external gauge fields. The existence of this anomaly led to an understanding of π^0 decay, and later on to the resolution of the U(1) problem[2] in QCD. These anomalies correspond to the breakdown of global axial symmetries, and their existence does not jeopardize unitarity or renormalizability. More dangerous anomalies appear whenever chiral currents are coupled to gauge fields. For example, in four dimensions we can consider V-A currents coupled to gauge fields as in the standard Weinberg-Salam model, and compute the same triangle diagram with V-A currents on each vertex.[3] Again one finds an anomaly, and unless the anomaly cancels after summing over all the fermion species, the theory will not be gauge invariant, implying a loss of renormalizability. If we recall the Feynman rules for non-Abelian gauge theories coupled to fermions in some representation T^a of the gauge group G, the anomaly for gauge currents is proportional to a purely group theoretic factor times a Feynman integral. The anomaly cancellation condition implies the vanishing of the sum of

*Supported in part by the National Science Foundation under Grant Number PHY82-15249, and by an Alfred P. Sloan Fellowship.

the group theoretic factor over the various representations. Explicitly

$$\sum_{T_L} \mathrm{STr}\, T_L^a T_L^b T_L^c - \sum_{T_R} \mathrm{STr}\, T_R^a T_R^b T_R^c = 0 \tag{1.1}$$

where T_L (T_R) stands for the representations of G carried by the left- (right-) handed fermions, and STr means the symmetrized trace over the group generators involved. Unless this condition is satisfied, the theory will not be gauge invariant. There are various ways of computing the group invariant factors in (1), and their generalizations to higher dimensions,[4] and we will spend part of Section II in presenting some efficient ways of computing the group theory factors appearing in various dimensions. Even though there are many ways of satisfying (1) in four dimensions, it is worth pointing out the remarkable cancellation of anomalies within each family of the standard Glashow-Weinberg-Salam model. If we write the known quarks and leptons in terms of left-handed Weyl fermions we find for each family the following SU(3) × SU(2) × U(1) quantum numbers:

$$(3,2)_{1/6} \oplus (\bar{3},1)_{-2/3} \oplus (1,2)_{-1/2} \oplus (\bar{3},1)_{1/3} \oplus (1,1)_{1} \,. \tag{1.2}$$

This representation is clearly complex, because if we complex conjugate (2), we do not get back to the same representation. The fact that the U(1) hypercharge anomaly only cancels after including the contributions of every member of the family plus the perversely chiral structure of (2) is indicative of a more profound structure which we have not been able to unravel. Understanding the quantum numbers of the known quarks and leptons displayed in (2) is one of the main motivations which led to the development of grand unified theories[5] and to the resurrection of Kaluza-Klein theories.[6]

Anomalies have also proven useful in providing constraints on the possible spectrum of composite models by requiring certain matching conditions on Green's functions of flavor currents between the elementary and composite levels of the theories.[7] Using the σ-model anomalies,[8] it is possible to understand the 't Hooft matching conditions from the low energy point of view.[9]

Soon after the discovery of the axial anomaly, it was realized[10] that the same triangle diagram with the vector currents substituted by energy momentum tensors is anomalous, implying that the axial vector symmetry is also violated in the presence of external gravitational fields. Similarly, for any U(1) gauge field coupled to a V-A current, we can consider the triangle diagram with two external gravitons, which is now proportional to the trace of Q [Q is the generator of U(1)]. If $\mathrm{Tr}\, Q \neq 0$, then one cannot maintain at the same time the conservation of the energy momentum tensor and the U(1) gauge symmetry.[11] Getting rid of this anomaly requires that Tr Q vanishes for the generator of any U(1) factor in the gauge group. The reader may easily check that the quantum numbers assigned in (2) satisfy this requirement. This condition will be automatically satisfied if the low energy gauge group comes from a larger simple (or semisimple) unified gauge group because the trace of any generator always vanishes for a compact simple group. The fact that the standard model satisfies this requirement automatically for each family, may be taken as some evidence for the existence of an underlying unified structure.

With the renewed interest in Kaluza-Klein theories in the late seventies and early eighties, some authors[4] analyzed the generalization of anomalies and their cancellation to higher dimensions. Again one can consider anomalies in both global and gauged currents. Since the anomalies in global currents do not spoil any consistency requirement in the theory, their presence can be interpreted in four dimensions. The anomalies in gauge currents however produce much more stringent requirements than in four dimensions.[2] Even though we will explain their computation in detail below, the main conditions are simple to explain. If one considers chiral fermions coupled to gauge fields in 2n dimensions, then the first diagram which might become anomalous is a polygon with $n+1$ sides. The usual requirement of Bose symmetry for the external gauge lines, plus the fact that each vertex contains a factor T^a implies that the group theory factor appearing in the anomaly is:

$$\sum_{T_L} \mathrm{STr}(T_L^{a_1} \ldots T_L^{a_{n+1}}) - \sum_{T_R} \mathrm{STr}(T_R^{a_1} \ldots T_R^{a_{n+1}}) \ . \tag{1.3}$$

Cancelling (3) for $n > 2$ is more complicated than cancelling (2), and consequently, the set of anomaly free chiral representations of fermions is considerably reduced with respect to the four dimensional case.

Once it is decided to consider higher dimensional theories, there is another problem to be taken into account. For chiral fermions in 4k+2 dimensions the energy momentum tensor may be anomalous.[11] In general, for any even number of dimensions, one can get anomalies in gauge currents (gauge) anomalies, anomalies in the energy momentum tensor (gravitational anomalies) and mixed anomalies corresponding to graphs containing the gravitons and gauge fields on the external lines. Requiring all these anomalies to cancel leaves a very small class of theories which are chiral and anomaly free. If one further requires the theory to be supersymmetric, there are basically two types of theories known exhibiting a non-trivial anomaly cancellation: $N=2$ chiral supergravity in ten dimensions,[11] and $N=1$ super Yang-Mills theory coupled to $N=1$ supergravity in ten dimensions with gauge groups $E_8 \times E_8$ or SO(32).[12,13] These theories can be obtained as the zero slope limit of certain string theories.[12,13] This non-trivial cancellation of anomalies, together with the very real possibility that these string theories provide an ultraviolet finite quantum theory of gravity, provide the main motivation for the recent interest in string theories.[14]

From a more mathematical standpoint, the study of anomalies has elicited very interesting applications of index theory[15,16] in quantum field theory. It was realized in the mid-seventies[17] that the global U(1) axial anomaly could be understood as a consequence of the Atiyah-Singer index theorem.[15] Loosely speaking, the expectation value of the divergence of the axial current can be related to the spectral asymmetry of the Dirac operator. Since the Dirac operator anticommutes with γ_5 (or the higher dimensional analogue), for any non-zero eigenvalue λ with eigenfunction ψ_λ, there is an eigenfraction $\gamma_5\psi_\lambda$ with eigenvalue $-\lambda$. Thus the "spectral asymmetry" is concentrated on the space of zero modes, some of whose properties are determined by index theorem arguments. Apart from obtaining the correct form and normalization of the U(1) anomaly using index theory, this approach provides a powerful way of quantitatively analyzing some of the subtleties of fermion-gauge field dynamics. Recently,[18-25] a more powerful form of the Atiyah-Singer index theorem[16] (the index theorem for families of elliptic operators) has been used to provide a global understanding of the non-Abelian anomaly as well as the gravitational anomalies. The advantage of this method is that it is very well suited for the study of the qualitative properties of the fermionic Hilbert space as one moves the gauge field over the space of all possible gauge fields. This form of the index theorem gives a rationale to the so-called "descent" equations of Stora and Zumino[26] which provide a rather elegant algebraic setting for the derivation of the anomaly and the Wess-Zumino-Witten[27,28] Lagrangian describing the anomalous low-energy interactions of Goldstone bosons in QCD-like theories.

Even though these results were derived within the Euclidean formulation of field theory, it is possible to understand these structures in the Hamiltonian formalism from an algebraic[29,30] or topological point of view.[31]

Finally, using spectral flows and index theory for families of operators,[15 16] Witten[32] in a remarkable series of papers has been able to analyze the possibility of obtaining global gauge and gravitational anomalies. The last part of these lectures will be dedicated to an introduction to Witten's results and the various tools needed to obtain them.

The outline of these lectures is as follows: Section II contains a quick review of spinors in Euclidean and Minkowski space, some other group theory results relevant for the computation of anomalies in various dimensions, and an exposition of the index theorem[15,16] in the way we will use it. Section III starts our analysis of fermion determinants and chiral effective actions by deriving the non-Abelian anomaly from index theory. Using the results of Section II, we will present the anomaly cancellation recently discovered by Green and Schwarz[12,13] in Section IV as well as the connection of these results of Section III with the descent equations[26] and the Wess-Zumino-Witten Lagrangians. Section V contains the generalization of anomalies to σ-models[8,9] and some of its application in string theory. Section VI will deal with the anomalies from the Hamiltonian point of view. An exact formula for the imaginary part of the effective action for chiral fermions in the presence of arbitrary external gauge and gravitational fields will be derived in Section VII, and used in Section VIII for the study of global anomalies.

There are some topics which are not covered in these lectures. One very interesting recent development has been the computation of the anomaly functional for supersymmetric gauge theories in the superfield formulation. The interested reader is encouraged to look up the entries listed in Ref. 33.

II. GENERALITIES AND HEURISTIC CONSIDERATIONS

A. This section contains some basic material we will use frequently in later parts of these lectures. Unless otherwise stated, we will be working in Euclidean space compactified to a sphere in order to avoid infrared problems. Since one of the basic ingredients in the study of anomalies is the presence of chiral fermions, we start by reviewing some basic properties concerning the spinor representations of the Lorentz group both in Euclidean and Minkowski space.[34] The starting point of any spinor analysis is the Clifford algebra in 2n dimensions.

$$\{\Gamma^m,\Gamma^n\} = 2\eta^{mn} \qquad m,n = 0,\ldots,2n-1 \tag{2.1}$$

η^{mn} is the flat metric with signature (t,s), i.e. η^{mn} is a diagonal matrix with t(+1) eigenvalues, and s(-1) eigenvalues. A Dirac spinor is a field which under an infinitesimal SO(t,s) transformation changes by

$$\begin{gathered}\delta\psi = -\tfrac{1}{2}\varepsilon_{mn}\Sigma^{mn}\psi \\ \Sigma^{mn} = -\tfrac{1}{4}[\Gamma^m,\Gamma^n] \\ \Gamma^{1\dagger} = \Gamma^1,\ldots,\Gamma^{t\dagger} = \Gamma^t,\Gamma^{t+1\dagger} = -\Gamma^{t+1},\ldots,\Gamma^{t+s\dagger} = -\Gamma^{t+s}\end{gathered} \tag{2.2}$$

In any even number of dimensions we can define the analogue of γ_5 : $\bar{\Gamma}$ a matrix which anticommutes with Γ^m for all m:

$$\bar{\Gamma} = \alpha\,\Gamma^0\Gamma^1\ldots\Gamma^{2n-1} \tag{2.3}$$

and α is a phase chosen so that $\bar{\Gamma}^2 = +1$, i.e.

$$\alpha^2 = (-1)^{(s-t)/2}, \qquad \alpha^* = (-1)^{(s-t)/2}\alpha \, . \tag{2.4}$$

Since $\{\bar{\Gamma},\Gamma^m\} = 0$, $\bar{\Gamma}$ commutes with the generators of the Lorentz group. In terms of $\bar{\Gamma}$ we can define two projection operators:

$$P_\pm = \tfrac{1}{2}(1 \pm \bar{\Gamma}) \tag{2.5}$$

and project ψ into the two irreducible representations

$$\psi_\pm = P_\pm \psi \, . \tag{2.6}$$

ψ_+ (ψ_-) is called a positive (negative) chirality Weyl spinor. $\bar{\Gamma}\psi_+ = \psi_+$, $\bar{\Gamma}\psi_- = -\psi_-$.

The Clifford algebra (2.1) has a unique faithful representation of dimension 2^n. This means that Γ^m, Γ^{m*}, $-\Gamma^{m*}$, Γ^{mT}, $-\Gamma^{mT}$ are related by a similarity transformation because all these matrices satisfy the same anticommutation relation (2.1). Hence there exists a matrix B such that

$$(\Sigma^{mn})^* = B\,\Sigma^{mn}B^{-1} \, . \tag{2.7}$$

Using B, we can define a charge conjugate Dirac spinor:

$$\psi^c = C\psi \equiv B^{-1}\psi^* \, . \tag{2.8}$$

Using (2.7), it is easy to show that ψ^c and ψ have the same Lorentz transformation properties:

$$\begin{gathered} \delta\psi^c = -\tfrac{1}{2}\varepsilon_{mn}\Sigma^{mn}\psi^c \\ [C,\Sigma^{mn}] = 0 \, . \end{gathered} \tag{2.9}$$

If B can be found such that $C^2 = +1$, we can define projection operators $(1 \pm C)/2$ and

$$\begin{aligned} \psi_M &= (1 + C)\psi / 2 \\ \psi_{\bar{M}} &= (1 - C)\psi / 2 \end{aligned} \tag{2.10}$$

are respectively Majorana and anti-Majorana fields.

It is also important to know whether C and $\bar{\Gamma}$ commute. In order to determine this, one can represent $\bar{\Gamma}$ in terms of the Σ^{mn}'s

$$\bar{\Gamma} = (-2)^n \alpha\,\Sigma^{01}\Sigma^{23}\ldots\Sigma^{2n-2,2n-1} \, . \tag{2.11}$$

Using (2.4) and the definition of C we get

$$C\bar{\Gamma} = (-1)^{(s-t)/2}\,\bar{\Gamma}C \, . \tag{2.12}$$

If $(s-t)/2$ is odd, C flips the chirality of the fermion. If $(s-t)/2$ is even, C does not change the chirality. For definiteness, take $t = 1$, $s = 2n-1$ (Minkowskian signature). Then in $4k$ dimensions $\{C,\bar{\Gamma}\} = 0$, whereas $[C,\bar{\Gamma}] = 0$ in $4k+2$ dimensions. We can get a more physical picture by considering the solutions to the massless Dirac equation in 2n dimensions.

Consider a free massless fermion with definite momentum moving in the direction 2n-1. Its momentum space wave function satisfies

$$\Gamma^0 u(p) = \Gamma^{2n-1} u(p) \ . \tag{2.13}$$

The helicity operator can be defined to be

$$h = \Sigma^{12}\Sigma^{34}\ldots\Sigma^{2n-3,2n-2} \sim \Gamma^1\Gamma^2\ldots\Gamma^{2n-2} \tag{2.14}$$

Thus

$$hu(p) = h\Gamma^0\Gamma^0 u(p) = \Gamma^0 h\Gamma^{2n-1} u(p) = \bar{\Gamma} u(p) \tag{2.15}$$

so helicity and chirality are tied up in four dimensions. This implies that in 4k-dimensional Minkowskian space charge conjugation flips helicity, and in 4k+2 dimensions it doesn't. In Euclidean space one gets that 4k-dimensional spinors behave like (4k+2)-dimensional Minkowskian spinors, and (4k+2)-dimensional spinors like those in 4k-dimensional Minkowskian space. In order to determine whether $C^2 = +1$, we need more detailed information about the matrix B. Since $-\Gamma^{m*}$ and Γ^m are related by a similarity transformation we have

$$\Gamma^{m*} = -B\,\Gamma^m B^{-1}$$

$$\Gamma^m = -B^*\Gamma^{m*}B^{*-1} = (B^*B)\Gamma^m(B^*B)^{-1} \tag{2.16}$$

and by Schur's lemma, we can choose B so that

$$B^*B = \varepsilon I \qquad |\varepsilon| = 1 \ . \tag{2.17}$$

In fact $\quad BB^* = \varepsilon^* I, \quad B^*B = \varepsilon I$

implies $\quad \varepsilon = \pm 1$.

Next, let us introduce the standard charge conjugation matrix by the similarity transformation relating $-\Gamma^{mT}$ and Γ^m

$$\Gamma^{mT} = -C\,\Gamma^m C^{-1} \ . \tag{2.18}$$

For definiteness, we will work in Minkowski space. We can now construct C in terms of B and Γ^0. Using $\Gamma^{m\dagger} = \Gamma^0\Gamma^m\Gamma^0$ and (2.16) we have:

$$\Gamma^{mT} = (\Gamma^{m*})^\dagger = -(B^{\dagger-1}\Gamma^0)\Gamma^m(B^{\dagger-1}\Gamma^0)^{-1}$$

$$\Gamma^{mT} = (\Gamma^{m\dagger})^* = -(B\Gamma^0)\Gamma^m(B\Gamma^0)^{-1} \ , \tag{2.19}$$

i.e. choosing our normalization convention appropriately

$$B^\dagger B = 1 \ . \tag{2.20}$$

Combining (2.17) and (2.20) we get

$$B^*B = \varepsilon, \qquad B^{-1} = B^\dagger = \varepsilon B^* \ ,$$

i.e. $$B^T = \varepsilon B \tag{2.21}$$
$$C = B\Gamma^0 \tag{2.22}$$

and finally:
$$C^T = (B\Gamma^0)^T = \varepsilon\Gamma^{0T} B = \varepsilon\Gamma^{0*} B = -\varepsilon B\Gamma^0 = -\varepsilon C$$

$$C^T = -\varepsilon C$$
$$B^T = \varepsilon B \;. \tag{2.23}$$

Since our arguments have been basis independent, we can now construct a particular basis for the Γ-matrices and check the value of ε for various dimensions. We simply take

$$\Gamma^0 = \sigma_x \otimes \overbrace{1 \otimes \ldots \otimes 1}^{n-1}$$
$$\Gamma^1 = i\sigma_y \otimes 1 \otimes \ldots \otimes 1$$
$$\Gamma^2 = i\sigma_3 \otimes \sigma_x \otimes 1 \ldots \otimes 1$$
$$\Gamma^3 = i\sigma_3 \otimes \sigma_y \otimes 1 \ldots \otimes 1$$
$$\Gamma^{2k-2} = \overbrace{i\sigma_3 \otimes \sigma_3 \otimes \ldots \otimes \sigma_3}^{k-1} \otimes \sigma_x \otimes 1 \ldots \otimes 1$$
$$\Gamma^{2k-1} = i\sigma_3 \otimes \sigma_3 \otimes \ldots \otimes \sigma_3 \otimes \sigma_y \otimes 1 \ldots \otimes 1$$
$$- - - - -$$
$$\Gamma^{2n-1} = i\sigma_3 \otimes \sigma_3 \otimes \ldots \otimes \sigma_y$$
$$\bar{\Gamma} = \sigma_3 \otimes \sigma_3 \otimes \ldots \otimes \sigma_3 \;. \tag{2.24}$$

It is now a simple exercise to check that

$$C = \sigma_y \otimes \sigma_x \otimes \sigma_y \otimes \ldots \tag{2.25}$$

where the number of σ_y's in C is k in 4k dimensions, and k+1 in 4k+2 dimensions. Therefore

$$\varepsilon = +1 \quad \text{for} \quad 2n = 2,4 \mod 8$$
$$\varepsilon = -1 \quad \text{for} \quad 2n = 0,6 \mod 8 \tag{2.26}$$

and the Majorana condition (2.10) can only be imposed in 2,4 mod 8 dimensions. Since $[C,\bar{\Gamma}] = 0$ only in 4k+2 dimensions, we learn that only in 2 mod 8 dimensions is it possible to simultaneously impose a Weyl-Majorana condition. These results are important when constructing minimal on-shell multiplet representations of higher dimensional supersymmetry and supergravity theories. For example in d = 10 dimensions a massless vector field represents 8 degrees of freedom (the little group of massless states in d = 10 is SO(8), and the states created by a ten-dimensional vector field transform like a vector under the little group SO(8)). If we want to find a supersymmetric multiplet containing a vector field, we have to find a fermion field with the same number of degrees of freedom. Notice that a Weyl fermion in $d = 10$ has $2^4 = 16$ degrees of freedom; thus once we impose both the Majorana and Weyl conditions, we end up with 8 degrees of freedom. In fact, it is possible to construct a $N = 1$ supersymmetric Lagrangian for any gauge group in $d = 10$ (for a review see Ref. 35).

$$S = \int d^{10}x \, \mathrm{Tr}\,[-\frac{1}{4} F_{mn} F^{mn} + \frac{i}{2} \bar{\lambda} \Gamma^m D_m \lambda] \tag{2.27}$$

$$D_m \lambda = \partial_m \lambda + ig[A_m, \lambda] \; .$$

λ is a Weyl-Majorana fermion in the adjoint representation of the gauge group. Constructing the $N = 1$ supergravity multiplet in $d = 10$ is more difficult. The field content of the theory contains a graviton g_{mn}, a Weyl-Majorana gravitino ψ_m (left-handed), a right-handed Weyl-Majorana fermion λ, a second-rank antisymmetric tensor B_{mn}, and a scalar ϕ. To convince the reader that the number of bosonic and fermionic degrees of freedom agree, let us count them. The graviton h_{mn} has $d(d+1)/2$ components. Imposing d coordinate conditions, transversality $\partial^m h_{mn} = 0$ and tracelessness $h^m{}_m = 0$ we get $(d-1)(d-2)/2 - 1$ degrees of freedom (symmetric traceless representation of the little group $SO(d-2)$. The two-form B_{mn} has $(d-2)(d-3)/2$ degrees of freedom, and ϕ just one. λ has eight degrees of freedom and a Weyl-Majorana gravitino will have $(d-3)2^{[d/2]}/4$ degrees of freedom after appropriate gauge fixing. For $d = 10$, we have $35 + 28 + 1 = 64$ bosonic degrees of freedom, and $56 + 8 = 64$ fermionic ones. In principle we have all the information we need to study the anomaly (gauge and gravitational) in ten dimensions. The detailed form of the Lagrangian which couples the $N = 1$ super Yang-Mills multiplet to the supergravity multiplet is mostly irrelevant (we will only need the form of the gauge field strength associated to B_{mn}. This plays a very important role in the anomaly cancellation mechanism discovered by Green and Schwarz.[12] This concludes our quick review of spinors.

The next piece of group theory that we will need is the computation of the group invariant traces (1.3). As remarked in the introduction, one needs to evaluate in general $\mathrm{STr}\, T^{a_1} \ldots T^{a_r}$. Rather than dealing with this expression, it is easier (and completely equivalent) to compute the quantity $\mathrm{Tr}_{(r)} F^n$ for an arbitrary matrix F in the representation r of the Lie algebra one is interested in. There are two cases which are very different qualitatively, namely n even or odd. When n is even $n = 2k$, $\mathrm{Tr}\, F^{2k}$ is positive definite, and therefore a theory containing only left-handed fermions will always be anomalous. For instance $N = 1$ super Yang-Mills in $d = 10$ is anomalous for any gauge group (for $d = 10$, $k = 6$ as we will see). If $n = 2k+1$, it is possible to have purely left-handed anomaly-free representations. In this case, we can get a fairly detailed general analysis of which groups and which representations may potentially be anomalous. Let us recall that in general there are three types of unitary representations (for a review on group theory see for example Ref. 36).

(a) Real symmetric (or simply real)

If R^a are the Hermitian generators of the representation, R^a is real if

$$R^{a^T} = - S^{-1} R^a S$$
$$S^T = S \ . \tag{2.28}$$

More explicitly, this means that the group matrices in this representation $\exp i R^a \alpha^a = U$ (α^a real numbers representing the group parameters) is equivalent to its complex conjugate $(\exp i R^a \alpha^a)^* = \exp - i R^{a^T} \alpha^a$ and that it is possible to find a basis where the R^a's are purely imaginary.

(b) Real antisymmetric (or pseudoreal)

The definition is the same as above, but now $S^T = -S$. This means again $U^* = S^{-1} U S$, but the antisymmetry of S implies that there is no basis where all the R^a's can be chosen as purely imaginary matrices. The simplest example is the two-dimensional representation of SU(2). The generators are Pauli matrices $\sigma^a/2$, $a = 1,2,3$; and $S = \sigma^2$, $\sigma^{a^T} = -\sigma^2 \sigma^a \sigma^2$, $S^T = -S$, and clearly there is no possible basis where all σ^a's can be chosen to be purely imaginary.

(c) Complex representations

This is the case when a representation U and its complex conjugate are not equivalent.

A simple consequence of these definitions, is that $\mathrm{Tr}\, F^{2k+1}$ vanishes for real or pseudoreal representation:

$$\mathrm{Tr}\, F^{2k+1} = \mathrm{Tr}(F^{2k+1})^T = \mathrm{Tr}(F^T)^{2k+1} = - \mathrm{Tr}(S^{-1} F S)^{2k+1} = 0 \ . \tag{2.29}$$

If we now analyze the simple groups, we can easily identify which groups and which representations are automatically safe in 4k dimensions.

SO(2n+1) has two basic representations: the vector representation (used to define SO(2n+1)) which is manifestly real, and a spinor representation. Since the spinor representation is unique (it is obtained by including $\bar{\Gamma}$ as Γ^{2n+1} and then defining the generators as in (2.9)), it must be equivalent to its complex conjugate and hence either real or pseudoreal. Since all the representations of SO(2n+1) can be obtained by suitable tensor products of these two representations, we obtain that all the representations of SO(2n+1) are real or pseudoreal, and thus SO(2n+1) is safe. The symplectic groups Sp(n), quaternionic unitary matrices, have the property that the fundamental representation is 2n-dimensional, pseudoreal, and all other representations can be obtained from it by tensor products. Similarly, the exceptional groups G_2, F_4, E_7, E_8 are safe; their fundamental representations have dimensions respectively 7 (real), 26 (pseudoreal), 56 (pseudoreal), 248 (real), and all other representations are either real or pseudoreal. The only dangerous possibilities are SO(2n), SU(n) and E_6. From our previous analysis of spinors in Euclidean space, and their behavior under charge conjugation, we know that the spinors of SO(4k) are not complex, and therefore SO(4k) is safe. For SO(4k+2) complex conjugation exchanges the spinor representations, thus indicating that they are complex. The computation of $\mathrm{Tr}\, F^n$ can be done in general by using the theory of group characters,[37] and will be explained below. The unitary groups SU(n) have anomalies in all even dimensions, and their cancellation is in general elaborate. Finally E_6 is a group of 78 generators, and its fundamental representations 27, $\overline{27}$ are complex. Even though the anomaly in the 27 representation cancels in four dimensions (this is easily checked by considering the embedding $E_6 \supset SO(10) \times U(1)$ so that $27 = 1(4) + 10(-2) + 16(1)$,[38] and computing the anomaly with respect to $SO(10) \times U(1)$), it does not in any 4k dimensions for $k \geq 2$.

The method of group characters is usually fairly efficient for the computation of $\mathrm{Tr}\, F^n$ for arbitrary n, and for the analysis of anomaly cancellation. Since the anomaly cancellation of Green and Schwarz for $E_8 \times E_8$ and SO(32) in $d = 10$ will be explained in a later section, we will do the computation explicitly for this case in particular. Given any Lie algebra G, we can always find a maximal subset of commuting generators (the Cartan subalgebra) H_i, $i = 1, \ldots, r = $ rank of G. The various representations are then labeled by the eigenvalues of the vector $\vec{H}$ on the states of the representation $\vec{H}|\vec{\mu}\rangle = \vec{\mu}|\vec{\mu}\rangle$. $\vec{\mu}$ is called the weight vector, and it has r components. Any group matrix V corresponding to a group element g is then unitarily equivalent to a matrix of the form $\exp i\,\vec{x}\cdot\vec{H}$, the vector $\vec{x}$ is an r-dimensional vector which corresponds to the "elementary" rotation angles required to describe the eigenvalues of V(g). Since V is a representation of G, the x's depend only on g and not on the representation considered. The character of g in the representation V is defined by

$$\mathrm{Tr}\, V(g) = \sum_{\vec{\mu}} \exp i\,\vec{x}\cdot\vec{\mu} \,. \tag{2.30}$$

Since any group element V can be represented as $\exp F$, computing (2.30) is equivalent to computing all the traces of the form $\mathrm{Tr}\, F^n$. Although it is possible to write a general formula for (2.30),[37] it is not very useful in order to analyze anomaly cancellations. What we can do instead is to compute the character (2.30) in terms of the character of the fundamental representation of G. In general $\mathrm{Tr}\, F^n$ for some representation R can be rewritten in terms of traces and product of traces in the fundamental representation (the traces in the fundamental representation will be denoted by lower case $\mathrm{tr}\, F^n$ following Green and Schwarz[12]):

$$\mathrm{Tr}\, F^n = A_1\, \mathrm{tr}\, F^n + A_2\, \mathrm{tr}\, F^2\, \mathrm{tr}\, F^{n-2} + \ldots \tag{2.31}$$

The number of independent coefficients A_i needed to be taken into account in the analysis of anomaly cancellation thus depends on the number of irreducible traces $\mathrm{tr}\, F^n$ (traces which cannot be factorized into products of lower-order traces), and this in turn is related to the Casimir operators of the group G. For example SU(3) has only two independent Casimirs related to $\mathrm{tr}\, F^2$, $\mathrm{tr}\, F^3$. Any other trace of the form $\mathrm{tr}\, F^n$, $n > 3$, can always be rewritten in terms of combinations of products of $\mathrm{tr}\, F^2$, $\mathrm{tr}\, F^3$. (In general, the number of independent traces for a simple Lie group G is given by the rank of G.) Thus $\mathrm{tr}\, F^5 \sim \mathrm{tr}\, F^3\, \mathrm{tr}\, F^2$, etc. For simple enough representations, we can almost always calculate the coefficients of (2.31) by a straightforward procedure. For instance, if we want to compute $\mathrm{Tr}\, F^n$ for the adjoint representation of SO(N) or Sp(N) in terms of similar traces for the fundamental representation, we first construct the adjoint representation from tensor products of the fundamental representation. If U_{ij} is a generic group matrix in the fundamental representation of SO(N) or Sp(N), it acts on the representation space by

$$t_i \to U_{ij}\, t_j \ .$$

For SO(N) the adjoint representation is equivalent to the antisymmetric tensor irrep. (irreducible representation). Thus

$$t_{ij} \to \tfrac{1}{2}(U_{ik}\, U_{j\ell} - U_{i\ell}\, U_{jk})t_{k\ell} \qquad (\mathrm{SO(N)}) \ . \tag{2.32}$$

On the other hand, for the symplectic group, due to the existence of an invariant, antisymmetric bilinear form, the adjoint representation is given by the symmetric second-rank tensor:

$$t_{ij} \to \tfrac{1}{2}(U_{ik}\, U_{j\ell} + U_{i\ell}\, U_{jk})t_{k\ell} \qquad (\mathrm{Sp(N)}) \ . \tag{2.33}$$

We can now compute the character of the adjoint in terms of the character of the fundamental irrep.

$$\mathrm{Tr}\, U = \tfrac{1}{2}\left((\mathrm{tr}\, U)^2 + \varepsilon\, \mathrm{tr}\, U^2\right) \tag{2.34}$$

$$\varepsilon = +1 \qquad \mathrm{Sp(N)}$$

$$\varepsilon = -1 \qquad \mathrm{SO(N)} \ .$$

Since $U = \exp F$ we simply expand both sides of (2.34) in power series in F to get the desired coefficients A_i:

$$\mathrm{Tr}\, F^2 = (N + 2\varepsilon)\, \mathrm{tr}\, F^2 \tag{2.35a}$$

$$\operatorname{Tr} F^4 = (N + 8\varepsilon)\operatorname{tr} F^4 + 3(\operatorname{tr} F^2)^2 \tag{2.35b}$$

$$\operatorname{Tr} F^6 = (N + 32\varepsilon)\operatorname{tr} F^6 + 15 \operatorname{tr} F^2 \operatorname{tr} F^4 \ . \tag{2.35c}$$

For E_8 we have to work a bit harder to get the analog of (2.35). One of the nicest ways to think about E_8 is in terms of its subgroup SO(16). The fundamental (248 equal to the adjoint rep) irrep. of E_8 decomposes under SO(16)

$$\underline{248} = \underline{120} + \underline{128}_{(+)}$$

120 is the adjoint of SO(16) and $128_{(+)}$ is the positive chirality spinor of SO(16). Since E_8 and SO(16) have both rank 8, the weights of their representations are eight dimensional vectors. If we choose the standard basis for $\mathbb{R}$, $e_i \cdot e_j = \delta_{ij}$, the weight of the fundamental irrep. of SO(16) are $\pm e_i$, $i = 1,8$; and in terms of them the weight of the adjoint representation becomes

$$\pm e_i \pm e_j \qquad i < j \tag{2.36}$$

plus 8 zeros corresponding to the Cartan subalgebra. And for the spinor:

$$\tfrac{1}{2} \sum_{a=1}^{8} \varepsilon_a e_a$$

$$\varepsilon_a = \pm 1, \qquad \prod_{a=1}^{8} \varepsilon_a = +1 \ . \tag{2.37}$$

Now, an arbitrary matrix F in the adjoint rep. of E_8 can be described in terms of 8 parameters $x_1,\ldots,x_8$ (the elementary rotations described above). Therefore the character of the 248 of E_8 is simply the sum of the characters of the 120 and 128 of SO(16):

$$\begin{aligned} \operatorname{Tr}_{E_8} U &= \sum_{i<j} \exp(\pm x_i \pm x_j) + \sum_{\{\varepsilon_b\},\pi\varepsilon_b=+1} \exp\left(\tfrac{1}{2} \sum_{a=1}^{8} \varepsilon_a x_a\right) \\ &= \tfrac{1}{2}\left[\left(\sum_i 2\cosh x_i\right)^2 - \sum_i 2\cosh 2x_i + \prod_{i=1}^{8} 2\cosh\frac{x_i}{2} + \prod_{i=1}^{8} 2\sinh\frac{x_i}{2}\right]. \end{aligned} \tag{2.38}$$

This computes the character of E_8. A tedious expansion now shows that for E_8:

$$\operatorname{Tr} F^4 = \frac{1}{100} (\operatorname{Tr} F^2)^2$$

$$\operatorname{Tr} F^6 = \frac{1}{7200} (\operatorname{Tr} F^2)^3 \ . \tag{2.39}$$

The relations (2.35,2.39) play a crucial role[12,13] as will be explained in Section IV. A similar analysis can be carried out for other exceptional groups by choosing suitable maximal subgroups. For example, for E_6 we can choose SO(10) × U(1) and for E_7, SO(12) × SU(2). The decomposition of the fundamental and adjoint irreps. are:[38]

$$27 = 1(4) + 10(-2) + 16(1)$$

$$78 = 45(0) + 1(0) + 16(-3) + \overline{16}(3) \qquad \text{for } E_6 \text{, and}$$

$$56 = (12,2) + (32,1)$$

$$133 = (1,3) + (66,1) + (32',2)$$

for E_7. (32, 32' are the two spinors of positive and negative chirality of SO(12) respectively.) It is left an exercise to the reader to compute the characters of E_6 and E_7 in the representations given above.

In the perturbative analysis of anomalies, we consider a set of chiral currents $\bar{\psi} T^a \Gamma^m P_+ \psi$ or energy momentum tensors $\bar{\psi}(D_m \Gamma_n + D_n \Gamma_m) P_+ \psi$ or both in the presence of external gauge and gravitational fields; one draws one-loop diagrams with these currents at the vertices, regulates the diagram and then checks for current conservation. Each amplitude obtained in this way can be divided into parity conserving and parity violating parts. Up to an overall constant, the real part of the amplitude is the same as the one obtained for the same process, but with Dirac fermions running around the loop, and in the representation of the gauge group as the original fermions. General arguments now guarantee that this amplitude is always gauge-invariant (and/or generally coordinate-invariant) by adding suitable counterterms to the effective action. The reason is simply that for Dirac fermions one can write down explicit gauge-invariant mass terms, and consequently the real part of the amplitude could be defined in terms of Pauli-Villars regulators in a gauge-invariant way. For the parity-violating part the conclusions are very different. There are two ways of defining the parity violating part of the amplitudes: The first (and most cumbersome from a calculational point of view) is to define the parity-violating part of the amplitude preserving Bose symmetry on the external lines. The second is to consider the same diagram with a single axial vector current in one vertex, and vector currents on the other, defining the diagram by requiring Bose symmetry and current conservation in the vector channels (the Adler-Rosenberg method). One can then check for current conservation in the axial channel. In more modern notation,[39] the first method leads to the consistent anomaly, and the second to the covariant anomaly, and there is a simple algorithm which interpolates between these two forms of the anomaly as will be shown in Section IV. Since the parity-violating part must necessarily contain the $\varepsilon_{m_1 \ldots m_{2n}}$ tensor, we can determine which diagrams are potentially anomalous. Since we want to check current conservation in the axial channel, the polarization vector of the gauge field incoming at the axial vertex is proportional to the incoming momentum P_μ. If there are k other vertices, there are $P_1, \ldots, P_k$ other momenta, and $\varepsilon_1, \ldots, \varepsilon_k$ polarization vectors of the incoming vector fields. The minimal value for k can be obtained by noting that momentum conservation requires $P + \Sigma P_i = 0$, and therefore the simplest diagram giving in principle a contribution to the anomaly $<\partial_m J^m_5>$ is a n+1 polygon, so that the amplitude contains n polarization vectors $\varepsilon_{1\,m_1}\ldots\varepsilon_{n\,m_n}$, and n independent momenta to saturate the ε-tensor; finally the group theory factor will clearly be equal to $(\pm)\mathrm{STr}\, T^{a_1}\ldots T^{a_{n+1}}$, the + (-) sign corresponds to positive (negative) chirality fermions: $\bar{\Gamma}\psi = +\psi\ (-\psi)$ running around the loop. Even though higher polygons also contribute to the anomaly, their coefficients are fixed by the Wess-Zumino consistency conditions[27] (see Section IV) once the (n+1)-polygon is known. Therefore if the first polygon is non-anomalous the higher polygons will accordingly be non-anomalous. This argument justifies some of the remarks made in the introduction, and also shows why the group theoretic tools developed in this section are relevant for anomaly computations.

The same argument applies to the gravitational anomalies,[11] because the graviton polarization vector is a second-rank symmetric tensor. From a physical point of view one can understand why the anomalies will appear in these one-loop diagrams. When we consider the standard Feynman rules for a fermion field ψ_+ running around the loop, we have to consider both the particles and antiparticles of the field inside the loop. In four dimensions, for example, the particles of a left-handed fermion field have positive helicity and the antiparticles negative helicity. Thus if we consider that there are only gravitons in the external lines, the fact that C flips the eigenvalues of $\bar{\Gamma}$ means that as far as the gravitational interactions are concerned, there is no difference between the particles and antiparticles, the degrees of freedom "look" vector-like. Consequently there should be no purely gravitational anomalies in four dimensions. The same argument applies to any 4k dimensions. There are no purely gravitational anomalies in 4k dimensions. In 4k+2 dimensions however $[C,\bar{\Gamma}] = 0$; the particles and antiparticles of a chiral fermion carry the same chirality, so that the interaction with the gravitational field will be genuinely chiral, and one expects to get anomalies in the energy momentum tensor (or in local Lorentz invariance, depending on how one sets up the problem (see Section IV)). Indeed explicit computations show[11] that 4k+2 dimensional chiral theories contain gravitational anomalies. The same argument can be applied to the gauge case, or the combined gauge and gravitational case.

Even though we have presented these arguments in Minkowski space, we could have started with our treatment in Euclidean space. When we perform the Wick rotation, the ε-tensor picks up a factor of i, so that the parity-violating part of the effective action becomes the imaginary part of the Euclidean effective action. In this case, the real part of the effective action is always invariant under infinitesimal gauge transformations[11] (those that can be reached continuously from the identity). If for simplicity we restrict our considerations to four-dimensional chiral gauge theories and label the effective action in some representation r by $\Gamma_r[A]$,

$$2\,\mathrm{Re}\,\Gamma_r[A] = \Gamma_r[A] + \Gamma_r[A]^* \,. \tag{2.40}$$

Since the complex conjugation changes the sign of the parity-violating amplitudes, $\Gamma_r[A]^*$ is the same as the effective action for a left-handed fermion in the complex conjugated representation $\Gamma_{\bar{r}}[A]$. If we remember that the functional integral defining $\Gamma_r[A]$ is gaussian in the fermion fields, we get:

$$2\,\mathrm{Re}\,\Gamma[A] = \Gamma_{r+\bar{r}}[A] \,. \tag{2.41}$$

However, in four dimensions a left-handed fermion in the representation $r+\bar{r}$ is equivalent to a Dirac fermion in the representation r, and we can regulate $\Gamma_{r+\bar{r}}[A]$ in a gauge-invariant way by using say Pauli-Villars fields or a ζ-function regularization.[40] Once again we find that only the imaginary part of $\Gamma_r[A]$ may be responsible for the anomaly. If the representation r is pseudoreal, then r+r is a real representation and $2\,\mathrm{Re}\,\Gamma_r[A]$ is gauge-invariant under infinitesimal gauge transformations. This does not necessarily mean that $\Gamma_r[A]$ will be invariant under gauge transformations which cannot be reached from the identity. The functional integral over the fermions gives $\exp-\Gamma_r[A]$, and in going from $\exp-2\Gamma_r[A]$, to $\exp-\Gamma_r[A]$ one has to extract a square root. The sign of the square root may not be defined under large gauge transformations. Indeed this is the origin of the SU(2) anomaly found by Witten,[32] and the first discovered example of a global anomaly. These types of anomalies are considered in Section VIII.

B. In this subsection we review some elementary ideas in index theory which will be used in the next few sections. There are several proofs in the literature of the Atiyah-Singer (A.S.) index theorem for manifolds without boundary[15] and for the Atiyah-Patodi-Singer (A.P.S.) index theorem for manifolds with boundary,[16] and we will not attempt to present any proofs of these fundamental results in modern mathematics.[41] We will simply formulate their content in a form useful for our later arguments. The reader interested in understanding how these results are obtained should consult the references given.

In order to present the statements of the index theorem, we have to quickly go over some basic results in Riemannian geometry[42] and the theory of characteristic classes.[43] We will follow the presentation of Ref. 20. Given some compact manifold M and a principal G bundle over it, there are two basic geometrical objects to be considered. The Riemannian metric on M, $g_{\mu\nu}$, and a connection on the bundle over M, $A = A_\mu\, dx^\mu$, a one form taking values on the Lie algebra of G (we take A to be antiHermitian). Given a metric $g_{\mu\nu}$, there is a unique connection which leaves $g_{\mu\nu}$ invariant and is torsion-free: the Riemannian connection. The standard way of writing it is in terms of the Christoffel symbol $\Gamma^{\mu}{}_{\nu\rho}$ as a one form:

$$\Gamma^{\alpha}{}_{\beta} = \Gamma^{\alpha}{}_{\beta\rho}\, dx^{\rho}$$

$$\Gamma^{\alpha}{}_{\beta\rho} = \tfrac{1}{2} g^{\alpha\sigma}(-\partial_{\sigma} g_{\beta\rho} + \partial_{\beta} g_{\sigma\rho} + \partial_{\rho} g_{\beta\sigma})$$

$$\nabla_{\mu} g_{\alpha\beta} = \partial_{\mu} g_{\alpha\beta} - \Gamma^{\rho}{}_{\mu\alpha} g_{\rho\beta} - \Gamma^{\rho}{}_{\mu\beta} g_{\rho\alpha} = 0 . \tag{2.42}$$

This is all one needs to define local Riemannian geometry if one is only interested in dealing with ordinary tensor fields. Since we want to consider spinor fields, we have to introduce a vielbein $\{e^{a}_{\mu}\}$ or a section of the bundle of orthonormal frames over M.

$$g_{\mu\nu} = e^{a}_{\mu} e^{b}_{\nu} \delta_{ab} . \tag{2.43}$$

The spin connection is a one-form with values in the Lie algebra of SO(n) (n = dim M). The torsion-free condition is used to define the spin connection $\omega^{a}{}_{b} = \omega_{\mu}{}^{a}{}_{b}\, dx^{\mu}$:

$$T^{a} = d\, e^{a} + \omega^{a}{}_{b}\, e^{b} = 0 \tag{2.44}$$

(The wedge product is suppressed) $e^{a} = e^{a}{}_{\mu}(x) dx^{\mu}$. Under a local frame redefinition $e^{a} \to (L^{-1})^{a}{}_{b}\, e^{b}$; L(x) an arbitrary orthogonal matrix, the connection transforms as a gauge field:

$$\omega \to L^{-1}(\omega + d)L \tag{2.45}$$

(SO(n) indices suppressed). The connection one-form defines parallel transport of orthonormal frames. For an infinitesimal displacement given by the vector ξ^{μ}, the frame rotates to $\omega^{a}{}_{b}(\xi) e^{b} = \omega_{\mu}{}^{a}{}_{b}\, \xi^{\mu} e^{b}$. We can use the vielbein $e^{a}{}_{\mu}$ and its inverse $(e^{-1})^{\mu}{}_{b}$ to refer any tensor to orthonormal frames:

$$\Sigma^{a_1 \ldots a_p}_{b_1 \ldots b_q} = e^{a_1}{}_{\mu_1} \ldots e^{a_q}{}_{\mu_q} (e^{-1})^{\nu_1}{}_{b_1} \ldots (e^{-1})^{\nu_q}{}_{\mu_q} \Sigma^{\mu_1 \ldots \mu_q}_{\nu_1 \ldots \nu_q} \tag{2.46}$$

and the covariant derivative acting on Σ can be written in analogy with the gauge covariant derivative:

$$D = d + [\omega,\ ] . \tag{2.47}$$

If $\Sigma^{a\ldots}_{b\ldots}$ is a p-form, the bracket is defined to act on the left on all upper indices, and to the right with sign $-(-1)^{p}$ on lower indices. For instance, if $\Sigma^{a}{}_{b}$ is a 0-form:

$$(D\Sigma)^{a}{}_{b} = d\Sigma^{a}{}_{b} + \omega^{a}{}_{c}\, \Sigma^{c}{}_{b} - \Sigma^{a}{}_{c}\, \omega^{c}{}_{b} . \tag{2.48}$$

Now we can treat the gauge field A and the spin connection ω in a similar manner geometrically. The curvature and gauge field strengths are defined by:

$$R = d\omega + \omega^2$$
$$F = dA + A^2 \tag{2.49}$$

and the Bianchi identities can be obtained by acting with d in (2.49), and using $d^2 = 0$

$$dR = d\omega\omega - \omega d\omega = R\omega - \omega R$$
$$dF = dAA - AdA = FA - AF$$

or

$$DR = dR + [\omega, R] = 0$$
$$DF = dF + [A,F] = 0 . \tag{2.50}$$

If one wants to include the gravitational effect of torsion, $T^a = de + \omega e$, there is one more Bianchi identity:

$$dT = d\omega e - \omega de = Re - \omega T . \tag{2.51}$$

For $T = 0$, this implies $Re = 0$, which is the cyclic identity for the curvature tensor. In components, (2.49-2.51) are

$$R^a{}_b = \tfrac{1}{2} R^a{}_{b\mu\nu}\, dx^\mu \wedge dx^\nu$$
$$F = \tfrac{1}{2}(\partial_\mu A_\nu - \partial_\nu A_\mu + [A_\mu, A_\nu]) dx^\mu \wedge dx^\nu \tag{2.49'}$$
$$\nabla_\lambda R^a{}_{b\mu\nu} + \nabla_\mu R^a{}_{b\nu\lambda} + \nabla_\nu R^a{}_{b\lambda\mu} = 0 \tag{2.50'}$$
$$D_\lambda F_{\alpha\beta} + D_\alpha F_{\beta\lambda} + D_\beta F_{\lambda\alpha} = 0$$
$$R^a{}_{bcd} + R^a{}_{cdb} + R^a{}_{dbc} = 0 . \tag{2.51'}$$

Under gauge transformations or frame redefinitions, A, ω, R, F, T change according to:

$$A \to g^{-1}(A + d)g$$
$$F \to g^{-1} F g$$
$$\omega \to L^{-1}(\omega + d)L$$
$$R \to L^{-1} R L$$
$$T \to L^{-1} T \quad . \tag{2.52}$$

The spin connection ω_{ab} can be computed once we have a choice of vielbein; de^a can be rewritten as

$$de^a = \tfrac{1}{2} \xi^a{}_{bc}\, e^b e^c$$
$$\omega^a{}_b = \omega^a{}_{b,c}\, e^c ,$$

and (2.44) becomes:

$$\xi_{a,bc} = \omega_{ab,c} - \omega_{ac,b} \,. \tag{2.53}$$

Adding cyclic permutations, we obtain

$$\omega_{ab,c} = \tfrac{1}{2}(\xi_{a,bc} + \xi_{b,ca} - \xi_{c,ab}) \tag{2.54}$$

If we use instead a coordinate basis and the Christoffel connection, all the formulae are essentially unchanged:

$$\begin{aligned} \nabla &= d + [\Gamma,\ ] \\ \Gamma^{\alpha}{}_{\beta} &= \Gamma^{\alpha}{}_{\gamma\beta}\, dx^{\gamma} \\ R^{\alpha}{}_{\beta} &= \tfrac{1}{2} R^{\alpha}{}_{\beta\lambda\mu}\, dx^{\lambda}\, dx^{\mu} = d\Gamma^{\alpha}{}_{\beta} + \Gamma^{\alpha}{}_{\gamma}\Gamma^{\gamma}{}_{\beta} \,. \end{aligned} \tag{2.55}$$

From the definition of the covariant derivatives, one obtains the Ricci identities:

$$\begin{aligned} \nabla^2 &= [R,\ ] \\ D^2 &= [F,\ ] \,. \end{aligned} \tag{2.56}$$

Since the Riemannian connections ω,Γ are uniquely determined once $e^{a}{}_{\mu}$ and $g_{\mu\nu}$ are given, they are related to each other. The relation stems from rewriting the torsion-free condition (2.44) in terms of Γ-covariant derivatives:

$$(e^{-1})_{c}{}^{\mu}\nabla_{b}\, e^{a}{}_{\mu} - (e^{-1})_{b}{}^{\mu}\nabla_{c}\, e^{a}{}_{\mu} + \omega^{a}{}_{c,b} - \omega^{a}{}_{b,c} = 0 \,.$$

Adding and subtracting appropriate cyclic permutations, we get:

$$\omega^{a}{}_{b,\mu} = -(e^{-1})_{b}{}^{\nu}\nabla_{\mu}\, e^{a}{}_{\nu} \tag{2.57}$$

where ∇_{μ} is only acting on the coordinate index of $e^{a}{}_{\nu}$. Using (2.57) one can easily interpolate between frame and coordinate covariant derivatives. For example,

$$D_{\mu}\, t^{a} = e^{a}{}_{\nu}\nabla_{\mu}\, t^{\nu} \tag{2.58}$$

and similar expressions for higher-order tensors. In theories describing the interactions of spin-1/2 particles with the gravitational field, it is necessary to introduce orthonormal frames because GL(n) does not have spinor representations. Before describing the geometry of spinors on a manifold M, one must first check whether M admits a spin structure. This requires us to check whether the frame bundle with structure group SO(n) lifts to a spin(n) bundle. The obstruction to this lifting is measured by the second Stieffel-Whitney class w_2.[43] Furthermore, if one wants to define chiral fermions, one has to be able to define smoothly $\bar{\Gamma} = \varepsilon_{\mu_1 \ldots \mu_{2n}} \Gamma^{\mu_1} \ldots \Gamma^{\mu_{2n}}/(2n)!$, which means that the manifold is orientable, i.e. there is a well-defined volume element on M. The obstruction to orientability is measured by the first Stieffel-Whitney class w_1.[43] Thus before we analyze spinor dynamics on M, we must check $w_1 = w_2 = 0$. The covariant derivative on spinors is defined in terms of the spin connection and the generators of SO(n) in the spinor representation (2.2)

$$D_\mu \psi_\pm = \partial_\mu \psi_\pm + \tfrac{1}{4} \omega_{\mu ab} \Sigma^{ab} \psi_\pm \tag{2.59}$$

$$\{\Gamma^a, \Gamma^b\} = 2\delta^{ab} \tag{2.60}$$

$$\Gamma^\mu = (e^{-1})_a{}^\mu \Gamma^a$$

and the Dirac action becomes:

$$\int d^2x \sqrt{g}\, \bar{\psi}\, i\, \Gamma^\mu D_\mu \psi \ . \tag{2.61}$$

In the analysis of gravitational theories with spinors, there are two types of invariances that one has to consider: general coordinate invariance, or invariance under the diffeomorphism group of M, Diff(M), and invariance under local frame redefinitions. Bardeen and Zumino[39] called the anomalies on Diff(M) Einstein anomalies, and Lorentz anomalies those appearing in frame redefinitions. We now discuss the action of general coordinate transformations on geometrical objects. Under a general coordinate transformation $x \to x'(x)$, the tensor valued forms $\Sigma^{\mu\ldots}_{\nu\ldots,\alpha\ldots}\, dx^\alpha \ldots$ are defined to transform as

$$\Sigma(x) \to \Sigma'(x') \equiv \Lambda^{-1}_{\text{upper}}(x)(\Sigma(x)dx)\Lambda_{\text{lower}}(x) \tag{2.62}$$

$$(\Lambda^{-1})^\mu{}_\nu = \partial x'^\mu / \partial x^\nu$$

$$\Lambda^\mu{}_\nu = \partial x^\mu / \partial x'^\nu \ .$$

Λ^{-1} acts on all the upper indices, and Λ on all the lower indices. The connection Γ transforms like a GL(n) gauge field:

$$\Gamma(x) \to \Gamma'(x') = \Lambda^{-1}(\Gamma + d)\Lambda\Big|_x \ . \tag{2.63}$$

Notice that in (2.62,2.63) the coordinate transformation is defined according to the "passive" point of view; we change the geometrical object $\Sigma(x)$ as well as its argument: $\Sigma(x) \to \Sigma'(x')$. Similarly the volume element will change as well $dV(x) \to dV(x')$ by the appropriate Jacobian determinant. There is also an "active" point of view in which the objects $\Sigma(x)$ are varied according to $\Sigma(x) \to \Sigma'(x)$, i.e. shifting back to the point x, and leaving the integration measure unchanged. Both of these representations of the action of the diffeomorphism group enter in the discussion of Einstein anomalies. Under an infinitesimal coordinate change:

$$x^\alpha \to x^\alpha - \xi^\alpha(x) \ . \tag{2.64}$$

we find from (2.62) the infinitesimal passive action

$$\delta'_{\xi}\Sigma = \Sigma'(x') - \Sigma(x) = -[V(\xi),\Sigma] \tag{2.65}$$

$$(V(\xi))^{\alpha}{}_{\beta} \equiv \frac{\partial\xi^{\alpha}}{\partial x^{\beta}} \ .$$

Similarly, the passive change in the connection Γ is given by

$$\delta'_{\xi}\Gamma = dV(\xi) - [V(\xi),\Gamma] = \nabla V(\xi) \ . \tag{2.66}$$

The prime on δ'_{ξ} stands for passive transformation. The infinitesimal active action of Diff(M) is essentially equivalent to the Lie derivative along ξ. Acting on the tensor valued form Σ, the Lie derivative is defined by

$$\begin{aligned}\mathcal{L}(\xi)\Sigma^{\alpha\ldots}_{\beta\ldots,\mu\ldots} &= \xi^{\lambda}\,\partial_{\lambda}\,\Sigma^{\alpha\ldots}_{\beta\ldots,\mu\ldots} \\ &\quad - \partial_{\lambda}\,\xi^{\alpha}\,\Sigma^{\lambda\ldots}_{\beta\ldots,\mu\ldots} - \ldots \\ &\quad + \Sigma^{\alpha\ldots}_{\lambda\ldots,\mu\ldots}\,\partial_{\beta}\,\xi^{\lambda} + \ldots \end{aligned} \tag{2.67}$$

or more concisely $\mathcal{L}(\xi) = \xi\cdot\partial - [V(\xi),\Sigma]$. The Lie derivative is one of Schouten's differential concommitants, namely an operator which can be defined on tensors without introducing the connection but transforming covariantly nonetheless. Geometrically (2.67) is obtained by constructing the integral curves of $dx^{\alpha}/dt = \xi^{\alpha}(x)$, $x^{\alpha}(t) = \phi^{\alpha}(x,t)$. ($x$ is the initial condition of the system of differential equations.) Next we pull back to x the value of Σ at $x^{\alpha}(t) = \phi^{*}_{t}\Sigma$. The Lie derivative is given by minus the t derivative of this one parameter family of Σ's at $t=0$. A simple property of (2.67) is that we can change the standard derivative ∂ by the Riemannian covariant derivative ∇, and the expression remains the same (i.e. $\mathcal{L}(\xi)\Sigma$ is covariant). Furthermore, the Lie derivative is a derivative on the tensor algebra over M, and generates a Lie algebra structure on the space of vector fields over M:

$$\mathcal{L}(\xi)(T_1 \otimes T_2) = (\mathcal{L}(\xi)T_1) \otimes T_2 + T_1 \otimes \mathcal{L}(\xi)T_2$$

$$[\mathcal{L}(\xi_1),\mathcal{L}(\xi_2)] = \mathcal{L}([\xi_1,\xi_2])$$

$$[\xi_1,\xi_2]^{\alpha} = \xi_1^{\lambda}\,\partial_{\lambda}\,\xi_2^{\alpha} - \xi_2^{\lambda}\,\partial_{\lambda}\,\xi_1^{\alpha} \ . \tag{2.68}$$

We thus see that the infinitesimal active change of tensor fields under $x^{\alpha} \to x^{\alpha} - \xi^{\alpha}(x)$ is directly given by the Lie derivative:

$$\delta_{\xi}\Sigma = \Sigma'(x) - \Sigma(x) = \mathcal{L}(\xi)\Sigma \ . \tag{2.69}$$

For example,

$$\mathcal{L}(\xi)g_{\alpha\beta} = \nabla_{\alpha}\,\xi_{\beta} + \nabla_{\beta}\,\xi_{\alpha} \ .$$

For tensor valued forms, we can write (2.69) in terms of operators naturally defined for forms: the inner multiplication of a p-form by a vector field ξ is defined by:

$$i(\xi)\omega_p = \frac{1}{(p-1)!}\xi^{\alpha_1}\omega_{\alpha_1\alpha_2\ldots\alpha_p}dx^{\alpha_2}\ldots dx^{\alpha_p}$$

$$\omega_p = \frac{1}{p!}\omega_{\alpha_1\ldots\alpha_p}dx^{\alpha_1}\ldots dx^{\alpha_p}\,. \tag{2.70}$$

Elementary properties of $i(\xi)$ are:

$$i(\xi)^2 = 0$$

$$[i(\xi),\mathscr{L}(\xi)] = 0$$

$$i(\xi)\Omega_p\Lambda_q = (i(\xi)\Omega_p)\Lambda_q + (-1)^p\Omega_p(i(\xi)\Lambda_q)\,. \tag{2.71}$$

Ω_p is a p-form, Λ_q is a q-form. For scalar valued forms ω_p, the Lie derivative is given by:

$$\mathscr{L}(\xi)\omega_p = (di(\xi) + i(\xi)d)\omega_p\,. \tag{2.72}$$

For tensor valued forms:

$$\begin{aligned}\mathscr{L}_\xi &= di(\xi) + i(\xi)d - [U(\xi),\ ]\\ U(\xi)^\alpha{}_\beta &= \frac{\partial\xi^\alpha}{\partial x^\beta}\,.\end{aligned} \tag{2.73}$$

Similarly, the change in the connection Γ from an active point of view is

$$\delta_\xi\Gamma = (di(\xi) + i(\xi)d)\Gamma + \nabla U(\xi)\,. \tag{2.74}$$

From (2.65) we see that on tensor valued forms:

$$\delta_\xi = di(\xi) + i(\xi)d + \delta'_\xi\,. \tag{2.75}$$

The spin connection on the other hand transforms like a scalar under passive transformations $\delta'_\xi\omega = 0$, and under an active transformation

$$\begin{aligned}\delta_\xi\omega &= \mathscr{L}(\xi)\omega = (di(\xi) + i(\xi)d)\omega\\ &= i(\xi)R - i(\xi)\omega\omega + \omega\, i(\xi)\omega + d(i(\xi)\omega)\\ &= i(\xi)R + d(i(\xi)\omega) + [\omega,i(\xi)\omega]\,.\end{aligned} \tag{2.76}$$

If we recall (2.52) for infinitesimal local Lorentz transformations,

$$\begin{aligned}\delta^L e &= -\alpha e\\ \delta^L\omega &= d\alpha + [\omega,\alpha] = D\alpha\end{aligned} \tag{2.77}$$

we get

$$\delta_\xi\omega = i(\xi)R + D(i(\xi)\omega)\,. \tag{2.78}$$

Similarly

$$\mathscr{L}(\xi)e^a = D\xi^a + \delta^L_{i(\xi)\omega}e^a\,. \tag{2.79}$$

The definition of the change in e and ω under coordinate transformations can always be modified by an arbitrary (ξ-dependent) local Lorentz transformation, and this allows one to write many different forms for the gravitational anomalies depending on the compensating Lorentz transformation chosen. A useful thing to notice is that for forms Σ of maximal degree (= dim M)

$$(di(\xi) + i(\xi)d)\Sigma = d(i(\xi)\Sigma) \ . \tag{2.80}$$

Thus acting on forms of maximal degree, (2.80) implies the equivalence of the passive and active points of view up to harmless total derivatives.

After this digression on notation we can start our study of characteristic classes. In general there are various homology and cohomology groups that can be introduced to study the topological complexity of a generic manifold M. The simplest ones by far are the De Rham cohomology groups; they are defined in terms of the harmonic forms on M. We have already encountered the exterior derivative d. Given two p-forms ω_1, ω_2, we can define a scalar product by:

$$\begin{aligned}(\omega_1, \omega_2) &= \int_M \omega_1 \wedge *\omega_2 \\ &= \frac{1}{p!}\int_M \omega_{i_1 \dots i_p}\, \omega^{i_1 \dots i_p}\end{aligned} \tag{2.81}$$

$*\omega$ is the Hodge duality operation, defined for orientable manifolds. If Λ^p denotes the space of p-forms over M_n,

$$*: \Lambda^p \to \Lambda^{n-p}$$

$$(*\omega)_{a_1 \dots a_{n-p}} = \frac{1}{p!}\, \varepsilon_{a_1 \dots a_{n-p}\, b_1 \dots b_p}\, \omega^{b_1 \dots b_p} \ . \tag{2.82}$$

The exterior derivative provides a map

$$\begin{aligned} d &= \Lambda^p \to \Lambda^{p+1} \\ d^2 &= 0 \ . \end{aligned} \tag{2.83}$$

We can define the adjoint of d with respect to (2.81)

$$\begin{gathered}(\omega_1, d\omega_2) = (d^*\omega_1, \omega_2) \\ d^* = \Lambda^p \to \Lambda^{p-1} \ ; \ d^{*2} = 0 \\ (d^*\omega)_{i_1 \dots i_{p-1}} = -\nabla^j\, \omega_{j i_1 \dots i_{p-1}} \ . \end{gathered} \tag{2.84}$$

A p-form ω is said to be closed if $d\omega_p = 0$, and it is exact if $\omega_p \equiv d\alpha_{p-1}$. Similarly ω is coclosed if $d^*\omega_p = 0$, and coexact if $\omega_p \equiv d^*\alpha_{p+1}$. The pth De Rham cohomology group $H_p(M, \mathbb{R})$ is defined by the equivalence classes of closed p-forms modulo exact p-forms, i.e. ω_1, ω_2 define the same class in $H_p(M, \mathbb{R})$ if $\omega_1 - \omega_2 = d\alpha_{p-1}$. In terms of d and d^* we can define a second-order elliptic semipositive definite operator: the Laplacian $\Box = dd^* + d^*d$; notice that $\Box: \Lambda_p \to \Lambda_p$. If M is compact without boundary, we can easily characterize the zeroes of $\Box$. If $\Box\omega_p = 0$, then

$$(\omega_p, \Box\omega_p) = (\omega_p, (d^*d + dd^*)\omega_p) = (d\omega_p, d\omega_p) + (d^*\omega_p, d^*\omega_p) = 0 \,. \tag{2.85}$$

Since every term is positive in (2.85), we conclude that

$$\Box\omega_p = 0 \leftrightarrow d\omega_p = 0,\ d^*\omega_p = 0 \,. \tag{2.86}$$

The zeroes of $\Box$ are called harmonic forms, and we have just proved that the space of harmonic forms is equivalent to the space of closed and coclosed forms. A basic result in De Rham cohomology is the Hodge decomposition theorem. It states that any p-form can be uniquely rewritten as:

$$\omega_p = d\alpha_{p-1} + d^*\beta_{p+1} + \gamma_p \tag{2.87}$$

and γ_p is harmonic. The proof of this theorem is elementary and can be found for example in Ref. 23. Hodge's theorem implies that every class in H_p can be represented by a harmonic form. Given a gauge or spin connection on M : ω, $\Omega = d\omega + \omega^2$, we can construct closed forms which in general will represent some class on $H_p(M, \mathbb{R})$. The construction of these forms is very simple.[43] Since ω is a connection for some bundle with structure group G, let $P(\alpha)$ be an invariant polynomial where α is an arbitrary element in the Lie algebra of G: $P(g^{-1}\alpha g) = P(\alpha)$. If we now write Ω instead of α in P, $P(\Omega)$ defines a closed form, and its integrals over appropriate closed submanifolds of M are independent of the connection ω (as long as the connections considered are in the same topological class. Before proving these results, let us recall that if we have a principle bundle P over M (or any of its associated vector bundles), the twisting of P is measured by its transition functions: let $\{U_i\}$ be a collection of good coordinate chart of M, the bundle P is topologically defined by giving the transition functions $g_{ij} = U_i \cap U_j \to G$ in the overlaps of two patches, i.e. the g_{ij}'s tell us how to identify the fibers over U_i with those over U_j. All the topological set-up is encoded in the collection of $\{g_{ij}\}$. A familiar example is given by the classification of bundles over spheres S^n. We can cover S^n by two coordinate patches D_+, D_- (upper and lower hemispheres), $D_+ \cap D_- = S^{n-1}$. Thus the transition function is a map $g = S^{n-1} \to G$. These maps are classified by $\pi_{n-1}(G)$ (n-1 st homotopy group of G). For more general manifolds the classification of G bundles is more complicated, and we refer the reader to the literature for further details on bundle classification and obstruction theory.[44] Given a connection ω on the bundle P, the characteristic polynomials $P(\Omega)$ define a map from the bundle P to the De Rham cohomology groups of M. We are now ready to prove the topological invariance of $P(\Omega)$. The computation can be simplified by considering instead monomials of the form:

$$P_m \equiv \mathrm{Tr}\,\Omega^m \ . \tag{2.88}$$

That P_m is closed follows from the Bianchi identity (2.50):

$$dP_m = d\,\mathrm{Tr}\,\Omega^m = m\,\mathrm{Tr}\,D\Omega\,\Omega^{m-1} = 0 \ . \tag{2.89}$$

Let ω_1, ω_2 be two connections on P with topologically equivalent transition functions, and let $\omega(t)$ $0 \le t \le 1$ be an interpolation between ω_1, ω_2, $\omega(t{=}0) = \omega_1$, $\omega(t{=}1) = \omega_2$. For each value of t we can define $P_m(t)$. Then

$$\frac{d}{dt}\Omega_t = d\left(\frac{d\omega(t)}{dt}\right) + \frac{d\omega(t)}{dt}\,\omega(t) + \omega(t)\,\frac{d\omega(t)}{dt}$$

$$= D_t\left(\frac{d\omega(t)}{dt}\right)$$

$$D_t M \equiv dM + [\omega(t), M] \ .$$

Thus

$$\frac{d}{dt}P_m(t) = m\,\mathrm{Tr}\,\frac{d\Omega(t)}{dt}\,\Omega^{m-1}(t)$$

$$= m\,\mathrm{Tr}\,D_t\left(\frac{d\omega}{dt}\right)\Omega^{m-1}(t)$$

$$= m\,d\,\mathrm{Tr}\,\frac{d\omega(t)}{dt}\,\Omega^{m-1}(t) \tag{2.90}$$

In the last step we have used the Bianchi identity (2.51). Integrating (2.90) from $t = 0$ to $t = 1$:

$$P_m(\Omega_2) - P_m(\Omega_1) = d\,m\int_0^1 dt\,\mathrm{Tr}\,\frac{d\omega(t)}{dt}\,\Omega^{m-1}(t)$$

$$= d\,Q^0_{2m-1}(\omega_t, \Omega_t) \ . \tag{2.91}$$

Q^0_{2m-1} is the Chern-Simons form. If we integrate (2.91) over a closed 2m-submanifold we get (using Stokes theorem):

$$\int_{M_{2m}} P_m(\Omega_2) = \int_{M_{2m}} P_m(\Omega_1) \ . \tag{2.92}$$

The numbers obtained this way are invariant not only under changes of ω, but also under deformations of M_{2m}. Let B_{2m+1} be a 2m+1 dimensional manifold with boundary $\partial B_{2m+1} = M_{2m} - M'_{2m}$ where M'_{2m} is a deformation of M_{2m}, and the minus sign is included so that M_{2m}, M'_{2m} have the same orientation. Since $dP_m = 0$, we can integrate dP_m over B_{2m+1} and get

$$\int_{M_{2m}} P_m = \int_{M'_{2m}} P_m ,$$

i.e. the numbers defined by (2.92 are characteristic numbers of the topology and differential structure on M.

There is a much richer structure of homology and cohomology groups which have been used in mathematics[43] to analyze the topology of M. Since we will not need these more exotic structures until the last section, we will not consider them here.

When the bundle group is U(n), Ω is an antiHermitian matrix of 2-forms, and the Chern classes are defined by the expansion of:

$$c(\Omega) = \det(1 + \frac{i}{2\pi}\Omega) = 1 + c_1 + c_2 + \dots \tag{2.93}$$

$$c_1 = \frac{i}{2\pi}\mathrm{Tr}\,\Omega, \text{ etc.}$$

Formally we can diagonalize the matrix $i\Omega/2\pi$. Let $x_i,\dots,x_n$ be its formal eigenvalues. Then

$$\begin{aligned} c(\Omega) &= \prod_i (1 + x_i) \\ c_k(\Omega) &= \sum_{i_1<\dots<i_k} x_{i_1},\dots x_{i_k} \; . \end{aligned} \tag{2.94}$$

It is easier to do the computations in terms of the x_i's, and at the end rewrite the answer in terms of the P_m's. (This is a ham-handed version of the Hirzebruck splitting principle.[43]) $c(\Omega)$ is known as the total Chern class, and it is multiplicative for direct sums of vector bundles $c(E \oplus F) = c(E)c(F)$. Another useful polynomial is the Chern character $ch(\Omega)$ defined to be additive with respect to direct sums of bundles:

$$ch(\Omega) = \mathrm{Tr}\, e^{i\Omega/2\pi} = 2n + ch_1(\Omega) + \dots \tag{2.95}$$

If the bundle group is SO(n), as in Riemannian geometry, we can define the Pontrjagin classes in a similar manner. Now the curvature Ω is an antisymmetric matrix of 2-forms. Again we can formally skew-diagonalize Ω:

$$\frac{\Omega}{2\pi} \sim \begin{bmatrix} 0 & x_1 & & & & \\ -x_1 & 0 & & & & \\ & & 0 & x_2 & & \\ & & -x_2 & 0 & & \\ & & & & \cdot & \\ & & & & & \cdot \\ & & & & & & \cdot \end{bmatrix} \tag{2.96}$$

The total Pontrjagin class of the bundle E is defined by:

$$p(E) = \det(1 + \frac{1}{2\pi}\Omega) = \prod_i (1 + x_i^2) = 1 + p_1 + p_2 + \dots \tag{2.97}$$

$$p_1 = \sum_i x_i^2 \; , \qquad p_2 = \sum_{i<j} x_i^2 x_j^2 \; \dots$$

Since $\Omega^T = -\Omega$, the Pontrjagin classes are different from zero only in 4k-dimensional spaces whereas the Chern classes can be different from zero in any even dimensional space.

Now we are in a position to formulate the index theorem. Since we are basically interested in the Dirac operator for fermions coupled to external gauge and gravitational fields in even dimensions, one has two vector bundles over M: $S_+ \otimes V$ and $S_- \otimes V$. V is the space carrying the representation of the gauge group, S_+ (S_-) is the space of positive (negative) chirality spinors. More explicitly, a field taking values on $S_+ \otimes V, \psi_{\alpha A}$, carries two indices: α runs over the components of the positive $\bar{\Gamma}$ spinor representation, and A runs over the dimension of the representation of G carried by the fermions. In this setting, the Weyl operator $D_+ = i\not{D} P_+$ (its adjoint is $D_- = i\not{D} P_-$) sends objects in $S_+ \otimes V$ into objects in $S_- \otimes V$

$$S_+ \otimes V \underset{D_-}{\overset{D_+}{\rightleftarrows}} S_- \otimes V \tag{2.98}$$

$D_\pm$ are given explicitly by:

$$D_\pm = i\gamma^\mu(\partial_\mu + \tfrac{1}{2}\omega_{\mu ab}\Sigma^{ab} + A_\mu)P_\pm \ . \tag{2.99}$$

The index of D_+ is defined to be the dimension of the kernel of D_+ minus the kernel of $D_+^\dagger = D_-$.

$$\text{ind}\, D_+ = \dim\ker D_+ - \dim\ker D_- \ . \tag{2.100}$$

In principle this is a problem in analysis. What Atiyah and Singer showed[15] is that ind D_+ is a number which only depends on the topological set up (2.98) and it is given by the integral over M of a particular characteristic class. In particular, for the operator (2.99) the answer is:

$$\begin{aligned} \text{ind}\, D_+ &= \int_M [\text{ch}(F)\hat{A}(M)]_{\text{vol}} \\ \hat{A}(M) &\equiv \prod_a \frac{x_a/2}{\sinh x_a/2} \\ \text{ch}(F) &= \text{Tr}\, e^{iF/2\pi} \ . \end{aligned} \tag{2.101}$$

$\hat{A}(M)$, the $\hat{A}$ or Dirac genus of M, is a polynomial in the two forms x_a which can be rewritten in terms of the P_m's. Since M is finite dimensional and the x_a's are the 2-forms (2.96), $\hat{A}(M)$ is a finite polynomial (same for ch(F). The subscript 'vol' in (2.101) means that one has to extract the form whose degree is equal to the dimension of M. The expansion of $\hat{A}(M)$ is rather cumbersome, and we will write it in terms of the P_m's and as a polynomial in the Pontrjagin classes. A careful expansion yields:[20]

$$\hat{A}(M) = 1 + \frac{1}{(4\pi)^2}\frac{1}{12}\operatorname{Tr} R^2 + \frac{1}{(4\pi)^4}\left[\frac{1}{288}(\operatorname{Tr} R^2)^2 + \frac{1}{360}\operatorname{Tr} R^4\right]$$

$$+ \frac{1}{(4\pi)^6}\left[\frac{1}{10368}(\operatorname{Tr} R^2)^3 + \frac{1}{4320}\operatorname{Tr} R^2 \operatorname{Tr} R^4 + \frac{1}{5670}\operatorname{Tr} R^6\right]$$

$$+ \frac{1}{(4\pi)^8}\left[\frac{1}{497664}(\operatorname{Tr} R^2)^4 + \frac{1}{103680}(\operatorname{Tr} R^2)^2 \operatorname{Tr} R^4 + \right.$$

$$\left. + \frac{1}{68040}\operatorname{Tr} R^2 \operatorname{Tr} R^6 + \frac{1}{259200}(\operatorname{Tr} R^4)^2 + \frac{1}{75600}\operatorname{Tr} R^8\right] + \ldots$$

$$= 1 + \frac{1}{2^2}\left(-\frac{1}{6}P_1\right) + \frac{1}{2^4}\left(\frac{7}{360}P_1^2 - \frac{1}{90}P_2\right)$$

$$+ \frac{1}{2^6}\left(-\frac{31}{15120}P_1^3 + \frac{11}{3780}P_1P_2 - \frac{1}{945}P_3\right)$$

$$+ \frac{1}{2^8}\left(\frac{127}{604800}P_1^4 - \frac{113}{226800}P_1^2P_2 + \right.$$

$$\left. + \frac{4}{14175}P_1P_3 + \frac{13}{113400}P_2^2 - \frac{1}{9450}P_4\right) + \ldots \qquad (2.102)$$

If $\dim V = r$, we have

$$\operatorname{ch}(F) = r + \frac{i}{2\pi}\operatorname{Tr} F + \frac{i^2}{2(2\pi)^2}\operatorname{Tr} F^2 + \ldots + \frac{i^n}{n!(2\pi)^n}\operatorname{Tr} F^n + \ldots \qquad (2.103)$$

Combining (2.102) and (2.103) we can compute the form of the index theorem in any dimension. For example, in $d = 4$ and $d = 8$:

$$\operatorname{ind} D_+ = \frac{1}{(2\pi)^2}\int_M \left(\frac{i^2}{2}\operatorname{Tr} F^2 + \frac{r}{48}\operatorname{Tr} R^2\right) \qquad d = 4$$

$$= \frac{1}{(2\pi)^4}\int_M \left(\frac{i^4}{24}\operatorname{Tr} F^4 + \frac{i^2}{96}\operatorname{Tr} F^2 \operatorname{Tr} R^2 + \frac{r}{4608}(\operatorname{Tr} R^2)^2 + \frac{r}{5760}\operatorname{Tr} R^4\right) \qquad d = 8$$

$$(2.104)$$

The other operators whose indices are of interest in the computation of anomalies are obtained by replacing V by some particular vector bundle. For instance for the graviton field V = TM, the tangent bundle over M. In this case A is simply the spin connection taking values on the Lie algebra of SO(2n) in the vector representation $(T^{ab})_{cd} = \delta^a{}_c\delta^b{}_d - \delta^a{}_d\delta^b{}_c$. Thus

$$\operatorname{Tr} e^{R_{ab}T^{ab}/4\pi} = \sum_a 2\cosh x_a \qquad (2.105)$$

In the standard quantization of a spin 3/2 field,[45] one has to add ghost fields to remove unphysical degrees of freedom. The constraints $k^a\psi_a = 0$ and $\psi_a \to \psi_a + k_a\chi$ remove two spin 1/2 degrees of freedom of the same chirality as ψ_a, and the constraint $\gamma^a\psi_a = 0$ removes one spin 1/2 degree of freedom of opposite chirality. Including the ghost field in the index theorem for a spin 3/2 field we get:

$$\operatorname{ind} i\not{D}_{3/2} = \int \hat{A}(M)(\operatorname{Tr} e^{R/2\pi} - 1)\operatorname{ch}(F) \qquad (2.106)$$

where the last factor accounts for the possibility that the spin 3/2 field carries some extra gauge index. Because of the dimensional dependence of $\mathrm{Tr}\exp R/2\pi$, we exhibit the quantity $\hat{A}(M)\mathrm{Tr}(e^{R/2\pi}-\mathbb{1})$. To order 16, this polynomial has the following form:

$$\begin{aligned}
\hat{A}(M)\mathrm{Tr}(e^{R/2\pi}-\mathbb{1}) &= -\frac{1}{(4\pi)^2}\, 2\,\mathrm{Tr}\,R^2 \\
&+ \frac{1}{(4\pi)^4}\left[-\frac{1}{6}(\mathrm{Tr}\,R^2)^2 + \frac{2}{3}\mathrm{Tr}\,R^4\right] \\
&+ \frac{1}{(4\pi)^6}\left[-\frac{1}{144}(\mathrm{Tr}\,R^2)^3 + \frac{1}{20}\mathrm{Tr}\,R^2\,\mathrm{Tr}\,R^4 - \frac{4}{45}\mathrm{Tr}\,R^6\right] \\
&+ \frac{1}{(4\pi)^8}\left[-\frac{1}{5184}(\mathrm{Tr}\,R^2)^4 + \frac{1}{540}(\mathrm{Tr}\,R^2)^2\,\mathrm{Tr}\,R^4 - \right. \\
&\left. -\frac{22}{2835}\mathrm{Tr}\,R^2\,\mathrm{Tr}\,R^6 + \frac{1}{540}(\mathrm{Tr}\,R^4)^2 + \frac{2}{315}\mathrm{Tr}\,R^8\right] + \ldots \\
&= \frac{1}{2^2}(4P_1) + \frac{1}{2^4}\left(\frac{2}{3}P_1^2 - \frac{8}{3}P_2\right) \\
&+ \frac{1}{2^6}\left(\frac{1}{30}P_1^3 - \frac{2}{15}P_1P_2 + \frac{8}{15}P_3\right) \\
&+ \frac{1}{2^8}\left(\frac{1}{1260}P_1^4 - \frac{16}{945}P_1^2P_2 - \frac{8}{189}P_1P_3\right. \\
&\left. + \frac{52}{945}P_2^2 - \frac{16}{315}P_4\right) + \ldots \qquad (2.107)
\end{aligned}$$

In the computation of the gravitational anomaly for chiral $N=2$ supergravity in ten dimensions[11] the anomaly also receives contributions from a four index antisymmetric tensor field $A_{\mu_1\mu_2\mu_3\mu_4}$ whose field strength is self-dual. In any even number of dimensions, the self dual representation of SO(2n) appears along with a number of anomaly-free representations[11] in the product of two chiral spinor representations. For that reason it is sufficient to consider the index theorem for a bispinor field $\phi_{\alpha\beta}$, i.e. a spinor with the extra spinor index in the representation $(T^{ab})_{\alpha\beta} = (\Sigma^{ab})_{\alpha\beta}$. The index theorem for these objects is related to a classic result in mathematics, the signature theorem.[43] In previous paragraphs we analyzed the De Rham complex and encountered the Hodge duality operation $*$. It is not difficult to check that $[*,\Box] = 0$, $[d,\Box] = [d^*,\Box] = 0$. Thus it is natural to understand the action of $*$ on the harmonic forms. If ω_p is a harmonic form, $*\omega_p$ is also harmonic. Thus we can divide the De Rham complex $\bigoplus_{p=1}^{2n} \Lambda^p = \Lambda^*(M)$ into self-dual and antiself-dual pieces. If $2n = 4k$, $*^2 = +1$ so that the eigenvalues of $*$ are ± 1. If $n = 4k+2$, $*^2 = -1$, and the eigenvalues are $\pm i$. Notice that for harmonic forms ω_p of degree $\neq n$, we can construct a self-dual and anti-self-dual form: $(\omega_p \pm *\omega_p)/2$, $2n = 4k$,

$(\omega_p \pm i{*}\omega_p)/2$, $2n = 4k+2$ and thus the trace of * over the harmonic forms is concentrated on the homology of the middle dimension $H_n(M, \mathbb{R})$. In fact $H_n(M)$ can be split into self-dual and antiself-dual spaces $H_n(M) = H_n^+(M) \oplus H_n^-(M)$ with obvious notation. The dimension of the vector space $H_p(M)$ is known as the pth Betti number of M, b_n. The signature of M is defined by sign $(M) = b_n^+ - b_n^-$. If we want to eliminate the constraint of working with harmonic forms directly, we can define the signature in terms of a heat kernel expansion:

$$\mathrm{sign}(M) = \mathrm{Tr} * e^{-\beta\Box} \; . \tag{2.108}$$

This number can be computed in terms of the AS index theorem (2.101) if we realize that the tensor product of two spinor representations decomposes into a direct sum of all the antisymmetric tensor representations of SO(2n), i.e. a bispinor $\phi_{\alpha\beta}$ is equivalent to the whole De Rham complex $\Lambda^*(M)$. Since $\phi_{\alpha\beta}$ has two spinor indices, there are various ways of defining the action of $\bar{\Gamma}$ on $\phi_{\alpha\beta}$. It turns out that the *-operation on $\Lambda^*(M)$ is equivalent to the action of $\bar{\Gamma} \otimes 1$ on $\phi_{\alpha\beta}$. Thus (2.108) is equivalent to considering an index problem for a Dirac-like operator interpolating between

$$S_+ \otimes (S_+ \oplus S_-) \to S_- \otimes (S_+ \oplus S_-) \; . \tag{2.109}$$

Using (2.101) and the additivity of Chern characters we get:

$$\begin{aligned} \mathrm{sign}(M) &= \int_M \mathrm{ch}(S_+ \oplus S_-)\hat{A}(M) \\ &= \int_M (\mathrm{ch}\, S_+ + \mathrm{ch}\, S_-)\,\hat{A}(M) \; . \end{aligned} \tag{2.110}$$

We now have to compute

$$\mathrm{Tr}\exp \frac{1}{2\pi} R_{ab}\,\Sigma^{ab} = \mathrm{ch}\, S_+ + \mathrm{ch}\, S_- \; ,$$

i.e. the character of SO(2n) in the rep. $S_+ \oplus S_-$. A straightforward exercise (which is strongly recommended to the reader) gives

$$\mathrm{ch}\, S_+ + \mathrm{ch}\, S_- = \prod_{a=1}^{n} 2\cosh\frac{x_a}{2}$$

and finally

$$\mathrm{sign}(M) = \int_M [L(M)]_{vol} \tag{2.111}$$

$$L(M) = \prod_a 2\frac{x_a/2}{\tanh x_a/2} = \left(\prod_a \frac{x_a}{\tanh x_a}\right)_{n^{th} \text{ order in } x\text{'s}}$$

L(M) is the Hirzebruch L polynomial and its expansion is given by:

$$L(M) = 1 - \frac{1}{(2\pi)^2}\frac{1}{6}\mathrm{Tr}\,R^2 + \frac{1}{(2\pi)^4}\left[\frac{1}{72}(\mathrm{Tr}\,R^2)^2 - \frac{7}{180}\mathrm{Tr}\,R^4\right]$$

$$+ \frac{1}{(2\pi)^6}\left[-\frac{1}{1296}(\mathrm{Tr}\,R^2)^3 + \frac{7}{1080}\mathrm{Tr}\,R^2\,\mathrm{Tr}\,R^4 - \frac{31}{2835}\mathrm{Tr}\,R^6\right]$$

$$+ \frac{1}{(2\pi)^8}\left[\frac{1}{31104}(\mathrm{Tr}\,R^2)^4 - \frac{7}{12960}(\mathrm{Tr}\,R^2)^2\,\mathrm{Tr}\,R^4 + \right.$$

$$\left. + \frac{31}{17010}\mathrm{Tr}\,R^2\,\mathrm{Tr}\,R^6 + \frac{49}{64800}(\mathrm{Tr}\,R^4)^2 - \frac{127}{37800}\mathrm{Tr}\,R^8\right] + \ldots$$

$$= 1 + \frac{1}{3}P_1 + \left(-\frac{1}{45}P_1^{\,2} + \frac{7}{45}P_2\right) + \left(\frac{2}{945}P_1^{\,3} - \frac{13}{945}P_1P_2 + \frac{62}{945}P_3\right)$$

$$+ \left(-\frac{1}{4725}P_1^{\,4} + \frac{22}{14175}P_1^2P_2 - \frac{71}{14175}P_1P_3 - \frac{19}{14175}P_2^{\,2} + \frac{127}{4725}P_4\right) + \ldots \tag{2.112}$$

Next, we would like to indicate quickly the changes in (2.101) when the manifold M_{2n} is compact with boundary $\partial M_{2n} = B_{2n-1}$. This is the A.P.S. index theorem.[16] It turns out that the boundary conditions one needs to impose are non-local. (Only for the computation of the Euler number in an index problem is it possible to impose local boundary conditions.)[46] The boundary conditions chosen by A.P.S. can be described as follows for the Dirac operator: close to ∂M_{2n} we can choose coordinates so that M_{2n} looks like $[0,\infty] \times B_{2n-1}$. The Dirac operator may be written in this neighborhood of B_{2n-1} as:

$$\Gamma^0\left(\frac{\partial}{\partial x^0} + \Gamma^0\, \not{D}^{(2n-1)}\right) . \tag{2.113}$$

x^0 is the direction normal to the boundary and $\Gamma^0\,\not{D}^{(2n-1)}$ is essentially the Dirac operator restricted to the boundary $H = \Gamma^0\,\not{D}^{(2n-1)}$ is elliptic and self-adjoint, so that it has a discrete spectrum. Let $\mathcal{U}_+$ be the projection operator which projects onto the positive part of the spectrum of H, and $\mathcal{U}_-$ the projection operator projecting onto the negative part of the spectrum of H. The A.P.S. boundary conditions require that the positive (negative) spinors satisfy on B_{2n-1} the condition $\mathcal{U}_+\psi_+ = 0$ ($\mathcal{U}_-\psi_- = 0$). This boundary condition is elliptic, and one can now ask what is the generalization of (2.101). What A.P.S. found is:

$$\mathrm{ind}\,\not{D} = \int_M \hat{A}(M)\,\mathrm{ch}(F) - \tfrac{1}{2}(\eta(0) + h) . \tag{2.114}$$

The first term in (2.114) is the familiar volume term, and the second piece is the boundary correction. $\eta(0)$ is a spectral invariant for the Dirac operator on B_{2n-1} defined as follows. Let $\{\lambda\}$ be the eigenvalues of $\not{D}^{(2n-1)}$, construct:

$$\eta(s) \equiv \sum_\lambda \frac{\mathrm{sgn}\lambda}{(\lambda)^s} . \tag{2.115}$$

$\eta(s)$ is convergent for Re s large enough, and it can be analytically extended as a meromorphic function on the complex s-plane. Furthermore A.P.S. showed that after analytic continuation $\eta(0)$ is finite. $\eta(0)$ measures the spectral asymmetry of $\not{D}^{(2n-1)}$; it is a regulated form of $\Sigma_{\lambda>0} 1 - \Sigma_{\lambda<0} 1$. Finally the term h in (2.114) counts the number of zero modes of $D^{(2n-1)}$ on the boundary B_{2n-1}. The A.P.S. index theorem plays a central role in the analysis of global anomalies (Sec. IX). (For more details and applications see for example Refs. 47, 48.)

III. THE ANOMALY AND THE INDEX THEOREM

We begin by recalling briefly the relationship between the Abelian (or global, axial U(1)) anomaly and the index theorem. This is most easily exhibited in Fujikawa's method[49] for computing anomalies. The effective action for a massless Dirac fermion in Euclidean 2n-dimensional space is given by a Gaussian functional integral:

$$e^{-\Gamma[A,g]} = \int d\psi\, d\bar{\psi} \exp - \int d^{2n} x \sqrt{g}\, \bar{\psi}\, i \not{D} \psi \; . \tag{3.1}$$

The classical action is invariant under global chiral rotations of the fermions

$$\begin{aligned} \psi &\to e^{i\alpha\bar{\Gamma}} \psi \\ \bar{\psi} &\to \bar{\psi}\, e^{i\alpha\bar{\Gamma}} \end{aligned} \tag{3.2}$$

because $\{\bar{\Gamma}, \not{D}\} = 0$. One way of defining (3.1) is by expanding $\psi, \bar{\psi}$ in terms of the eigenfunctions of $i\not{D}$:

$$\begin{aligned} \psi(x) &= \sum_n a_n \psi_n(x) \\ \bar{\psi}(x) &= \sum_n \bar{b}_n \psi_n^{\dagger}(x) \\ i\not{D}\psi_n &= \lambda_n \psi_n \\ \int d^{2n} x \sqrt{g}\, \bar{\psi}\, i\not{D} \psi &= \sum_n \lambda_n \bar{b}_n a_n \; . \end{aligned} \tag{3.3}$$

The measure in (3.1) now becomes $\prod_n d\bar{b}_n\, da_n$, and one has to keep in mind that a_n and $\bar{b}_n$ are independent Grassmann numbers. If we make a change of variables:

$$\begin{aligned} \psi &\to \psi + i\alpha(x)\bar{\Gamma}\psi \\ \bar{\psi} &\to \bar{\psi} + \bar{\psi}\, i\, \bar{\Gamma}\, \alpha(x) \; . \end{aligned} \tag{3.4}$$

The action changes by:

$$\int (dx) \bar{\psi}\, i\not{D}\psi \to \int (dx)\bar{\psi}\, i\not{D}\psi + \int d^{2n} x \sqrt{g}\, \alpha(x)\, (\nabla_\mu j_5^\mu) \tag{3.5}$$

$$j_5^\mu = \bar{\psi}\, \Gamma^\mu \bar{\Gamma}\, \psi$$

$$(dx) \equiv d^{2n}x \sqrt{g} \; .$$

Expanding in power of $\alpha(x)$, and taking into account the invariance of $\Gamma[A]$ under changes of variables, gives naive Ward identities implying conservation of the axial current at the quantum level. What Fujikawa realized is that in performing all these manipulations, we have to worry about the Jacobian factor induced in the measure by the change of variables (3.4). The Jacobian is divergent, and therefore the computation of the anomaly reduces to computing the change in the measure in an appropriately regulated fashion. If we take α to be small, the Jacobian factor multiplies the measure by:

$$\exp - 2i\int (dx)\alpha(x) \sum_n \psi_n^\dagger(x)\bar{\Gamma}\,\psi_n(x) = \exp - 2i\int (dx)\alpha(x)\sigma(x) \ . \tag{3.6}$$

We can now regulate the sum by using for example a Gaussian cut-off. Since we want to preserve gauge invariance, the natural choice is:

$$\begin{aligned} J &= 2\int (dx)\alpha(x) \sum_n \psi_n^\dagger(x)\,\bar{\Gamma}\,\psi_n(x) \equiv \\ &\equiv \lim_{M\to\infty} 2\int (dx)\alpha(x) \sum_n \psi_n^\dagger(x)\,\bar{\Gamma}\,\psi_n(x) e^{-\lambda_n^2/M^2} \\ &= \lim_{M\to\infty} 2\int (dx)\alpha(x) \sum_n \psi_n^\dagger(x)\,\bar{\Gamma}\, e^{-(i\not{D})^2/M^2}\,\psi_n(x) \ . \end{aligned}$$

In the limit when $\alpha(x)$ is almost constant (zero momentum Ward identities), we can factorize α out of the integral, and we are left with

$$J = 2 \lim_{M\to\infty} \alpha \,\mathrm{Tr}\,\bar{\Gamma}\, e^{-(i\not{D})^2/M^2} \ . \tag{3.7}$$

Since $\{\bar{\Gamma},\not{D}\} = 0$, given $i\not{D}\psi_\lambda = \lambda\psi_\lambda$, $\lambda \neq 0$, we can always construct $\bar{\Gamma}\psi_\lambda$ with $i\not{D}\bar{\Gamma}\psi_\lambda = -\lambda\bar{\Gamma}\psi_\lambda$ and the trace only receives contributions from the zero mode sector. In this case, we can write $\mathrm{Tr}\,\bar{\Gamma}\exp-(i\not{D})^2/M^2 = n_+ - n_- = \mathrm{ind}\,\not{D}$. Using (2.101) we can represent $n_+ - n_-$ in terms of a local quantity, and conclude that

$$\int (dx)\alpha(x)\langle\nabla_\mu j_5^\mu\rangle = 2\int_M \alpha(x)\,[\hat{A}(M)\,\mathrm{ch}(F)]_{\mathrm{vol}} \ . \tag{3.8}$$

Substituting (2.102-2.104) we get the standard form of the axial anomaly in the presence of both gauge and gravitational fields. In $2n = 4$ for instance, we get:

$$\int_M \langle\nabla_\mu j_5^\mu\rangle = \frac{1}{(2\pi)^2}\int_M (-\,\mathrm{Tr}\,F^2 + \frac{r}{24}\,\mathrm{Tr}\,R^2) \ . \tag{3.9}$$

Had we started with a Weyl fermion instead, the answer in (3.8) would have to be divided by 2. Even though we have been slightly cavalier about regularization of (3.1), a more careful analysis yields the same result.[49] Notice also that with the definition we have given of the measure, the effective action can be elegantly expressed as

$$e^{-\Gamma[A]} = \det i\not{D}(A) \ . \tag{3.10}$$

(3.10) can be regulated in a gauge invariant way by using for example Pauli-Villars fields:

$$\det i\not{D}(A) \rightarrow \frac{\det(i\not{D}(A))}{\det(i\not{D}(A)+iM)} \tag{3.11}$$

or through a ζ-function regularization

$$\not{D}^2\psi_\lambda = \lambda^2\psi_\lambda$$

$$\zeta(s) \equiv \sum_\lambda \frac{1}{(\lambda^2)^s}$$

$$\det i\not{D}(A) \equiv e^{-\frac{1}{2}\frac{d\zeta}{ds}\Big|_{s=0}} \quad . \tag{3.12}$$

If we consider chiral fermions instead, it is not possible to define the effective action as $\det D_+ = \det i\not{D}\, P_+$. We simply have to observe that D_+: $S_+ \otimes V \rightarrow S_- \otimes V$ and therefore D_+ does not have a well defined eigenvalue problem. The operator whose determinant should be identified with $\exp - \Gamma_r[A]$ can be identified by considering the perturbative evolution of the fermionic effective action. Gauge invariance for the effective action under infinitesimal gauge transformations means that

$$\Gamma[A-Dv] - \Gamma[A] = \int dx\, v^a(x)\, D_\mu \frac{\delta\Gamma[A]}{\delta A_\mu{}^a} \tag{3.13}$$

since

$$e^{-\Gamma[A]} = \int d\lambda\, d\bar{\lambda} \exp - \int dx\, \bar{\lambda}\, i\not{D}_+ \lambda$$

$$\frac{\delta\Gamma[A]}{\delta A_\mu{}^a} = \langle\bar{\lambda}\, \Gamma_\mu\, P_+\, T^a \lambda\rangle_A \quad . \tag{3.14}$$

(3.13) and (3.14) imply that gauge invariance is equivalent to current conservation as pointed out in the introduction. From Section II we know that for complex representations of fermions, $\Gamma_r[A]$ will generically be anomalous, and now we want to relate the anomaly to detailed properties of the chiral fermion determinant. (As a side remark, an equivalent way of rephrasing the arguments leading to (2.40) is to say that if it is not possible to write a gauge invariant mass term, then the theory is potentially anomalous. If such a mass term existed, we could regulate the theory using Pauli-Villars fields. Note however that in 4 dimensions for example, the existence of a mass term requires that in the Clebsch-Gordon decompositions of $r \times r$ there should be a singlet and that the Clebsch-Gordon coefficient should be symmetric under the exchange of fermions. Using the definitions of the different types of representations of a group (2.28), these conditions are equivalent to the reality of r.)

There is a simple way of defining $\Gamma_r[A]$ in terms of a determinant.[19] Consider a basis of Γ-matrices where $\bar{\Gamma}$ is diagonal. Instead of considering the operator $i\not{D} P_+ = D_+$, we consider the operator:

$$\hat{D} \equiv i\Gamma^\mu(\partial_\mu + A_\mu P_+) \tag{3.15}$$

(We are going to work explicitly with the gauge case for the time being. The extension to the gravitational case will be considered later on.) (3.15) is an elliptic operator acting on Dirac rather than Weyl fermions, and therefore has an eigenvalue problem. Since $\hat{D}$ is not self-adjoint, its eigenvalues are complex, and one has to simultaneously consider left and right eigenfunctions:

$$\begin{aligned} \hat{D}\phi_n &= \lambda_n \phi_n \\ \chi_n^\dagger \hat{D} &= \lambda_n \chi_n^\dagger \\ (\chi_n, \phi_m) &= \delta_{nm} \ . \end{aligned} \tag{3.16}$$

(Remember that we compactified Euclidean space to a sphere, this plus the ellipticity of $\hat{D}$ imply the discreteness of the spectrum of $\hat{D}$.) We can define:

$$e^{-\Gamma_r[A]} \equiv \det \hat{D}(A) \ . \tag{3.17}$$

Since only the left-handed fermions couple to the gauge field, up to a field independent normalization constant, the perturbative expansion for $\not{D} P_+$ and $\hat{D}$ is identical. In the $\bar{\Gamma}$-diagonal basis, we can write D as:

$$\begin{bmatrix} 0 & D_+ \\ \partial_- & 0 \end{bmatrix} \tag{3.18}$$

where $\partial_- \equiv i\not{\partial} P_-$. Since ∂_- has no zero modes on S^{2n}, it follows that the only possible zero modes of $\hat{D}$ come from D_+. Thus, if we want to consider instanton effects, we again obtain the equivalence of (3.17) with the original theory. We can calculate the anomaly using Fujikawa's method. The action

$$\int (dx)\bar{\psi}\,\hat{D}\,\psi \tag{3.19}$$

is invariant under

$$\begin{aligned} \psi &\to g P_+ \psi + P_- \psi \\ \bar{\psi} &\to \bar{\psi}_- g^{-1} + \bar{\psi}_+ \ . \end{aligned} \tag{3.20}$$

Now we expand ψ, $\bar{\psi}$ in terms of ϕ_n, $\chi_n^\dagger$ defined in (3.16), $\psi = \Sigma_n a_n \phi_n$, $\bar{\psi} = \Sigma_n \bar{b}_n \chi_n^\dagger$. The measure is $\prod_n d\bar{b}_n da_n$, and (3.19) becomes $\Sigma_n \lambda_n \bar{b}_n a_n$ and (3.17) follows after a gaussian integration. Under an infinitesimal gauge transformation $g \approx 1+v$, we can repeat the steps leading to (3.7), and obtain for the Jacobian factor:

$$\delta_v\Gamma = \int v^a D_\mu \frac{\delta\Gamma[A]}{\delta A_\mu{}^a} = \int \lim_{\substack{M\to\infty \\ y\to x}} dx \,\mathrm{Tr}\, v(x)\bar{\Gamma}\, e^{-(i\hat{D})^2/M^2}\, \delta(x-y) \ . \tag{3.21}$$

Evaluation of (3.21) in a plane wave basis is tedious but straightforward. In four dimensions it gives the usual result:

$$\delta_v \Gamma[A] = \frac{1}{24\pi^2} \int d^4x \, \mathrm{Tr}\, v\, \varepsilon^{\lambda\mu\alpha\beta} \partial_\lambda (A_\mu \partial_\alpha A_\beta + \tfrac{1}{2} A_\mu A_\alpha A_\beta)$$

$$= \frac{1}{24\pi^2} \int \mathrm{Tr}\, v\, d(A\, dA + \tfrac{1}{2} A^3) \quad . \qquad (3.22)$$

In higher dimensions one can use the Wess-Zumino consistency conditions[27] to generate the rest of the terms once the leading term is known. The leading term in the number of derivatives in (3.21) is

$$\frac{i^{n+2}}{(2\pi)^n (n+1)!} \mathrm{Tr}[v(dA)^n] \quad . \qquad (3.23)$$

These arguments prove that one can define $\Gamma_r[A]$ in terms of $\det \hat{D}(A)$. One more property to check is that $|\det \hat{D}(A)|$ is gauge invariant. This can be done by the following formal argument:

$$|\det \hat{D}|^2 = \det \hat{D}^\dagger \hat{D} = \det \begin{bmatrix} \partial_+ \partial_- & 0 \\ 0 & D_- D_+ \end{bmatrix} = \det \partial_+ \partial_- \det D_- D_+ \quad . \qquad (3.24)$$

$\det \partial_+ \partial_-$ is the gauge field independent constant we mentioned before, and $\det D_- D_+ = \det i\not{D}$. This last result can be seen by decomposing the eigenfunctions of $i\not{D}$ in terms of left and right components.

$$D_+ e_{+n} = \lambda_n e_{-n}$$

$$D_- e_{-n} = \lambda_n e_{+n} \quad . \qquad (3.25)$$

Then

$$D_- D_+ e_{+n} = \lambda_n^2 e_{+n}$$

$$D_+ D_- e_{-n} = \lambda_n^2 e_{-n} \qquad (3.26)$$

so that $\det i\not{D} = \det D_- D_+$. This shows that (3.17) satisfies the general criteria that only the phase of $\det \hat{D}$ may be gauge non-invariant. Of course the argument leading to (3.24) needs some regularization. However, one can show by using ζ-functions or Pauli-Villars regulators that (3.24) is correct up to local counterterms; and this is enough to show that $\mathrm{Re}\,\Gamma[A]$ cannot be anomalous.

After these perturbation theory arguments, we would like to prove that the form of the anomaly (3.22) follows from an index theorem in 2n+2 dimensions, or as a consequence of the index theorem for families of operators. The connection between the 2n-dimensional anomaly and the 2n+2 index theorem can be formulated as follows. Consider a one-parameter family of gauge transformations:

$$g(\theta,x) = S^1 \times S^{2n} \to G$$

$$g(0,x) = g(2\pi,x) = \mathbf{1} \; . \tag{3.27}$$

As long as $\pi_1 G = 0$, $\pi_{2n} G = 0$, these maps are classified by $\pi_{2n+1} G$. Next we define a one-parameter family of gauge field configurations:

$$A^\theta \equiv g^{-1}(A + d)g \; . \tag{3.28}$$

Then going from A^θ to $A^{\theta+\delta\theta}$ involves making an infinitesimal gauge transformation with gauge parameter $g^{-1}\,\delta_\theta\, g$:

$$A^{\theta+\delta\theta} = A^\theta + \delta\theta\, D_{A^\theta}\,(g^{-1}\partial_\theta\; g)$$

$$D_{A^\theta}\, v = dv + [A^\theta, v] \; . \tag{3.29}$$

If we assume for simplicity that $\not{D}_A$ has no zero modes, then $\det \hat{D}(A^\theta)$ (properly regulated) defines a complex function of θ. Using the gauge invariance of $|\det \hat{D}(A^\theta)|$, we can write:

$$\det \hat{D}(A^\theta) = \sqrt{\det i\not{D}(A)}\; e^{iw(A,\theta)} \; . \tag{3.30}$$

The anomaly one gets by the infinitesimal transformation (3.29) is

$$-\delta_\theta \Gamma[A^\theta] = i\, \frac{dw(A,\theta)}{d\theta} \; . \tag{3.31}$$

Hence the function $e^{iw(A,\theta)}$ defines a map from $S^1 \to S^2$, and the anomaly measures the local winding number of this function. The total winding number is equivalent to the integrated anomaly along the one-parameter family of gauge transformations (3.27):

$$-\int_0^{2\pi} d\theta\, \frac{d\Gamma[A^\theta]}{d\theta} = i\int_0^{2\pi} d\theta\, \frac{dw(A,\theta)}{d\theta} = 2\pi n \; . \tag{3.32}$$

n is the winding number of $\exp i\,w(A,\theta)$. Our strategy will be to find an appropriate 2n+2-dimensional Dirac operator whose index equals the winding number (3.32). The integral form for the index then generates an expression for the anomaly in the form (3.22).

In order to compute n, let us consider instead a two-parameter family of gauge fields. Take for simplicity tA^θ, $0 \le t \le 1$, but in principle any other parameterization leads to the same result. Note that in this parameterization it is crucial that A defines a trivial gauge field (its Pontrjagin class vanishes). Otherwise we cannot interpolate between A^θ and 0. Geometrically this two-parameter family looks like a 2-disc in the space $\mathfrak{A}$ of all gauge field configurations on S^{2n} in a fixed (trivial) topological class. (The transition functions for the various gauge fields are homotopically equivalent.) The space $\mathfrak{A}$ is topologically trivial; it is an infinite dimensional affine space, and one can easily construct a deformation retract that shrinks $\mathfrak{A}$ to a point: Let A_1, A_2 be two configurations in $\mathfrak{A}$; define $A_t = A_1 + t(A_2 - A_1)$, $0 \leq t \leq 1$. If we fix A_1 and vary A_2 over $\mathfrak{A}$, the function A_t is the deformation contracting $\mathfrak{A}$ to A_1. The gauge invariant dynamics however is not defined on $\mathfrak{A}$, but on the space of gauge orbits $\mathfrak{A}/\mathfrak{g}_0$, $\mathfrak{g}_0 = \{g: S^{2n} \to G,\ g(x_0) = 1\}$, x_0 is some particular point on S^{2n} which corresponds to infinity in the uncompactified theory. The restriction to $\mathfrak{g}_0$ is required by Gauss' law. In the canonical formation, the infinitesimal generator of gauge transformations is given by the operator appearing in Gauss' law $Q^a = (D_i E_i)^a + \rho^a$ (ρ^a = fermionic non-Abelian charge, E^a = non-Abelian electric field). In quantizing the theory one has to impose the constraint $Q^a|\psi\rangle = 0$ ($|\psi\rangle$ is any physical state). If we want to construct finite gauge transformations by exponentiating Gauss' law, the only gauge transformations that one can represent in terms of Q^a are those corresponding in the Euclidean case to $\mathfrak{g}_0$. If Gauss' law cannot be consistently defined quantum mechanically, the longitudinal degrees of freedom of A_μ do not decouple from the theory and gauge invariance is lost. The anomaly from this point of view is related to the possibility of consistently restricting the dynamics from $\mathfrak{A}$ to $\mathfrak{A}/\mathfrak{g}_0$. Coming back to our previous discussion, the disc $A^{(t,\theta)}$ in $\mathfrak{A}$ projects down to $\mathfrak{A}/\mathfrak{g}_0$ to become a 2-sphere. This 2-sphere is non-contractible if the one-parameter family (3.27) generates a non-trivial homotopy class in $\pi_{2n+1}G$, otherwise the homotopy deformation which shrinks the sphere to a point can be used to deform the loop of gauge transformations $g(\theta,x)$ to a point. This is essentially the same construction used in monopole theory to show that $\pi_2(G/H) = \ker(\pi_1(H) \to \pi_1(G))$, G,H Lie groups, and H a subgroup of G. We can go even further, and use the exact homotopy sequence for bundles; after all, the space $\mathfrak{A}$ is the total space for a bundle with basic space $\mathfrak{A}/\mathfrak{g}_0$, and fiber $\mathfrak{g}_0$. Thus:

$$\ldots \to \pi_k(\mathfrak{A}) \to \pi_k(\mathfrak{A}/\mathfrak{g}_0) \to \pi_{k-1}(\mathfrak{g}_0) \to \pi_{k-1}(\mathfrak{A}) \quad \ldots \tag{3.33}$$

since $\mathfrak{A}$ is contractible, $\pi_k(\mathfrak{A}) = 0$ for all k, and

$$\pi_k(\mathfrak{A}/\mathfrak{g}_0) \approx \pi_{k-1}(\mathfrak{g}_0) \tag{3.34}$$

for $k = 2$, $\pi_2(\mathfrak{A}/\mathfrak{g}) \approx \pi_1(\mathfrak{g}_0)$; and the elements in $\pi_1(\mathfrak{g}_0)$ are loops of gauge transformations. We have constructed a non-contractible two-sphere in $\mathfrak{A}/\mathfrak{g}_0$ in a slightly singular way, similar to choosing the strong gauge in monopole theory, where all the "action" occurs at the south pole. This can be remedied by choosing two discs in $\mathfrak{A}$ and patching them together at the equator with the loop $g(\theta,x)$. The fermion determinant gives a complex number at each point of this two-sphere. If we denote by A_+ (A_-) the two-parameter family of gauge potentials describing the upper (lower) hemisphere, we have two functions $\det \hat{D}(A_+)$, $\det \hat{D}(A_-)$. At the equator $D_+ \cap D_- = S^1$, $A_+ = g^{-1}(\theta,x)\,(A_- + d)g(\theta,x)$, and the transition function is given by

$$\frac{\det \hat{D}(A^\theta)}{\det \hat{D}(A)} = e^{iw(A,\theta)} \quad . \tag{3.35}$$

In other words, the fermion determinant defines a complex line bundle over the 2-sphere whose transition function is the imaginary part of the effective action $\exp iw(A,\theta)$. These bundles are classified by the winding number of the transition function (3.32). This construction is very reminiscent of the theory of charged particles in the presence of a monopole with $eg = n$. The topological analysis is therefore concerned with finding a representative of the "monopole" field, i.e. the first Chern class of the determinant line bundle. Finding a representative of the first Chern class in a topologically non-trivial situation is equivalent to computing a particular form of the anomaly.

Since the winding number (3.32) is invariant under small deformation, in the 2-disc $A^{(t,\theta)}$, we can start shrinking the circle on the boundary $t = 1$ into the interior. If $\not{D}(A^{t,\theta})$ has no zero modes anywhere on (t,θ), the winding number will be zero and we cannot say anything in this situation. Let us imagine instead that at some set of point (t_i,θ_i), $\not{D}(A^{t_i,\theta_i})$ has zero modes. Since by asumption $\not{D}(A)$ has no zero modes, the index of $\not{D}(A^{t_i,\theta_i})$ vanishes ($\not{D}(A^{t,\theta})$ is a smooth deformation of $\not{D}(A)$, and the index is invariant under smooth deformations). This means that $\not{D}(A^{t_i,\theta_i})$ will have both a left- and a right-handed zero mode, and generically only one of each because we are only considering a two-parameter family of gauge fields. If $\nu(t_i,\theta_i)$ denotes the winding number of the phase $w(A,t,\theta)$ around a very small circle arount t_i,θ_i, we find

$$\int_0^{2\pi} d\theta \, \frac{dw(A,\theta)}{d\theta} = 2\pi \sum_i \nu(t_i,\theta_i) \quad . \tag{3.36}$$

Regarding the determinant of $\hat{D}$ as a regularized product of complex eigenvalues, its winding number around any contour is equal to the sum of the winding numbers of the individual eigenvalues. By continuity and discreteness of the winding number, it is clear that for those eigenvalues which do not vanish in the (t,θ)-disc, the winding number vanishes, and (3.36) is determined entirely by the winding number of the smallest eigenvalues in the neighborhood of the zeroes. We have thus reduced the computation of (3.36) to the behavior of the smallest eigenvalues of $D(A^{t,\theta})$ around their zeroes. This can be computed in terms of ordinary Schrödinger perturbation theory. We first compute the value of the winding number around each zero, and then we will show that the existence of zeroes of det $\hat{D}(A^{t,\theta})$ plus their winding numbers are determined for purely topological reasons. Using (3.18) $\hat{D} = \partial_- + D_+$, we deduce

$$\det\hat{D} = \det(\partial_- D_+) \ .$$

This follows from breaking up the eigenfunctions of $\hat{D}$ into positive and negative chirality projections $e_\pm$ satisfying $D_+e_+ = \lambda e_-$, $\partial_- e_- = \lambda e_+$, so that $\partial_- D_+ e_+ = \lambda^2 e_+$, $\lambda \neq 0$. The non-zero eigenvalues of $\hat{D}$ come in pairs $\pm\lambda$ because $\Gamma\hat{D} = -\hat{D}\Gamma$. We are left with doing perturbation theory of the non-Hermitian operator $N = \partial_- D_+(A^{t,\theta})$ around (t_i,θ_i). Let $N_i = N(t=t_i, \theta=\theta_i)$, and $f_n^{(i)}, e_n^{(i)}$ be left and right eigenfunctions of N_i:

$$\begin{aligned} f_n^{(i)} N_i &= \lambda_n f_n^{(i)} \\ N_i e_n^{(i)} &= \lambda_n e_n^{(i)} \\ (f_n^{(i)}, e_m^{(i)}) &= \delta_{nm} \ . \end{aligned} \tag{3.37}$$

Then using the standard methods of quantum mechanics, the first order correction to the eigenvalues of $N_i + \delta N$ is:

$$\delta\lambda_n = (f_n^{(i)}, \delta N\, e_n^{(i)}) / (f_n^{(i)}, e_n^{(i)}) \tag{3.38}$$

In the case at hand, we have

$$\begin{aligned} i\not{D}(A^{t_i,\theta_i})\psi_\pm^i &= 0 \\ D_+(A^{t_i,\theta_i})\psi_+^i &= 0 \\ D_-(A^{t_i,\theta_i})\psi_-^i &= 0 \ . \end{aligned} \tag{3.39}$$

Consequently $\psi_-^{i\dagger} D_+(A^{t_i,\theta_i}) = 0$, and the left and right zero eigenfunctions of N_i are $f_0^{(i)} = \psi_-^{\dagger}(\partial_-)^{-1}$, $e_0^{(i)} = \psi_+^i$. Choosing coordinates on the disc (ϕ_1, ϕ_2) so that t_i, θ_i corresponds to $\phi_i = 0$, $i = 1,2$, the smallest eigenvalue of N close to $\phi = 0$ is

$$\delta\lambda = (\psi_-^{\dagger}\partial_-^{-1}, \delta(\partial_- D_+)\psi_+) = (\psi_-^{\dagger}, \delta i\not{A}\psi_+) \equiv z_1\phi_1 + z_2\phi_2 \tag{3.40}$$

$$z_a^{(i)} = \left(\psi_-^{(i)}, \left(\frac{\partial}{\partial\phi^a}\not{A}\right)_{\phi=0} \psi_+^{(i)}\right) \qquad a = 1, 2$$

If we write $z_a^{(i)} = x_a^{(i)} + iy_a^{(i)}$, and $\delta\lambda = \delta\lambda_1 + i\delta\lambda_2$, we can write

$$\begin{pmatrix} \delta\lambda_1 \\ \delta\lambda_2 \end{pmatrix} = \begin{pmatrix} x_1^{(i)} & x_2^{(i)} \\ y_1^{(i)} & y_2^{(i)} \end{pmatrix} \begin{pmatrix} \phi_1 \\ \phi_2 \end{pmatrix} \tag{3.41}$$

Choosing coordinates so that $\mathrm{Re}\, z_1^* z_2^{(i)} = 0$, the winding number of $\delta\lambda$ is given by the sign of the determinant of the 2×2 matrix in (3.41), i.e.

sign Im $z_1^* z_2^{(i)}$:

$$\nu(t_i,\theta_i) = \text{sign Im}\, z_1^{*(i)} z_2^{(i)} \tag{3.42}$$

Now we want to show that for topological reasons $\not{D}(A^{t,\theta})$ must vanish for some (t_i,θ_i) and that the sum (3.36) is given by a 2n+2 dimensional index theorem.[19] Let us enlarge 2n space-time S^{2n} to a 2n+2 dimensional space by including (t,θ) as coordinates: $D_2^+ \times S^{2n}$, defining the 2n+2 gauge field $(0,0,A^{(t,\theta)})$. The first two zeroes refer to the components of the gauge field in the (t,θ) direction. In order to avoid using the A.P.S. theorem for the time being, we consider another 2-disc $D_2^- \times S^{2n}$. The gauge field in 2n+2 dimensions is thus:

$$\begin{array}{lll} D_2^+ \times S^{2n} & A_+ = t\, g^{-1}(A+d)g & 0 \leq t \leq 1 \\ D_2^- \times S^{2n} & A_- = A - s\, d_\theta\, g\, g^{-1} & 0 \leq s \leq 1 \end{array} \tag{3.43}$$

On $(D_2^+ \times S^{2n}) \cap (D_2^- \times S^{2n}) = S^1 \times S^{2n}$ $(t=1,\ s=1)$ one has:

$$A_+ = g^{-1}(\theta,x)(A_- + d + d_\theta)g(\theta,x) \tag{3.44}$$

$$d_\theta = d_\theta \frac{\partial}{\partial\theta}\,.$$

Now we want to study the Dirac operator in 2n+2 dimensions.

$$i\not{D}_{2n+2} = i\sum_{a=1}^{2n+2}(\partial_a + A_a)\tilde{\Gamma}^a \tag{3.45}$$

where we choose the following convenient representation for $\tilde{\Gamma}^a$:

$$\begin{aligned} \tilde{\Gamma}^\mu &= \sigma_1 \otimes \Gamma^\mu \\ \Gamma^{2n+1} &= \sigma_2 \otimes 1 \\ \Gamma^{2n+2} &= \sigma_1 \otimes \bar{\Gamma} \\ \bar{\bar{\Gamma}} &= \sigma_3 \otimes 1\,. \end{aligned} \tag{3.46}$$

Applying the index theorem (2.101)

$$\text{ind}\, i\not{D}_{2n+2} = \frac{i^{n+1}}{(2\pi)^{n+1}(n+1)!}\int_{S^2\times S^{2n}} \text{Tr}\mathscr{F}^{n+1} \tag{3.47}$$

$\mathscr{F}$ is the 2n+2-dimensional gauge field strength

$$\mathscr{F} = (d_t + d_\theta + d)A + A^2 \tag{3.48}$$

On each patch D_+, D_- we can write using (2.91):

$$\begin{aligned} \text{Tr}\mathscr{F}_+^{n+1} &= \tilde{d}\, Q_{2n+1}^0(A_+) \\ \text{Tr}\mathscr{F}_-^{n+1} &= \tilde{d}\, Q_{2n+1}^0(A_-) \end{aligned} \tag{3.49}$$

where $\tilde{d} = d_t + d_\theta + d$.

A simple computation yields:

$$\operatorname{ind} i\not{D}_{2n+2} = \frac{(-1)^n\, i^{n+1}}{(2\pi)^{n+1}} \frac{n!}{(2n+1)!} \int_{S^1 \times S^{2n}} \operatorname{Tr}(g^{-1}dg)^{2n+1} \quad . \tag{3.50}$$

For groups with $\pi_{2n+1}(G) = \mathbb{Z}$, every 2n+1-sphere in G is a multiple of a basic sphere S_0. The right-hand side of (3.50) is then an integer proportional to the number of times the map $g(\theta,x)$ wraps around S_0. By the index theorem (3.47), this integer also gives the net difference between the number of positive and negative chirality zero modes of $i\not{D}_{2n+2}$. Choosing $g(\theta,x)$ to be the map generating $\pi_{2n+1}(G)$, we conclude that $\operatorname{ind} i\not{D}_{2n+2} = 1$, and by a small perturbation we can always choose our 2n+2-dimensional gauge configuration such that $i\not{D}_{2n+2}$ has a single left-handed zero mode. Next we deform the operator $i\not{D}_{2n+2}$ to $i\not{D}^{\varepsilon}_{2n+2}$ by making the size of the discs $D_\pm$ enormous relative to S^{2n}. In the upper patch for example:

$$i\not{D}^{\varepsilon}_{2n+2} = \frac{1}{\varepsilon} i \sum_{i=1}^{n} D_\mu \tilde{\Gamma}^\mu + i\not{D}_2 \quad . \tag{3.51}$$

Since this is a smooth deformation of the operator $\not{D}_{2n+2}$, the index does not change. When ε is very small, we can calculate the zero mode of $\not{D}^{\varepsilon}_{2n+2}$ in the adiabatic approximation. Let us then assume that $\operatorname{ind} \not{D}^{\varepsilon}_{2n+2} = +1$, and consider:

$$H_\varepsilon \equiv (i\not{D}^{\varepsilon}_{2n+2})^2 = \frac{1}{\varepsilon^2} (i\, D_\mu \Gamma^\mu)^2 + (i\, D_i \Gamma^i)^2 + \frac{i^2}{\varepsilon} \Gamma^i \Gamma^\mu (D_i\, D_\mu - D_\mu D_i) \quad . \tag{3.52}$$

A 2n+2 dimensional Weyl spinor has 2^n components, so it is equivalent to a Dirac fermion in 2n dimensions. In $D_2^+ \times S^{2n}$, we can explicitly write:

$$H_\varepsilon = \frac{1}{\varepsilon^2}\, 1 \otimes (i\not{D}_{2n}(A^{t,\theta}))^2 - \partial_i^2 + \frac{1}{\varepsilon} \Gamma^i \Gamma^\mu\, i\, \partial_i (i\, A_\mu^{t,\theta}) \quad . \tag{3.53}$$

If the eigenvalues of $i\not{D}_{2n}(A^{t,\theta})$ are strictly different from zero for all (t,θ), then $[i\not{D}_{2n}(A^{t,\theta})]^2$ will be strictly greater than zero, and its eigenvalues grow like $1/\varepsilon^2$ as $\varepsilon \to 0$. Since $-\partial_i^2$ is positive definite and the third term in (3.53) is bounded, choosing ε small enough, we can (under this assumption) make H_ε strictly positive which is a contradiction, because $\operatorname{ind} \not{D}^{\varepsilon}_{2n+2} = +1$. Thus a zero mode of H_ε then requires that $(i\not{D}_{2n}^{t,\theta})^2$ have vanishing eigenvalues for some values of (t,θ). As pointed out before, these zeroes come in pairs of positive and negative chirality. Generically, these zeroes are also isolated as will become clear in the next few paragraphs. Let (t_i,θ_i) be one of the zeroes, and $\psi_\pm^{(i)}$ the two zero modes of $i\not{D}_{2n}^{(t,\theta)}$. Close to (t_i,θ_i) we again choose some (ϕ_a) coordinates, and:

$$\delta(i\not{D}_{2n}) = \sum_{a=1}^{2} i\, \partial_a \not{A}^{(t,\theta)} \phi_a \quad . \tag{3.54}$$

In the $(\psi_+^{(i)}, \psi_-^{(i)})$ basis, the matrix elements of (3.54) are:

$$\begin{bmatrix} 0 & \sum_a z_a^* \phi_a \\ \sum_a z_a \phi_a & 0 \end{bmatrix} \tag{3.55}$$

with the z_a's defined as in (3.40). Notice that the vanishing of (3.55) requires the vanishing of a complex function of two parameters. This justifies the claim made earlier that the zeroes of $\not{D}(A^{t,\theta})$ are isolated on $D_+ \times S^{2n}$. The eigenvalues of (3.55) are simply:

$$\lambda(\phi_i) = \pm |z_1 \phi_1 + z_2 \phi_2| \ .$$

In the basis $\operatorname{Re} z_1^* z_2 = 0$

$$\lambda(\phi_i) = |z_1|^2 \phi_1^2 + |z_2|^2 \phi_2^2 \ . \tag{3.56}$$

In the adiabatic approximation, we can construct the zero mode of (3.52) as a linear combination of $\psi_\pm^{(i)}$:

$$\psi(x,t,\theta) = f_+(t,\theta)\psi_+^{(i)} + f_-(t,\theta)\psi_-^{(i)} \ . \tag{3.57}$$

This is justified in the $\varepsilon \to 0$ limit, because the eigenfunctions will be concentrated close to the zeroes of the potential of (3.52). If we denote by χ the chirality of the zero modes of $\not{D}_{2n+2}^{\varepsilon}$ in the 2n+2 dimensional sense, one gets after some simple algebra that H_ε applied to (3.57) takes the form of a 2×2 matrix operator

$$\left[-\frac{\partial^2}{\partial\phi_1^2} - \frac{\partial^2}{\partial\phi_2^2} + \frac{1}{\varepsilon^2}\left(|z_1|^2\phi_1^2 + |z_2|^2\phi_2^2\right)\right]\mathbb{1} + \frac{1}{\varepsilon}\begin{pmatrix} 0 & \chi z_1^* + i z_2^* \\ \chi z_1 - i z_2 & 0 \end{pmatrix} . \tag{3.58}$$

The first term in (3.58) is a harmonic oscillator Hamiltonian with ground state energy $(|z_1| + |z_2|)/\varepsilon$. Including the effect of the third term in (3.58) we obtain in the $\varepsilon \to 0$ limit the following two lowest eigenvalues for H_ε:

$$\lambda_\pm = \frac{1}{\varepsilon}(|z_1| + |z_2|) \pm \frac{1}{\varepsilon}\left[|z_1|^2 + |z_2|^2 + 2\chi \operatorname{Im} z_1^* z_2\right]^{1/2}$$
$$\operatorname{Re} z_1^* z_2 = 0 \ . \tag{3.59}$$

Hence the only way for H_ε to have a zero eigenvalue is that

$$|z_1||z_2| = \chi \operatorname{Im} z_1^* z_2 \ . \tag{3.60}$$

In particular

$$\operatorname{sign} \operatorname{Im} z_1^* z_2 = \chi \ . \tag{3.61}$$

The reader may now remember that in (3.42) the only thing we needed to know to determine $\nu(t_i,\theta_i)$ was the sign of $\mathrm{Im}\, z_1^* z_2$. We have now completed the argument:

$$\sum_i \nu(t_i,\theta_i) = \mathrm{ind}\, \not{D}_{2n+2} \ . \tag{3.62}$$

Notice that in the adiabatic approximation we have only considered the behavior of $\not{D}_{2n+2}$ in the upper hemisphere $D_+ \times S^{2n}$. This is justified in the adiabatic approximation, because in the $\varepsilon \to 0$ limit, the form of the Dirac operator in $D_- \times S^{2n}$ takes the form:

$$\not{D}^{(2)} + \frac{1}{\varepsilon} 1 \otimes \not{D}(A) \ . \tag{3.63}$$

Since $\not{D}(A)$ has no zero eigenvalues by assumption, as $\varepsilon \to 0$ (3.63) will not have any zero modes. This means that on the lower hemisphere we take the zero mode of H_ε to be zero. Since the wave function (3.57) of the zero mode vanishes away from the zeroes (t_i,θ_i) as $\exp(-1/\varepsilon)$; in this approximation it is consistent to take the wave function to be zero in $D_- \times S^{2n}$. Now combine (3.32), (3.49) and (3.63) to obtain the integrated form of the anomaly. Using formulae of the previous section, (2.91), with $\omega_1 = 0$, $\omega_2 = A$, we can explicitly write down Q^0_{2n+1} for $\omega(s) = sA$:

$$Q^0_{2n+1} = (n+1)\int_0^1 ds\, \mathrm{Tr}\, A\,(s dA + s^2 A^2)^n \ . \tag{3.64}$$

In four dimensions, $n = 2$:

$$Q^0_5 = 3\int_0^1 ds\, s^2\, \mathrm{Tr}\, A(dA + sA^2)^2 = \mathrm{Tr}(A F^2 - \frac{1}{2}A^3 F + \frac{1}{10}A^5) \ . \tag{3.65}$$

Using (3.65) and (3.43), we obtain

$$-\int d\theta\, \frac{d\Gamma[A^\theta]}{d\theta} = 2\pi\, \frac{1}{(2\pi)^3 2!}\int_{S^2\times S^4} \mathrm{Tr}\mathscr{F}^3 = \frac{1}{24\pi^2}\int d\theta \int_{S^4} \mathrm{Tr}\, g^{-1}\partial_\theta g\, d(A^\theta dA^\theta + \frac{1}{2} A^{\theta^3}) \tag{3.66}$$

which agrees with (3.22) term by term. In 2n-dimensions it is possible to give a rather simple formula for the integrand (3.66). For A_+ we have:

$$F_+ = (d + d_t + d_\theta)(tA^\theta) + (tA^\theta)^2 = F^{(t,\theta)} + dt\, A^\theta - t\, D v_\theta$$

$$F^{(t,\theta)} = d\, A^{(t,\theta)} + (A^{t,\theta})^2$$

$$D v_\theta = dv + v_\theta A^\theta + A^\theta v_\theta \ , \quad v_\theta = g^{-1} d_\theta g \ . \tag{3.67}$$

Then:

$$\mathrm{STr}\, F_+^{n+1} = -\, n(n+1)\mathrm{STr}\left\{(t\, D v_\theta)(dt\, A^\theta)(F^{t,\theta})^{n-1}\right\}$$

integrating over t; and using the Bianchi identity for $F^{t,\theta}$:

$$\int_0^1 dt\, \mathrm{STr}\, F_+^{n+1} = n(n+1)\int_0^1 dt\,(1-t)\mathrm{STr}\left\{v_\theta\, d\left(A^\theta (F^{t,\theta})^{n-1}\right)\right\}$$
$$-\,(n+1)\int_0^1 dt\, \frac{\partial}{\partial t}\, \mathrm{Tr}[v(F^{t,\theta})^n] \ . \tag{3.68}$$

In the lower patch

$$F_- = (d_s + d_\theta + d)A_- + A_-^2 =$$

$$= F + d_\theta g g^{-1} ds + s\, d(d_\theta g g^{-1}) - s(A\, d_\theta g g^{-1} + d_\theta g g^{-1} A) ,$$

i.e.

$$\int_0^1 ds \operatorname{Tr} F_-^{n+1} = (n+1)\operatorname{Tr}(d_\theta g g^{-1} F^n)$$

cancelling the last term in (3.68). In other words, we can write

$$d\theta\, i \frac{dw}{d\theta}(A,\theta) = \frac{i^{n+2}}{(2\pi)^n (n-1)!} \int_0^1 dt\,(1-t) \operatorname{STr}\left\{ v\, d\, [A^\theta (F^{t,\theta})^{n-1}] \right\} . \quad (3.69)$$

As we will see in the next section, (3.69) automatically satisfies the Wess-Zumino consistency conditions. In order to get (3.69) we needed to use the fact that ind $\not{D}_{2n+2} \neq 0$, and in our particularly simple setting $S^2 \times S^{2n}$, this meant that $\pi_{2n+1}G = \mathbb{Z}$. In general however one can find G's and n's with $\pi_{2n+1}G \neq \mathbb{Z}$. The simplest example is a U(1) gauge theory coupled to a Weyl Fermion. This theory is anomalous and $\pi_{n+1}U(1) = 0$ for $n \geq 1$. (It is interesting to note that in four dimensions, all the anomaly-free groups have $\pi_5 G = 0$, which is necessary in order to make contact between the global and local analysis of the anomaly.) From the previous argument one might conclude that the topological analysis is not strong enough to detect the non-Abelian anomaly in all possible cases. There are two attitudes one can take with respect to this problem. The first, and most pragmatic, is to realize that by using the topological analysis (in those cases where non-trivial topological structures are available) we have obtained the form of the anomaly that one would obtain by using perturbation theory. Since in perturbation theory the difference in the various groups lies in the algebraic factors $\operatorname{Tr} T^{a_1} \ldots T^{a_r}$, we can simply take (3.69) as the anomaly functional for any gauge group, because after all, (3.69) essentially "computes" all the anomalous diagrams. A better way of presenting this argument is in terms of local (rather than global) cohomology. Even though the functional (3.69) may not have a meaning in terms of global topology when we compactify space-time to S^{2n}, one may still ask whether (3.69) is the gauge variation of a local functional. In this sense, (3.69) defines a non-trivial class in the local cohomology of the gauge fields. This point of view (stressed mainly by R. Stora and B. Zumino) is very close to the usual principles of field theory, because if we cannot add a counterterm to the action whose variation cancels the gauge variation of the fermion determinant, then the anomaly is genuine, and cannot be removed. It has recently been shown by O. Alvarez and B. Zumino[50] that (3.69) and similar anomaly functionals including gravitational effects, are non-trivial in the sense of local cohomology.

The other point of view one can take is to realize that one is after all trying to reproduce a purely perturbative answer. Thus instead of compactifying on S^{2n}, one can compactify on more complicated manifolds so that the index theorem in the corresponding 2n+2 dimensional space is not zero. This for instance can be implemented for anomalous U(1) theories by compactifying on tori, and using monopole-like connections to generate a non-trivial index theorem. Similar constructions can be carried out for other groups apart from U(1) (for instance $\pi_7(SU(3)) = \mathbb{Z}_6$, even though there is a non-Abelian anomaly in six dimensions for any group).

Let us now recapitulate the geometrical meaning of our manipulations. Taking space-time to be some 2n-dimensional manifold M, we consider the space of all possible gauge fields $\mathfrak{A}$ on M within a given topological class (for simplicity a class with zero index). The space of gauge-invariant configurations is $\mathfrak{A}/\mathfrak{g}_0$, and the fermion determinant defines geometrically a complex line bundle over $\mathfrak{A}$. By our construction, we have computed a section of this determinant line bundle, and under certain circumstances shown that the line bundle was non-trivial. Line bundles are classified by their first Chern class (monopole number); an element of $H^2(M,\mathbb{Z})$. In our case, we have a line bundle over $\mathfrak{A}/\mathfrak{g}_0$, and $H^2(\mathfrak{A}/\mathfrak{g}_0) \approx H^1(\mathfrak{g}_0)$, and the index theorem in 2n+2 dimensions has provided a representative of the first Chern class of the line bundle. Recall that we explicitly constructed the 2-cycle generating $H^2(\mathfrak{A}/\mathfrak{g}_0)$ in terms of a 2-parameter family of gauge fields. Then:

$$c_1 = \frac{1}{(2\pi)^{n+1}(n+1)!} \int_{M_{2n}} \mathrm{Tr}\, F^{n+1} \tag{3.70}$$

is a 2-form, and a representative of $H^2(\mathfrak{A}/\mathfrak{g}_0, \mathbb{R})$. (If the manifold M_{2n} is sufficiently complicated, we have to modify the integrand of (3.70), and we will do so later.) The 2-cycle in $\mathfrak{A}/\mathfrak{g}_0$ on which we integrated c_1 is a 2-sphere.

In mathematics what we have done goes under the name of the index theorem for families of operators. We can make this relation more precise as follows: Fix the space-time manifold M and consider a gauge theory with gauge group G. For a fixed topological sector, we can consider all possible gauge connections $A \in \mathfrak{A}$. For each A we consider the Weyl operators $D_+(A)$, $D_-(A)$. Thus we get an infinite parameter family of Dirac operators parameterized by A. Since $D_\pm(g^{-1} A g) = g^{-1} D_\pm(A) g$, we can in principle parameterize the family by $\mathfrak{A}/\mathfrak{g}_0$. We can consider the two operators $D_-(A)D_+(A)$ and $D_+(A)D_-(A)$. These operators are elliptic and self-adjoint, and their eigenfunctions span Hilbert spaces $\mathcal{H}_+(A)$, $\mathcal{H}_-(A)$ respectively. Thus we have a Hilbert bundle over $\mathfrak{A}/\mathfrak{g}$. Since we have a fixed topological sector, we know that ind D_A = constant independent of A. However ind D_A is the simplest topological invariant we can define out of this infinite dimensional family of operators. Consider a finite (dimensional) subset of $\mathfrak{A}$ which projects to some compact set in $\mathfrak{A}/\mathfrak{g}_0$. On this finite dimensional parameter space P we have two Hilbert bundles $\mathcal{H}_{+p}$, $\mathcal{H}_{-p}$, and we may ask whether they are topologically non-trivial. As we move A_p in P, the eigenvalues of $N_+(p) = D_-(A_p)D_+(A_p)$ and $N_-(p) = D_+(A_p)D_-(A_p)$ will change as a function of p. Since P and M are compact, if we choose some fiducial configuration A_{p_0}, say without any zero eigenvalues, at most a finite number of eigenvalues can go through zero as p moves over P; if we order the eigenvalues $\lambda_0(p) \le \lambda_1(p) \le \ldots \le \lambda_n(p) \le \ldots$ we can split $\mathcal{H}_+(A_p)$ and $\mathcal{H}_-(A_p)$ into two pieces $\mathcal{H}^0_\pm(A_p)$ and $\mathcal{H}^1_\pm(A_p)$. $\mathcal{H}^0_\pm(A_p)$ contains the eigenfunctions whose eigenvalues may vanish somewhere on P. Notice that if λ^+_p is an eigenfunction of $N_+(p)$ $\lambda^+_p \neq 0$, $p \in P$,

$$N_+(p)\phi^+_p = \lambda^+_p \phi^+_p \tag{3.71}$$

then

$$D_+(A_p)\phi^+_p \tag{3.72}$$

is an eigenfunction of $N_-(p)$ with the same eigenvalue, i.e. $D_+(A_p)$ provides an isomorphism between $\mathcal{H}^1_+(A_p)$ and $\mathcal{H}^1_-(A_p)$. By definition $D_+(A_p) : \mathcal{H}^1_+(A_p) \to \mathcal{H}^1_-(A_p)$ is invertible. Since $\mathcal{H}^1_+(A_p)$ and $\mathcal{H}^1_-(A_p)$ are isomorphic, there is no twisting between $\mathcal{H}^1_+(A_p)$ and $\mathcal{H}^1_-(A_p)$, and any relative twisting between $\mathcal{H}_+(p)$, $\mathcal{H}_-(p)$ must come exclusively from $\mathcal{H}^0_\pm(A_p)$. Since $\mathcal{H}^0_\pm(A_p)$ are finite dimensional we now have two finite dimensional vector bundles over P. In order to make the notation simpler, let $V_\pm$ be the vector bundles obtained this way. We can characterize $V_\pm$ by their Chern characters $ch(V_+)$, $ch(V_-)$. From our computation of the anomaly we know that $ch(V_+)$or $ch(V_-)$ are not the relevant quantities. The anomaly came from the relative twisting of $\psi^0_+(t,\theta)$ and $\psi^0_-(t,\theta)$, the eigenfunctions of $\not{D}^{(t,\theta)}$ which vanished at (t_0,θ_0). In fact the winding number (3.42) measures the twisting of ψ_- and ψ_+ (see (3.40)) around (t_0,θ_0), and this is simply another way of defining the first Chern class of a line bundle. Define a section of the line bundle (a representative at each value p of the parameter space), and find those points where it vanishes. The $\int c_1$ over any 2-cycle is given by the sum of winding numbers (3.42) around each vanishing point. In order to understand the relative twisting of V_+ and V_-, we have to introduce mathematically the difference bundle $V_+ \ominus V_-$. We know how to define the direct sum of vector bundles $E \oplus F$; at each point of the base manifold $x \in M$, we simply consider a fiber which is the direct sum of the fiber E_x and F_x of the bundles E and F. By the definition of Chern classes, we have $ch(E \oplus F) = ch(E) + ch(F)$, and $c(E \oplus F) = c(E) \cdot c(F)$. The definition of $E \ominus F$ can be made in the context of K-theory[51] (for a readable account of some of the properties of K-theory and the index of families of operators, see Refs. 8, 51). We can consider real or complex K-theory. The K-groups classify bundles over a manifold. Let us consider complex vector bundles (real K-theory is fairly subtle) over some manifold M. By the direct sum of bundles we can define addition on K[M]. In order to make K[M] into a group with respect to addition, we have to be a bit more careful about the definition of K[M]. The construction of $E \ominus F$ follows naturally the construction of the integers $\mathbb{Z}$ from the natural numbers N (positive integers). Recall that an arbitrary integer is defined by pairs of positive integers (n,m) with the equivalence relation $(n+k, m+k) \sim (n,m)$, k any positive integer. The set of equivalence classes: $(N \times N/\sim)$ is identical to the integers, and $(n,m) \equiv n-m$; $(n,m) = -(m,n)$. Similarly, let $\mathcal{B}$ be the collection of classes of vector bundles over M. We can define pairs of them $(E,F) \in \mathcal{B} \times \mathcal{B}$, and define the equivalence relation $(E,F) \approx (E+S, F+S)$. Then K[M] is defined to be $[(\mathcal{B} \times \mathcal{B})/\approx]$. In this sense $(E,F) = \mathrm{ch}(E) - \mathrm{ch}(F)$. Similarly, one can define a multiplication operation in K[M] using the tensor product of bundles $E \otimes F$ defined by the tensor product of the fibers at each point of M; then $\mathrm{ch}(E \otimes F) = \mathrm{ch}(E) \cdot \mathrm{ch}(F)$. Technically, K[M] has the algebraic structure of a ring (induced by addition and multiplication). Now we can give a reasonably precise presentation of the index for families of operators, and reinterpret the computations done before in this context. In the familiar form of the index theorem introduced in Section II, one considers some operator (which we will always take to be the Dirac operator) D between two vector bundles: $S_+ \otimes V \to S_- \otimes V$ with base manifold M. We then compute $\dim\ker D$, $\dim\ker D^+$, and the index is the difference:

$$\mathrm{ind}\, D = \dim\ker D - \dim\ker D^+ \ . \tag{3.73}$$

If instead of a single operator D, we have a family of them D_p parameterized by some manifold P, $p \in P$, we can distinguish two cases. The first and simpler corresponds to the case that $\dim\ker D_p$ is constant over P. Since $\mathrm{ind}\, D_p$ is a topological invariant, this means that $\dim\ker D_p^+$ is also constant. Thus $\ker D_p$ and $\ker D_p^+$ are well defined bundles over P, and in the K-theory sense $\mathrm{Ind}\, D \equiv \ker D \ominus \ker D^+$ is another vector bundle over P, the index bundle, and we can compute its Chern classes. The second case is slightly more subtle. It may happen that $\dim\ker D_p$ jumps at various subsets of P. Since $\mathrm{ind}\, D_p$ is invariant under smooth deformations, $\ker D_p^+$ will also have jumps in dimension so that even though $\ker D_p$ and $\ker D_p^+$ do not define smooth bundles over P, Ind D is still well defined in the sense of K-theory. One way we can define Ind D in this case, is by using the bundles $V^{\pm}$ introduced previously. If one can characterize Ind D, then the standard index theorem is obtained as the simplest Chern character of

Ind D, i.e. ch_0 ind D = dim ker D − dim ker D^+. In paper number IV of Ref. 15, Atiyah and Singer were able to give a complete topological characterization of Ind D, and gave representatives of the Chern characters of Ind D. This is a rather profound result in mathematics, and the anomaly can be viewed as a simple corollary of the index theorem for families of elliptic operators. For the case we are interested in, we can write down ch(Ind D) as follows:[21] Let Z be the manifold which is locally P × M (we have considered the simplest case Z ≡ P × M. In general however M is fibered over P, and Z is the total space of this fibration, with P as base space and M as fiber), and let $\hat{A}(Z)$ be the Dirac genus defined in (2.101). Then

$$ch(\mathrm{Ind}\, D) = \int_M \hat{A}(Z)\, ch(V) \; . \tag{3.74}$$

In the first part of this section we have "proven" the simplest case of (3.74), with $M = S^{2n}$, $P = S^2$, and the bundle is constructed with transition function $g(\theta,x)$ along the equator of S^2. We hope that this proof makes the general result (3.74) plausible. The reader interested in the full-fledged form of the index theorem for families of operators is referred to paper IV of Ref. 15.

We can now apply (3.74) to the gravitational anomalies. In this case the vector bundle V is purely "geometrical". For spin 1/2 fields V is 0, for spin 3/2 fields V is the tangent bundle TM, and for self-dual tensor gauge fields (which also contribute to the gravitational anomalies, see Ref. 11 for details) $V = S^-$ (negative chirality spinors). This means that the characteristic polynomial of (3.74) will be a polynomial in Pontrjagin classes. Since the Pontrjagin classes are forms of degree divisible by four, we conclude that ch (Ind D) = c (Ind D) will not vanish only if M has dimension 4k-2 (so that Z has dimension 4k). This corroborates the heuristic arguments presented in Section II that gravitational anomalies only exist in 4k-2 dimensions. As pointed out in Section II (and explored in detail in Section IV) in the gravitational case we want to preserve invariance under diffeomorphisms and local Lorentz transformations in order to have a well-defined theory of chiral fermions in the presence of a gravitational field. Thus it might seem that there are two different types of anomalies. From the topological arguments, it follows that if there is a non-trivial index theorem for a 2-parameter family of Dirac operators with varying metric, the characteristic polynomial (3.74) which gives the anomaly will be the same. In other words, the anomaly cancellation conditions are the same for diffeomorphism or Lorentz anomalies. What is not so obvious from the purely topological argument is that they are in fact the same, or more precisely, one can shift the anomaly from diffeomorphisms to local Lorentz transformations by adding a local counterterm to the effective action (see Section IV for more details). If we concentrate on diffeomorphisms for the moment, it is not so easy to exhibit a two-parameter family of Dirac operators with a non-trivial family index theorem. If such an example can be found in any 4k-2 dimensions, then we know that the characteristic polynomial (3.74) gives the anomaly. After all, we are only trying to reproduce the perturbative anomaly, and we are free to look at a favorable case where the global topology permits us to infer the local form of the anomaly. M. F. Atiyah[53] found an example of a two-parameter family of two-dimensional Dirac operators with a non-zero index. The example is based on the moduli space of Riemann surfaces of genus higher than two. This example involves fibering a Riemann surface M over another Riemann surface M' in such a way that the signature τ of the total space (Z) is not equal to zero. Using this example one can obtain an example in any number of dimensions by including products of M with powers of any 4-manifold with non-vanishing index (for instance the K3 surface). If one tries to look at pure diffeomorphism anomalies (i.e. anomalies in $\nabla_\mu T^{\mu\nu}$), there are some technical difficulties in extending the argument we presented for the gauge theory case. If we tried to do so, the analogs of $\mathfrak{A}$ and $\mathfrak{G}_0$ are $\mathcal{M}$; the space of Riemannian metrics on M, and $\mathrm{Diff}_0(M)$; the groups of diffeomorphisms which reduce to the identity at some particular point (infinity for example). Thus the analog of $\mathfrak{A}/\mathfrak{G}_0$ is $\mathcal{M}/\mathrm{Diff}_0(M)$. Once again $H_2(\mathcal{M}/\mathrm{Diff}_0,\mathbb{R}) \approx H_1(\mathrm{Diff}_0(M))$. The anomaly (first Chern class of the determinant line bundle) defines a 1-form in $\mathrm{Diff}_0(M)$. As in our previous construction, we would take $P = S^2$ in $\mathcal{M}/\mathrm{Diff}_0(M)$; now $D_+ \times M$ and $D_- \times M$ are related by a 1-parameter family of diffeomorphisms $\phi(\theta,x)$. If we can find a manifold M such that $\pi_1(\mathrm{Diff}_0(M)) \approx \mathbb{Z}$, then the argument for the gauge case would translate automatically word for word to conclude that the integrand in (3.74) gives the anomaly. However as pointed out in Ref. 21, if we take $M = S^{4k-2}$, it is known that for $k = 1$ and $k \geq 6$ the first cohomology group $H_1(\mathrm{Diff}_0(S^{4k-2}),\mathbb{R}) = 0$, and it is believed to be true for any k. It is not clear what would happen for manifolds other than S^{4k-2}. Thus one would have to either rely on Atiyah's example, in explicit computations,[11] or in arguments based on local cohomology[26,50] (which shows that (3.74) generates a non-trivial form in the sense of local cohomology) to show that the diffeomorphism anomaly (3.74) is genuine. If on the other hand one concentrates on the local Lorentz anomaly, then $\pi_{4k-1}SO(4k-2)$ contains the integers, and we get exactly the same anomaly as in a SO(4k-2) gauge theory. Either way, we obtain the same conclusion. The invariant polynomial which determines the gravitational anomaly is given by (3.74). Let us now write it explicitly for some interesting cases:

(a) For spin 1/2, $V = 0$, $\mathrm{ch}\, V = 1$, $\hat{A}(Z)$ is as usual (2.101):

$$I_{\frac{1}{2}} = \Pi \frac{x_i/2}{\sinh x_i/2} \qquad (3.75)$$

(the x_i's are as in Section II the skew eigenvalues of $R/2\pi$) and we have to expand $I_{\frac{1}{2}}$ to order 2k in order to obtain the 4k-form characterizing the gravitational anomaly for spin 1/2 in 4k-2 dimensions.

(b) For spin 3/2 we have to be careful with the ghost fields appearing in the quantization procedure (see Ref. 45 for a detailed description of the quantization of a spin 3/2 field in the presence of an external gravitational field, and for references to earlier literature). The final answer is that one needs to include two spin 1/2 ghost fields of opposite chirality with respect to the spin 3/2 field, and one spin 1/2 ghost of the same chirality. This means that in (3.74) we have to substitute ch(V) for ch(V)-1. In the spin 3/2 case V = TM (4k-2 dimension tangent bundle). Consequently:

$$I_{3/2} = \left(\Pi \frac{x_i/2}{\sinh x_i/2}\right)\left(\mathrm{Tr}(e^{R/2} - \mathbf{1}) + 4k-3\right) . \qquad (3.76)$$

(c) For the self-dual tensor gauge field, the bundle V is S^- (negative chirality spinors). Recall that in Euclidean space $S^+ \otimes S^-$ contains the self-dual odd forms while $S^- \otimes S^-$ contains the anti-self-dual even forms, and $D : S^+ \otimes S^- \to S^- \otimes S^-$. Since we will not deal with this field any more in these lectures, we will simply quote the form of $I_{s.d.}$ (for details see Refs. 11, 21):

$$I_{s.d.} = -\frac{1}{8}\prod_i \frac{x_i}{\tanh x_i} . \qquad (3.77)$$

(The minus sign is because self-dual fields must be quantized like bosons.) One way of making (3.77) plausible is to notice that in $d = 2$ a self-dual tensor gauge field is simply a right moving scalar, and therefore it is the bosonized form of a right-handed Weyl fermion. Expanding (3.77) and (3.75) it is easy to see that the anomalies agree if we include the factor of 1/8 in (3.77). (Topologically this is the same as saying that the index of $\not{D}$ on a compact 4-manifold M without boundary is related to the signature of M $\tau(M)$ by ind $\not{D} = -\tau(M)/8$. See Section II for the definition of $\tau(M)$ and the index theorem related to it ((2.108-2.112)). Equations(3.75-3.76) and (3.74) will be the starting point of our analysis of the Green and Schwarz anomaly cancellation mechanism[12, 13] at the end of Section IV.

Let us recapitulate this section with some questions that remain open. We have shown that the non-Abelian and gravitational anomalies can be understood topologically in terms of the non-trivial topology of the space $\mathcal{O}\!\mathcal{l}/\mathfrak{g}$. Using the index theorem for families of Dirac operators we showed that the anomaly can be interpreted in terms of the first Chern class of the determinant line bundle $c_1 \in H^2(\mathcal{O}\!\mathcal{l}/\mathfrak{g})$. The simplest situation corresponds to groups such that $\pi_{2n+1}G = \mathbb{Z}$. In this case we explicitly constructed the

2-sphere generating $H^2(\mathcal{A}/\mathcal{G})$, and the first Chern class. Let us for a moment imagine that $G = SU(N)$ with N large. Then one has $\pi_3 SU(N) = \pi_5 SU(N) = \ldots = \pi_{2n-1} SU(N) = \mathbb{Z}$. Consider also for definiteness that $d = 4$. Using $\pi_{k+1}(\mathcal{A}/\mathcal{G}) \approx \pi_{d+k} G$, (3.33-3.34), with appropriate conditions, for $d+k-1 \leq 2n-1$, we can consider a k-parameter family of gauge transformations $g(\theta_1, \ldots, \theta_k, x)$ generating $\pi_{d+k} G$, and thus construct a representative of the generator of $\pi_{k+1}(\mathcal{A}/\mathcal{G})$ by $tA^{g(\theta,x)}$ (again A is in the trivial topological sector). If $g(\theta,x) : S^k \times S^4 \to G$ has winding number equal to 1, then the index of the Dirac operator on the d+k+1 dimensional space determined by (t,θ,x) is equal to 1. Using the adiabatic approximation again, this means that at some value $(t_0, \vec{\theta}_0)$ (k+1) eigenvalues of the Dirac operator go through zero, and the relative twisting of their eigenfunctions around (t_0, θ_0) characterizes the non-triviality of the (k+1)+4 index. In the $N \to \infty$ limit this remains true for any odd k. Thus these higher homotopy groups give us information about the submanifold of $\mathcal{A}/\mathcal{G}$ where $\det D_A$ vanishes. It is a very interesting and open question to understand whether these higher homotopy groups may become relevant in the non-perturbative understanding of the gauge field - fermion dynamics. These higher invariants cannot be interpreted as new anomalies. Once the anomaly in the phase of the fermion determinant is cancelled, it follows from Fermi statistics that the Green's functions with any number of fermions will behave covariantly under gauge transformations. It may be that these higher spheres have something to do with the breaking of global chiral symmetries. We simply do not know yet.

IV. THE DESCENT EQUATIONS AND WESS-ZUMINO LAGRANGIANS

In this section we relate the topological arguments developed in the last section with the descent equation of Stora and Zumino.[26] This approach to the study of anomalies is essentially based on local cohomology and the BRS[54] (Becchi-Rouet-Stora) transformations. To understand the origin of the anomalies from this point of view, we start with the Wess-Zumino consistency conditions.[27] The anomaly equation under infinitesimal gauge transformations of the fermionic effective action can be implicitly written as:

$$\delta_v \Gamma[A] = \int \mathrm{Tr}\, v\mathcal{A}(A) = W[v,A] \tag{4.1}$$

where $\mathcal{A}(A)$ is a local functional (the anomaly). Since the group of gauge transformations is a group, the commutator of two infinitesimal gauge transformations $[\delta_u, \delta_v]$ is another infinitesimal gauge transformation:

$$[\delta_u, \delta_v] = \delta_{[u,v]} \ . \tag{4.2}$$

Applying (4.2) to $\Gamma[A]$, and using (4.1) we get

$$\delta_u W[v,A] - \delta_v W[u,A] = W([u,v],A) \ . \tag{4.3}$$

Equation (4.3) is the statement of the Wess-Zumino consistency conditions. It clearly imposes constraints on the local form of the anomaly. Notice incidentally that any local functional of the gauge field A automatically satisfies (4.3) so that we do not have to worry about the various counter-terms one needs to add to $\Gamma[A]$ in order to make it finite. The consistency conditions (4.3) can be written in a more transparent way in terms of differential forms. To do so, let us go back to the definition of the Chern-Simons forms. From (2.90) using a linear interpolation tA (on a single coordinate patch, we get

$$\mathrm{Tr}\, F^{n+1} = (n+1)d\int_0^1 \mathrm{Tr}\, A(tdA + t^2A^2)^n \ . \tag{4.4}$$

Since in the index theorem we had to include a factor of $i^{n+1}(2\pi)^{-n}/(n+1)!$ in order to get the anomaly in the previous section, we will define for convenience the Chern-Simons form to include such factors:

$$\begin{aligned} Q_{2n+1}(A,F) &\equiv N_{n+1}(n+1)\int_0^1 dt\, \mathrm{Tr}\, A\,(tdA + t^2A^2)^n \\ N_{n+1} &= \frac{i^{n+1}}{(2\pi)^n(n+1)!} \end{aligned} \tag{4.5}$$

Using Cartan's homotopy formula,[26] we can write down a rather simple way to derive (4.5) or more general expressions. Let A_t be a one-parameter family of gauge fields, and define an antiderivation operator ℓ_t as follows:

$$\begin{aligned} \ell_t A_t &= 0 \\ \ell_t F_t &= \delta A_t = \delta t\, \frac{\partial A_t}{\partial t} \ . \end{aligned} \tag{4.6}$$

If we extend ℓ_t to act on forms as an antiderivation:

$$\ell(\Sigma_p \Lambda_q) = (\ell\, \Sigma_p)\Lambda_q + (-1)^p\, \Sigma_p\, \ell\, \Lambda_q$$

It can be shown that on any polynomial function of A_t and F_t one has

$$(d\ell_t + \ell_t d)S(A_t,F_t) = \delta S(A_t,F_t) \tag{4.7}$$

As a consequence of (4.6):

$$\begin{aligned} (d\ell_t + \ell_t d)A_t &= \ell_t(F_t - A_t^2) = \ell_t F_t = \delta A_t \\ (d\ell_t + \ell_t d)F_t &= d(\delta A_t) + A_t\delta A_t + \delta A_t A_t = D_t\delta A_t = \delta F_t \end{aligned} \tag{4.8}$$

If we integrate (4.7) from 0 to 1 in t, we get Cartan's homotopy formula:

$$(dk_{01} + k_{01}d)S = S(A_1,F_1) - S(A_0,F_0) \tag{4.9}$$

and

$$k_{01} = \int_0^1 \delta t\, \ell_t \ . \tag{4.10}$$

In particular, if S is closed, we have:

$$d(k_{01}S) = S(A_1,F_1) - S(A_0,F_0) \tag{4.11}$$

and (4.4) becomes a particular application of (4.11) when $A_0 = 0$ and $S = \mathrm{Tr}\, F^{n+1}$. Particularly useful applications of this formalism can be obtained when one considers "bundle connections", i.e. gauge fields of the form $A^g = g^{-1}(A+d)g$, with g being a multiparameter family of gauge transformations. Let δ be the analog of the d operator acting on the parameters appearing in g. Then $\delta^2 = d\delta + \delta d = 0$, and we have

$$\delta A^g = -d(g^{-1}\delta g) - A^g g^{-1}\delta g - g^{-1}\delta g\, A^g = -D_{A^g}(g^{-1}\delta g) \ . \tag{4.12}$$

In other words, varying with respect to δ is equivalent to making an infinitesimal gauge transformation with parameter $g^{-1}\delta g = v_a d\theta^a$. If δv_a denotes the ordinary variation of A^g with respect to θ^a, then

$$\delta A^g = d\theta^a \wedge D_{A^g} v_a = -D_{A^g}(g^{-1}\delta g)$$

$$D_{A^g} v_a = d v_a + [A^g, v_a] \ . \tag{4.13}$$

Using (4.9), we can easily obtain the variation of the Chern-Simons form under a finite gauge transformation. To do this, simply take

$$A_t = g^{-1} dg + g^{-1} tAg \qquad 0 \le t \le 1 \ . \tag{4.14}$$

Then

$$(d k_{01} + k_{01} d)Q_{2n+1} = Q_{2n+1}(A^g, F^g) - Q_{2n+1}(g^{-1}dg, 0) \ . \tag{4.15}$$

Since $F^g = g^{-1} F g$, and $dQ_{2n+1} = N_{n+1}\, \mathrm{Tr}\, g^{-1} F^{n+1} g = N_{n+1}\, \mathrm{Tr}\, F^{n+1}$, we have $k_{01} dQ_{2n+1} = Q_{2n+1}(A,F)$, and therefore:

$$Q_{2n+1}(A^g,F^g) - Q_{2n+1}(A,F) - Q_{2n+1}(g^{-1}dg,0) = d\, k_{01} Q_{2n+1} \equiv d\alpha_{2n}(A, dgg^{-1}) . \tag{4.16}$$

If we consider Γ as a function of A^g formally, and with $\delta(g^{-1}\delta g) = -(g^{-1}\delta g)^2$, for a two-parameter family of gauge transformations, $g^{-1}\delta g = v_1 d\theta^1 + v_2 d\theta^2$, the consistency condition (4.3)can be elegantly written as[26]

$$\delta W[g^{-1}\delta g, A^g] = 0 \ . \tag{4.17}$$

If we evaluate all the variations implied in (4.17) at $g = 1$, we automatically reproduce (4.3). From this point of view δ is just the generator of a BRS transformation. From an algebraic point of view, we can consider

$g^{-1}\delta g \equiv c$ the Faddeev-Popov ghost field, and then in the BRS language (4.17) becomes $\delta W[c,A] = 0$, and we can try to solve this equation using a local functional consistent with power counting, dimensional analysis, and the fact that the anomaly is parity violating. This provides an alternative way of obtaining the form of the anomaly functionals obtained using index theory in the previous section. In fact by a rather elegant set of formal minipulations using the formulae derived in this section one can get the complete form of the anomaly plus a proof of the consistency conditions. To do this, let A^g again be $g^{-1}(A+d)g$, d = exterior derivative in space-time, and introduce

$$\mathscr{A} \equiv g^{-1}(A + d + \delta)g$$

$$\mathscr{F} = \Delta\mathscr{A} + \mathscr{A}^2 \quad , \qquad \Delta \equiv d + \delta \ . \tag{4.18}$$

A simple computation shows that

$$\mathscr{F} = g^{-1}Fg \ , \qquad F = dA + A^2 \ . \tag{4.19}$$

Next construct the Chern-Simons forms for both A and $\mathscr{A}$, $Q_{2n+1}(\mathscr{A},\mathscr{F})$, $Q_{2n+1}(A,F)$. Then

$$\begin{aligned} N_{n+1}\,\mathrm{Tr}\,\mathscr{F}^{n+1} &= (d+\delta)Q_{2n+1}(\mathscr{A},\mathscr{F}) \\ &= d\,Q_{2n+1}(A^g,F^g) \end{aligned} \tag{4.20}$$

as a consequence of (4.19). Define $v = g^{-1}\delta g$, and $\bar{A} \equiv A^g$, $\bar{F} \equiv F^g$, and expand $Q_{2n+1}(\mathscr{A},\mathscr{F})$

$$Q_{2n+1}(\bar{A}+v,\ \bar{F}) = Q^0_{2n+1}(\bar{A},\bar{F}) + Q^1_{2n}(v,\bar{A}) + \ldots + Q^{2n+1}_0 \ . \tag{4.21}$$

The superscript denotes the number of powers of v appearing in the form $Q^p_{2n+1-p}(v,\bar{A},\bar{F})$. Substituting (4.21) in (4.20) gives the descent equations:[26]

$$\begin{aligned} \delta Q^0_{2n+1} + dQ^1_{2n} &= 0 \\ \delta Q^1_{2n} + dQ^2_{2n-1} &= 0 \\ \cdots\cdots\cdots\cdots\cdots\cdots \\ \delta Q^{2n+1}_0 &= 0 \ . \end{aligned} \tag{4.22}$$

In the previous section, we only considered one-parameter families $g(\theta,x)$, and in this case (4.21) stops at $Q^1_{2n}(v,\bar{A})$. Notice however that the second equation in (4.22) immediately implies that $Q^1_{2n}(v,A)$ can be identified as an anomaly functional in 2n-dimensions, satisfying the consistency conditions:

$$\delta W[v,A] = \int \delta Q^1_{2n}(v,\bar{A}) = -\int dQ^2_{2n-1} = 0 \tag{4.23}$$

suggesting that $Q^1_{2n}(v,\bar{A})$ coincides with (3.69). To prove it we only have to keep in mind that $\bar{A}$ is a 1-form in space-time, and that for a 1-parameter family of g's, $v^2 = 0$. Then, defining $\bar{A}_t = t\bar{A}$

$$
\begin{aligned}
\frac{1}{N_{n+1}} Q_{2n+1}(\mathscr{A},\mathscr{F}) &= (n+1)\int_0^1 dt\, \mathrm{Tr}\mathscr{A}\left(t(d+\delta)\mathscr{A} + t^2\mathscr{A}^2\right)^n \\
&= (n+1)\int_0^1 dt\, \mathrm{STr}(\bar{A}+v)[\bar{F}_t - t(1-t)\{A,v\}]^n \\
&= (n+1)\int_0^1 dt\, \mathrm{STr}\, v\,\bar{F}_t^{\,n} \\
&\quad - n(n+1)\int_0^1 dt\, \mathrm{STr}(\bar{A},\bar{F}_t^{n-1}\cdot\bar{A}v+v\bar{A})t(1-t) \ .
\end{aligned} \tag{4.24}
$$

There is no term independent of v, because $\bar{A},\bar{F}$ are a 1-form and a 2-form respectively in space-time, and the v-independent term is a 2n+1 form in 2n dimensional space-time. In the symmetrized trace in the last line of (4.24) one has to be careful because one is symmetrizing in $\bar{A},\bar{F}_t$ and $\{\bar{A},v\}$ but not in $\bar{A},\bar{F}_t$ and v, so that the presence of the anticommutator is necessary. Since $t(\bar{A}v+v\bar{A}) = D_t v - dv$, we get after an integration by parts and use of the Bianchi identity $D_t F_t = 0$:

$$
\begin{aligned}
&(n+1)\int_0^1 dt\, \mathrm{STr}\, v\,\bar{F}_t^n \\
&+ n(n+1)\int_0^1 dt\, (1-t)\mathrm{STr}\, v\, d(\bar{A}\,\bar{F}_t^{n-1}) \\
&- n(n+1)\int_0^1 dt\, \mathrm{STr}(v\, D_t\, \bar{A}\,\bar{F}_t^{n-1})\,(1-t) \ .
\end{aligned} \tag{4.25}
$$

Using $\frac{\partial}{\partial t}\bar{F}_t = D_t\bar{A}$ and integrating the last term in (4.25) by parts in t we finally obtain:

$$
Q_{2n}^1(v,\bar{A}) = N_{n+1}\, n(n+1)\int_0^1 dt\, (1-t)\mathrm{STr}\, v\, d(\bar{A}\,\bar{F}_t^{n-1}) \tag{4.26}
$$

which is identical to (3.69). Finally:

$$
-\delta_v\Gamma[A] = i\delta W(A) = i\int Q_{2n}^1(v,A) \ . \tag{4.27}
$$

From this more algebraic point of view, we have used the index theorem to fix the normalization of the anomaly.

The methods developed thus far permit a simple derivation of the Wess-Zumino Lagrangian which describes the low energy dynamics of Goldstone bosons coupled to gauge fields. This phenomenological Lagrangian summarizes the effect of the anomalies and allows a calculation of their physical consequences. These Lagrangians will be quite useful in Section V (σ-model anomalies). The Wess-Zumino Lagrangian basically gives a local action in terms of gauge fields and Goldstone bosons whose variation under gauge transformations reproduce the anomaly. This is very useful when one wants to consider the weak interactions of the pseudoscalar octet. The weak group SU(2) × U(1) is a subgroup of the global flavor symmetry of the massless QCD Lagrangian, and in computing the weak decays of mesons (as well as other weak processes), one does not need to know very much about the substructure of mesons and kaons. Thus it is very important to have a Lagrangian which describes at low energies the Goldstone bosons of the breaking of the chiral symmetry in QCD and that reproduces at low energies the flavor anomalies of the underlying quark theory. Thus for definiteness we consider in analogy with QCD an underlying theory with an exact global symmetry $G_L \times G_R$ spontaneously broken down to the diagonal subgroup G. The possible vacuum states are therefore in one to one correspondence with the points of the G manifold. Choosing the identity as our vacuum, the low energy dynamics of the Goldstone bosons can be conveniently described via the matrix $U(x) = \exp(2\pi^i\lambda^i/f_\pi)$ where the $\pi^i(x)$ are the "pion" fields. The matrix U(x) is a map from space-time into the group G, and transforms under the $G_L \times G_R$ symmetry as:

$$U \to g_L^{-1}\, U\, g_R, \quad g_L, g_R \in G \; . \tag{4.28}$$

(4.28) defines a non-linear realization of the broken symmetries in the theory. Imagine now that we gauge part or all of G_L. Then we can introduce A_L gauge fields, and write the standard kinetic term:

$$-\frac{f_\pi^2}{16}\int \mathrm{Tr}(D_\mu U^{-1} D^\mu U) \tag{4.29}$$

$$D_\mu = \partial_\mu + A_{\mu_L}$$

which is invariant under

$$U \to g_L^{-1} U$$

$$A_L \to g_L^{-1}(A_L + d)g_L \; . \tag{4.30}$$

If we take the space-time manifold to be S^{2n}, and assume that $\pi_{2n}(G) = 0$, we can find a one-parameter deformation $U_t(x)$, $0 \le t \le 1$, $U_0(x) = 1$, $U_1(x) = U(x)$, i.e. we have a map $U_t : I \times S^{2n} \to G$; and $I \times S^{2n}$ is topologically a ball B_{2n+1} in G whose boundary is the image under U(x) of S^{2n}.

The Wess-Zumino Lagrangian is defined as:[27,28]

$$W[A_L, U] = i\int_{B_{2n+1}} Q_{2n+1}(A_L^{U_t} + v_t, F_L^{U_t}) \tag{4.31}$$

$$A_L^{U_t} \equiv U_t^{-1}(A_L + d)U$$
$$F_L^{U_t} = (d + d_t)(A_L^{U_t} + v_t) + (A_L^{U_t} + v_t)^2 = U_t^{-1} F_L U_t .$$
$$v_t = U_t^{-1} d_t U_t \; .$$

Under a small deformation of U_t keeping the value of $U_t(x)$ at $t=1$ fixed (4.31) does not change, because $(d+d_t)Q_{2n+1} = \mathrm{Tr}(U_t^{-1} F_L U_t)^{n+1} = 0$. ($F_L$ is a 2-form only in the dx's.) The normalization factor included in Q_{2n+1} is very helpful, because even though $\pi_{2n}G = 0$, it may happen that $\pi_{2n+1}G = Z$ (for example, in $d=4$, $G = SU(3)$, $\pi_5 SU(3) = \mathbb{Z}$), and one can choose two different ways of interpolating between $U(x)$ and 1. The difference between the two actions (4.31) is the integral over the 2n+1 sphere generated by gluing together the two B_{2n+1} balls through their common boundary of the integrand of (4.31). This integral (by the index theorem discussed in Section III) equals 2π times the winding number on S^{2n+1} of the corresponding pion configuration. Since quantum mechanics only cares about $\exp i W[A_L, v]$, the change $W \to W + 2\pi n$ does not affect quantum mechanics. To show that (4.31) has the correct transformation law under gauge transformations, we notice by the definition of Q^1_{2n} that

$$Q_{2n+1}(A_L^{U_t} + U_t^{-1} d_t U_t, F_L^{U_t}) = Q^1_{2n}(v_t, A_L^{U_t}) \; . \tag{4.32}$$

If we recall (4.27), and integrate it along the one-parameter family of gauge transformations U_t, we get

$$-(\Gamma[A_L^U] - \Gamma[A_L]) = i\int Q^1_{2n}(v_t, A_L^{U_t}) \; . \tag{4.33}$$

Since under a gauge transformation (4.30)

$$A_L^{U_L} \to (A_L^{g_L})^{g_L^{-1} U_L} = A_L^{U_L}$$

is gauge invariant, the change of (4.31) under a gauge transformation (4.30) is given by the gauge variation of $\Gamma[A_L]$. Thus (4.31) has the correct transformation properties under gauge transformations. Using (4.16) we have:

$$W[A_L, U] = \int_{I \times S^{2n}} Q_{2n+1}(U_t^{-1}(d+d_t)U_t, 0) + \int_{S^{2n}} \alpha_{2n}(A_L, dU\, U^{-1}) \; . \tag{4.34}$$

In four dimensions the result is:

$$\begin{aligned} W[A_L, U] = {} & \frac{1}{240\pi^2} \int_{B_5} \mathrm{Tr}\left(U_t^{-1}(d+d_t)U_t\right)^5 \\ & - \frac{1}{48\pi^2} \int_{S^4} \mathrm{Tr}\Big[dU\, U^{-1}(A_L\, dA_L + dA_L\, A_L + A_L^3) \\ & - \frac{1}{2}\, dU\, U^{-1} A_L\, dU\, U^{-1} A_L - (dU\, U^{-1})^3 A_L\Big] \; . \end{aligned} \tag{4.35}$$

Since $\mathrm{Tr}(g^{-1}dg)^{2n+1}$ is a closed form, and B_{2n+1} is contractible, the first term in (4.35) is only a function of the pion fields in space time. Next we want to gauge the right-handed gauge fields A_R associated with G_R.

The covariant derivative is now:

$$D_\mu U = \partial_\mu U + A_{\mu L} U - U A_{\mu R} \tag{4.36}$$

and the kinetic term (4.29) is gauge invariant. Now we want to extend $W[A_L, U]$ to a functional $W[A_L, A_R, U]$ with the correct anomalous variations under A_L, A_R. Under infinitesimal left-right gauge transformations:

$$\delta_v^{L,R} A_{L,R} = D_{L,R}, v = dv + [A_{L,R}, v]$$

$$\delta_v^L U = -vU \qquad \delta_v^R U = Uv$$

$$\delta_v^L A_R = \delta_v^R A_L = 0 . \tag{4.37}$$

From (4.31 - 4.33) we easily get:

$$\delta_v^L W[A_L, U] = -\int Q_{2n}^1(v, A_L)$$

$$\delta_v^R W[A_L, U] = \int Q_{2n}^1(v, A_L^U) \tag{4.38}$$

because A_L^U transforms like A_R under right gauge transformations. To shift these variations to a more standard form, we construct Bardeen's counterterm[55] as follows. Define

$$R[A_1, A_0] = k_{01} Q_{2n+1}(A_t) \tag{4.39}$$

using the homotopy operator for $A_t = A_0 + t(A_1 - A_0)$. For variations δ_v such that $\delta_v A_{0,1} = dv + [A_{0,1}, v]$, A_t satisfies $\delta_v A_t = dv + [A_t, v]$, implying

$$\delta_v k_{01} Q_{2n+1} = k_{01} \delta_v Q_{2n+1}(A_t) = k_{01} \, d Q_{2n}^1(v, A_t)$$

$$= Q_{2n}^1(v, A_1) - Q_{2n}^1(v, A_0) - d k_{01} Q_{2n}^1(v, A_t) . \tag{4.40}$$

In integrated form:

$$\delta_v R[A_1, A_0] = \int Q_{2n}^1(v, A_1) - \int Q_{2n}^1(v, A_0) . \tag{4.41}$$

Now we can shift the anomalies (4.38) to standard form

$$W_{LR}[A_L, A_R, U] = W[A_L, U] + R[A_R, A_L^U] \tag{4.42}$$

with $A_t = A_L^U + t(A_R - A_L^U)$, and $\delta_v^R A_t = dv + [A_t, v]$, $\delta_v^L A_t = 0$. Thus:

$$\delta_v^L W[A_L, A_R, U] = -\int Q_{2n}^1(v, A_L)$$

$$\delta_v^R W[A_L, A_R, U] = +\int Q_{2n}^1(v, A_R) . \tag{4.43}$$

The difference in sign corresponds to the sign difference for the anomaly in left- and right-handed currents in the underlying theory. The relative sign in (4.43) implies that the diagonal subgroup G is free of anomalies. This suggests another form of the Wess-Zumino functional, called the VA form, which is invariant under vector gauge transformations, and all the anomaly is shifted to the axial gauge transformation:

$$\delta_v^V = \delta_v^L + \delta_v^R$$

$$\delta_v^A = \delta_v^L - \delta_v^R \tag{4.44}$$

and

$$\delta_v^V W_{LR} = \int \left(Q_{2n}^1(v,A_R) - Q_{2n}^1(v,A_L)\right) \tag{4.45}$$

using Bardeen's counterterm

$$\delta_v^V R[A_L,A_R] = Q_{2n}^1(v,A_L) - Q_{2n}^1(v,A_R) \tag{4.46}$$

and the functional we are looking for is:

$$W_{VA}[A_L,A_R,U] = W_{LR}[A_L,A_R,U] + \int R[A_L,A_R] \ . \tag{4.47}$$

If we restrict the computation to four dimensions, $R(A_1,A_0)$ is:

$$R[A_1,A_0] = \frac{1}{48\pi^2}\,\mathrm{Tr}[(F_0+F_1)(A_1A_0-A_0A_1) - (A_1^3A_0-A_0^3A_1) + \tfrac{1}{2}\,A_1A_0A_1A_0] \tag{4.48}$$

and after a long computation, the form of W_{LR} becomes:[27,28,56]

$$\begin{aligned} W_{LR}[A_L,A_R,U] = {} & \frac{1}{240\pi^2}\int_{B_5} \mathrm{Tr}[U_t^{-1}(d+d_t)U_t]^5 \\ & - \frac{1}{48\pi^2}\int_{S^4}\Big\{\mathrm{Tr}[A_L(dU\,U^{-1})^3 - \text{p.c.}] \\ & + \mathrm{Tr}[(A_L\,dA_L + dA_L\,A_L + A_L^3)(U\,A_R\,U^{-1} + U\,d\,U^{-1}) - \text{p.c.}] \\ & + \mathrm{Tr}(dU\,U^{-1}\,dA_L\,U\,A_R\,U^{-1} - \text{p.c.}) \\ & + \tfrac{1}{2}\,\mathrm{Tr}(A_L\,dU\,U^{-1}\,A_L\,dU\,U^{-1} - \text{p.c.}) \\ & + \mathrm{Tr}(U\,A_R\,U^{-1}\,A_L\,dU\,U^{-1}dU\,U^{-1} - \text{p.c.}) \\ & - \mathrm{Tr}(A_R\,U^{-1}\,dU\,A_R\,U^{-1}\,A_L\,U - \text{p.c.}) \\ & - \tfrac{1}{2}\,\mathrm{Tr}(A_R\,U^{-1}\,A_L\,U\,A_R\,U^{-1}\,A_L\,U)\Big\} \ . \end{aligned} \tag{4.49}$$

p.c. means parity conjugate: $A_L \to A_R$, $U \to U^{-1}$. Finally we will write for completeness the form of the anomaly in VA form, varying W_{VA}, and defining

$$V = \tfrac{1}{2}(A_L + A_R) \qquad A = \tfrac{1}{2}(A_L - A_R)$$

$$F_V = \tfrac{1}{2}(F_L + F_R) = dV + V^2 + A^2$$

$$F_A = \tfrac{1}{2}(F_L - F_R) = dA + AV + VA \,. \tag{4.50}$$

We get after some tedious algebra:

$$\delta_v^V W_{VA} = 0$$

$$\delta_v^A W_{VA} = \frac{1}{4\pi^2} \operatorname{Tr} v[F_V^2 + \tfrac{1}{3} F_A^2 - \tfrac{4}{3}(A^2 F_V + A F_V A + F_V A^2) + \tfrac{8}{3} A^4] \,. \tag{4.51}$$

(4.49) can be used to calculate low energy processes involving pseudoscalar mesons including the weak and electromagnetic interactions. These are included by gauging the $SU(2)_L \times U(1)_R$ subgroup of the global chiral symmetry. The Lagrangian (4.49) will have anomalies under this group, which cancel only after one includes the contributions from the lepton sector of the theory.

There is one more topic which we would like to cover before moving to the treatment of the gravitational anomalies from the point of view of the descent equations. This has to do with the so-called "covariant" and "consistent" forms for the anomaly.[39] In Section II we agreed that there are two ways of defining the anomalous diagrams. The simplest way of doing the computation, is to insert a single axial current, and the rest vector currents. Then one requires current conservation and Bose symmetry in the vector channels, obtaining the anomaly in the axial vector channel. This is equivalent to computing the anomaly using the naive form of Fujikawa's method[49] as applied to the non-Abelian anomaly. This corresponds to defining the functional integral for Weyl fermions by expanding ψ in terms of the eigenfunctions ϕ_n of $i\not{D}_- i\not{D}_+$, $\phi_n = \lambda_n \phi_n$, and $\bar{\psi}$ in terms of the eigenfunctions $\chi_n = |\lambda_n|^{-\frac{1}{2}} i\not{D}_+ \phi_n$ of $i\not{D}_+ i\not{D}_-$. This definition makes $\exp - \Gamma[A]$ formally equivalent to $[\det i\not{D}_- i\not{D}_+]^{\frac{1}{2}}$. Making the infinitesimal gauge transformation $A \to A + Dv$ and a compensating change of variables $\psi \to \Psi - v\psi$, $\bar{\psi} \to \bar{\psi} + \bar{\psi} v$, we get an anomalous Jacobian factor as in (3.6):

$$\log \mathcal{J} = \operatorname{Tr}_+ v_- - \operatorname{Tr}_- v \tag{4.52}$$

where the traces $\operatorname{Tr}_\pm$ run over the eigenfunctions of $i\not{D}_- i\not{D}_+$ and $i\not{D}_+ i\not{D}_-$ respectively. Regulating as in (3.7) by means of a gaussian cut-off and then evaluating the trace on a plane-wave basis, we get

$$\log \mathcal{J} = \lim_{M\to\infty} \operatorname{Tr} \bar{\Gamma} v\, e^{-(i\not{D})^2/M^2} = \frac{i^{n+2}}{(2\pi)^n n!} \int \operatorname{Tr} v F^n \tag{4.53}$$

or

$$\delta_v W[A] = \frac{i^{n+2}}{(2\pi)^n n!} \int \mathrm{Tr}\, v\, F^n \,. \tag{4.54}$$

(4.54) is known as the covariant form of the anomaly. It does not satisfy the consistency conditions (4.3) and it does not reduce to the Gross-Jackiw form of the anomaly (3.66) for $d = 4$. We know however that this method of computing the anomaly violates Bose symmetry from the beginning because we are considering all vertices but one to emit vector gauge fields, and in the original theory, the gauge fields couple to chiral currents so that each interaction vertex contains a factor of $(1+\bar{\Gamma})/2$, i.e. the gauge field A couples with equal strength to vector and axial vector currents. The full Bose symmetric procedure of computing the diagrams amounts to determining how the current J^c defined by

$$-\int \mathrm{Tr}\, v\, D_\mu J^{c\mu} = \frac{i^{n+1}}{(2\pi)^n n!} \int \mathrm{Tr}\, v\, F^n \tag{4.55}$$

differs from the automatically Bose symmetric current J appearing in the correct computation $\langle\bar{\psi}\gamma_\mu T^a P_+ \psi\rangle_A$. This can easily be determined as follows: we want to compute ΔJ_μ as a local function of A,F such that:

$$\mathrm{Tr}\, v\, D_\mu \Delta J^\mu = \frac{i^n}{(2\pi)^n n!} \int \mathrm{Tr}\, v\, F^n - Q^1_{2n}(v,A) \,. \tag{4.56}$$

ΔJ can be computed using the homotopy operator formulae. Using (4.7), we have (with $A_t = tA$):

$$\begin{aligned} Q_{2n+1}(A+\delta A, F+\delta F) - Q_{2n+1}(A,F) &= (\ell d + d\ell) Q_{2n+1}(A,F) \\ &= N_{n+1}\, \ell\, \mathrm{Tr}\, F^{n+1} + d\,\ell\Big[(n+1)\int_0^1 \delta t\, \mathrm{STr}(A, F_t^n)\Big] \\ &= N_{n+1}\, (n+1)\,\mathrm{Tr}\,\delta A\, F^n - d\, n(n+1)\int_0^1 \delta t\, \mathrm{STr}\, t(A, \delta A, F_t^{n-1}) \\ &= N_{n+1}\, (n+1)\,\mathrm{Tr}\,\delta A\, F^n + d\, n(n+1)\int_0^1 \delta t\, \mathrm{STr}(\delta A, A_t, F_t^{n-1}) \,. \end{aligned}$$

If $\delta A = -Dv = -dv - [A,v]$, (v is not a 1-form now)

$$\begin{aligned} -dQ^1_{2n}(v,A) &= -(n+1)d(\mathrm{Tr}\, v\, F^n)N_{n+1} - d\, n(n+1)\int_0^1 \delta t\, \mathrm{STr}(Dv, A_t, F_t^{n-1})N_{n+1} \\ &= -N_{n+1}(n+1)d(\mathrm{Tr}\, v\, F^n) + dn\, (n+1)N_{n+1}\int_0^1 \delta t\, \mathrm{STr}\big(v, D(A_t, F_t^{n-1})\big) \,. \end{aligned} \tag{4.57}$$

The last term in (4.57) can be used now to define the current ΔJ.

$$\Delta J = \int_0^1 \delta t\, n(n+1) N_{n+1}\, s(A_t, F_t^{n-1}) \tag{4.58}$$

where s is the symmetrized polynomial in its arguments defined similarly to the symmetrized trace, but without the trace. Thus, in explicit computations one can compute the covariant form of the anomaly (which is easier), and then use (4.55-4.58) to obtain the consistent form of the anomaly.

This completes our study of the consistency conditions and descent equations for the gauge case. Let us now turn to the gravitational case. We know from Section II that any theory containing spinors coupled to gravity has two types of invariances at the classical level: coordinate transformations and local Lorentz transformations. On the other hand, we know from the topological analysis that there is only one invariant polynomial associated with the anomaly. This makes one suspect that the anomalies in Lorentz transformations and diffeomorphisms are essentially one and the same thing. Since the invariant polynomial is the same for both anomalies, it is clear that cancelling any one of them will automatically make the other anomaly vanish. In fact even for anomalous theories, one can construct counterterms which shift the anomaly from diffeomorphisms to local Lorentz transformations and vice versa.[39] The derivation of the descent equations (4.22) is identical as in the gauge theory case. We start with some invariant polynomial P(R), and then follow the same steps that led to (4.22). We only have to be careful from the beginning whether we express R in terms of the spin connection ω or in terms of the Christoffel connection Γ (going from ω to Γ corresponds to making a GL(n) gauge transformation, where the GL(n) is the vielbein (2.57)). If we choose to write $R = d\omega + \omega^2$ the gauge variations δ in (4.22) correspond to local Lorentz transformations $\delta^L \omega = D\alpha = d\alpha + [\omega, \alpha]$. (Notice that the relation (2.57) between ω and Γ holds also in the presence of torsion. Thus everything we will derive below works with or without torsion.) From this point of view, everything is identical to the gauge theory case, including the consistency conditions $[\delta^L_{\alpha_1}, \delta^L_{\alpha_2}] = \delta^L_{[\alpha_1, \alpha_2]}$ for two infinitesimal local Lorentz transformations. Hence $Q^1_{2n}(\alpha, \omega, R)$ is a solution to the Lorentz consistency conditions, where Q^1_{2n} is derived from the polynomial $P_{2n+2}(R)$ given by the appropriate index theorem. Similarly, if we choose to write $R = d\Gamma + \Gamma^2$, and consider passive coordinate transformations δ'_ξ, Γ transforms like a GL(n) gauge connection $\Gamma \to \Lambda^{-1}(\Gamma + d)\Lambda$, $\Lambda^\mu_\nu = (\partial x'^\mu / \partial x^\nu)$; again the derivation of (4.22) goes through and $Q^1_{2n}(v_\xi, \Gamma, R)$ $(v_\xi{}^\alpha{}_\beta = \partial\xi^\alpha / \partial x^\beta)$ satisfies the consistency condition for passive coordinate transformations: $[\delta'_{\xi_1}, \delta'_{\xi_2}] = \delta'_{[\xi_1, \xi_2]}$. Before writing down the complete set of consistency conditions, let us understand how these anomalies appear when analyzing the dynamics of chiral fermions in the presence of external gravitational fields. Let $\Gamma[e, \omega]$ be the effective action for the fermions in the presence of an external gravitational field. The energy momentum tensor (or rather its expectation value) is given by:

$$\frac{\delta\Gamma}{\delta e^a_\mu} = e\langle T^\mu{}_a\rangle\,, \qquad e = \det e^a_\mu \quad . \tag{4.59}$$

If we make an infinitesimal local Lorentz transformation:

$$\delta^L_\alpha \Gamma = - \int dx\, \alpha^a{}_b\, e^b{}_\mu \frac{\delta\Gamma}{\delta e^a{}_\mu}$$

$$= - \int dx\, e\, \alpha^{ab} <T_{ab}> \ . \tag{4.60}$$

Thus if the expectation value of the energy momentum tensor is not symmetric, we will find a local Lorentz anomaly under an infinitesimal coordinate transformation $x^\mu \to x^\mu - \xi^\mu(x)$

$$\delta_\xi e^a{}_\mu = \xi^\nu \nabla_\nu e^a{}_\mu + e^a{}_\nu \nabla_\mu \xi^\nu \tag{4.61}$$

and

$$\delta_\xi \Gamma = \int d^{2n} x\, e\, \xi^\nu (\nabla_\mu <T_\nu{}^\mu> + \omega_{ab,\nu} <T^{ab}>) \ . \tag{4.62}$$

If we choose a gauge which makes T^{ab} manifestly symmetric, then the second term vanishes and we are left with a pure Einstein anomaly (using Zumino's nomenclature), i.e. the induced energy-momentum tensor is not conserved. For explicit computations (or for other reasons depending on the problem) we have to select a particular form of the local Lorentz and coordinate transformations for the basic fields $e^a{}_\mu$, $\omega_{\mu ab}$, ψ. Depending on which choice we make, the final form of the anomaly will be different as can easily be understood from (4.60-4.62). For example we can choose a gauge for local Lorentz transformations making the vielbein $e^a{}_\mu$ symmetric. This is always possible. By the polar decomposition theorem, any real matrix can be written as the product of a symmetric and an orthogonal matrix. Choosing the orthogonal matrix to be equal to 1 completely fixes the local Lorentz gauge. In this case the energy momentum tensor is symmetric, and we only have a pure Einstein anomaly, i.e. we can write the effective action Γ as a functional of $g_{\mu\nu}$: $\delta\Gamma/\delta g_{\mu\nu} = \sqrt{g}\, T_{\mu\nu}/2$, $\delta_\xi g_{\mu\nu} = \nabla_\mu \xi_\nu + \nabla_\nu \xi_\mu$ and

$$\delta_\xi \Gamma = - \int dx \sqrt{g}\, \xi_\mu \nabla_\nu T^{\mu\nu} \ . \tag{4.63}$$

In terms of the vielbein however, we have to modify the coordinate transformation role (4.61) by a compensating local Lorentz transformation which makes $\delta_\xi e$ symmetric again. A different prescription for example would be to make (4.61) covariant under local Lorentz transformations. ((4.61) is not because the $\nabla_\nu e^a{}_\mu$ piece is missing the spin connection term.) If we introduce a compensating local Lorentz transformation $\delta^L{}_\alpha$ with gauge parameter $\alpha = i(\xi)\omega$, then we get from (2.77-2.79):

$$\delta_\xi e^a{}_\mu = D_\mu \xi^a$$

$$\delta_\xi \omega = i(\xi) R \ . \tag{4.64}$$

These are the two choices used in the perturbative computation of the gravitational anomaly in Ref. 11. Depending on which choice one makes, the consistency conditions will look different and the anomaly functional can be quite different as well. (This can also be understood in terms of the operator formalism. Depending on which definition we choose for $T^R_{\mu\nu}$, we may have $\nabla^\mu T^R_{\mu\nu}$, which only receives contributions from the regulator, which may or may not be conserved.)

As pointed out before, let us start by setting up the consistency conditions for passive coordinate transformations. Later on we will see what happens when we consider active transformations. The passive transformations are easier to deal with because they behave formally like GL(n) gauge transformations on e and Γ. Then

$$\begin{aligned} \delta^L{}_\alpha e &= - \alpha e \\ \delta^L{}_\alpha \omega &= D\alpha \\ \delta'_\xi e^a{}_\alpha &= e^a{}_\beta (v_\xi)^\beta{}_\alpha \\ \delta'_\xi \Gamma &= \nabla v_\xi \\ \delta'_\xi \omega &= 0 \ . \end{aligned} \tag{4.65}$$

$$(v_\xi)^\alpha{}_\beta = \partial\xi^\alpha/\partial x^\beta \ .$$

The consistency conditions are then:

$$\begin{aligned} [\delta^L{}_\alpha, \delta^L{}_{\alpha'}] &= \delta^L{}_{[\alpha,\alpha']} \\ [\delta^L{}_\alpha, \delta'_\xi] &= 0 \\ [\delta'_{v_{\xi_1}}, \delta'_{v_{\xi_2}}] &= \delta'_{[v_{\xi_1}, v_{\xi_2}]} \ . \end{aligned} \tag{4.66}$$

If we choose to have the anomalies in Lorentz transformations, then we will have

$$\delta^L{}_\alpha \Gamma = Q^1_{2n}(\alpha,\omega) \ . \tag{4.67}$$

Since $\delta'_\xi \omega = 0$, and Q^1_{2n} satisfies the descent equations, Q^1_{2n} satisfies the consistency conditions (4.66). Equation (4.67) is the anomaly one would obtain by choosing a gauge where all the anomalies appear under local Lorentz transformations. Similarly, if we choose a symmetric gauge for $e^a{}_\mu$, the anomaly is a pure Einstein anomaly, and $Q^1_{2n}(v_\xi,\Gamma)$ satisfies (4.66) because $\delta^L{}_\alpha \Gamma = 0$, and the descent equations for passive coordinate transformations are also satisfied. Thus we have two anomaly functionals,

$Q^0_{2n+1}(\omega,R)$ and $Q^0_{2n+1}(\Gamma,R)$. Since we obtain one or the other by simply changing the gauge on local Lorentz transformations, let us now show explicitly that the two functionals $Q^0_{2n+1}(\omega,R)$ and $Q^0_{2n+1}(\Gamma,R)$ differ by the variation of a local counterterm.[39] Since ω and Γ are related by a GL(n) gauge transformation, the vielbein, e^a_μ, let us choose a one-parameter family e_t, $0 \leq t \leq 1$, so that at $t=0, e_0 = 1$, and at $t=1, e_1 = e$. This can always be done in a local coordinate patch. (The vielbein plays a role analogous to the Goldstone bosons in a Wess-Zumino Lagrangian describing the breaking of GL(n) to its SO(n) subgroup.) We now simply integrate up the GL(n) "anomaly" with infinitesimal parameter $\hat{v}_t = e_t^{-1} d_t e_t$. The local functional obtained is:

$$S = \int_{M_{2n}\times[0,1]} Q^0_{2n+1}(\omega^{e_t} + \hat{v}_t)$$

$$\omega^{e_t} = e_t^{-1}(\omega + d)e_t$$

$$\hat{v}_t = e_t^{-1} d_t e_t \quad . \qquad (4.68)$$

Thus
$$S = \Gamma[\Gamma,g] - \Gamma[\omega,\delta_{ab}]$$

and by adding or subtracting S we can shift the anomaly from Lorentz to Einstein anomalies and vice-versa. Notice that even though S is non-polynomial in e, it contains a finite number of derivatives, and thus it is a local counterterm. Finally, in order to discuss the anomalous conservation law of the energy momentum tensor, we must now consider the infinitesimal generator of diffeomorphisms in active form. From (2.71-2.75) we have:

$$\delta_\xi e^a_\alpha = (i_\xi\, d e)^a{}_\alpha + e^a{}_\beta (v_\xi)^\beta{}_\alpha$$

$$\delta_\xi \Gamma = (d\, i_\xi + i_\xi d)\Gamma + \nabla v_\xi \quad . \qquad (4.69)$$

The consistency conditions become[57]

$$[\delta_{\xi_1}, \delta_{\xi_2}] = \delta_{[\xi_2,\xi_1]}$$

$$[\delta^L_\alpha, \delta^L_{\alpha'}] = \delta^L_{[\alpha,\alpha']}$$

$$[\delta_\xi, \delta^L_\alpha] = -\delta^L_{(\xi\cdot\partial)\alpha} \quad . \qquad (4.70)$$

Now $\delta_\xi\omega = (i(\xi)d + d\,i(\xi))\omega$; and from (2.75)

$$\delta_\xi = d\,i(\xi) + i(\xi)d + \delta'_\xi \qquad (4.71)$$

for tensor valued forms. If we consider the pure Lorentz form of the anomaly $\Gamma[\omega,e]$, it is easy to check that (4.70) is satisfied.[20,39] For the purely Einstein form, we have to do a bit more work:

$$\delta_\xi \Gamma = - \int \xi_\mu \nabla_\nu \langle T^{\mu\nu} \rangle \equiv \int \mathrm{Tr}\, v_\xi \mathcal{a}_c(\Gamma) \tag{4.72}$$

$$\mathrm{Tr}\, v_\xi \mathcal{a}_c(\Gamma) = Q^1_{2n}(v_\xi, \Gamma) \ . \tag{4.73}$$

We want to show that

$$\delta_{\xi_1} \int \mathrm{Tr}\, v_{\xi_1} \mathcal{a}_c - \delta_{\xi_2} \int \mathrm{Tr}\, v_{\xi_1} \mathcal{a}_c = \int \mathrm{Tr}\, v_{[\xi_2,\xi_1]} \mathcal{a}_c \tag{4.74}$$

using (2.75) and the fact that $Q^1_{2n}(v_\xi,\Gamma)$ satisfies the consistency conditions for passive coordinate transformations. Using also that $\mathcal{a}_c$ is a form of maximal degree, we can rewrite the left-hand side of (4.74):

$$\delta_{\xi_1}\delta_{\xi_2}\Gamma = \delta_{\xi_1} \mathrm{Tr}\, v_{\xi_2} \mathcal{a}_c = \mathrm{Tr}\, v_{\xi_2}\big(i(\xi_1)d + d\, i(\xi_1)\big)\mathcal{a}_c + \delta'_{\xi_1} \mathrm{Tr}\, v_{\xi_2} \mathcal{a}_c$$

$$= \delta'_{\xi_1}\delta'_{\xi_2}\Gamma + \mathrm{Tr}\, v_{\xi_2}\big(i(\xi_1)d + d\, i(\xi_1)\big)\mathcal{a}_c \ ,$$

i.e.

$$[\delta_{\xi_1},\delta_{\xi_2}]\Gamma = \int \mathrm{Tr}[v_{\xi_1},v_{\xi_2}]\mathcal{a}_c - \int \mathrm{Tr}\big(d\, v_{\xi_2}\, i(\xi_1) - d\, v_{\xi_1}\, i(\xi_2)\big)\mathcal{a}_c$$

$$= \int \mathrm{Tr}\Big([v_{\xi_1},v_{\xi_2}] - \big(i(\xi_1)\cdot d\, v_{\xi_2}\big) + \big(i(\xi_2)\cdot d\, v_{\xi_1}\big)\Big)\mathcal{a}_c \ .$$

Since

$$\Big([v_{\xi_1},v_{\xi_2}] - \big(i(\xi_1)\cdot d\, v_{\xi_2}\big) + i(\xi_2)\cdot d\, v_{\xi_1}\Big)^\alpha{}_\beta =$$

$$= \partial_\beta[(\xi_2\cdot\partial)\xi_1^\alpha - (\xi_1\cdot\partial)\xi_2^\alpha] = (v_{[\xi_2,\xi_1]})^\alpha{}_\beta \ ,$$

we conclude that (4.74) is satisfied Q.E.D. Thus $Q^1_{2n}(v_\xi,\Gamma)$ gives the consistent form of the Einstein anomaly. Notice that a perfectly valid anomaly functional is $\alpha Q^0_{2n+1}(\omega) + \beta Q^0_{2n+1}(\Gamma)$, $\alpha+\beta = 1$, and it should be emphasized again that depending on the conventions chosen to perform the computations, the answers may look quite different even though we know *a priori* that they are equivalent.

Finally in the same way as with the gauge anomalies, one can define covariant and consistent anomalies. Since the derivations are identical to those in the gauge case (the algebra is slightly more involved), we will not present the corresponding results. The interested reader is referred to the literature for further details.[20,39]

As an application of the material presented in this and the previous section we would like to outline the Green and Schwarz anomaly cancellation mechanism. So far we only know of two non-trivial cancellations of gauge and gravitational anomalies between fields of different spins. The first one was found in Ref. 11 for the chiral $N = 2$ supergravity theory in

$d = 10$. In this theory there are no gauge fields, and the fields contributing to the gravitational anomaly are the gravitinos, a spin 1/2 field, and a bosonic field: a self-dual tensor gauge field. Although the cancellation of anomalies is rather striking, the possible phenomenological applications of this theory are not very promising (in all known Kaluza-Klein compactifications it leads to a vector-like four-dimensional theory, and the gauge group cannot be very big). The second non-trivial anomaly cancellation was recently discovered by Green and Schwarz for type I superstring theories with gauge group SO(32).[12,13] Green and Schwarz also noticed that in the field theory limit of the string theory (and $N = 1$ super Yang-Mills coupled to $N = 1$ supergravity in $d = 10$) the anomalies cancel also for $E_8 \times E_8$. Soon after the discovery of this anomaly cancellation, the Princeton group[58] constructed the so-called "heterotic" string with gauge group $E_8 \times E_8$. This theory is now being considered rather intensely and it may have non-trivial phenomenological implications. In order to compute the anomalies in this theory we have to first know the field content. The $N = 1$ supergravity multiplet contains the graviton $e^a{}_\mu$, a left-handed Weyl-Majorana gravitino ψ_μ, a right-handed Weyl-Majorana field λ, a 2-index antisymmetric tensor $B_{\mu\nu}$ and the dilaton ϕ. The $N = 1$ super-Yang-Mills multiplet contains simply the gauge field $A^a{}_\mu$ and a left-handed Weyl-Majorana gluino λ^a where $a = 1, \ldots, \dim G = N$ (G = gauge group). Let us shift all the gravitational anomalies to the local Lorentz transformations for simplicity. From the formulae in Section III we can write down immediately the 12-form which determines the anomaly through the descent equations. The characteristic polynomial we have to expand is in dimension d:

$$\begin{aligned}
&\hat{A}\left(\frac{R}{2\pi}\right) \operatorname{Tr} e^{iF/2\pi} && \text{(gluinos)} \\
&\hat{A}\left(\frac{R}{2\pi}\right) \left[\operatorname{Tr}(e^{R/2\pi} - \mathbb{1}) + d - 1\right] && \text{(gravitino)} \\
-&\hat{A}\left(\frac{R}{2\pi}\right) && \text{(right-handed Weyl-Majorana field)}
\end{aligned} \tag{4.75}$$

For $d = 10$ we have to extract the 12-form piece of (4.75). If we factor out various common numerical factors, we get after some algebra:

$$\begin{aligned}
I_{12} = &-\frac{1}{15} \operatorname{Tr} F^6 + \frac{1}{24} \operatorname{Tr} F^4 \operatorname{Tr} R^2 - \frac{\operatorname{Tr} F^2}{960} \left(5(\operatorname{Tr} R^2)^2 + 4 \operatorname{Tr} R^4\right) \\
&+ \frac{N-496}{7560} \operatorname{Tr} R^6 + \left(\frac{N-496}{5760} + \frac{1}{8}\right) \operatorname{Tr} R^4 \operatorname{Tr} R^2 \\
&+ \left(\frac{N-496}{13824} + \frac{1}{32}\right) (\operatorname{Tr} R^2)^3
\end{aligned} \tag{4.76}$$

where the trace over the gauge field F is taken in the adjoint representation. By computing the Chern-Simons form associated with I_{12}, and going through the descent equations, we will get the anomaly under combined gauge and local Lorentz transformations. If we only included the fermion field contribution to the anomaly, it is clear that (4.76) does not vanish for any gauge group and we would have to throw away the theory. The non-trivial step taken by Green and Schwarz is to analyze under what circumstances the anomaly induced by (4.76) can be cancelled by counterterms. Let us now repeat their argument. If we consider the leading terms in I_{12}, i.e. the term $\mathrm{Tr}\, F^6$ and $\mathrm{Tr}\, R^6$, the anomalies they will induce are respectively $Q^1_{10}(\omega)$ and $Q^1_{10}(A)$, and we know from previous arguments that neither can be obtained as the variation of a local functional. Thus if we require that the $\mathrm{Tr}\, R^6$ term vanish, we obtain N = 496, i.e. the gauge group must have 496 generators. Next we require that the group in the adjoint representation should not have a 6th order irreducible Casimir, i.e. we require that

$$\mathrm{Tr}\, F^6 = \alpha\, \mathrm{Tr}\, F^2\, \mathrm{Tr}\, F^4 + \beta(\mathrm{Tr}\, F^2)^3 \quad . \tag{4.77}$$

In this case the anomalous variation of the effective action will be proportional to $Q^1_2\, \mathrm{Tr}\, F^4$ and $Q^1_6\, \mathrm{Tr}\, F^2$, and there may be hope of finding a local counterterm cancelling this variation (for instance terms like $Q^0_3\, Q^0_7$ will under gauge variations generate $Q^1_2\, \mathrm{Tr}\, F^4$ and $\mathrm{Tr}\, F^2\, Q^1_6$). A crucial role in constructing the appropriate counterterms is played by the two form $B = B_{\mu\nu}\, dx^\mu \wedge dx^\nu/2$ appearing in the supergravity multiplet. A three-form field strength is formed from this potential:

$$H = dB + Q_{3L} + k\, Q_{3y}$$

$$d\,Q_{3L} = \mathrm{Tr}\, R^2 \quad d\,Q_{3y} = \mathrm{Tr}\, F^2 \quad . \tag{4.78}$$

The introduction of the Lorentz Chern-Simons form is an important modification of the definition of H in the N = 1 d = 10 supergravity Lagrangian.[60] Since

$$\delta Q_{3y} = -\, d\, Q^1_{2y}$$

$$\delta Q_{3L} = -\, d\, Q^1_{2L} \quad . \tag{4.79}$$

H is gauge and local Lorentz invariant only if

$$\delta B = Q^1_{2L} + k\, Q^1_{2y} \quad . \tag{4.80}$$

What Green and Schwarz[12] proved (see also Ref. 13 for the presentation we are following here) is that a local counterterm exists which cancels the anomalies whenever I_{12} can be factorized in the form:

$$I_{12} = (\mathrm{Tr}\, R^2 + k\, \mathrm{Tr}\, F^2)\, X_8 \tag{4.81}$$

where X_8 is an invariant 8-form constructed in terms of F and R. Straightforward, but somewhat tedious algebra shows that the factorization (4.81) occurs only if

$$\operatorname{Tr} F^6 = \frac{1}{48} \operatorname{Tr} F^2 \operatorname{Tr} F^4 - \frac{1}{14400} (\operatorname{Tr} F^2)^3 \tag{4.82}$$

$$k = \frac{-1}{30}$$

and

$$X_8 = \frac{1}{24} \operatorname{Tr} F^4 - \frac{1}{7200} (\operatorname{Tr} F^2)^2 - \frac{1}{240} \operatorname{Tr} F^2 \operatorname{Tr} R^2 + \frac{1}{8} \operatorname{Tr} R^4 + \frac{1}{32} (\operatorname{Tr} R^2)^2 . \tag{4.83}$$

Using (2.35) it is easy to show that (4.79) is satisfied by SO(32). Moreover, if we use (2.39) and the fact that the gluinos belong to the representation (248,1) + (1,248) of $E_8 \times E_8$ (so that $\operatorname{Tr}(F_1 + F_2)^6 = \operatorname{Tr} F_1^6 + \operatorname{Tr} F_2^6$, etc.). It can be checked that (4.79) also holds for $E_8 \times E_8$, and that these two groups have 496 generators. Once the factorization (4.81) takes place, then the counterterm that cancels the anomaly induced by I_{12} is simply:

$$S_c = \int [4(Q_{3L} - \frac{1}{30} Q_{3y})X_7 - 6 B X_8] \tag{4.84}$$

$$dX_7 = X_8 .$$

(X_8 is a closed form because it is an invariant polynomial.) For the conscientious reader who will check this computation, we should add one more technical detail: given an invariant polynomial of the form $\operatorname{Tr} F^n$, using the descent equations, it follows that Q^1_{2n-2} will be the anomaly. If, however, we have terms like $\operatorname{Tr} F^2 \operatorname{Tr} F^4$, when we compute its "Chern-Simons" form, we could have in general $\hat{Q}(\alpha,\beta) = \alpha Q^0_3 \operatorname{Tr} F^4 + \beta \operatorname{Tr} F^2 Q^0_7$ with $\alpha+\beta = 1$. Various choices of $\hat{Q}(\alpha,\beta)$ differ only by a total derivative, i.e. $\hat{Q}(\alpha,\beta) - \hat{Q}(\alpha',\beta') = d\Omega_{2n-2}$, so that one can go from the form of the anomaly given by $\hat{Q}(\alpha,\beta)$ to that induced by $\hat{Q}(\alpha',\beta')$ by simply varying a local counterterm. Thus which choice one makes of α and β is essentially irrelevant. There is however a canonical choice selected by Bose symmetry (of the external gluon or graviton lines in the Feynman diagrams contributing to the anomaly). Using Bose symmetry it is easy to show that the correct Chern-Simons form one would get for a term of the form $\operatorname{Tr} F^p \operatorname{Tr} F^q$ is $\alpha Q^0_{2p-1} \operatorname{Tr} F^q + \beta \operatorname{Tr} F^p Q^0_{2q-1}$ with $\alpha = p/(p+q)$, $\beta = q/(p+q)$. It is quite remarkable that in this context the anomaly cancellation pins down the gauge group to just two choices. This together with the possibility that superstring theories with these gauge groups are very likely finite in the ultraviolet (and therefore provide a consistent theory including quantum gravity non-trivially), make it very worthwhile trying to understand what makes these theories so special. (See Ref. 59 for some of the recent attempts to make contact between superstring theory and low-energy phenomenology.)

V. ANOMALIES IN σ-MODELS

There is an extension of the treatment of anomalies presented in Section III which applies to the coupling of fermions to arbitrary σ-model fields. These type of anomalies where introduced in Ref.8 where a thorough topological analysis is presented in terms of K-theory. The general conclusions concerning the topological nature of these anomalies are the same as those presented in Section III, but there are several interesting twists and applications that we would like to present. Although we will start by giving the general set-up for these anomalies, we will consider mostly applications to ordinary chiral models (G/H σ-models),[9] as well as some recent applications in extended supergravity theories and string theories.

A general bosonic σ-model is defined by a set of bosonic fields ϕ^i which geometrically represent maps from space time S^{2n} into some riemannian manifold M_d endowed with a metric $g_{ij}(\phi)$. The ϕ fields behave as coordinates on M_d with respect to coordinate reparametrizations of M_d. The standard minimal Lagrangian describing the dynamics of the ϕ fields is:

$$\mathscr{L} = \frac{1}{2} g_{ij}(\phi)\partial_\mu\phi^i\partial^\mu\phi^j. \tag{5.1}$$

Lagrangians like (5.1) are used for example to describe the low energy dynamics of the Goldstone bosons corresponding to the breaking of some symmetry group G to a subgroup H. In fact a large part of the formalism in this section is an elaboration of the classic papers of Coleman-Wess and Zumino[60] and Callan, Coleman, Wess, and Zumino.[61] [We will refer to these papers collectively as CCWZ). In the general case, the metric may or may not admit isometries. If we only have bosonic fields, (5.1) is usually the starting point for the analysis and there are no anomalies of any kind. If we want to couple fermions to the ϕ-field in a geometrically meaningful way, we have to introduce a connection one form ω_i, and its pull-back to space-time through the field ϕ: $\omega_\mu = \omega_i\partial_\mu\phi^i$. The connection ω_i is a one form with values in the Lie algebra of the holonomy group. The holonomy group of a connection ω for a connected manifold M_d, is defined as follows.[42] Given any point ϕ_0 on M_d, we consider all the closed loops at ϕ_0. Let γ be one of these loops. Given any vector v^a at ϕ_0, we can parallel transport v^a around γ with the connection ω. When one returns to ϕ_0, we get back v^a rotated by some matrix $m^a{}_b(\gamma)$ which only depends on the path γ and the connection ω, but not on v. Using the standard composition of paths, it follows that $H(\gamma_1{}^0\gamma_2) = H(\gamma_1)H(\gamma_2)$, and that $H(\gamma^{-1}) = H(\gamma)$ where γ^{-1} is the path γ traversed in the opposite direction. This shows that the set of matrices $H(\gamma)$ form a group; the holonomy group $\Psi(M,\omega,\phi_0)$ with connection ω and base point ϕ_0. If the manifold M is connected, the holonomy groups based at different ϕ_0's are all isomorphic (to any closed path at ϕ_0 we can associate a closed path at ϕ_0' by using the path joining ϕ_0 to ϕ_0'). We will simply use the notation $\Psi(M,\omega)$ for the holonomy group of the connection ω. It is sometimes useful to introduce the restricted holonomy group $\sigma(M,\omega) \subset \Psi(M,\omega)$ defined exactly on Ψ, but with the restriction that the path considered should be homotopically trivial. A simple consequence of the definition of Ψ, and the definition of curvature, is that the Lie algebra of $\sigma(M,\omega)$ is determined by the curvature R of the connection ω. If we consider an infinitesimal quasiparallelogram with sides $\delta\xi^i,\delta\zeta^j$, the infinitesimal element of the holonomy group is $R^a{}_{bij}\delta\xi^i\delta\zeta^j$. The holonomy group is useful because it summarizes most of the local geometric properties of a connection ω, and it provides a rather simple method of determining which objects are going to be covariantly constant. If $t^{a_1\ldots,a_n}$ is an invariant tensor with respect to the group $\Psi(M,\omega)$

$$H^{a_1}{}_{b_1}\cdots H^{a_n}{}_{b_n}\, t^{b_1\cdots b_n} = t^{a_1\cdots a_N} \tag{5.2}$$

we can define a covariantly constant tensor on M by simply choosing $t^{a_1\cdots a_N}(\phi_0) = t^{a_1\cdots a_N}$, and constructing $t^{a_1\cdots a_N}(\phi)$ by parallel transport from ϕ_0. The path used to parallel transport t from ϕ_0 to ϕ is irrelevant as a consequence of (5.2). The simplest example corresponds to a manifold $(M_{d,g})$ with the riemannian connection associated to the metric g. If we work in vielbein frames, the metric is represented by δ_{ab}. Since the metric is covariantly conserved with respect to the riemannian connection, we get that $H^TH=1$, for any $H \in \Psi\,(M,\omega)$, thus in this case $\Psi \subset SO(d)$ (we assume M_d to be orientable, otherwise $\Psi \subset O(d)$). Another interesting case appears when one considers Kähler manifolds.[42] A Kähler manifold $(M_{2d,g})$ is a riemannian manifold which is complex (i.e. it can be covered with a system of complex coordinates with holomorphic transition functions in the overlaps between coordinate patches), hermitean, and such that the complex structure is covariantly constant. In equations this means the following: A complex structure is determined by a tensor $J^i{}_j$ satisfying $J^i{}_kJ^k{}_j = -\delta^i{}_j$ (i.e. J defines a representation of $\sqrt{-1}$ on the tangent space) and such that the eigenvectors of J with eigenvalue +i define a holomorphic coordinate system; namely they can be written as $\partial/\partial Z^i$ $i=i,d$. The three conditions mentioned above can be expressed by:

$$J^i{}_kJ^k{}_j = -\delta^i{}_j \tag{5.3}$$

$$g_{ij}J^i{}_kJ^j{}_\ell = g_{k\ell}$$

$$D_kJ^i{}_j = 0 \tag{5.3}$$

where the covariant derivative is taken with respect to the riemannian connection. From the holonomy group point of view, $\Psi \to SO(2d)$ the condition (5.3) can be succinctly expressed by saying that $\Psi \subset U(d) \subset SO(2d)$. In vielbein frames the metric is simply the unit matrix $H^TH = 1$; and the complex structure satisfies $J^2 = -1$, $HJ = JH$ i.e. H is one of the U(d) matrices in the canonical embedding of U(d) in SO(2d). The reason for going at length into the theory of Kähler manifolds is that they provide some of the most interesting examples of the σ-model anomalies appearing frequently in low energy supersymmetric theories.[62] To conclude this brief interlude on holonomy groups it should be said that the possible holonomy groups of irreducible manifolds (manifolds whose holonomy group acts irreducibly on the tangent space) have been classified by M. Berger,[63] and the possible choices for non-homogeneous spaces is limited.

Going back to our main theme, once we are given a metric g_{ij} and a connection ω_i, the fermion fields are introduced as taking values on some vector bundle B over M with structure group given by the holonomy group H of the connection introduced in M. For minimally coupled fermions, the lagrangian is in 2n dimensions:

$$\mathscr{L} = \frac{1}{2} g_{ij}(\phi)\partial_\mu\phi^i\partial^\mu\phi^j + \delta_{AB}\bar{\Psi}^A\gamma^\mu P_L(D_\mu\Psi)^B + \delta_{A'B'}\bar{\Psi}^{A'}\gamma^\mu P_R(D_\mu\Psi)^{B'}$$

$$D_\mu\Psi^A = \partial_\mu\Psi^A + \omega_i^a\partial_\mu\phi^i(H^a)^A{}_B\Psi^B \tag{5.4}$$

where $(H^a)^A{}_B$ $a = 1, \dim H$ are the antihermitian generators of the holonomy group in the representation carried by the left-handed fermions (the A', B' indices correspond to right-handed fermions). One can also add non-minimal terms to (5.4) which need not concern us at this moment. In analogy with the gauge or gravitational cases treated in Section III, for each bosonic configuration $\phi: S^{2n} \to M_d$, we can integrate the fermions and define an effective action:

$$e^{-\Gamma_{eff}[\phi]} . \tag{5.5}$$

As before, defining this determinant requires a careful analysis of the regularization procedure because one has to compare two Weyl operators at two different configurations ϕ_0 and ϕ and this always implies some ambiguity (indeed, this ambiguity is the basic reason for the anomaly).

We may ask now what does it mean to have an anomaly in (5.5)? As pointed out in Ref.8, in these general circumstances there is no notion of $\mathcal{A}$ and $\mathcal{G}_0$ separately, one is given at first sight only the analog of $\mathcal{A}/\mathcal{G}_0$ represented by the space of maps $\mathcal{C} = \{\phi : S^{2n} \to M_d\}$. A more detailed analysis however,[8,9] shows that this is not so. One wants to end up with a theory where the various objects have an intrinsic geometrical meaning. Since the fermion fields take values on a bundle B over M (or rather its pull-back over space-time through the bosonic configuration $\phi(x)$), they will have intrinsic geometrical meaning if the effective action is invariant under bundle redefinitions. Since the bundle group is the holonomy group H, a bundle redefinition will be a map from M_d to H; i.e. an H-gauge transformation. We can express essentially the same fact by noticing that for non-trivial M_d, one needs several coordinate patches to cover M_d. In setting up the bundle B, one has to prescribe the bundle transition functions whenever two patches U_α, U_β overlap. These transition functions $h_{\alpha\beta}$ are H-gauge transformations, and therefore studying whether (5.5) has an intrinsic geometrical meaning is equivalent to computing the change of $\exp - \Gamma[\phi]$ in going from one patch to another. Since going from one patch to the other is the same as making an H-gauge transformation we can immediately conclude using the arguments of Section III that (5.5) will be anomalous whenever the fermions belong to anomalous representations of the holonomy group H. This can be exhibited more intrinsically by again analyzing the behavior of $\exp - \Gamma[\phi]$ as we move ϕ over $\mathcal{C}$. This brings us back to classifying line bundles over $\mathcal{C}$, identifying the line bundle over $\mathcal{C}$ defined by $\exp - \Gamma[\phi]$ and computing its first Chern class using the index theorem for families of Dirac operators (see Section III). This procedure identifies again a 2n+2 index theorem characterizing the triviality or non-triviality of the first Chern class of the index bundle defined by the fermions (5.4-5). The conclusions are the same as those one would obtain by extrapolating the analysis performed for the gauge case as indicated above. These two methods provide local and global ways of exhibiting the anomaly. If $\exp - \Gamma[\phi_\alpha]$ $\alpha = 1,2$ represents the value of (5.5) for a configuration ϕ in the intersection of two coordinate patches, and if $h(\phi(x))$ is the pull-back of the H-transition function, $\exp - \Gamma[\phi_1]$ is related to $\exp - \Gamma[\phi_2]$ by the analog of the Wess-Zumino action. If $\tilde{h}(t, \phi(x))$ is a one parameter family of H-transformations interpolating between h=1 and h (assuming we have taken a "good" cover of M_d,[8,9] so that we do not have to worry about a possible topological non-triviality of $h(\phi(x))$, then

$$e^{-(\Gamma[\phi_1]-\Gamma[\phi_2])} = e^{i\int_{B_{2n+1}}\Omega^0_{2n+1}(\tilde{h},\omega)} \tag{5.6}$$

with notation borrowed from Section IV.

So far we have maintained a very general discussion of the phenomenon of σ-model anomalies. These anomalies can be exhibited in the physically interesting case of fermions coupled to a Goldstone boson manifold describing the low energy sector of a theory with a global symmetry group G broken down to some subgroup H.[8,9] In this case the more meaningful question to ask is whether the Ward identities manifesting the low energy consequences of the global symmetry G are maintained after one loop corrections induced by chiral fermions are taken into account. If the possible anomalies induced by the fermions cannot be removed, then there would be no consistent way of coupling such fermion fields so as to preserve the G/H current algebra predictions. Asking that $\Gamma[\phi]$ (or some modifications to be defined below) be well defined under the action of G-isometries automatically eliminates the anomalies of a more general nature considered before in the case of G/H spaces. The conditions for the absence of anomalies under global G-transformations is a bit more subtle than in the gauge case, because the Goldstone boson fields can be used to construct counterterms which can cancel the anomalies induced by the fermions. In order to formulate the G-anomalies, and their cancellation in detail, we briefly review the CCWZ formalism in a way which makes its geometrical significance more apparent. This formalism has been used very recently in the discussion of Kaluza-Klein compactifications.[6]

Given a group G and a subgroup H, let T_A be the complete set of generators of G, H_i the generators of H, and X_a the broken generators of G, all of them taken to be antihermitian. If H is compact, the Lie algebra of G admits a reductive splitting.[64] This means that commutation relations can be written in the form:

$$\begin{aligned} [H_i,H_j] &= f_{ij}{}^{k}H_k \\ [H_i,X_a] &= f_{ia}{}^{b}X_b \\ [X_a,X_b] &= f_{ab}{}^{i}H_i + f_{ab}{}^{c}X_c . \end{aligned} \tag{5.7}$$

A homogeneous space G/H is called symmetric if it is possible to arrange $f_{ab}{}^{c} = 0$ (a simple example is given by the spheres $S^n = SO(n+1)/SO(n)$). In the local formulation of G/H, we start by choosing in each coordinate patch a coset representative $\ell(\phi) \in G$ for each class in G/H. Since dim(G/H) = dimG - dimH, there are as many coordinates ϕ as broken generators in G. This is equivalent to choosing a section of the H-bundle with total space G and base G/H. The left action of G on $\ell(\phi)$ is

$$g\ell(\phi) = \ell(\phi')h(\phi,g) \ . \qquad (5.8)$$

$h(\phi,g)$ is an element of H, usually called the compensating H-transformation, which is needed to bring $g\ell(\phi)$ back to our choice of coset representatives. This determines both ϕ' and $h(\phi,g)$. Under infinitesimal g transformations:

$$\begin{aligned} g &\approx 1 + \varepsilon^A T_A \\ h(\phi,g) &\approx 1 + \varepsilon^A \Omega_A^i(\phi) H_i \qquad (5.9) \\ \phi'^\alpha &= \phi^\alpha + \varepsilon^A k_A^\alpha(\phi), \ \alpha = 1, \dim(G/H) = N. \end{aligned}$$

The $k_A^\alpha(\phi)$'s define a vector field on G/H for each A. They are by definiton the Killing vectors of the left G action on G/H. Substituting (5.9) in (5.8) we get

$$\begin{aligned} k_A^\alpha \frac{\partial}{\partial\phi^\alpha} \ell(\phi) &= T_A \ell(\phi) - \ell(\phi)\Omega_A(\phi) \\ \Omega_A &\equiv \Omega_A^i H_i, \qquad (5.10) \end{aligned}$$

The vielbein and connection on G/H follow from the CCWZ prescription: Decompose the one-form $\bar{\ell}^1(\phi)d\ell(\phi)$ $(d = d\phi^\alpha\ \partial/\partial\phi^\alpha)$ along X and H

$$\bar{\ell}^1 d\ell = (\omega^i_{\ \alpha}(\phi)H_i + e^a_\alpha(\phi)X_a)d\phi^\alpha = \omega^i H_i + e^a X_a. \qquad (5.11)$$

$e_i^a(\phi)$ represents the vielbein and ω^i is an H-connection (the canonical connection) as we will see momentarily. We can give explicit formuli for k_A and Ω_A in terms of e,ω and the elements $D_A^{\ B}(g)$ of the adjoint representation of G:

$$g^{-1} T_A g = D_A^{\ B}(g) T_B, \qquad (5.12)$$

multiplying (5.10) by $\bar{\ell}^1$ and using (5.11), (5.12) we obtain

$$k_A^\alpha(e_\alpha^a X_a + \omega_\alpha^i H_i) = D_A^{\ B}(\ell) T_B - \Omega_A \qquad (5.13)$$

or

$$\begin{aligned} k_A^\alpha e_\alpha^a(\phi) &= D_A^{\ a}(\ell) \\ \Omega_A^i &= D_A^i(\ell) - k_A^\alpha(\phi)\omega_\alpha^i. \qquad (5.14) \end{aligned}$$

Finally, we need to know the transformation properties of e^a and ω^i under isometries. This can be done by using directly the formula for the Lie derivative $\mathscr{L}(k_A) = i(k_A)d + di(K_A)$, or by direct substitution of (5.8) in (5.11):

$$\begin{aligned} \ell(\phi) &\to \ell(\phi') = g\ell(\phi)h^{-1}(\phi,g) \\ e \equiv e^a X_a &\to h\, e\, h^{-1} \qquad (5.15) \\ \omega \equiv \omega^i H_i &\to h(\omega + d)h^{-1}, \end{aligned}$$

which in the infinitesimal version reads:

$$\mathscr{L}(\vec{k}_A)e = [\Omega_A, e]$$

$$\mathscr{L}(\vec{k}_A)\omega = -(d\Omega_A + [\omega, \Omega_A]). \tag{5.16}$$

Thus the isometries act on e,ω as H-gauge transformation (e transforms like a field in the adjoint representation, and ω like a H gauge connection). The bosonic kinetic term for the φ-fields is now (5.1) with

$$g_{\alpha\beta}(\phi) = f_\pi \, \mathrm{Tr}\,(e_\alpha e_\beta^\dagger), \tag{5.17}$$

making (5.1) invariant under G-transformations, and f_π is the analog of the pion decay constant. The fermionic matter is chosen to transform according to some representation of H. If $\mathscr{H}_i$ denotes the generators of H in this representation, the fermion lagrangian is (for left-handed fermions)

$$\mathscr{L} = \bar{\psi}\, i\gamma^\mu(\partial_\mu + \omega_\mu)P_+\psi$$

$$\omega_\mu = \omega^i_\alpha \partial_\mu \phi^\alpha \mathscr{H}_i. \tag{5.18}$$

$\mathscr{L}$ is invariant under the global G-transformations (5.15) if Ψ transforms according to $\Psi \to h\Psi$, where h is the compensating H-transformation h(φ,g) in the representation $\mathscr{H}_i$. Depending on the representation $\mathscr{H}_i$ one may be able to add extra Yukawa couplings to the action. Using Neother's theorem, the global G-currents are

$$J^\mu_A = f_\pi g_{\alpha\beta} k^\beta_A \partial^\mu \phi^\alpha + i\bar{\psi}\gamma^\mu \mathscr{H}_i P_+ (\omega^i_\alpha k^\alpha_A(\phi) + \Omega^i_A(\phi)), \tag{5.19}$$

and the naive Ward identities follow from $\partial_\mu J^\mu{}_A = 0$. In some cases, G is not the largest isometry group of G/H. If the normalizer of H in G, $N(H) = \{g \in G, \text{ such that } gHg^{-1} \subset H\}$ is larger than H, there are also isometries associated to the right action of N(H)/H on G/H. N(H)/H is a group because H is a normal subgroup of N(H). The elements of H do not generate non-trivial right isometries because they only redefine the coset representatives. Then for an element $g \in N(H)$ we have:

$$\ell(\phi)g = \ell(\phi')\hat{h}(\phi,g). \tag{5.20}$$

This quantity is well defined, because for some other choice of the φ-coset $\ell(\phi)h'$, $\ell(\phi)h'g = \ell(\phi)gh'' = \ell(\phi')\hat{h}h''$ because for $g \in N(H)$, $h'g = gh''$, $h', h'' \in H$. For the infinitesimal right action:

$$\hat{k}^\alpha_A \frac{\partial}{\partial\phi^\alpha}\ell(\phi) = \ell(\phi)\hat{T}_A - \ell(\phi)\hat{\Omega}_A(\phi). \tag{5.21}$$

($\hat{T}_A$ labels the generators of G in N(H) but not in H). Multiplying by ℓ^{-1} and keeping in mind that $\hat{\Omega}^i{}_j = 0$, we obtain

$$\hat{k}^{\alpha}_{a} = e^{\alpha}_{a} \qquad \hat{k}^{\alpha}_{i} = 0$$

$$\hat{\Omega}^{i}{}_{a} = \omega^{i}{}_{a} e^{\alpha}{}_{a} \qquad \hat{\Omega}^{i}{}_{j} = 0. \tag{5.22}$$

Since H is a normal subgroup of N(H), the Lie algebra of N(H) splits into two pieces whose generators we can label H, K with $[H,H] \subset H$, $[K,K] \subset K$, $[K,H] = 0$. Then the rest of the broken generators, L, can be chosen so that $[K,L] \subset L$. For infinitesimal right transformations $g \approx 1 + \varepsilon^{A}K_{A}$, we get:

$$\mathscr{L}(\hat{k}_{A})\omega = -(d\hat{\Omega}_{A} + [\omega,\hat{\Omega}_{A}])$$

$$\mathscr{L}(\hat{k}_{A})e = -[K,e] + [\Omega_{A},e]. \tag{5.23}$$

The fermionic lagrangian is invariant under N(H)/H transformations because:

$$\ell(\phi) \to \ell(\phi)\, g\hat{h}^{-1}(\phi,g)$$

$$\omega \to \hat{h}(\omega+d)\hat{h}^{-1}$$

$$\Psi \to \hat{h}\Psi, \tag{5.24}$$

leaving (5.18) invariant. A good example of G/H with nontrivial N(H) is the squashed seven sphere SO(5)/SO(3),[6] where the SO(3) is embedded in one of the SO(3)'s of the SO(3) x SO(3) subgroup of SO(5). Then N(H)/H = SO(3), and the isometry group is SO(5) x SO(3).

Now we can consider the "isometry anomalies" induced by the fermion determinant. Given some bosonic configuration $\ell(\phi)$: $S^{2n} \to G/H$ (which we may take in the trivial homotopy class for the moment) as a background for the fermions, we can perform the fermionic functional integral which depends explicitly only on ω_{μ}:

$$e^{-\Gamma[\omega]} = \int d\psi d\bar{\psi}\; e^{-\int dx\,\bar{\psi}(i\gamma^{\mu}\,\partial_{\mu}+\omega_{\mu})P_{+}\Psi} \tag{5.23}$$

If the fermions belong to a complex representation of H, one cannot regulate (5.25) in a gauge invariant way under H-gauge transformations, and any regulator will inevitably break H-gauge invariance. According to (5.15,16) the connection ω_{μ} transforms like an H-gauge field under an infinitesimal G-isometry with gauge parameter $\varepsilon^{A}\Omega_{A}$. Translating now the formulae of Sections III-IV to this case we get

$$\delta_{\varepsilon}\Gamma[\omega] = \int_{S^{2n}} Q^{1}_{2n}(\varepsilon^{A}\Omega_{A},\omega). \tag{5.26}$$

(Ω_{A},ω takes values in the matrices $\mathscr{H}_{i}$ of the fermion representation of H). The anomalous change (5.26) results from the local change in the fiber basis (given by our choice of coset representatives) induced by the G-action, because the symmetry under such local frame rotations is spoiled by a quantum anomaly. This has the effect of breaking the invariance of $\exp-\Gamma[\phi]$ under the action of G. This anomaly is different from anomalies in currents coupled to gauge fields. Its existence does not render the theory inconsistent from a quantum mechanical point of view. It means that we are not realizing the G symmetry in a way that makes the ϕ-fields behave like Goldstone bosons. The H symmetry of course is still manifest because no compensating gauge transformation is required. If the fermions transform under an anomaly free representation of H, (5.26) vanishes and the usual Ward identities can be implemented as usual. This is however not the only way of cancelling the anomalies. We can construct counterterms which under certain conditions, cancel the anomaly (5.26). Adding such counterterms make the combined Fermi-Bose theory free of G-anomalies. The effect of these counterterms in the current algebra of the model may be quite important, as exemplified by the inclusion of the Wess-Zumino term in the standard $SU(N)_L \times SU(N)_R/SU(N)_V$ current algebra[27,28] which affects the low energy phenomenology of the pseudoscalar octet. A trivial case where the anomaly always vanishes corresponds to G/H with dim G/H < dim S^{2n}. This is the case of the CP^1 supersymmetric σ-model in four (or higher) dimensions.

The natural counterterm suggested by (5.26) consists of constructing a Wess-Zumino lagrangian with the coset representative $\ell(\phi)$ and the Cartan connection ω. To do this, imagine that $\ell(\phi)$ maps S^{2n} into a single patch of M_d, (the more complicated case where several patches are involved, will be considered later on, and will show that the topological and local analysis in this case are basically identical); and take ℓ and ω in some representation of G. In analogy with Section IV, we can define an interpolation ℓ_t t=0 ℓ=1, t=1 $\ell_1 = \ell(\phi)$, and construct $\tilde{\omega}^{\ell_t^{-1}} = \ell_t(\tilde{\omega}+d)\ell_t^{-1}$. (The twiddle indicates that ω takes values in the chosen representation of G). Then

$$W(\ell,\tilde{\omega}) = -\int_{0(t)}^{1}\int_{S^{2n}} Q^1_{2n}(\ell_t d_t \ell_t^{-1}, \tilde{\omega}^{\ell_t^{-1}}), \tag{5.27}$$

which is the same as the integrated anomaly in (4.31), under a g transformation:

$$\tilde{\omega}^{\ell_t^{-1}} \to (\tilde{\omega}^{\bar{h}^1})^{h\ell_t^{-1}\bar{g}^1} = (\tilde{\omega}^{\bar{\ell}^1_t})^{\bar{g}^1}, \tag{5.28}$$

and using the Equation (4.38) the variation of $W(\ell,\tilde{\omega})$ gives

$$\delta_\varepsilon W(\ell,\tilde{\omega}) = \int_{S^{2n}} Q^1_{2n}(\varepsilon^A \tilde{\Omega}_A, \tilde{\omega}). \tag{5.29}$$

Since (5.29) is proportional to the anomaly in the chosen representation of G: M^A, if we restrict A to run only over H, we get the condition for (5.29) to cancel the anomalous variation of the fermion effective action:

$$\mathrm{S\,Tr}M^{i_1}\cdots M^{i_{n+1}} = \mathrm{S\,Tr}\mathscr{H}^{i_1}\cdots\mathscr{H}^{i_{n+1}} \tag{5.30}$$

in other words, the G-anomaly induced by (5.29) must match the $\mathscr{H}$-anomaly induced by the fermions. If this condition is met, adding $W(\ell,\tilde{\omega})$ to the action (with an $\hbar$ coefficient) makes the theory G-invariant. A simple example is given by the grassmannian manifolds $G/H = U(p+q)/U(p) \times U(q)$. The representation $(p,1) \oplus (1,q)$ of $U(p) \times U(q)$ is allowed as long as the relative "strength" of the U(p), U(q) charges are identical. The anomalies then match those of the (p+q) representation of U(p+q) when restricted to the subgroup $U(p) \times U(q)$. Another example is provided by $CP^N = SU(N+1)/SU(N) \times U(1)$. The representation N of SU(N) with unit charge under U(1) is also allowed (i.e. there is a counterterm (5.27) cancelling the anomaly. In this case we simply take the fundamental of SU(N+1) in (5.27)). Note that in this example, the fermions would not combine to form a representation of G = SU(N+1).

As pointed out in the introduction, these anomalies in G/H are reminiscent of the 't Hooft anomaly matching conditions.[7] The G/H σ-model may be considered as the effective low energy theory describing the massless modes which result from the dynamical breaking of G in an underlying higher energy theory with a flavor G symmetry. The fermions Ψ transforming under the unbroken group H are those fermions which remain in the low energy theory, kept massless presumably by some unbroken chiral symmetry. In the absence of massless fermions, a flavor G-anomaly of the underlying theory can be reproduced only in the low energy theory by Goldstone bosons. The underlying fermions would then have to have been anomaly free under the subgroup H, since unbroken currents do not produce Goldstone bosons from the vacuum and hence cannot give any contributions to the low energy G-anomaly to match that of the underlying G-symmetry. If there are remaining massless fermions, then they may transform anomalously under H as long as their anomalies match those of the underlying $H \subset G$. The non-trivial result is that massless chiral fermions can be consistently coupled to a G/H σ-model only when they could have resulted from such a dynamical breakdown. The exception to this interpretation occurs when G is non-compact, then there is no underlying theory with a linearly realized G-symmetry, even though σ-models like $SU(m,n)/SU(m) \times SU(n) \times U(1)$ can be well defined as long as the H-representations are again chosen to have an anomaly matching that of some representation of G when restricted to the H generators.

Another interesting application of these anomalies happens in the context of supersymmetric preon models. It is known [72] that a supersymmetric σ-model in four dimensions requires the manifold M to be Kähler. If we consider locally symmetric spaces then G/H is Kähler if G and H have the same rank, and H contains at least a U(1) factor. Examples are

$$SU(p+q)/Su(p)\times SU(q)\times U(1),\quad SU(N+1)/SU(N)\times U(1)$$
$$E_6/SO(10)\times U(1),\quad E_7/SU(5)\times SU(3)\times U(1) \tag{5.31}$$
$$E_8/SO(10)\times SU(3)\times U(1)$$

supersymmetric models based on these spaces are all anomalous under the U(1) part of H, in a way that cannot be cancelled by adding counterterms. The reason is that the fermion content of the theory is fixed by supersymmetry to lie on the holomorphic tangent bundle of G/H. For instance, in the CP^N case, we obtain the transformation properties of the fermions by decomposing the adjoint of SU(N+1) with respect to SU(N) x U(1); getting

$$(N+1)^2-1 = (N^2-1)_0 + N_{N+1} + \bar{N}_{-N-1} + 1_0 \tag{5.32}$$

with the lower subscripts indicating the U(1) charges. Equation (5.32) indicates that the fermions transform like the N of SU(N) but with charge (N+1) under the U(1), and this differs from the U(1) assignment of the (N+1) representation of SU(N+1) under SU(N) x U(1). For the exceptional group cases in (5.31) the argument is even easier, because E_6, E_7, E_8 are anomaly free in four dimensions, and for example the fermions of $E_6/SO(10)\times U(1)$ transform like a 16 of SO(10) with charge -3 under U(1), hence the U(1) anomaly cannot be cancelled.

We showed before that in these cases where N(H) ≠ H, there are additional isometries. The question now is whether the counterterm (5.27) is enough to cancel the anomalies under N(H)/H transformations in the effective action:

$$\delta_\varepsilon\Gamma[\omega] = \int_{S^{2n}} Q'_{2n}(\varepsilon^A\hat{\Omega}_A,\omega). \tag{5.33}$$

Fortunately, (5.33) is cancelled by the variation of (5.27) under a N(H)/H transformation, as a consequence of

$$\tilde{\omega}^{\bar{\ell}^1_t} \to (\tilde{\omega}^{\hat{h}^{-1}})^{\hat{h}g^{-1}\bar{\ell}^1_t} = \tilde{\omega}^{\bar{\ell}^1_t}$$

and thus $\delta W[\ell,\omega]$ cancels (5.33) if (5.30) is satisfied.

Finally we have to check that the addition of (5.27) to $\Gamma[\omega]$ makes the theory well defined as we move from patch to patch is the space of bosonic configurations $\{\phi: S^{2n} \to G/H\}$. When G/H is topologically non-trivial, we used several patches to cover G/H, $\{U_\alpha, \ell_\alpha\}$ where ℓ_α gives the coset representative in the patch U_α. For points in $U_\alpha \cap U_\beta$, ℓ_α, ℓ_β are related by a right H transformation: $\ell_\beta = \ell_\alpha h_{\alpha\beta}$. When we look at a map from S^{2n} into G/H, it will in general hit several patches, and we have to compute the effective action in each patch $\exp-\Gamma_\alpha[\phi]$; and $\exp-\Gamma_\alpha$, $\exp-\Gamma_\beta$ must agree on the overlaps. Since going from U_α to U_β requires making a H-gauge transformation $h_{\alpha\beta}(\phi(x))$ on the fermions as well:

$$\Psi_\beta(x): \quad h^{-1}_{\alpha\beta}(\phi(x))\Psi_\alpha(x)$$

we get

$$\Gamma_\alpha[\phi] - \Gamma_\beta[\phi] = \int_I \int_{S^{2n}} Q^1_{2n}(h^{-1}_{\alpha\beta}(t) d_t h_{\alpha\beta}, \omega^{h_{\alpha\beta}(t)}). \tag{5.34}$$

$h_{\alpha\beta}(t)$ interpolates between 1 and $h_{\alpha\beta}(\phi(\alpha))$ in $U_\alpha \cap U_\beta$. This, however, is cancelled by the counterterm, because in the same way as $\Gamma[\phi]$, we have to define the counterterm patchwise. Thus for each α we have $W_\alpha[\omega, \ell_\alpha]$. Since W_α is a Wess-Zumino term, we know from the previous section, that under a finite gauge transformation it changes in exactly the same way as the fermionic effective action, (5.34). This if (5.30) is satisfied, the combination $\exp(-\Gamma_\alpha[\phi] - W_\alpha[\omega_\alpha, \ell_\alpha])$ is globally well-defined. This also proves that the anomaly from the global point of view is going to be captured by an index theorem for a two parameter family of Dirac operators. Rather than producing the appropriate changes of the arguments of Section III in order to apply them to this case, we leave it as an exercise to the reader, or refer to Ref. 8 for further details.

An interesting application of σ-model anomalies appears in N=8 supergravity where the seventy scalars in the supergravity multiplet are in the coset space $E_{7,7}/SU(8)$. Since the 8 Weyl gravitinos transform like the 8 of SU(8), and the spin 1/2 fermions like the 56 of SU(8), one might be tempted to conclude that the theory has a σ-model anomaly.[65] A more careful analysis[66] indicates that when the effects of the 28 vector fields in the N=8 multiplet are properly taken into account, the theory is free of SU(8) anomalies. The anomaly cancellation in this case is obtained by combining the 28 gauge field strengths $F^{IJ}_{\mu\nu}$, I,J =1,8 with their dual: $F^{IJ} - i{*F}^{IJ}$ to generate a 28 of SU(8) (the 28 F^{IJ} transform only in the adjoint of SO(8)). Details can be found in Ref. 66.

As a final comment, σ-model anomalies have been used in Ref. 67 to analyze general N = 1/2 supersymmetric σ-models in two dimensions. These field theories are interesting because they underlie the formulation of the heterotic string.[58] What Hull and Witten show is that the transformations required by Green and Schwarz for the field b_{ij} in order to cancel the anomalies in the ten dimensional theory (see Section III), can be discovered using the σ-model anomalies in the analysis of the N = 1/2 supersymmetric theories in two dimensions (the string world sheet).

If one wants to analyze the anomalies with respect to the holonomy group for an arbitrary σ-model, the topological and local analysis remains unchanged, and the only question that remains is to find general criteria for the cancellation of these anomalies. Although some criteria can be formulated [9] which generalize the notion of the Wess-Zumino counterterm used in this section, we have not found simple examples different from the examples considered in this section; where such generalized counterterms could be exhibited.

VI. THE HAMILTONIAN APPROACH

In the preceeding sections we have concentrated on the euclidean functional integral approach to the fermion determinant and its gauge variation. We want to reinterpret now our results from the hamiltonian point of view. Since there are already several good accounts on the algebraic approach to the hamiltonian form of the anomalies[29,30], we will mainly concentrate on the topological analysis[31], and sketch the algebraic approach based on the theory of cocycles and ray representations.

For simplicity, we will carry out the discussion in the four dimensional case even though the formalism can be extended to higher dimensions and to include gravitational anomalies. As before, we start by considering a Weyl fermion in the presence of an arbitrary external gauge field. In order to present the hamiltonian formulation in the clearest way, we choose the $A^0=0$ gauge, so that the hamiltonian looks like the euclidean action for a three dimensional gauge theory:

$$H[A] = \int d^3x\, \lambda^+ \, i\sigma^i D_i(A)\lambda \tag{6.1}$$

$$i = 1,2,3, \qquad D_i = \partial_i + A_i .$$

σ^i $i=1,2,3$ are the Pauli matrices. In this case (6.1) is self-adjoint, and λ behaves like an SU(2) spinor under three dimensional rotations, so that the issues regarding self-adjointness of the euclidean action for a complex Weyl fermion are absent. For fixed A, (6.1) describes a perfectly well-defined quantum theory. In first quantization, we solve the three-dimensional Dirac operator $\sigma\cdot D$, and divide the spectrum in positive and negative energy states:

$$\sigma\cdot D(A)\phi_E = E\phi_E \qquad E>0$$

$$\sigma\cdot D(A)\psi_E = E\psi_E \qquad E<0, \tag{6.2}$$

as usual we associate annihilation operators to the positive energy states, and creation operators to the negative energy states, so that

$$\lambda(x) = \sum_{E>0} a_E\phi_E + \sum_{e<0} b_E^+\psi_E \tag{6.3}$$

(we are forgetting zero modes for the moment). The operators a_E, b_E do depend on the gauge field A. Since (6.2) is covariant under three dimensional gauge transformations, we can in principle define covariant transformation rules for a_E, b_E under gauge transformations. Let $\mathcal{H}[A]$ be the Hilbert space generated by the wave functions ϕ_E, ψ_E, and $\mathcal{H}^+[A]$, $\mathcal{H}^-[A]$ the subspaces of $\mathcal{H}[A]$ generated respectively by ϕ_E, ψ_E: $\mathcal{H}[A] = \mathcal{H}^+[A] \oplus \mathcal{H}^-[A]$. The second quantization of this system requires that we construct the Fock space generated by a_E, b_E. This necessitates the definition of the Fock vacuum $|0,A\rangle$ for each gauge field configuration A. Using Dirac's old methods, we can define the Fock vacuum by filling in all the negative energy states. Once $|0\rangle$ is defined, we can construct the rest of Fock space by acting on $|0\rangle$ with arbitrary polynomials in a_E^+, b_E^+. This way, we have a perfectly well defined quantum theory for fixed A. The important question is what happens when we consider this quantum theory as a function of A. In other words, in the hamiltonian formulation we have a Hilbert bundle $\mathcal{H}[A]$ over the space $\mathfrak{A}^{(3)}$, or more precisely, on $\mathfrak{A}^{(3)'}$, where $\mathfrak{A}^{(3)'}$ is $\mathfrak{A}^{(3)}$ minus the variety $V \subset \mathfrak{A}^{(3)}$ for which $\sigma\cdot D(A)$ has zero modes. For each such A, we can construct fermionic wave functionals $\Psi[A]$ which would have components in the various multiparticle levels in Fock space. A gauge transformation therefore corresponds to a map from $\mathcal{H}[A]$ into $\mathcal{H}[A^g]$, $A^g = g^{-1}(A+d)g$, and g is a map from S^3 (compactified space) into the gauge group G: $U_g: \mathcal{H}[A] \to \mathcal{H}[A^g]$. Since the "first quantized" theory (6.2) has no problems, we can define the gauge transforms of $a_E(A)$ and $b_E(A)$ by $U_g a_E(A) U_g^+ = a_E(A^g)$ and $U_g b_E(A) U_g^+ = b_E(A^g)$. Since any state in Fock space can be constructed in terms of $|0\rangle$, a_E^+, b_E^+, gauge covariance will be achieved if we can construct the Fock vacuum $|0,A\rangle$ consistent with Gauss' law. Topologically, the Fock vacuum is again a line bundle over $\mathfrak{A}^{(3)'}$. We are trying to see whether this line bundle is trivial, and whether we can reduce the vacuum line bundle from $\mathfrak{A}^{(3)}$ to $\mathfrak{A}^{(3)}/\mathcal{G}_0$ in a way consistent with Gauss' law. For any wave functional, and in particular for the Fock vacuum, this means that

$$U_g\Psi[A] = \Psi[A^g] \tag{6.4}$$

$$U_{g_1} U_{g_2} = U_{g_1 g_2} \tag{6.5}$$

in infinitesimal form, (6.4) means that

$$Q^a \Psi = \mathrm{Tr} T^a D_i \left(\frac{\delta}{\delta A_i} - i\lambda^+ T^a \lambda \right) \Psi = 0 \tag{6.6}$$

and (6.5) in infinitesimal form implies that

$$[Q_u, Q_v] = Q_{[u,v]} \tag{6.7}$$

$$Q_u = \int d_x^3 \, u^a(x) Q^a(x). \tag{6.8}$$

Equation (6.7) is the local integrability condition for (6.6). We can re-express (6.4-5) by saying that the space of physical functionals provide a faithful representation of the gauge group $\mathfrak{g}_0$. For anomalous theories we will find that (6.4) and (6.5) get extra phases i.e. we get a ray representation, and that the phase which changes the right hand side of (6.5) is given by $Q_3^2(u,v,A)$ (see (4.21-22)). Using the families index theorem we will show that

$$U_{g_1} U_{g_2} = \exp i Q_3^2(g_1, g_2 A) \; U_{g_1 g_2}.$$

This translates into a Schwinger term in the commutator of (6.7) which is the hamiltonian representation for the anomaly. In recent work, Faddeev and coworkers[29] showed that the descent equations of Stora and Zumino can be reinterpreted from the point of view of the cocycles that characterize the ray representations of the gauge group. This provides an algebraic approach to the results we will obtain using topological methods. Some recent work along the lines of Faddeev's ideas can be found in Ref. [68]. Let us quickly review some definitions concerning the theory of projective representations.

One starts with a space M with points $a \in M$, and a group G acting on M. The action of $g \in G$ on $a \in M$ will be represented by ag. Given a function of n+1 arguments $\omega_n(a; g_1, \ldots, g_n)$, one can introduce the boundary operator as follows:

$$(\delta\omega_n)(a; g_1 \ldots, g_{n+1}) = \omega_n(ag_1; g_2, \ldots, g_{n+1})$$
$$- \omega_n(a; g_1 g_2, g_3, \ldots, g_{n+1}) + \cdots + (-1)^i \omega_n(a; \ldots, g_i g_{i+1} \ldots, g_{n+1})$$
$$\cdots + (-1)^{n+1} \omega_n(a; g_1, \ldots, g_n). \tag{6.9}$$

It is easy to check that $\delta^2 = 0$, and we can define cohomology groups in terms of the operator δ. The functions ω_n are called cochains. If $\omega_n = \delta\Lambda_{n-1}$, one says that ω_n is a coboundary. On the other hand, if $\delta\omega_n = 0$, we say that ω_n is a cocycle. We can define the cohomology groups associated with δ: H_n = {n-cocycles modulo n-coboundaries}. (These generalizations of the cohomology groups allow for the definition of cohomology even for finite groups). 1- and 2-cocycles play an important role in the theory of representations of the group G. Thus 1-cocycles may appear in the representations of G in the spaces of functions over M:

$$U(g)f(a) = \exp i\omega_1(a;g)f(ag) \tag{6.10}$$

it follows that $U(g_1)U(g_2) = U(g_1g_2)$ if $\delta\omega_1 = 0$. The 2-cocycles independent of "a" characterize the projective representations $U(g_1)U(g_2) = \exp i\alpha_2(g_1,g_2)$ $U(g_1,g_2)$. Associativity holds if $\delta\alpha_2 = 0$. The case we are interested in is a bit more general, we have a set of functions on M (M is $\mathfrak{A}$) valued in some vector space (fermionic Fock space), so that the group action is

$$U(g)f(a) = V(a;g)f(ag), \tag{6.11}$$

and the U-matrices satisfy:

$$V(a;g_1)V(a;g_2) = \exp i\omega_2(a;g_1,g_2)V(a;g_1\,g_2) \tag{6.12}$$

so that

$$U(g_1)U(g_2)f(a) = \exp i\omega_2(a\,;\,g_1g_2)U(g_1g_2)f(a) \tag{6.13}$$

$$\delta\omega_2 = 0.$$

Using the descent Equations (4.20-22), Faddeev then proceeds to identify Q_3^2 with the anomalous commutator of the gauge charges. Let us now show that topological arguments lead naturally to identifying Q_3^2 as the 2-cocycle in (6.13). The Stora-Zumino descent Equations (4.101) then guarantee that ω_2 found this way satisfied $\delta\omega_2 = 0$. Since the arguments use the families index theorem and are very similar to those presented in Section III, we will simply sketch the main steps in the derivation.

We have shown before that the basic issue with gauge invariance of the infinite dimensional family of fermionic Fock spaces parametrized by $\mathfrak{A}^{(3)}$ is concerned with the possibility of defining a Fock vacuum $|0,A\rangle$ which can be smoothly restricted from $\mathfrak{A}^{(3)}$ to $\mathfrak{A}^{(3)}/\mathfrak{G}_0^{(3)}$. To analyze this issue, we can take a particular point $A \in \mathfrak{A}^{(3)}$ such that $i\sigma\cdot D(A)$ has no zero modes so that $|0,A\rangle$ can be defined without ambiguities by filling in all the negative energy states. Next, we consider the fiber of $\mathfrak{A}^{(3)}$ over A, and study the parallel transport of this line bundle along the fiber to check whether our construction of $|0,A\rangle$ is equivariant under three dimensional gauge transformations i.e. whether Gauss' law is satisfied. Since $|0,A\rangle$ defines a line bundle on $\mathfrak{G}_0$ for given A. We can restrict the line bundle to the generators of $H_2(\mathfrak{G}_0)$. As in Section III, we can choose $G = SU(N)$, $N \geq 3$, and compactify space to S^3. Since we are considering a two parameter family of gauge transformations in $\mathfrak{G}_0$, for every point (θ,ϕ) in the two parameter family the winding number of the map $g_{\theta,\phi}(x)$: $S^3 \to G$ is zero. Since $\pi_2 G = 0$, we find again that $\pi_2(\mathfrak{G}_0) \approx \pi_5\,SU(N) = Z$. (In more complicated cases we may have to compactify on a different three manifold, and the generator of $H_2(\mathcal{G}_0)$ may not be a 2-sphere, but the flavor of the argument is the same, and for pedagogical reasons we will avoid these complications). The operator defining the parallel transport on the surface of this S^2 will be by construction the operator U_g which implements a gauge transformation on $|0,A\rangle$. In order to test for the triviality of this line bundle, we fill in the 2-sphere (θ,ϕ) to make a three ball $B_3 \subset \mathfrak{A}^{(3)}$, $\partial B_3 = S^2$; if the north pole of S^2 is taken to be A, we can interpolate adiabatically inside B_3 between A and $A^{(\theta,\phi)}$. This gives a two parameter family of lines joining A with $A^{(\theta,\phi)}, (\theta,\phi) \in S^2$. This way we produce a section of the Fock vacuum bundle by comparing the vacuum at A with that at (θ,ϕ). If $g(\theta,\phi)$ is the generator of $\pi_2(\mathcal{G}_0)$, we are going to show that somewhere inside B_3 (t_0,θ_0,ϕ_0), two eigenvalues of $\sigma\cdot D(A_t^{(\theta,\phi)})$ cross. Before we prove this, let us recall some simple facts about spectral flows. Fix (θ,ϕ) and let A_t be the 4-dimensional gauge field interpolating slowly betwen A and $A^{g(\theta,\phi)}$. Instead of considering $i\sigma\cdot D(A_t)$, let us consider the d = 4 Weyl equations:

$$D_\pm\Psi = \frac{\partial}{\partial t} \pm i\sigma^i D_i(A_t)\,\Psi. \tag{6.14}$$

Let us also assume that D_+ (D_-) has a zero mode Ψ_+ (Ψ_-). Since we are interpolating very slowly, we can solve $D_+\Psi_+ = 0$ $(D_-\Psi_-\phi = 0)$ using the adiabatic approximation. Since Ψ_+ (Ψ_-) is normalizable, we can expand $\Psi_\pm(t,x) = \sum_n f_\pm^{(n)}(t)\phi_n^t(\vec{x})$; $\phi_n^+(x)$ are the simultaneous eigenfunctions of $i\sigma^i D_i(A_t)$ for fixed t with eigenvalues $\lambda_n(t)$. In terms of $f_\pm^{(n)}$ (6.14) becomes

$$\frac{d}{dt} f_\pm^{(n)} \pm \lambda_n(t) f_\pm^{(n)}(t) = 0, \tag{6.15}$$

and

$$f_\pm^{(n)}(t) = \text{const. } \exp \mp \int_0^t \lambda_n(\tau)d\tau \tag{6.16}$$

for $\Psi_\pm$ to be normalizable, we must require that

$$\int_{-\infty}^{-\infty} |f_\pm^{(n)}(t)|^2\, dt < \infty \tag{6.17}$$

Notice, however, that if $\lambda_n(t)$ is always either positive or negative along the interpolation, (6.17) cannot be satisfied, thus for such n's, $f_\pm^{(n)} = 0$. The only way (6.17) can be satisfied is if one of the lowest eigenvalues crosses zero. For $f_+^{(n)}(t)$ $(f_-^{(n)}(t))$ normalizability requires that at $t\to\infty$ $\lambda_n(t) \to \lambda_n(\infty) > 0$ $(\lambda_n(\infty) < 0)$ and at $t\to-\infty$ $\lambda_n(t) \to \lambda_n(-\infty) < 0$ $(\lambda_n(-\infty) > 0)$. This result, familiar from instanton arguments,[69] means that the existence of zero modes of the d=4 Weyl operators can be interpreted as spectral flows of the one parameter family of three dimensional Dirac operators $i\sigma^i D_i(A_t)$. Returning to our problem, we have an interpolating field $A_t^{(\theta,\phi)}$ for any point in S^2. As $t\to-\infty$ $A_t^{(\theta,\phi)}\to A$, $t\to+\infty$ $A_t^{(\theta,\phi)}\to A^{(\theta,\phi)}=g^{-1}(A+d)g$. This defines a 2-parameter family of four-dimensional Dirac operators $\not{D}(A_t^{\theta,\phi})$. If (θ,ϕ) is close to the north pole, we know by continuity that $i\sigma^i D_i(A_t)$ will have no zero modes for any value of t because we choose A such that $i\sigma^i D_i$ has no zero modes. Hence ind $\not{D}(A_t^{\theta,\phi})=0$ $(\theta,\phi)\in S^2$. This however does not mean that the index bundle Ind $\not{D}(\theta,\phi)$ does not have a non-trivial 1st Chern class. We can compute $c_1(\mathrm{Ind}_{S^2}\not{D})$ using the adiabatic approximation along the lines of Section III. We have a 3-parameter family of 3-dimensional Dirac operators $i\sigma^i D_i(A_t^{\theta,\phi})$ $(t,\theta,\phi)\in B_3^+$. We construct a well defined index problem as in Section III by including a "lower cap" B_3^-, $A_s^{(\theta,\phi)}$ such that in the overlap $(B_3^+\times S^3)\cap(B_3^-\times S^3)=S^2\times S^3$ the transition function is $g(\theta,\phi,\vec{x})$. Let $\mathscr{F}$ be the six dimensional gauge field constructed this way. Then using the same arguments we get that a representative of the 1st Chern class of the Fock line bundle is

$$c_1(\theta,\phi)=\int_{S^3} Q_5^0(\mathscr{A},\mathscr{F})$$

$$\mathscr{A}=g^{-1}(A+d+\delta)g$$

$$\delta=d\theta\nabla_\theta+d\phi\nabla_\phi \tag{6.18}$$

(we leave it as an exercise to the reader to adapt the arguments of Section III to our case to prove (6.18)). Since

$$\int_{S^2} c_1=1. \tag{6.19}$$

($g(\theta,\phi)$ is the generator of $\pi_2(\mathcal{G}_0)$ with winding number 1), this means generically that for some (θ_0,ϕ_0) D_+,D_- will have a jump in their kernels i.e. D_+ and D_- has one zero eigenvalue at (θ_0,ϕ_0). Using our previous arguments borrowed from instanton theory, we know that the two lowest eigenvalues (in absolute value) will cross at some value of t.

Let $\lambda_1<0<\lambda_2$ be the two eigenvalues closest to zero of $i\sigma^i D_i(A)$. As we interpolate along this particular line (θ_0,ϕ) $\lambda_1(t)$ will evolve like a kink between λ_1 and λ_2, and $\lambda_2(t)$ as an antikink (also between λ_2 and λ_1). Therefore $\lambda_1(t),\lambda_2(t)$ will cross at some particular value of t. This shows that the Fock vacuum line bundle is twisted. If we neglect the states whose eigenvalues do not go through zero, and concentrate on the two level system defined by ϕ_1, ϕ_2: the eigenfunctions with eigenvalues λ_1, λ_2 respectively, the Fock vacuum at A is defined so that ϕ_1 is filled and ϕ_2 empty. If we follow these two states adiabatically along the line (θ_0,ϕ_0), we will find that after the interpolation is completed, ϕ_2 is occupied and ϕ_1 is empty. The vacuum has "rolled over" into a particle-antiparticle state, and by the index theorem this is unavoidable. The hole as usual has to be interpreted as an antiparticle (it is the absence of a particle of negative energy -E, and negative helicity in the representation N of SU(N). Recall that we have restricted our computation to G = SU(N) and a Weyl fermion in the fundamental representation). The state so obtained contains a particle-antiparticle pair with zero projection onto the Fock vacuum i.e. the Fock bundle behaves exactly like a charged particle in the presence of a monopole, with the monopole string going through S^2 at (θ_0,ϕ_0). This ties in rather nicely with Berry's phase[70] (a modification of the adiabatic theorem) which is equivalent as B. Simon showed,[70] to the holonomy of the connection of a line bundle. Let P be any point of the 2-sphere $S(\theta,\phi)$, and let $|P\rangle$ be the Fock vacuum there. Let Q and R be any other two points on S^2. Since on S^2 parallel transport is given by the operator implementing Gauss' law, on S^2 we can parallel transport from P to Q to R and back to P. Going from P to Q corresponds to making a gauge transformation g, and going from Q to R by making a gauge transforamtion h. Finally, we go back from R to P by making a gauge transformation $(hg)^{-1}$. Since we have a non-trivial curvature for the Fock line bundle, after we parallel transport around the spherical triangle $P\to Q\to R$ (staying always on S^2), we do not get back to $|P\rangle$, we pick up a phase given by the holonomy of the gauge connection which determines c_1 i.e.

$$U^{-1}_{hg}U_h U_g|P\rangle = e^{i\int_\Omega c_1}|P\rangle, \tag{6.20}$$

where Ω is the spherical triangle determined by (P,Q,R). For Q,R infinitesimally closed to P, we can use (6.5), (6.7), and (6.18) to obtain finally that

$$[Q_{u_1},Q_{u_2}] = iQ_{[u_1,u_2]} + Q^2_3(v,\mathcal{A})$$

$$v = d\alpha^1(g^{-1}\partial_1 g) + d\alpha^2(g^{-1}\partial_2 g) = d\alpha^1 u_1 + d\alpha^2 u_2. \tag{6.21}$$

u_1 and u_2 are the infinitesimal gauge transformations required to go from P (A^g) to Q and R, and α_1,α_2 parameterize the infinitesimal lines joining P with Q and R respectively. More explicitly, we go from P to Q (R) by making an infinitesimal gauge transformation with gauge parameter $u_1 = g^{-1}\partial_1 g\delta\alpha_1$ $(u_2 = g^{-1}\partial_2 g\delta\alpha_2)$. Restricting (6.21) to the north pole A where $g^2=1$, we obtain after some simple algebra:

$$[Q^a(\vec{x}),Q^b(\vec{y})] = if^{abc}\,Q^c(\vec{x})\,\delta^3(\bar{x}-\bar{y}) + \frac{1}{12\pi^2 i}\,d^{abc}\,\varepsilon^{ijk}\,\partial_i\,A^c_j\,\partial_k\,\delta^{(3)}(\vec{x}-\vec{y})$$

$$d^{abc} = \mathrm{Tr}\,T^a\{T^b,T^c\}\ . \qquad (6.22)$$

The phase appearing in (6.20) can also be obtained directly by an explicit computation of Berry's phase using perturbation theory in the Hamiltonian formulation. This has indeed been done in a very nice recent paper by H. Sonoda.[71]

VII. NON-PERTURBATIVE CONSIDERATIONS

So far we have been dealing exclusively with perturbative anomalies, i.e. we have computed the change of the fermionic effective action under gauge transformations that can be continuously deformed to the identity. In Section VIII we will introduce the global anomalies which require the evaluation of the change of $\mathrm{Im}\,\Gamma[A]$ under a gauge transformation (or diffeomorphism) which cannot be continuously connected to the identity. These questions can not be answered by perturbative methods, and we need to develop a non-perturbative understanding of the effective action for chiral fermions in the presence of gauge and gravitational fields. In the following we will present an exact formula for the imaginary part of the fermion determinant for chiral fermions.[73-76]

One of the most interesting open problems in field theory is the study of chiral gauge theories: gauge theories with fermions in complex representations of the gauge group. It is poorly understood how the global symmetries of these systems are realized after confinement except for the possibilities allowed by the 't Hooft anomaly matching conditions.[7] See Ref. 77 for the heuristic arguments based on the Maximal Attractive Channel (MAC) or instanton arguments. The problem has two sides to it. First, there is no way of making reliable computations in the continuum limit taking into account non-perturbative effects. This problem is aggravated because we lack a method of writing a consistent lattice approximation to these theories as a consequence of the generic doubling problem for lattice fermions.[78] Second, we do not have reliable qualitative methods of analyzing the dynamical subtleties inherent in the chiral structure of the theory. In vector-like theories like QCD, there are more powerful theoretical tools which provide a qualitative (and sometimes quantitative) understanding of the intricacies of the fermion-gauge field dynamics: lattice analysis, large N limit,[79] qualitative estimates,[80] ... We will dedicate this section to an exact computation of the imaginary part of the effective action $\mathrm{Im}\,\Gamma[A]$. The formulae to be derived are simple enough that this may ameliorate the present situation concerning chiral gauge theories.

(For some recent ideas based mainly on effective Lagrangians and extrapolations of the known behavior in QCD see Ref. 81.) The results presented were motivated by previous work on the parity anomaly in odd dimensions[47] and on Witten's computation of global gravitational anomalies.[32] (See Section VIII.)

As before we will work in Euclidean space and compactify space-time to either S^n or $(S^1)^n = T^n$; moreover we will do the computations mostly in the gauge case, even though some of our results extend to the gravitational and to the combined gauge and gravitational case. Apart from simplicity, this restriction is justified because one ultimately envisages the application of the results to the lattice analysis of chiral gauge theories. We know from Sections II and III that only $\mathrm{Im}\,\Gamma[A]$ may be gauge non-invariant, and that $\mathrm{Re}\,\Gamma[A]$ is gauge-invariant. We can say furthermore, we know that $2\,\mathrm{Re}\,\Gamma_r[A]$ is identical to the effective action for a vector-like theory with Dirac fermions in the representation r of G (gauge group). Thus $\mathrm{Re}\,\Gamma[A]$ is essentially vector-like, while the imaginary part is sensitive to the fact that the fermions are chiral. In previous sections we have described the recent progress made in understanding the gauge non-invariant part of $\mathrm{Im}\,\Gamma[A]$. We know however from ordinary perturbation theory that only the first few diagrams in the diagramatic evaluation of $\mathrm{Im}\,\Gamma[A]$ may produce an anomaly, and that $\mathrm{Im}\,\Gamma[A^g] - \mathrm{Im}\,\Gamma[A]$ is a local polynomial on A and $dg\,g^{-1}$ given by the Wess-Zumino Lagrangian (see Section IV). In spite of the elegance and generality of the arguments in Sections III-IV, it is clear that with them one is only probing the first few diagrams in $\Gamma_r[A]$. Furthermore, when the anomalies cancel, all the Wess-Zumino terms cancel, $\mathrm{Im}\,\Gamma[A]$ is gauge-invariant, and we know (except in $d = 2$) that there is a gauge-invariant non-local piece in $\mathrm{Im}\,\Gamma_r[A]$ carrying the information about the chiral nature of the fermionic representation. Before getting into the details of how to exactly compute $\mathrm{Im}\,\Gamma_r[A]$, it seems worthwhile to review why the standard methods fail in the analysis of chiral gauge theories. Let us first consider the large-N limit.[79] For QCD-like theories (color SU(N) and fermions in the fundamental representation), standard arguments show[79] that in the $N \to \infty$ limit one can neglect internal fermion loops, i.e. we can "quench" $\det \not{D}(A)$ and set it equal to 1, and define the expectation value of products of fermion operators in terms of gauge field averages of products of fermionic Green's functions $\langle x|\left(i\not{D}(A)\right)^{-1}|y\rangle$. This is unfortunately an unacceptable distortion of the theory when there are chiral fermions present. The easiest way to see this is by considering an example. The simplest SU(N) anomaly-free chiral multiplet contains a two-index antisymmetric left-handed tensor ψ^{ij} field and (N-4) fields in the $\bar{N}$ representation ϕ_i^a, $a = 1, N-4$; $i = 1, N$. (This is the straightforward extension to $N > 5$ of the SU(5) family structure $\bar{5}+10$.) In this theory, the number of gluons is $N^2 - 1$, and the number of fermionic states running around loops is $N(N-1)/2$ for ψ^{ij} plus $(N-4)\text{-}N$ for ϕ^a_i, i.e. $3N(N-3)/2$. Hence even though one may be able to define a large N approximation, the standard arguments about the suppression of fermion loops are manifestly wrong. Essentially, throwing away the determinant would imply throwing away the chiral properties of the theory. Similarly, the arguments based on large N for the breaking of global chiral symmetries[82] break down for the same reason. If one tries to use the arguments of Ref. 80 in order to get non-trivial qualitative properties of these theories, one cannot even get started because for chiral fermions the measure on the functional integral is not only not positive-definite, it is complex. Finally, the lattice version of the standard fermion action is afflicted with the doubling problem:[78] any attempt to write down a local and gauge-invariant action for the fermions seems to be doomed to give a vector-like spectrum in the continuum limit. Without giving any detailed proofs, there is a simple qualitative reason why this should happen. In the usual approach to the lattice formulation of chiral gauge theories, one writes down an action and a measure which are both gauge-invariant (and regulated). This is not very satisfactory, because in the continuum the action is indeed gauge-invariant, but the measure (which defines the quantum mechanics) is not gauge-invariant. In general it will have anomalies. By characterizing the imaginary part of the effective action exactly, one may be able to find a lattice prescription for it, and therefore add it to the measure on the lattice in order to put in "by hand" some of the chiral properties of the fermions. It is not clear yet whether such a program can be implemented.

The main result of this section can be understood quite easily. Let A be a gauge field defining a connection on a trivial bundle over space-time (if we had an instanton-like configuration, we would define the fermion determinant with the zero modes removed), and let A_0 be some fiducial configuration in the same topological sector. If M is space-time, we consider $\mathbb{R} \times M$ ($\mathbb{R}$ is the real line), and an "instanton"-like interpolation A_t between A_0 and A, i.e. at $t \to -\infty$, $A_t \to A_0$; $t \to +\infty$, $A_t \to A$. For simplicity we will take $A_0 = 0$. On $\mathbb{R} \times M$, we define the (2n+1)-dimensional Dirac operator:

$$H = i\,\bar{\Gamma}\frac{\partial}{\partial t} + i\not{D}(A_t) \; . \tag{7.1}$$

For arbitrary A, (7.1) has a spectral asymmetry (see Section II). Using suitable boundary conditions to be defined later, we can show that $\text{Im}\,\Gamma[A]$ is exactly given by $\pi\eta(0)$ + terms which vanish for anomaly-free theories. We know that theories with fermions in real representations

$\text{Im}\,\Gamma[A] = 0$. In this case we can concentrate only on $\pi\eta(0)$ and neglect the other terms. As a consistency check, we want to show that $\pi\eta(0)$ vanishes in this case mod $2\pi n$. This is a consequence of the fact that for real representations $A^* = -S^{-1} A S$, and that the spinor representation is unique. If $H\phi_\lambda = \lambda\phi_\lambda$, we can complex conjugate the equation and use $A^* = -S^{-1} A S$, $\Gamma^* = -B^{-1}\Gamma B$, to show that $S^{-1} B^{-1} \phi^*_\lambda$ is an eigenfunction of H with eigenvalue $-\lambda$. Thus the spectrum is symmetric around zero and $\eta(0) = 0$. (If there are h zero modes of H, we have to change $\pi\eta(0)$ to $\pi(\eta(0)+h)$.) Similarly, if H has a continuous spectrum, the spectral density is also symmetric around zero as long as the fermion representation is real. In order to get acquainted with spectral asymmetries, let us compute a few examples. The advantage of the examples we will consider is that they contain all the basic ingredients without any of the technical complications. Let M(t) be an arbitrary $n \times n$ matrix, and consider the operator:

$$i\frac{d}{dt} + M(t) \qquad 0 \leq t \leq 1$$

$$M^+(t) = M(t) \;. \tag{7.2}$$

With the boundary condition on eigenfunctions:

$$\Psi(1) = U\Psi(0)$$

$$U^+ U = 1 \;. \tag{7.3}$$

We compute the spectral asymmetry of (7.2) by changing variables $\Psi(t) = G(t)\,\xi(t)$ with

$$\left(i\frac{d}{dt} + M(t)\right)G(t) = 0 \;,$$

$$G(0) = 1 \;. \tag{7.4}$$

Then the eigenvalue problem for (7.2) becomes:

$$i\frac{d}{dt}\xi(t) = E\,\xi(t)$$

$$\xi(1) = G^{-1}(1)U\,\xi(0) \;. \tag{7.5}$$

Since $G^{-1}(1)U$ is a unitary matrix, we can diagonalize it. Let the eigenvalues be $\exp{-i\delta_i}$, $i = 1,n$. Since $G^{-1}(1)U$ is time-independent, we can redefine ξ again by the unitary transformation which diagonalizes $G^{-1}(1)U$. Now (7.5) is reduced to n-independent 1-dimensional problems, and the eigenvalues of H are simply:

$$\lambda = \sum_{i=1}^{n} \lambda_i \;,$$

$$\lambda_i = 2\pi n_i + \delta_i \;. \tag{7.6}$$

Thus the spectral asymmetry of (7.2) is given by

$$\eta(0) = \lim_{\varepsilon\to 0} \sum_{i=1}^{n} \left(\sum_{m_i \in \mathbb{Z}} \operatorname{sign}(2\pi m_i + \delta_i) e^{-\varepsilon|2\pi m_i + \delta_i|} \right) \tag{7.7}$$

which modulo even integers becomes:

$$\eta(0) = n - \frac{1}{\pi} \sum_i \delta_i \pmod{2\mathbb{Z}} \tag{7.8}$$

and notice that

$$-\sum_i \delta_i = -i \log \det G^{-1}(1) U \ . \tag{7.9}$$

Now we can re-express (7.5) in terms of U and M(t) explicitly as follows:

$$\log \det G^{-1}(1)U - \log \det U = \int_0^1 dt \frac{d}{dt} \log \det \left(G^{-1}(t) U\right)$$

$$= +\int_0^1 dt \operatorname{Tr} G^{-1}(t) \frac{dG}{dt} = +i \int_0^1 dt \operatorname{Tr} M(t) \ . \tag{7.10}$$

Hence

$$\eta(0) = n + \frac{1}{\pi} \int_0^1 dt \operatorname{Tr} M(t) - \frac{i}{\pi} \log \det U \ . \tag{7.11}$$

In (7.11) it is important to notice that there are two distinct contributions. The first one involves integration over t of Tr M(t), and it could be called the volume term, while the last term log det U is exclusively determined by the boundary conditions. Thus in general when M(t) is substituted by something like the Dirac operator, the spectral asymmetry of (7.1) will depend on the type of boundary conditions imposed on $\mathbb{R} \times M$ or $I \times M$, $I = [0,1]$. A simple generalization of (7.11) which is closer to the case we are interested in (if we are cavalier about regularization) is the next example.

$$H = i\Gamma \frac{d}{dt} + M(t)$$
$$[M(t), \Gamma] = 0 \ . \tag{7.12}$$

Since $[M,\Gamma] = 0$ we can split the eigenvalues of M(t) into $e_{+n}(t)$, $e_{-n}(t)$, $\Gamma e_{\pm n}(t) = \pm e_{\pm n}(t)$. We now impose independent boundary conditions on the $\Gamma = \pm 1$ components of the eigenfunctions:

$$\psi_\pm(1) = U_\pm \psi_\pm(0)$$
$$U_\pm U_\pm^\dagger = 1 \ . \tag{7.13}$$

(M and U are $2n \times 2n$ matrices, and we assume that M(t) has no zero modes for any t.) Repeating the previous computation for each sector, we get:

$$\eta(0) = \frac{1}{\pi} \int_0^1 dt \operatorname{Tr} \Gamma M(t) - \frac{i}{\pi} (\log \det U_+ - \log \det U_-) \pmod{2\mathbb{Z}} \ . \tag{7.14}$$

More complicated cases can be dealt with in a similar way. Given a Hermitian operator on some space M, there are several useful representations of the spectral asymmetry. Some of them are listed now because we will need them later.

$$\eta(s) = \frac{2}{\Gamma\left(\frac{1+s}{2}\right)} \int_0^\infty d\beta\, \beta^s \,\mathrm{Tr}\, H\, e^{-\beta^2 H^2} \tag{7.15a}$$

$$= \mathrm{Tr}\, H\,(H^2)^{-(s+1)/2} \quad . \tag{7.15b}$$

Choosing s large and positive, (7.15a) and (7.15b) are well-defined, and $\eta(s)$ can be analytically continued to a meromorphic function of the complex s-plane. As discussed in Section II, $\eta(0)$ exists for the Dirac operator on a compact spin manifold. Even though (7.15a) and (7.15b) are in general impossible to compute, sometimes the variations of $\eta(0)$ with respect to some parameter are relatively easy to compute. If H_ε is a one-parameter family of elliptic operators, a simple computation shows that:

$$\frac{d\eta_\varepsilon(0)}{d\varepsilon} = \lim_{\beta\to 0} \frac{2}{\sqrt{\pi}}\, \beta\, \mathrm{Tr}\frac{dH}{d\varepsilon}\, e^{-\beta^2 H^2}$$

$$= -\lim_{s\to 0} s\, \mathrm{Tr}\frac{dH}{d\varepsilon}\,(H^2)^{-(s+1)/2} \tag{7.16}$$

After this digression on general features of η-invariants, let us get back to our main problem, i.e. computing $\mathrm{Im}\,\Gamma[A]$ for an arbitrary gauge field A. Since we want to obtain a non-perturbative definition of $\mathrm{Im}\,\Gamma[A]$ which agrees with the perturbative evaluation of $\mathrm{Im}\,\Gamma[A]$, we first start by assuming that A is a perturbative configuration. By this we mean first of all that $\not{D}(A)$ has no zero modes, i.e. A is topologically trivial, and furthermore that we can choose a path from $A = 0$ (or some reference configuration A_0) to A, A_t so that $\not{D}(A_t)$ has no zero modes for any t. Thus on $\mathbb{R}\times M$ we define a gauge field A_t such that for $t \to -\infty$, $A_t \to A_0$, and for $t \to +\infty$, $A_t \to A$. As a five (or 2n+1) dimensional gauge field we take $A_5 = 0$. The operator (7.1) is now well-defined and elliptic acting on square-integrable functions on $\mathbb{R}\times M$. Its spectrum however may not be discrete. On the space of square-integrable functions we can define the η-invariant for the operator H (see Eq. (7.1)) using (7.15b) for s large enough.

$$\eta(s) = \mathrm{Tr}\, H\,(H^2)^{-(s+1)/2} \quad . \tag{7.17}$$

We want to prove that (7.17) is exactly $\mathrm{Im}\,\Gamma[A] - \mathrm{Im}\,\Gamma[A_0]$ up to local Chern-Simons forms. We must give a precise prescription of how to compute the fermion determinant. For simplicity of exposition we will present the proof using a Pauli-Villars regularization.[73] This derivation is not very rigorous. If one wants to be mathematically precise, it is better to start by defining the determinant using ζ-functions (this is done in detail in Ref. 74). Since we are assuming that A is a perturbative configuration, the operators $\hat{D}_A$, $\hat{D}_{A_t}$ are invertible for all values of t. Then the ζ-function definition of the determinant is:

$$\log\det \hat{D}_A \equiv -\frac{1}{2}\frac{d\zeta(s)}{ds}\bigg|_{s=0}$$

$$\zeta(s) = \mathrm{Tr}(\hat{D}_A^2)^{-s} \ . \tag{7.18}$$

Powers of operators like those appearing in (7.18) can be defined in terms of contour integrals.[83] (See ref. 74 for more details and references.) Since the symbol of $\hat{D}_A^2$ (the term with the highest number of derivatives) is positive-definite, we know that the spectrum of $\hat{D}_A^2$ (except for a finite number of eigenvalues) is concentrated on a wedge about the real axis. Choose any contour C running from $-\infty - i\varepsilon$ to the origin, circles the origin, and goes back to $-\infty + i\varepsilon$. Then

$$\mathrm{Tr}(\hat{D}_A^2)^{-s} = \frac{1}{2\pi_i}\int_C d\lambda\, \lambda^{-s}\frac{1}{\lambda - \hat{D}_A^2} \ . \tag{7.19}$$

(In order to make the writing shorter, we will ignore the possibility that some eigenvalues may have to be treated separately.) The choice of contour C is important because it determines the choice of cut in the logarithm. We make a choice so that $\overline{(\lambda^s)} = \overline{(\lambda)}^s$. Otherwise the definition of the determinant obtained would differ by a local piece proportional to ζ(0). The two choices give the same physics. They simply correspond to different subtraction schemes. The definition (7.18) can be made more general by including an independent right-handed gauge field:

$$D_{AB} = i(\partial\!\!\!/ + A\!\!\!/P_+ + B\!\!\!/P_-) \tag{7.20}$$

and eventually take $B = 0$. The advantage of our choice of contour in (7.19) is that the imaginary part of $\log\det D_{AB}$ can be expressed in a rather symmetrical way:

$$\mathrm{Im}\log\det D_{AB} = -\frac{1}{4}\frac{d}{ds}[\zeta(D_{AB}^2,s) - \zeta(D_{BA}^2,s)]\bigg|_{s=0} \ . \tag{7.21}$$

Relating (7.21) to the η-invariant rigorously is technically rather involved (see Ref. 74), and it is beyond the scope of this lecture. From now on we will sacrifice rigor to clarify and define the determinant using Pauli-Villars regulators, and concentrate on the four-dimensional gauge case, and set $B = 0$ in (7.20). What we will do is show explicitly that the imaginary part of the sum of all the one-loop Feynman diagrams with an arbitrary number of external gauge fields add up to the η-invariant. Choosing suitable Pauli-Villars coefficients, the free fermion propagator becomes:

$$S_M(p) = \frac{\not p}{p^2} \frac{M^2}{p^2 + M^2} \tag{7.22}$$

thus regulating both the real and imaginary parts of $\Gamma[A]$. If S_M denotes the regulated propagator, the graphical expansion of the effective action is

$$\Gamma[A] = \sum_{n=1}^{\infty} \frac{(-1)^n}{n} \mathrm{Tr}(i \not A P_+ S_M)^n \ . \tag{7.23}$$

Graphs in (7.23) with more than four external lines are convergent and one could take the $M \to \infty$ limit. The regulator is essential in so far as it makes the first few diagrams finite. The imaginary part of (7.23) is given by

$$\mathrm{Im}\,\Gamma[A] = \frac{1}{2} \sum \left[\frac{(-1)^n}{n} \mathrm{Tr}(i \not A P_+ S_M)^n - (P_+ \leftrightarrow P_-)\right] \ . \tag{7.24}$$

Starting from (7.24), one can compute the change of $\mathrm{Im}\,\Gamma[A]$ along the path chosen to join A_0 with A, A_u: $d\,\mathrm{Im}\,\Gamma[A_u]/du$. Since in the diagrams with more than four powers of the gauge field one can shift the loop momenta without introducing ambiguities, it is straightforward to check that those diagrams are identical to the corresponding ones occuring in the perturbative expansion of

$$\frac{1}{2} \mathrm{Tr}\, \gamma_5 \frac{d\not D_u}{du} \not D_u^{-1} (1 + \not D_u^2/M^2)^{-1} \tag{7.25}$$

where $\not D$ is the full Dirac operator. Problems appear in the first few graphs. Due to the divergent nature of the original Feynman integrals, the shift in momenta needed in going from (7.24) to (7.25) introduces ambiguities which must be dealt with directly. This is most easily done if one notices that the variation of $\mathrm{Im}\,\Gamma[A]$ along a path is the same as computing the induced fermion current $\langle j \rangle_u$ in the presence of the external gauge field A_u. In (7.24) we computed the induced current preserving the Bose symmetry of the original theory, whereas in (7.25) one is computing the induced current in a Bose non-symmetric treatment of the lowest order diagrams. Since the difference between (7.24) and (7.25) can be traced back to the regulators, the difference can be written in terms of a local current constructed in terms of A_u. On the other hand, this current has already been computed in the literature; it is the current which allows one to go from the covariant to the consistent form of the anomaly.[39] Indeed, if we compute the anomaly from (7.24) we obtain the consistent anomaly, while (7.25) produces the covariant form of the anomaly as a direct consequence of the covariant properties of the Dirac operator under gauge transformations. Thus

$$\mathrm{Im}(\Gamma[A] - \Gamma[A_0]) = \frac{1}{2} \int_0^1 du \left[\mathrm{Tr}\, \gamma_5 \frac{d\not D_u}{du} \not D_u^{-1} (\not D_u^2/M^2 + 1)^{-1} - \mathrm{Tr}\, \frac{dA_u^\mu}{du} J_\mu(A_u)\right] \ . \tag{7.26}$$

Explicit formulae for $J(A_u)$ can be found in Ref. 39. For the derivation of our final result we do not need to use any explicit representation for it.

Let us now look at the spectral asymmetry of the operator (7.1). One of the nice features about the η-function, is that on a compact manifold it changes by a local amount when one makes small changes in the gauge field. This local amount is in fact the Chern-Simons form.[16] For non-compact manifolds, one has to be a bit more careful in the derivation of the variation of η in order to keep track of boundary contribution. To show that (7.26) is related to the spectral asymmetry, we consider a one-parameter family of five-dimensional operators H_u defined by a two-parameter family of gauge fields $A(t,u)$ with the following properties: for $t = -\infty$ to $-T$, $A(t,u) = A_0$ (the reference configuration); for $t = -T$ to T, $A(t,u)$ smoothly interpolates between A_0 and A_u and from $t = T$ to ∞, $A(t,u) = A_u$. Since we are assuming that both A_0 and A are perturbative configurations, $A(t,u)$ can be chosen so that $D(A_{a,u})$ has no zero modes. Our aim is to compute $\eta_{u=1}$. This is done by computing the variation of η with respect to u, and then integrating. It is simpler to evaluate the variations of η rather than η itself. Since we have used Pauli-Villars regularization to define the four-dimensional fermionic effective action, it is convenient to use a similar definition for η. If in any odd number of dimensions one computes the fermion effective action in the presence of an arbitrary external gauge field by means of a regularization procedure which preserves gauge invariance, contrary to intuition, one encounters an induced imaginary part given exactly by $i\pi\eta(0)/2$.[47] Using this result we can write:

$$\pi\eta_u = - \operatorname{Im} \log \det \frac{H_u - iM}{H_u + iM} \tag{7.27}$$

in the limit as M goes to infinity. (As it stands, (7.27) is not fully regulated. A complete regularization would require making one more Pauli-Villars subtraction. Since the results are the same after one includes the second Pauli-Villars regulator, we have not written it out explicitly in order to avoid the cluttering of subsequent equations. Finally, one has to choose a cut for the logarithm. We choose it along the positive imaginary axes to treat along the lines of Ref. 47.) We can write (7.27) equivalently as

$$\eta_u(0) = - \frac{1}{\pi} \operatorname{Im} \operatorname{Tr} \ln \frac{H_u - iM}{H_u + iM} \tag{7.28}$$

where the trace is over $R \times M_4$. We can keep track of the non-compactness of the manifold by first cutting off the integral over R from $-L$ to $L, L \gg T$, and then letting L go to infinity. This is implemented inserting the function $P_L(t) = \theta(t+L)(1-\theta(t-L))$ inside the trace of (7.28). Then

$$\frac{d\eta_u^{(L)}}{du} = -\frac{1}{\pi}\,\mathrm{Im}\int_{-M}^{M} i\,dz\,\mathrm{Tr}\,P_L \frac{dH}{du}(H-iz)^{-2} + \frac{1}{\pi}\int_{-M}^{M} dz\,\mathrm{Tr}\,\gamma_5 \frac{dP_L}{dt}(H_u-iz)^{-1}\frac{dH}{du}(H_u-iz)^{-2}\ . \tag{7.29}$$

The first term in (7.29) gives a purely local contribution while the second one is concentrated on the boundary, and it is in fact the one which reproduces (7.25). The first term can be computed using a plane wave basis. A short computation yields

$$\frac{1}{(2\pi)^2 2!}\,\mathrm{Tr}\,\frac{dA}{du}F_u^2 = 2\pi\frac{d}{du}Q_5^0(A) + \text{total derivative}\ . \tag{7.30}$$

The second term is more interesting, and can be rewritten as:

$$-\frac{1}{\pi}\int_{-M}^{M} dz\int dx\,\mathrm{Tr}\,\gamma_5\left(\frac{H_u+iz}{H_u^2+z^2}\frac{d\not{D}_u}{du}\frac{(H_u+iz)^2}{(H_u^2+z^2)^2}\right)(Lx,Lx) \tag{7.31}$$

$$-\ (L \to -L)$$

The final step is to show that we can substitute H in (7.31) by H_0, the t-independent operator for $t>L$, in the L going to infinity limit.

Let us concentrate on the term at $t=L$. The same considerations apply to the $t=-L$ term. For $t>L$, we have that $H_u-H_0 = i(A(t,u)-A(u)) = 0$. Furthermore, $dH_u/du - dH_0/du$ only differs from zero in the region $-T$, T. Since we are assuming for the time being that H_u has no zero eigenvalues, the propagators for both H_u and H_0 decay exponentially at large distances, with exponential decay determined by the lowest eigenvalue of $D(A_{t,u})$ in the first case, and the lowest eigenvalue of $D(A_u)$ in the second. Now we can rewrite the integrand in (7.31) as:

$$\begin{aligned}(H_u-iz)^{-1}\frac{dH_u}{du}(H_u-iz)^{-2} - (H_0-iz)^{-1}\frac{dH_0}{du}(H_0-iz)^{-2} &=\\ = \left[(H_u-iz)^{-1}-(H_0-iz)^{-1}\right]&\frac{dH_u}{du}(H_u-iz)^{-2}\\ +(H_0-iz)^{-1}\left[\frac{dH_u}{du}-\frac{dH_0}{du}\right]&(H_u-iz)^{-2}\\ +(H_0-iz)^{-1}\frac{dH_0}{du}&\left[(H_u-iz)^{-2}-(H_0-iz)^{-2}\right]\ .\end{aligned} \tag{7.32}$$

To show that the right-hand side of (7.32) vanishes exponentially in the limit as L goes to infinity, it is crucial to use the information that H_u and H_0 are identical for $t>T$, and then rewrite

$$\frac{1}{H-iz}-\frac{1}{H_0-iz} = \frac{1}{H-iz}(H-H_0)\frac{1}{H_0-iz} \tag{7.33}$$

$$\frac{1}{(H-iz)^2}-\frac{1}{(H_0-iz)^2} = \frac{1}{(H-iz)^2}(H-H_0)\frac{1}{(H_0-iz)} + \frac{1}{H-iz}(H-H_0)\frac{1}{(H_0-iz)^2}$$

Substituting (7.33) in the right-hand side of (7.32), and using the exponential decay of the propagators $G(t,t')$ and $G_0(t,t')$ as $|t-t'|$ becomes large, one can show after a rather lengthy computation that (7.32) vanishes exponentially as L goes to infinity. Thus in (7.31) we can substitute H_u by H_0 as L becomes large (and a similar substitution in the term where L is replaced by -L). Since H_0 is t-independent, (7.31) becomes

$$-\frac{1}{\pi}\int_{-M}^{M}dz\int_{-\infty}^{+\infty}\frac{dE}{2\pi}\,\mathrm{tr}\,\gamma_5\frac{\not{D}_u+iz+E\Gamma}{\not{D}_u^2+E^2+z^2}\frac{d\not{D}_u}{du}\left(\frac{\not{D}_u+iz+E\Gamma}{\not{D}_u^2+E^2+z^2}\right)^2 = \frac{1}{2\pi}\,\mathrm{tr}\,\gamma_5\frac{d\not{D}_u}{du}\not{D}_u^{-1}\,(1+\not{D}_u^2/M^2)^{-1} \tag{7.34}$$

where tr is the four-dimensional operator trace. (Note that at $u=0$, $\eta_u(0)=0$, the gauge configuration is time-independent and the spectrum of the associated five-dimensional operator is symmetric around zero.) We can summarize our results so far in the following expression:

$$\pi\frac{d\eta_u}{du} = \frac{1}{2}\,\mathrm{tr}\,\gamma_5\frac{d\not{D}_u}{du}\not{D}_u^{-1}(1+\not{D}_u^2/M^2)^{-\frac{1}{2}} + \frac{1}{(2\pi)^2 2!}\int_{R\times M}\mathrm{Tr}\frac{dA_u}{du}F_u^2 \;. \tag{7.35}$$

Comparing (7.34) with (7.26), we can immediately conclude that for non-anomalous theories the spectral asymmetry exactly reproduces $\mathrm{Im}(\Gamma[A]-\Gamma[A_0])$. For anomalous theories however there is a correction to the spectral asymmetry given by the second term in the right-hand side of (7.35) and minus the contribution due to the current $J(A_u)$ in (7.26). These two terms combine after one integrates over u to produce the Chern-Simons form $Q_5^0(A_{u=1})$ as follows from the relation between the covariant and the consistent anomaly.[39] Thus, for $A_0=0$ (as in ordinary perturbation theory):

$$\mathrm{Im}\,W[A] = \pi\eta(0) - 2\pi\,Q_5^0(A_t) \;. \tag{7.36}$$

Another way to understand the appearance of the Chern-Simons term in (7.36) is to notice that (7.36) should be independent of the path A_t chosen to interpolate between A_0 and A. Since two such paths only differ in a compact region, we can apply the standard formulae for the variation of $\eta(0)$,[16] which is equal to the difference between the two Chern-Simons forms for the two different paths, thus cancelling neatly the variation of the Q-term in (7.36). Using this property one could have guessed from the beginning the form of the correction to $\eta(0)$ in order to reproduce $\mathrm{Im}\,\Gamma[A]$, by requiring path independence.

If it happens that the five-dimensional operator H has a non-trivial kernel (but with a gap between the zero eigenmodes and the continuum spectrum), the only change in (7.36) is that we have to add to $\eta(0)$ the dimension of the kernel of H, i.e.

$$\mathrm{Im}\,W[A] = \pi\big(\eta(0)+\dim\ker H\big) - 2\pi\,Q_5^0(A_t) \;. \tag{7.37}$$

This is because the existence of a zero mode of H forces two eigenvalues of $D(A_u)$ to go through zero at some value of u as we interpolate between A_0 and A. In order to obtain (7.37), one has to modify slightly the definition of $\eta(0)$:

$$\eta = \lim_{\lambda \to 0} \mathrm{Tr}\, H\,(H^2 + \lambda^2)^{-(s+1)/2} . \tag{7.38}$$

Similarly Eqs. (7.25), (7.34) do not make sense as they stand. This is fixed by separating out the two eigenvalues which go through zero at some value of u, and treat them separately. Using then an argument similar to Witten's treatment of the SU(2) anomaly,[16] one easily gets that for every zero crossing, $\mathrm{Im}\,\Gamma[A]$ picks up a factor of $\pm\pi$. This is most easily derived in the case of a nonanomalous chiral gauge theory. Since $\mathrm{Im}\,\Gamma[A]$ is only defined modulo 2π, we end up with (7.37).

A more difficult situation appears when we try to extend these considerations to the instanton sectors. In this case the non-vanishing of the index of the Dirac operator throughout the interpolation implies that the continuum spectrum of the five-dimensional operator extends all the way down to zero and it is essential to define $\eta(0)$ through a limiting procedure as the one indicated in (7.38). There is however a more subtle problem one has to deal with, and this is the contribution to $\mathrm{Im}\,\Gamma[A]$ of the zero-mode eigenfunctions. Since in the non-zero instanton sector the determinant vanishes identically, the only meaningful way of defining $\mathrm{Im}\,\Gamma[A]$ is to either define a truncated determinant or to infer the form of the phase by computing the expectation value of a string of fermionic operators. This is basically the method followed by Atiyah and Singer[18] to define the fermion determinant in such a situation. Let $Y_+(u)$ and $Y_-(u)$ be the kernel and cokernel spaces for the Weyl operator respectively at the point u in the interpolation. We can construct two complex line bundles by taking the highest exterior powers of Y_+ and Y_-: $L_+(u) = \Lambda Y_+(u)$, $L_- = \Lambda Y_-(u)$. Each of these line bundles is not smoothly defined throughout the space of gauge fields (there will be jumps in their dimensions); however their "ratio" defined in the sense of K-theory, is well defined and one can associate a phase to it. By Fermi statistics this is in fact the contribution to the phase of the fermion determinant coming from the zero-mode sector. A very drastic simplification takes place if we only consider non-anomalous theories. Then the curvature of the ratio of these line bundles is zero, and the only contribution to the change in the phase of the fermion determinant coming from the zero modes as we interpolate between A_0 and A in the instanton sector is given by the holonomy of this flat line bundle. Even though the connection of the line bundle is flat, its holonomy may contain non-trivial information, as Witten has showed in his study of global gravitational anomalies.

A different way of obtaining the same results consists of interpolating in a finite interval rather than the real line. This was the approach takin in Ref. 73a, but the difference with the approach taken here is that one has to impose non-local boundary conditions in the generic case that A_0 and A are not gauge equivalent. The boundary conditions take into account the fact that the Hilbert spaces of $D(A_0)$ and D(A) are rotated by a non-local unitary operator.

In the next section we will apply the results obtained to the study of global gravitational anomalies. An interesting open question independent of anomalies is whether the η-invariant admits a lattice representation. If this were the case, one could probably obtain new insights in the non-perturbative structure of chiral gauge theories.

VIII. GLOBAL ANOMALIES

In this last section we would like to give a review of some recent work of Witten[32] on the study of global gravitational and gauge anomalies. The problem is easy to formulate, but the computations require a good deal of ingenuity and the use of rather powerful results in mathematics. We can illustrate the problem first in a very simple example. Consider a three-dimensional gauge theory with group SU(N) and a single fermion in the fundamental representation. This theory is known to be the classic example of an odd dimensional anomaly[84,47] and it also illustrates some of the ideas involved in the study of global anomalies. We will follow in this introduction the approach of Ref. 47. In any odd number of dimensions there are no local anomalies simply because there is no chirality, and also because the spinor representation of SO(2n+1) is real or pseudoreal, hence by the arguments of Sections II and III, the effective action $\Gamma[A]$ is always well defined under small gauge transformations. However, there may be "large gauge transformations" under which $\Gamma[A]$ changes in a non-trivial way.

In three dimensions, depending on whether one wants to look at the problem from a Euclidean or Hamiltonian point of view, the space of gauge invariant configurations is either $\mathfrak{A}^{(3)}/\mathfrak{G}^{(3)}$ or $\mathfrak{A}^{(2)}/\mathfrak{G}^{(2)}$. If we look at the problem from the Hamiltonian point of view and compactify space to S^2, the group of gauge transformations is given by maps from $S^2 \to SU(N)$, such that $g(x_0) = 1$ ($x_0 \equiv$ south pole, the point at infinity in $\mathbb{R}^2$). In this case, one finds in contrast with 3+1 dimensions, that there is no θ-vacuum structure, because $\pi_2(SU(N)) = 0$, or the number of components of $\mathfrak{G}_0^{(2)}$ is 1, i.e. $\pi_0\mathfrak{G}^{(2)} = \pi_2(SU(N))$. There are no large gauge transformations, and consequently no θ-parameters may arise. This might lead one to think that Gauss' law in the presence of fermions can be imposed trivially. In other words, one can integrate $Q\Psi = 0$ without any problem. (Q is the generator of gauge transformations, and Ψ is any physical wave functional.) This is not so, because $\mathfrak{G}_0^{(2)}$ is not simply connected. In fact, $\pi_1 \mathfrak{G}_0^{(2)} \simeq \pi_3 SU(N) = \mathbb{Z}$. There are non-contractible loops in $\mathfrak{G}_0^{(2)}$, and it may happen that depending on the fermion representation, the vacuum wave functional may not be well defined when one considers the action of these one-parameter families of gauge transformations generating the non-contractible loops in $\mathfrak{G}_0^{(2)}$. This can also be seen from the Euclidean point of view because in this case $\mathfrak{G}_0 = \{g: S^3 \to G,\ g(x_0) = 1\}$ and $\pi_0(\mathfrak{G}_0) = \mathbb{Z}$, i.e. the space of gauge-invariant configurations $\mathfrak{A}/\mathfrak{G}_0$ is not simply connected $\pi_1(\mathfrak{A}/\mathfrak{G}_0) \approx \pi_0(\mathfrak{G}_0) = \mathbb{Z}$. If we look at trivial gauge transformations (those in the trivial sector of $\pi_0(\mathfrak{G}_0)$), the absence of anomalies in odd dimensions implies that $\Gamma[A]$ is well defined. $\Gamma[A^g] = \Gamma[A]$ if $\nu(g) = 0$. (ν = the winding number of $g: S^3 \to SU(N)$.) If we want to consider "large" gauge transformations with $\nu(g) = n \neq 0$, we have to be a bit more careful. If we want to compare $\Gamma[A^g]$ and $\Gamma[A]$ we cannot do it by interpolating from A to A^g along the gauge fiber because we are considering a "disconnected" gauge transformation. Since $\mathfrak{A}$ is topologically trivial however, we can interpolate between A and A^g by some path in $\mathfrak{A}$. For instance $A + t(A^g - A) = A_t$. In $\mathfrak{A}/\mathfrak{G}_0$ A_t represents a non-contractible loop, and we ask what happens to the fermion determinant as a function of t from $t = 0$ to $t = 1$. In any odd number of dimensions, the Dirac operator does not have a spectrum symmetric around zero: the positive and negative non-zero eigenvalues are not paired because there is no analog of γ_5. If we define the fermion determinant using ζ-functions (any other method gives the same conditions), we find that $\Gamma[A]$ develops an imaginary part similar to the spectral asymmetry of the Dirac operator $\eta(0)$ introduced in Section II in the formulation of the Atiyah-Patodi-Singer (APS) index theorem.[16] This is proven as follows:

$$\not{D}_A \phi_\lambda = \lambda\, \phi_\lambda$$

$$\begin{aligned} \zeta(s) &= \sum_\lambda \frac{1}{\lambda^s} \\ &= \sum_{\lambda>0} \frac{1}{|\lambda|^s} + \sum_{\lambda<0} \frac{(-1)^s}{|\lambda|^s} \ . \end{aligned} \tag{8.1}$$

Taking the derivative of $\zeta(s)$ and choosing a branch for $(-1)^s$ close to $s = 0$ which agrees with the perturbative evaluation of the fermion determinant, we have

$$\zeta'(s) = -\sum_{\lambda>0} \frac{\ln|\lambda|}{|\lambda|^s} - \sum_{\lambda<0} (-1)^s \frac{\ln|\lambda|}{|\lambda|^s} + i\pi \sum_{\lambda<0} (-1)^s \frac{1}{|\lambda|^s} \ . \tag{8.2}$$

If we consider a one-parameter family A_t interpolating slowly between A and A^g, we have a four-dimensional gauge configuration A_t which is essentially an instanton in the temporal gauge. Standard arguments[69] show that a consequence of the zero mode of the four-dimensional Dirac operator is that for some t_0, the three-dimensional Dirac operator has a zero. More precisely, if $\nu(g) = n$, there is a spectral flow of n eigenvalues. We know that the eigenvalues of $D(A_0)$ and $D(A_1)$ are identical. If we label them from $-\infty$ to $+\infty$ by $\lambda_n \ldots < \lambda_n < \lambda_{n+1} < \ldots$, having a spectral flow of n eigenvalues means that as a function of t, the eigenvalues of $\not{D}(A_t)$, $\lambda_k(t)$, are such that $\lambda_k(1) = \lambda_{k-n}(0)$. This means in particular, that n eigenvalues changed sign from negative to positive (or vice-versa). Thus if we compare the effective actions as we move along t, we see from (8.2) that

$$\frac{e^{-\Gamma[A^g]}}{e^{-\Gamma[A]}} = e^{i\pi\nu(g)} \quad . \tag{8.3}$$

Hence if we choose $\nu(g) = 2k+1$, the effective action changes sign, and we lose gauge invariance. Even though this can be fixed in this case by adding a Chern-Simons term with the appropriate coefficient,[84] this example makes the point clear. Even though the theory may be free of perturbative anomalies, there may still be an anomaly under large or "global" gauge transformations. If $\mathfrak{G}_0 \times \mathrm{Diff}_0(M)$ represents the group of gauge transformations and diffeomorphisms of a theory defined on some manifold M, the number of components of this group $\pi_0(\mathfrak{G}_0 \times \mathrm{Diff}_0(M))$ is a discrete group, and gives the possible global gauge transformations. If the theory is free from perturbative anomalies, we want to know whether $\exp - \Gamma[A,g]$ provides a non-trivial one-dimensional representation of $\pi_0(\mathfrak{G}_0 \times \mathrm{Diff})$. Were this the case, the theory would have a global gauge and/or gravitational anomaly.

In Ref. 32 Witten analyzed the question of global gravitational anomalies for the $N = 2$, $d = 10$ anomaly-free chiral supergravity theory[11] and the $N = 1$, $d = 10$ theory of supergravity coupled to super Yang-Mills theories with groups SO(32) and $E_8 \times E_8$ which are free of perturbative anomalies. The first question one has to ask is how many components of the $\mathrm{Diff}_0(M)$ are there in ten dimensions (diffeomorphisms which become the identity at infinity). Even though the answer is not known in general, one can consider a special class of diffeomorphisms which are different from the identity only in some ball B_{10}, and become the identity outside. These diffeomorphisms are the same as diffeomorphisms on S^{10}, and thus it is natural to ask about $\pi_0(\mathrm{Diff}\ S^{10})$, and whether the theories mentioned above are invariant under the action of these disconnected diffeomorphisms. Fortunately there is a lot of information in the mathematics literature (see Ref. 32 for references) about $\pi_0(\mathrm{Diff}\ S^{10})$ which can be used to address these questions. This is related to the existence of exotic S^{11}, i.e. manifolds which are homeomorphic to S^{11} but not diffeomorphic. The relation between exotic S^{n+1} spheres and non-trivial diffeomorphisms $\pi: S^n \to S^n$ can be seen as follows: take the standard S^{n+1} sphere and cut it in half at the equator, then glue the two halves back together after transforming the boundary of the northern hemisphere by π. The manifold obtained this way S^{n+1}_π is topologically equivalent to S^{n+1}, but it may not be diffeomorphic to S^{n+1}. It can be shown that S^{n+1} and S^{n+1}_π are diffeomorphic only if π is topologically trivial, and furthermore that any exotic n+1 sphere is S^{n+1}_π for some π.[85] Hence exotic n+1 spheres and topological classes of diffeomorphisms of the n-sphere are in 1-1 correspondence. In the study of the exotic sphere one goes one step further and considers a manifold B_{n+2} whose boundary is S^{n+1}_π. This manifold always exists, because the conditions for a manifold M_n to bound some (n+1)-dimensional manifold is that the Stiefel-Whitney and Pontrjagin numbers vanish for M_n. This is indeed the case for the spheres S^{n+1}_π. Since we are mainly interested in $d = 10$, there is a nice argument in Ref. 32 which is very helpful in understanding some of the features of the exotic spheres. We know from Section II that for compact manifolds without boundary, the integrals of products of characteristic classes are topological invariants. In particular if B is a 12-dimensional manifold, the quantities:

$$I_1 = \int_B (\mathrm{Tr}\, R^2)^3$$
$$I_2 = \int_B \mathrm{Tr}\, R^2\, \mathrm{Tr}\, R^4$$
$$I_3 = \int_B \mathrm{Tr}\, R^6 \qquad (8.4)$$

are independent of the metric and the connection. If B has a boundary $\partial B = M$, in general I_i, $i = 1, 2, 3$, are not topological invariants. However, if on the boundary of M one can solve $\mathrm{Tr}\, R^2 = dH$, then I_1, I_2 can be generalized to

$$\bar{I}_1 = \int_B (\mathrm{Tr}\, R^2)^3 - \int_{\partial B} H(\mathrm{Tr}\, R^2)^2$$
$$\bar{I}_2 = \int_B \mathrm{Tr}\, R^2\, \mathrm{Tr}\, R^4 - \int_{\partial B} H\, \mathrm{Tr}\, R^4 \quad . \qquad (8.5)$$

By choosing the transformations of H under a change in the metric appropriately, $\bar{I}_1$, $\bar{I}_2$ can easily be shown to be topological invariants. Furthermore, the ambiguity in choosing $H \to H + d\Lambda$ does not affect $\bar{I}_1$, $\bar{I}_2$. I_3 on the other hand does not admit a similar generalization. Similarly, in eight dimensions P_1^2 admits a generalization to a manifold with boundary and on the boundary $\mathrm{Tr}\, R^2 = dH$ can be solved. From the formulae of Section II we know:

$$\text{ind}\, i\not{D} = \frac{1}{5760}(7P_1^2 - 4P_2)$$

$$\sigma = -\frac{1}{45}P_1^2 + \frac{7}{45}P_2 \tag{8.6}$$

i.e.
$$\text{ind}\, i\not{D} = \frac{1}{896}(P_1^2 - 4\sigma)\ . \tag{8.7}$$

For a compact 8-manifold the right-hand side of (8.7) is therefore an integer. For a manifold with boundary $\partial B = M$, such that P_1^2 admits a generalization similar to $\bar{I}_1$, one can define a topological invariant:

$$\lambda(M) = \frac{1}{896}(P_1^2 - 4\sigma) \bmod 1 \tag{8.8}$$

which only depends on M. To prove this, consider two manifolds B,B' with boundary M, and glue them together with the same orientation: $X = B+(-B')$. This is a manifold without boundary, and $\lambda(B)-\lambda(B') = \text{ind}\,\not{D}(X)$ = integer. Thus $\lambda(M)$ depends only on M. If M is the exotic seven sphere S_π^7, B has signature equal to 8, and $P_1 = 0$. (See Ref. 32.) Thus $\lambda(S_\pi^7) = -1/28$. Since $\lambda = 0$ for the standard S^7, S_π^7 must be an exotic sphere. In fact, if we take connected sums of S_π^7 with itself, we can generate 27 exotic spheres. It is not so easy to prove[85] that these are all of the exotic 7 spheres. A similar computation in eleven dimensions[32] shows that

$$\lambda(M) = (\alpha P_1^3 + \beta P_1 P_2 - \frac{\sigma}{8.992})\ . \tag{8.9}$$

Thus if $M = S_\pi^{11}$, $\lambda(M) = -1/992$, and we obtain 991 exotic spheres. Again these are all of the exotic eleven spheres.

Let us now review Witten's computations[32] in the case of $N=2$ chiral supergravity in ten dimensions. This theory is known to be anomaly-free,[11] and in fact the absence of anomalies is equivalent to a rather intriguing relation between the Hirzebruck signature σ, and the index of the Rarita-Schwinger and Dirac operators in twelve dimensions, namely

$$\frac{\sigma}{8} = \text{ind}(R-S) - 3\,\text{ind}\, i\not{D}\ . \tag{8.10}$$

In $d=10$ dimensions, the basic spinors in Minkowski space are Weyl-Majorana spinors. Thus the effective action $\Gamma[g]$ for a Weyl-Majorana spinor, is half of $\Gamma_D[g]$ for a Weyl spinor. What we want to compute is the difference $\Gamma_D[g^\pi] - \Gamma_D[g]$. To do this, we proceed in analogy with the gauge case explained at the beginning of this section, and consider a one-parameter family of metrics g_t which interpolate between g and g^π. Since the anomaly only affects the imaginary part of the effective action, we want to compute

$$\Delta\Gamma_D \equiv i\int_0^1 dt\, \frac{d}{dt}\,\text{Im}\,\Gamma[g_t]\ . \tag{8.11}$$

Arguments entirely similar to those presented in Section VII lead to

$$\Delta\Gamma_D = i\pi\eta_D(0) \tag{8.12}$$

where $\eta_D(0)$ is the spectral asymmetry for the Dirac operator (see Section II) computed on the manifold $(M \times S^1)_\pi$. This is a cylinder with base M and "side" the interval [0,1], but with the top and bottom identified with the twist π. In other words, we have fibered the manifold M over the circle S^1 with transition function equal to π. If we have N_D Weyl-Majorana fermions, the total change in the action is:

$$\Delta S_D = N_D \frac{i\pi}{2} \eta_D(0) \; . \tag{8.13}$$

The same argument applies to the Rarita-Schwinger field. We only have to notice that a Rarita-Schwinger field in n+1 dimensions decomposes to a Rarita-Schwinger field plus a spinor in n dimensions. Keeping track of the Weyl-Majorana condition, we get for the change in the action:

$$\Delta S_R = N_R \frac{i\pi}{2} \left(\eta_R(0) - \eta_D(0)\right) \; . \tag{8.14}$$

N_R = number of spin 3/2 Weyl-Majorana fields and in η_R one includes the effect of ghosts. For the self-dual tensor field one has to be slightly more careful. In this case the change in the effective action in the interpolation from g to g^π is given by the spectral asymmetry of an operator introduced in Ref. 16 for the treatment of the signature theorem for manifolds with boundary. This operator is *d acting on even forms on $(M \times S^1)_\pi$. The result for N_S of these fields is:[32]

$$\Delta S_A = -\frac{i\pi}{4} \eta_A(0) \tag{8.15}$$

and the minus sign is due to the different statistics with respect to the fermions. Adding (8.12)-(8.15) we get

$$\Delta S = \frac{i\pi}{2} \left(N_D \eta_D + N_R (\eta_R - \eta_D) - \frac{N_S}{2} \eta_A\right) \; . \tag{8.16}$$

Now we use the APS index theorem formulated at the end of Section IIB. If B is a 12-manifold with boundary $(M \times S^1)_\pi$:

$$\begin{aligned} \tfrac{1}{2}\eta_D &= -\operatorname{ind} \not{D} + \int_B \hat{A}(B) \\ \tfrac{1}{2}\eta_R &= -\operatorname{ind}(R-S) + \operatorname{ind}(i\not{D}) + \int_B [K(B) - \hat{A}(B)] \\ \eta_S &= -\sigma(B) + \int_B L(B) \end{aligned} \tag{8.17}$$

where K(B) is the characteristic polynomial for the index of the R − S operator (2.106),(2.107), and we subtract the contribution from a spin 1/2 index in $\eta_R/2$, because an R - S field in 12 dimensions again decomposes into a Rarita-Schwinger plus a Dirac field on the boundary. Using (8.17) in (8.16):

$$- \Delta S = 2\pi i \left(\frac{1}{2} N_D \text{ind } i\not{D} + \frac{1}{2} N_R \ (\text{ind}(R-S) - 2 \text{ ind } i\not{D}) - \frac{1}{8} N_S \sigma \right)$$

$$- 2\pi i \int_B \left[\frac{N_D}{2} \hat{A}(B) + \frac{N_R}{2} \left(K(B) - 2\hat{A}(B)\right) - \frac{1}{8} N_S L(B) \right] . \qquad (8.18)$$

For the N = 2 chiral supergravity theory, $N_D = -2$, $N_R = 2$, $N_S = 1$, and the perturbative cancellation of anomalies[11] implies that the integrand in (8.18) vanishes identically. Since in d = 12 Euclidean space charge conjugation does not change the chirality of spinors (see Section II), we conclude that ind $\not{D}$ and ind(R-S) are always even integers. Hence modulo $2\pi i$ they do not contribute to ΔS. We thus conclude that

$$\Delta S = 2\pi i \frac{\sigma(B)}{8} . \qquad (8.19)$$

If the theory is formulated in S^{10}, we know from Milnor's results[85] that $\sigma(B) = 0 \bmod 8$, and $\Delta S = 0 \bmod 2\pi i$. Thus in this case the theory is free of global anomalies.

A similar analysis can be carried out in the N = 1, d = 10 case with gauge group $E_8 \times E_8$ or SO(32). The analysis is more complicated because one has to include the effect of the Green-Schwarz counterterms, and the effect of regulator fields. The interested reader is referred to Ref. 32 for more details.

IX. CONCLUSION

We have tried to present a reasonably thorough overview of some of the recent developments in the relation between geometry, topology and field theory. The study of anomalies, both local and global provides a rather fertile ground for the interplay between various fields of physics and mathematics. It seems to us quite likely that these ideas and techniques will eventually generate new qualitative methods for dealing with non-perturbative aspects of quantum field theory. Due to space limitations, some of the issues covered in these lectures have not been treated in as much detail as many readers might wish. It is hoped that the list of references may help the interested reader in finding more information about the topics covered in these lectures.

Acknowledgements

I have learned a great deal about the subject of these lectures from many colleagues and collaborators. I would like to thank M. F. Atiyah, J. Bagger, L. Baulieu, R. Bott, H. Braden, A. Cohen, S. Coleman S. Della

Pietra, E. D'Hoker, E. Farhi, P. Frampton, P. Ginsparg, R. Jackiw, A. Manohar, G. Moore, P. Nelson, A. Niemi, H. Osborn, G. Semenoff, I. M. Singer, R. Stora, E. Witten, and B. Zumino for sharing their insights with me. I would finally like to thank G. Velo and A. Wightman for giving me the opportunity to present part of this material at the Erice Summer School in Problems in Gauge Theory. Subsequently, these lectures were presented at the Summer Workshop of the Ecole Normale Superieure. I would like to thank E. Cremmer, J. Iliopoulos, B. Julia and Ph. Meyer for their invitation and their kind hospitality. I am especially grateful to M. Goulian for many comments which helped improve the final form of these lectures.

REFERENCES

1. S. Adler, Phys. Rev. 177, 2426 (1964), and in *Lectures on Elementary Particles and Quantum Field Theory*, ed. S. Deser *et al.* (M.I.T. Press, 1970); J. Bell and R. Jackiw, Nuovo Cimento 60A, 47 (1969); R. Jackiw, in *Lectures on Current Algebra and Its Applications* (Princeton Univ. Press, 1972); S. L. Adler and W. Bardeen, Phys. Rev. 182, 1517 (1969); W. A. Bardeen, Phys. Rev. 184, 1848 (1969); R. W. Brown, C. C. Shi and B. L. Young, Phys. Rev. 186, 1491 (1969); J. Wess and B. Zumino, Phys. Lett. 37B, 95 (1971); A. Zee, Phys. Rev. Lett. 29, 1198 (1972).
2. J. Steinberger, Phys. Rev. 76, 1180 (1944); J. Schwinger, Phys. Rev. 82, 664 (1951); L. Rosenberg, Phys. Rev. 129, 2786 (1963); R. Jackiw and K. Johnson, Phys. Rev. 182, 1459 (1969); S. Adler and D. G. Boulware, Phys. Rev. 184, 1740 (1969); S. L. Adler, B. W. Lee, S. B. Treiman and A. Zee, Phys. Rev. D4, 3497 (1971); R. Aviv and A. Zee, Phys. Rev. D5, 2372 (1972); M. V. Terentiev, JETP Lett. 14, 140 (1971); A. M. Belavin, A. M. Polyakov, A. S. Schwarz and Yu. S. Tyupkin, Phys. Lett. 59B, 85 (1975); G. 't Hooft, Phys. Rev. Lett. 37, 8 (1976), Phys. Rev. D14, 3437 (1976); C. Callan, R. Dashen and D. J. Gross, Phys. Lett. 63B, 334 (1976); R. Jackiw and C. Rebbi, Phys. Rev. Lett. 37, 172 (1976).
3. D. J. Gross and R. Jackiw, Phys. Rev. D6, 477 (1972); C. Bouchiat, J. Iliopoulos and Ph. Meyer, Phys. Lett. 38B, 519 (1972); H. Georgi and S. Glashow, Phys. Rev. D6, 429 (1972).
4. P. H. Frampton and T. W. Kephart, Phys. Rev. Lett. 50, 1343, 1347 (1983); P. K. Townsend and G. Sierra, Nucl. Phys. B222, 493 (1983); B. Zumino, W. Y.-Shi and A. Zee, Nucl. Phys. B239, 477 (1984); P. H. Frampton and T. W. Kephart, Phys. Rev. D28, 1010 (1983); P. H. Frampton, J. Preskill and H. V. Dam, Phys. Lett. 124B, 209 (1983).
5. For a review on GUT theories see for example, P. Langacker, Phys. Rep.

72, 185 (1981); A. Zee in *Unity of Forces in the Universe*, Vol. I,II, (World Scientific, 1982).

6. For a review on Kaluza-Klein theories and references to earlier literature, see A. Salam and J. Strathdee, Ann. of Phys. 141, 316 (1982); P. Van Nieuwenhuizen, "An Introduction to Simple Supergravity and the Kaluza-Klein Program" in *Relativity Groups and Topology - II*, Ed. R. Stora (North-Holland, 1985).
7. G. 't Hooft in *Recent Developments in Gauge Theories"*, Eds. G. 't Hooft *et al.* (Plenum Press, N.Y., 1980); A. A. Ansel'm, JETP Lett. 32, 138 (1980); A. Zee, Phys. Lett. 95B, 290 (1980); Y. Frishman, A. Schwimmer, T. Banks and S. Yankielowicz, Nucl. Phys. B177, 157 (1981); S. Coleman and B. Grossman, Nucl. Phys. B203, 205 (1982); G. R. Farrar, Phys. Lett. 96B, 273 (1980); S. Weinberg, Phys. Lett. 102B, 401 (1981); C. H. Albright, Phys. Rev. D24, 1969 (1981); I. Bars, Phys. Lett. 109B, 73 (1982); T. Banks, S. Yankielowicz and A. Schwimmer, Phys. Lett. 96B, 67 (1980); A. Schwimmer, Nucl. Phys. B198, 269 (1982).
8. G. Moore and P. Nelson, Phys. Rev. Lett. 53, 1519 (1984) and Comm. Math. Phys. 100, 83 (1985).
9. A. Manohar, G. Moore and P. Nelson, Phys. Lett. 152B, 68 (1985); L. Alvarez-Gaumé and P. Ginsparg, Nucl. Phys. B262, 439 (1985); J. Bagger, D. Nemeschanski, S. Yankielowicz, SLAC-PUB-3588; E. Cohen and E. Gomez, Nucl. Phys. B254, 235 (1985); P. DiVecchia, S. Ferrara and L. Girardello, Phys. Lett. 151B, 199 (1985).
10. R. Delbourgo and A. Salam, Phys. Lett. 40B, 381 (1972); T. Eguchi and P. Freund, Phys. Rev. Lett. 37, 1251 (1976).
11. L. Alvarez-Gaumé and E. Witten, Nucl. Phys. B234, 269 (1983).
12. M. B. Green and J. H. Schwarz, Phys. Lett. 149B, 117 (1984).
13. M. B. Green, J. H. Schwarz and P. C. West, Nucl. Phys. B254, 377 (1984).
14. For some useful reviews of string theories before the construction of the heterotic string[58] see J. Scherk, Rev. Mod. Phys. 47, 123 (1975); J. H. Schwarz, Phys. Rep. 89, 223 (1982); M. B. Green, Surveys in H.E.P. 3, 127 (1983); L. Brink, "Superstrings", Lectures at the 1984 Bonn Summer School on Supersymmetry, CERN preprint TH 4006/84.
15. M. F. Atiyah and I. M. Singer, Ann. of Math. 87, 485, 546 (1968), 93, 1, 119, 139 (1971); M. F. Atiyah and G. B. Segal, Ann. of Math. 87, 531 (1968).
16. M. F. Atiyah, V. I. Patodi and I. M. Singer, Math. Proc. Camb. Phil. Soc. 77, 43 (1975), 78, 405 (1975), 79, 71 (1976).
17. R. Jackiw, C. Nohl and C. Rebbi in *Particles and Fields*, Eds. D. Boch and A. Kamal (Plenum Press, N.Y. 1978); N. K. Nielsen, H. Römer, B. Schroer, Nucl. Phys. B136, 478 (1978); for a recent review see R. Jackiw

in *Relativity Groups and Topology II*, Eds. B. S. DeWitt and R. Stora (North Holland, 1984).

18. M. F. Atiyah and I. M. Singer in Proc. Nat. Acad. Sci. USA 81, 2597 (1984).
19. L. Alvarez-Gaumé and P. Ginsparg, Nucl. Phys. B243, 449 (1984).
20. L. Alvarez-Gaumé and P. Ginsparg, Ann. of Phys. 161, 423 (1985).
21. O. Alvarez, I. M. Singer and B. Zumino, Comm. Math. Phys. 96, 409 (1984).
22. C. Gomez, Salamanca preprint (1984).
23. S. Goldberg, *Curvature and Homology* (Dover, N.Y., 1982).
24. T. Sumitani, UT-KOMABA 84-7 (1984).
25. J. Lott, Comm. Math. Phys. 93, 533 (1984).
26. R. Stora, Lecture at the 1983 Cargese Summer School, in *Progress in Gauge Theory*, Ed. G. 't Hooft *et al.* (Plenum Press, N.Y., 1984); B. Zumino, Lectures at the 1983 Les Houches Summer School, *Relativity, Groups and Topology II*, Eds. B. S. DeWitt and R. Stora (North Holland, 1984); B. Zumino, Y. S. Wu and A. Zee, Nucl. Phys. B239, 477 (1984); L. Baulieu, Nucl. Phys. B241, 557 (1984).
27. J. Wess and B. Zumino, Phys. Lett. 37B, 95 (1971).
28. E. Witten, Nucl. Phys. B223, 422 (1983).
29. L. D. Faddeev, Phys. Lett. 145B, 81 (1984); L. D. Faddeev and S. Shatashvili, Math. Phys. 60, 206 (1984).
30. B. Zumino, Nucl. Phys. B253, 477 (1985); R. Jackiw, MIT preprint CTP-1298, to appear in Comments in Nuclear and Particle Physics.
31. P. Nelson and L. Alvarez-Gaumé, Comm. Math. Phys. 99, 103 (1985).
32. I. M. Singer, MIT preprint, to appear in the Proceedings of the Conference in honor of E. Cartan, June 1984. E. Witten, Phys. Lett. 117B, 324 (1982), Comm. Math. Phys. 100, 197 (1985).
33. N. K. Nielsen, Nucl. Phys. B244, 499 (1984); O. Piguet and K. Sibold, Nucl. Phys. B247, 484 (1984); G. Girardi, R. Grimm and R. Stora, Annecy preprint LAPP-TH-130 (1985); S. Ferrara, L. Girardello, O. Piguet, and R. Stora, CERN preprint TH 41134/85; L. Bonora, P. Pasti and M. Tonin, Padova preprint DFPD 20/84 (1985); H. Itoyama, V. P. Nair, H. C. Ren, Columbia-Princeton preprint (March 1985); Guadagnini, Konishi and M. Mintoker, Pisa preprint IFHP-TH-10/85; Zi Wang, Youg-Shi Wu, Univ. of Utah preprint (August 1985); D. S. Hwang, LBL preprint LBL-20133 (September 1985).
34. See for example F. Gliozzi, J. Scherk and D. Olive, Nucl. Phys. B122, 253 (1977); E. Cartan, *Theory of Spinors* (Dover, N.Y., 1981); C. Wetterich, Nucl. Phys. B211, 177 (1982).
35. E. Cremer, Lectures at the Trieste School on Supergravity, 1981.
36. H. Georgi, *Lie Algebras in Particle Physics* (Benjamin, 1982).

37. H. Weyl, *Classical Groups* (Princeton Univ. Press, 1939).

38. For useful and practical information on group theory for model building and group embeddings, see D. Slanski, Phys. Rep. 79, 1 (1981).

39. W. Bardeen and B. Zumino, Nucl. Phys. B244, 421 (1984); see also Ref. 20.

40. For general results and definitions of ζ-functions, see R. Seeley, Proc. Symp. Pure Math. 10, 288 (1967), Am. J. Math. 91, 889 (1969), 91, 963 (1969). For physical applications of the ζ-function definition of determinants, see J. Dowker and R. Critchley, Phys. Rev. D13, 3224 (1976); S. W. Hawking, Comm. Math. Phys. 55, 133 (1977).

41. For a derivation of the index theorem more accessible to physicists, see L. Alvarez-Gaumé, Comm. Math. Phys. 90, 161 (1983), J. Phys. A16, 4177 (1983); E. Getzler, Comm. Math. Phys. 92, 163 (1983); D. Friedan and P. Windey, Nucl. Phys. B235, 395 (1984); B. Zumino, LBL preprint 17972, to appear in the Proceedings of the Shelter Island Conference - II (June 1983).

42. See for instance, A. Lichnerowicz, *General Theory of Connections and The Holonomy Group* (Noorhooff, 1976); Goldberg, *Curvature and Homology* (Dover, N.Y.); R. Bott and L. Tu, *Differential Form in Algebraic Geometry* (Springer-Verlag, 1982).

43. S. S. Chern, *Complex Manifolds without Potential Theory* (Van Nostrand, 1967); J. Milnor and J. Stasheff, *The Theory of Characteristic Classes* (Princeton Univ. Press, 1974); T. Eguchi, P. B. Gilkey and A. Hanson, Phys. Rep. 66, 243 (1980); F. Hirzebruch, *Topological Methods in Algebraic Geometry* (Springer-Verlag, 1966).

44. N. Steenrod, *The Topology of Fiber Bundles* (Princeton Univ. Press, 1951); D. Husemoller, *Fiber Bundles* (Springer-Verlag, 1966).

45. For a detailed analysis and references to earlier literature, see P. Van Nieuwenhuizen, Phys. Rep. 68, 189 (1981).

46. M. F. Atiyah and R. Bott in *Differential Analysis*, Bombay Colloquium (Oxford Univ. Press, 1964).

47. L. Alvarez-Gaumé, S. Della Pietra and G. Moore, Ann. of Phys. 163, 288 (1985).

48. A. Niemi and G. Semenoff, Phys. Rev. Lett. 55, 927 (1985). See also "Index Theorems on Open Infinite Manifolds", IAS preprint (1985).

49. K. Fujikawa, Phys. Rev. Lett. 42, 1195 (1979), 44, 1733 (1980), Phys. Rev. D21, 2848 (1980), D22, 1499(E) (1980), D23, 2262 (1981), D29, 285 (1984); M. B. Einhorn and D. R. T. Jones, Phys. Rev. D29, 331 (1984); A. P. Balachandran, G. Marno, V. P. Nair, and C. G. Trahern, Phys. Rev. D25, 2713 (1983).

50. O. Alvarez and B. Zumino, LBL preprint in preparation.

51. M. F. Atiyah, *K Theory* (Benjamin, N.Y., 1967).

52. O. Alvarez, I. M. Singer and B. Zumino, LBL preprint in preparation.
53. M. F. Atiyah, *The Signature of Fiber Bundles*, in Collected Mathematical Papers in Honor of K. Kodaira (Tokyo Univ. Press, Tokyo, 1969).
54. C. Becchi, A. Rouet and R. Stora, Ann. of Phys. (N.Y.) 98, 287 (1976).
55. W. A. Bardeen, Phys. Rev. 184, 1848 (1969).
56. C. K. Chan, G. H. Ying, W. Ke, Phys. Lett. B134, 67 (1984); A. Manohar and G. Moore, Nucl. Phys. B243, 55 (1984); C. Callan and E. Witten, Nucl. Phys. B239, 161 (1984); K. Chou, H. Guo, X. Li, K. Wu and X. Song, AS-ITP-84-018; H. Kawai, H. Tye, Phys. Lett. B140, 403 (1984); J. L. Mañez, LBL-17318 (1984).
57. See R. Stora's Cargese Lectures quoted in Ref. 26.
58. D. Gross, J. Harvey, E. Martinec and R. Rohm, "The Heterotic String I-II", Princeton preprints to appear in Nucl. Phys. B.
59. P. Candelas, G. Horowicz, A. Strominger and E. Witten, Princeton preprint, and in the Proceedings of the Argonne Meeting on Anomalies, Geometry and Topology, March 1985; W. A. Bardeen and A. White Eds. (World Scientific, 1985).
60. S. Coleman, J. Wess and B. Zumino, Phys. Rev. 177, 2239 (1969).
61. C. Callan, S. Coleman, J. Wess and B. Zumino, Phys. Rev. 177, 2247 (1969).
62. W. A. Bardeen and V. Visnjic, Nucl. Phys. B149, 422 (1982); W. Büchmuller, S. Love, R. Peccei, T. Yanagida, Phys. Lett. 124B, 67 (1983); W. Büchmuller, R. Peccei, T. Yanagida, Nucl. Phys. B227, 503 (1983), B231, 53 (1984).
63. M. Berger, Comptes Rendus Acad. Sci.,Paris, 262A, 1316 (1966).
64. A. Lichnerowicz, *Geometry of Groups of Transformations* (Noordhoff Ed. Holland, 1976).
65. E. Cohen and C. Gomez, Nucl. Phys. B254, 235 (1985); P. Di Vecchia, S. Ferrara and L. Girardello, Phys. Lett. 151B, 151 (1985).
66. N. Marcus, Phys. Lett. 157B, 383 (1985).
67. C. Hull and E. Witten, Princeton preprint (June 1985).
68. Proceedings of the Argonne Workshop on *Anomalies, Topology and Geometry*, Eds. W. A. Bardeen and A. White (World Scientific, 1985).
69. See for example R. Jackiw and C. Rebbi, Phys. Rev. Lett. 37, 172 (1976); C. Callan, R. Dashen and D. Gross, Phys. Rev. D17, 2717 (1978).
70. M. Berry, Proc. Royal Soc.,London A392, 45 (1984); B. Simon, Phys. Rev. Lett. 51, 2167 (1983).
71. H. Sonoda, Caltech preprints 68-1271, 68-1242.
72. B. Zumino, Phys. Lett. 87B, 203 (1979).
73. (a) L. Alvarez-Gaumé and S. Della Pietra in *Recent Developments in Quantum Field Theory* (Niels Bohr Centennial Conference), Eds. J. Amb-

jøorn, B. J. Durhuus and J. L. Petersen (Elsevier Science Pub. B.V., 1985); (b) L. Alvarez-Gaumé, S. Della Pietra and V. Della Pietra, Harvard preprint HUTP-85/A034, to appear in Phys. Lett. B.

74. L. Alvarez-Gaumé, S. Della Pietra and V. Della Pietra, Harvard preprint HUTP-85/A083, to appear in Comm. Math. Phys.

75. R. C. Ball and H. Osborn, DAMTP85-10.

76. D. Freed and J. Bismut, D. R. Acad. Sc. Paris t.301, serie I, No. 14, 54 (1985), and paper in preparation.

77. For a comprehensive review, see M. Peskin in *Recent Developments in Field Theory and Statistical Mechanics*, Eds. J. B. Zuber and R. Stora (North Holland, 1982).

78. N. K. Nielsen and N. Ninomiya, Nucl. Phys. B185, 20 (1981). See also J. Kogut's contribution in the volume in Ref. 77.

79. See for instance, S. Coleman in *Aspects of Symmetry* (Cambridge Univ. Press, 1985).

80. D. Weingarten, Phys. Rev. Lett. 51, 1830 (1983); N. Nussinov, Phys. Rev. Lett. 51, 2081 (1983); E. Witten, Phys. Rev. Lett. 51, 2351 (1983); C. Vafa and E. Witten, Nucl. Phys. B234, 173 (1984).

81. See H. Georgi, Lectures presented at the 1985 Les Houches Summer School.

82. S. Coleman and E. Witten, Phys. Rev. Lett. 45, 100 (1980).

83. R. Seeley in *Proc. Sym. Pure Mathematics*, Vol. X (American Math. Soc., 1967).

84. A. N. Redlich, Phys. Rev. Lett. 52, 1 (1984); see also Ref. 11.

85. J. Milnor, Ann. of Math. 64, 399 (1956); M. Korvaire and J. Milnor, Ann. of Math. 77, 504 (1963).

CONSTRUCTIVE GAUGE THEORY*

T. Balaban** and A. Jaffe***

** Department of Mathematics
Northeastern University
Boston, MA 02115

***Harvard University
Cambridge, MA 02138

CONTENTS

---

*Supported in part by the National Science Foundation under Grants PHY 82-03669, PHY 85-13554 and DMS 84-01989.

I. BASIC CONCEPTS AND MODELS

Here we provide some insight into mathematical methods to analyze quantized gauge theories. This approach is being used to establish existence as well as to prove properties of gauge field models. Up to now, only abelian gauge fields are known to exist, namely the U(1) Higgs model on $\mathbb{R}^2$ and $\mathbb{R}^3$ and electrodynamics on the tori T^2 and T^3. In these lectures we study infrared properties of some of these models. The non-Abelian gauge models are interesting for additional reasons: First, they have a natural geometric interpretation. Furthermore, because of asymptotic freedom, non-Abelian models have better local regularity properties than Abelian theories. Furthermore, they provide a rich analytic structure.

In these notes we discuss ultraviolet properties of the $d = 3$ Yang-Mills theory. We use the Euclidean point of view throughout. The connection to real time follows from established results [1].

1. Models

(i) The φ^4 Scalar Field. A nonlinear field model with a stable ground state arises from the Euclidean action

$$A = A(\varphi) = \int \left[\frac{1}{2} |\nabla\varphi|^2 + \lambda^2 |\varphi|^4 + \sigma |\varphi|^2 \right] dx \quad , \tag{1.1}$$

where σ and λ are real parameters. Here φ is vector valued. This model is "simple" because the action density is a polynomial of low degree and because its interpretation involves neither topology nor geometry.

This φ^4 model has been studied extensively for the real or complex valued φ. A general reference to such results is [2]. For space-time dimension $d < 4$, existence of a Wightman field theory is known, and many more detailed properties of the fields have been established. Correspondingly, for $d > 4$, a nonexistence theorem has been established. The case $d = 4$ is extremely delicate, although most people also expect nonexistence for $d = 4$.

The model has been proved to extend to complex λ in a small, pie-shaped sector for $d = 1,2,3$ of the form

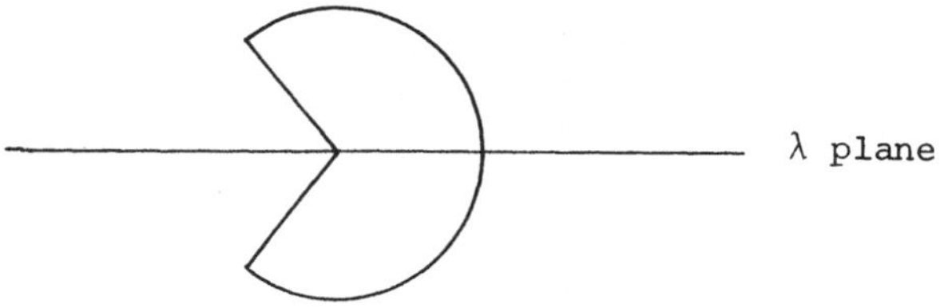

and in $d = 4$ to the region

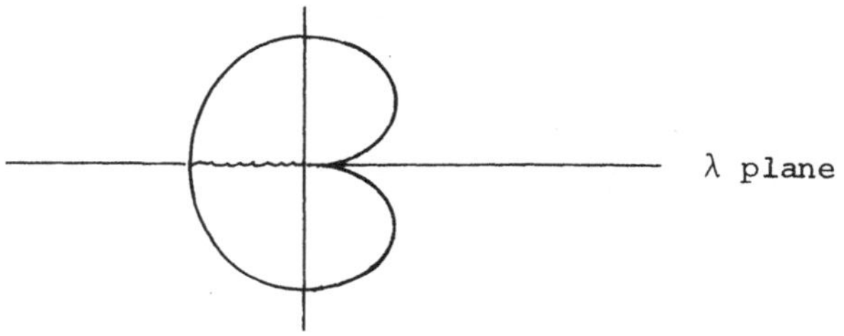

This region in $d=4$ excludes any point on the positive real axis, and so does not contain a candidate for a field theory with a stable ground state. In fact, the $d<4$ analyticity presumably extends to the union of these regions, namely to a small circle with a cut along the negative axis,

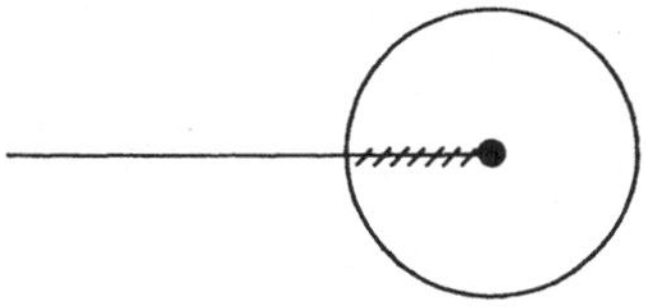

though this has not been established. These analyticity results, as well as other properties of the model, yield Borel summability at $\lambda = 0$ for $d=1,2,3$. In other words, the full theory can be recovered from the perturbation series in the coupling constant.

(ii) Gauge Theories. Gauge theory provides a generalization of electromagnetism. The potentials A in gauge theory take values in the Lie algebra of a compact Lie group G, the gauge group. The fields F are related to the potential A by

$$F = dA + gA\wedge A \quad \text{or in components} \quad F_{ij} = \partial_i A_j - \partial_j A_i + g[A_i, A_j] \quad .$$

Geometrically, A is a connection defining a covariant differentiation on a space with curvature F. The action for the Yang-Mills theory is

$$\mathscr{A} = \|F\|^2 = \frac{1}{4}\,\mathrm{Tr}\sum_{ij}\int F_{ij}^{*}\, F_{ij}\, dx \quad . \tag{1.2}$$

Here g is the gauge coupling constant.

Electromagnetism is the case $G=U(1)$ and $g=e$, the electric charge. The Lie algebra in this case is $\mathbb{R}^1$ and the trace is the identity. The variational equations for $\mathscr{A}$ characterize the critical points of $\mathscr{A}$ and correspond in this case to F being harmonic. For the general Yang-Mills action the critical points can also be interpreted in this way and yield the classical approximation for quantum fields.

(iii) The Higgs Model. The Higgs model combines the φ^4 model with gauge theory. The simplest case is the coupling of electromagnetism to a charged (complex) scalar φ^4 theory. The coupling is taken to be "minimal," as defined by the covariant gradient $|d_A\varphi|^2 = |d\varphi - iA\varphi|^2$. (By convention, A is chosen to be real.) The action is then

$$\mathscr{A} = \frac{1}{4e^2}\int F_{ij}^2\, dx + \frac{1}{2}\int |d_A\varphi|^2\, dx + \int (\lambda^2|\varphi|^4 - |\varphi|^2)\, dx \quad . \tag{1.3}$$

The physics interpretation of φ and the nontrivial geometry associated with critical points of $\mathscr{A}$ depend on the negative coefficient for the quadratic term, as we discuss in Section I.7 on infrared behavior.

2. Regularization

Euclidean field theory requires the construction of a measure with density $\exp(-\mathscr{A})$ on field space. Such an integral is extremely singular, so regularization is necessary. The two standard problems are local regularity and infinite volume, and generally they can be separated. The local regularity is associated with ultraviolet divergences and renormalization counter terms.

Even though some supersymmetric models do not exhibit ultraviolet divergences, taking advantage of this symmetry from a mathematical point of view has not--up to now--been possible. Most quantum field models exhibit ultraviolet divergences in perturbation theory. They require an approximation, parameterized, for example, by ε. The approximate model needs to be analyzed uniformly in ε; finally the limit $\varepsilon \to 0$ must yield to the desired theory.

Often models whose Lagrangians exhibit a basic symmetry are less divergent than might otherwise be expected. Gauge invariance and supersymmetry fall in this category, leading to algebraic identities whose consequence is the cancellation of certain divergences. To exhibit this desired increase in regularity, the regularization procedure must preserve the symmetry. A regularization which preserves gauge invariance alone (much less supersymmetry) is hard to find.[1] In fact the only known regularization which preserves both gauge invariance and reflection positivity (the property which yields a Hilbert space and a quantum mechanics without ghosts) is the lattice approximation proposed by Wilson [3]. Lattice approximation have substantial disadvantages:

---

[1]Some supersymmetric models do not exhibit ultraviolet divergences at all--they are "finite." However, taking advantage of this symmetry from a mathematical point of view has not been possible up to now. No regularization preserving supersymmetry, as well as reflection positivity has been used, even for non-gauge models. Without the supersymmetry, the models are formally more singular, and the $\varepsilon \to 0$ limit becomes more difficult (or impossible) to control. For example, the simplest supersymmetric field theory is the $\lambda^2\varphi^4 + \lambda\bar{\psi}\psi\varphi$ coupling of a scalar to a fermion with a Yukawa interaction. In two dimensions, this model exists by the classical methods of constructive field theory, which do not rely on supersymmetry. In fact, the ultraviolet divergences cancel identically (for the specified ratio of Yukawa to φ^4 coupling constants) and this cancellation has an interpretation in terms of supersymmetry. However, the cancellation has not been used in the mathematical construction. In fact, it is not even known whether the ground state of the existing theory is supersymmetric--although, in principle, existing methods should provide a framework for proving this result. Thus one should analyze supersymmetry within the framework of constructive field theory (possibly using cluster expansion methods) to determine whether the supersymmetry of the ground state is implemented or is broken. Until such a mathematical analysis of supersymmetry is carried out--and is extended to provide analytic control over regularizations of supersymmetric actions--we must conclude that "finite models are more difficult than infinite ones!" Perhaps a new continuum regularization would lead to a solution of this dilemma.

D1) Such an approximation is analytically extremely "messy" and unaesthetic, and

D2) Geometrical insight is greatly obscured.

Nevertheless, it is possible to proceed without further ado to the study of specific models using lattice approximations. Moreover, the lattice regularization has a number of specific advantages:

A1) It preserves the physical (reflection) positivity. This property is an infinite set of nonlinear inequalities. They have never been established (and appear extremely difficult to establish) except in cases where they are preserved by the regularization.

A2) The lattice approximation is compatible with gauge invariance, which is necessary to exhibit cancellation of many divergences.

A3) Certain topological interpretations of configurations of lattice gauge models have been worked out.

A4) Natural geometric concepts (such as orbits of the gauge group or critical points of the action functional) can be formulated in the lattice approximation. Sobolev spaces on the lattice provide a framework to study such questions uniformly as the lattice spacing tends to zero. We describe results along these lines in later sections.

A5) Because of the compact domain of integration on a finite lattice the lattice approximation provides a natural framework to investigate problems associated with gauge fixing and the Faddeev-Popov procedure, etc.

In the infinite dimensional case, these concepts have only a formal character. We must deal with the gauge fixing problem in order to pass to the infinite dimensional limit. The reason is that we require uniform estimates to control the $\varepsilon \to 0$ convergence. The estimates which can be established without gauge fixing are not uniform.

It is well-known that the ultraviolet behavior of gauge theories is gauge dependent. For example, the classical solutions (critical points of the action functional $\mathscr{A}$) have different regularity properties in different gauges. The same is true for Green's functions of the covariant Laplacian, which is the fundamental object for establishing properties in perturbation theory.

The method to deal with questions of local regularity without defining a global cross section (i.e., without global gauge fixing) is to employ local gauge fixing. The notion of selecting which gauge degrees of freedom to fix based on localization in phase space cells is quite natural for a non-perturbative field theory expansion. Degrees of freedom are chosen to correspond to given momentum ranges and given space-time localization properties. Such an organization is characteristic of "phase cell cluster expansions" and can naturally be implemented using renormalization group methods explained in these lectures.

3. Lattice Approximations, Gauge Invariance and Reflection Positivity

The gauge potential is a connection which defines the notion of parallel transport. This concept motivates the definition of the basic lattice variables as parallel transport operators along bonds of a lattice. The bonds $b = (b_-, b_+)$ connect nearest neighbor sites. The variables

U(b) have values in a compact Lie group G, the gauge group. For simplicity we assume that for some n, $G \subset U(n)$, the group of $n \times n$ unitary matrices. Let $\mathcal{A}$ denote the space of gauge field, i.e. the space of functions from bonds to G.

We consider configurations U which satisfy $U(b)^{-1} = U(b^{-1})$, where

$$b^{-1} = (b_-, b_+)^{-1} \equiv (b_+, b_-) \quad . \tag{1.4}$$

An ordered sequence of bonds $b_1, \ldots, b_n$ define a contour $\Gamma = b_1 \circ b_2 \circ \cdots \circ b_n$ which traverses the bonds in order. We assume $(b_j)_+ = (b_{j+1})_-$. Define the reversed contour Γ^{-1} by

$$\Gamma^{-1} = b_n^{-1} \circ \cdots \circ b_1^{-1} \tag{1.5}$$

and the contour variable $U(\Gamma)$ by

$$U(\Gamma) = U(b_1) \cdots U(b_n) \quad . \tag{1.6}$$

This variable defines parallel transport along Γ.

Parallel transport around the boundary ∂p of an elementary square p is especially important. Let Tr denote the matrix trace normalized so $\mathrm{Tr}(I) = 1$. If $\partial p = b_1 \circ \cdots \circ b_4$, note that $\mathrm{Tr}\, U(\partial p)$ is invariant under cyclic permutation of $b_1, \ldots, b_4$ and hence a function of p. K. Wilson proposed that an appropriate action functional for the lattice theory is

$$A^{\varepsilon}(U) = \sum_{p \in T_\varepsilon^d} \varepsilon^{d-4} [1 - \mathrm{Re}\, \mathrm{Tr}\, U(\partial p)] \tag{1.7}$$

where T_ε^d denotes the set of bonds on a d-dimensional toroidal lattice with spacing ε. Pointwise, as $\varepsilon \to 0$,

$$\lim_{\varepsilon \to 0} A^{\varepsilon}(U) = \int \mathrm{Tr}|F|^2 \quad . \tag{1.8}$$

Here we consider a smooth connection A with curvature $F = dA + A \wedge A$. Then $A^{\varepsilon}(U)$ is defined by restricting A to the ε-lattice and defining U(b) on that lattice as

$$P \exp\left(ig \int_b A\right)$$

where P denotes the path ordered integral. More precisely, this is the solution $U(\varepsilon)$ to the differential equation

$$\frac{d}{ds} U(s) = igA(s)U(s), \qquad U(o) = I \quad ,$$

where $s \in [o, \varepsilon]$ parameterizes the bond b. Approximately $U(\varepsilon)$ is equal to $\exp(ig\varepsilon A(b))$ where A(b) is the component of A along b, evaluated at b_-. For b along the μ-th coordinate axis and e_μ a corresponding unit vector,

$$A_\mu(x) = \lim_{\varepsilon\to 0} (ig\varepsilon)^{-1} \ln U(x, x+\varepsilon e_\mu) \quad . \tag{1.9}$$

Let $h: T_\varepsilon \to G$ be a G-valued function on the lattice. Such a function defines a gauge transformation of $\{U(b)\}$ as follows:

$$U(b) \to U^h(b) = h(b_-)^{-1} U(b) h(b_+) \quad . \tag{1.10}$$

These transformations form a group, since

$$(U^{h_1})^{h_2} = U^{h_1 h_2} \quad . \tag{1.11}$$

Let $\mathfrak{G}$ denote this group, and define $\mathfrak{A}/\mathfrak{G}$ as the space of gauge orbits. For every such h,

$$A^\varepsilon(U^h) = A^\varepsilon(U) \quad , \tag{1.12}$$

so we have a "gauge invariant lattice approximation." In other words, A^ε is a function on the spaces of gauge orbits $\mathfrak{A}/\mathfrak{G}$.

Physics suggests that all observable functions f are gauge invariant, i.e. f is defined on $\mathfrak{A}/\mathfrak{G}$. Fundamental objects to study are expectation values

$$\langle f \rangle_{\mathfrak{A}} = \frac{1}{Z} \int_{\mathfrak{A}} f(U) e^{-\frac{1}{g^2} A^\varepsilon(U)} dU \quad . \tag{1.13}$$

Here dU denotes the product of normalized Haar measures $dU(b)$,

$$\prod_b dU(b) \tag{1.14}$$

and Z is the normalizing constant $\int \exp(-A^\varepsilon(U))\, dU$. The integral (1.13) is well-defined, since (1.14) is a finite product of Haar measures on a compact group.

Since both f and A^ε are functions on $\mathfrak{A}/\mathfrak{G}$ and since the Haar measure dU is translation invariant, the integral (1.13) should really be expressed as an integral over $\mathfrak{A}/\mathfrak{G}$ rather than over $\mathfrak{A}$. Here we can use either point of view, since the integral (1.13) is defined. Symbolically, we can draw the space of gauge configuration $\mathfrak{A}$ as a union of orbits:

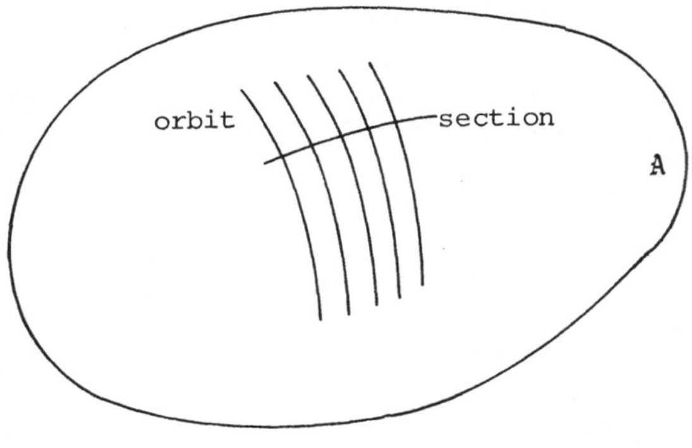

Fig. 1. Gauge orbits in the space $\mathfrak{A}$ of gauge configurations and a section in $\mathfrak{A}/\mathfrak{G}$.

The choice of one point on each orbit is the choice of a global section. In the physics literature such a choice is called the choice of gauge. The definition of a gauge invariant integral over $\mathfrak{A}/\mathfrak{G}$ then reduces to the definition of an integral over such a global cross-section. In other words, we may wish to replace (1.13) by an integral

$$\langle f\rangle_{\mathfrak{A}/\mathfrak{G}} = \frac{1}{Z}\int_{\mathfrak{A}/\mathfrak{G}} f(U)e^{-\frac{1}{g^2}A^{\varepsilon}(U)}\, dW(U) , \qquad (1.15)$$

where "dW(U)" would be a natural restriction of dU to $\mathfrak{A}/\mathfrak{G}$. The definition of such a measure is called *gauge fixing*.

Why do we care? Since (1.13) is well-defined, it seems unnecessary to complicate life with (1.15). As an additional argument against trying to define (1.15) is that in many cases it is impossible to do it in a nice way, i.e. by extending the section in Figure 1 to a global section. This result was established for continuum gauge theories by Singer. More precisely [4]:

> No continuous global cross-section of a gauge theory on the sphere S^4 (or S^3) with a non-abelian, compact, Lie Gauge group exists.

(A similar result presumably applies on the lattice.) This fact depends on the topology of the underlying space S^4. On the Euclidean space $\mathbb{R}^4$, there is a continuous global section. This is given by axial gauge for which $A_1 = 0$, $A_2(x_1=0)$, $A_3(x_1=x_2=0) = 0,\cdots,A_d(x_1=x_2=\cdots=x_{d-1}=0) = 0$. The same gauge is also good, for example, for a cube or a ball in $\mathbb{R}^d$ with Neumann boundary conditions.

The real reason that we care about gauge fixing is the possibility to formulate perturbation expansions in powers of the coupling constant g. Through such a perturbation theory we make contact with standard physics, especially questions of renormalizability. In the following section we explain why perturbation theory forces an analysis of gauge fixing.

Let us now turn to reflection positivity,sometimes called physical positivity or Osterwalder-Schrader positivity [5]. It is this property which provides the connection to the transfer matrix of statistical mechanics (on the lattice) and to the Hilbert space and Hamiltonian of quantum theory (in the continuum limit).

Reflection positivity is defined in terms of a reflection plane, perpendicular to one lattice direction (which we can interpret as the time direction). For convenience, we choose this plane Π midway between two lattice planes. This plane divides the torus into two pieces, T_+ and T_-, as illustrated in Figure 2. We define two spaces of gauge invariant functions $\mathscr{E}_\pm$ depending on bond variables $U(b)$ with $b \in T_\pm$. We also define an antilinear map

$$\Theta:\ \mathscr{E}_\pm \to \mathscr{E}_\mp$$

by

$$(\Theta f)(U) = \bar{f}(rU)$$

where r denotes reflection of bonds in the plane Π, and

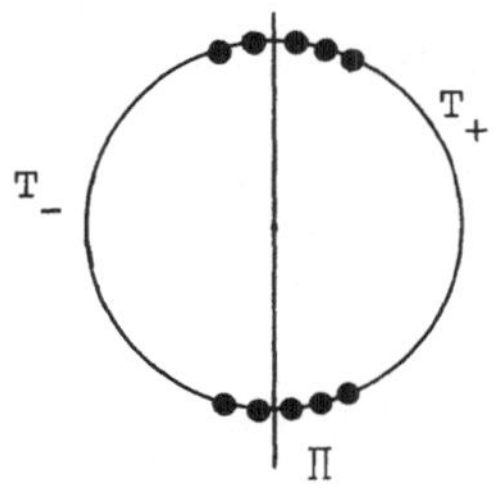

Fig. 2. The reflection plane Π divides the torus T into T_+ and T_-.

$$rU(b) = U(rb) \quad .$$

Mathematically, Θ is the composition of complex conjugation with the lift of r from the space of bonds to $\mathscr{E}$.

THEOREM (Reflection Positivity). *With the expectation* $\langle\,\rangle$ *defined by (1.13), and* $f \in \mathscr{E}_+$,

$$0 \leqslant \langle(\Theta f) f\rangle \quad .$$

Proof. First we fix the gauge so that $U(b) = I$ for all bonds b which intersect Π. Then we write

$$\mathscr{A} = \mathscr{A}_+ + \mathscr{A}_- + \mathscr{A}_{+,-} \quad ,$$

where $\mathscr{A}_\pm \in \mathscr{E}_\pm$ and

$$\mathscr{A}_{+,-} = \sum_{b \cap \Pi \neq \emptyset} -\varepsilon^{d-4} \operatorname{Re} \operatorname{Tr} U(b) U(rb)^{-1}$$

as illustrated in Figure 3. Note that $\exp(-\mathscr{A}_-) = \Theta \exp(-\mathscr{A}_+)$, so $\exp(-\mathscr{A}_+)$ can be considered part of f. The exponential $\exp(-\mathscr{A}_{+,-})$ is written as a product over plaquettes. Expand each term as a power series. Each term in the resulting expansion has the form of a positive constant times a product of terms of the form

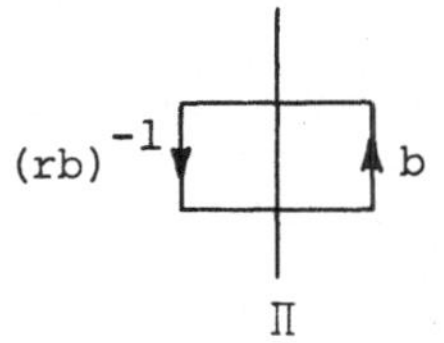

Fig. 3. Plaquettes which intersect Π contribute to $\mathscr{A}_{+,-}$. Since $U((rb)^{-1}) = U(rb)^{-1}$, and the bonds intersecting Π are gauge fixed, the formula for $\mathscr{A}_{+,-}$ follows.

$$U_{jk}(b)\overline{U_{jk}(rb)} = (\Theta U_{jk}(b))U_{jk}(b) , \qquad b \in T_+ ,$$

or the same expression with $U_{jk}(b)$ replaced by $\overline{U_{jk}(b)}$. Thus again, each term can be represented as a positive constant times $(\Theta f')f'$ with $f' \in \mathscr{E}_+$. With these preliminaries, the expectation can be written

$$\langle(\Theta f)f\rangle = \sum_{f' \in \mathscr{E}_+} c(f') \int(\Theta f')f' \, dU$$

where $c(f') \geqslant 0$. But

$$\int(\Theta f')f' \, dU = \int \Theta f' \, dU \int f' \, dU$$

$$= \left|\int f' \, dU\right|^2 \geqslant 0 ,$$

which completes the proof.

The physical positivity condition defines a Hilbert space $\mathscr{H}$ of states using the sesquilinear form on $\mathscr{E}_+$,

$$\langle f,g\rangle_{\mathscr{H}} = \langle(\Theta f)g\rangle .$$

Dividing by the kernel of this form in the standard way, we obtain an inner product, satisfying

$$0 \leqslant \langle f,f\rangle_{\mathscr{H}} , \qquad f = 0 \Leftrightarrow \langle f,f\rangle = 0 .$$

The definition of the transfer matrix K, $0 \leqslant K \leqslant I$ automatically follows, assuming we can take (and have taken) the limit of an infinite volume in the t-direction. Define K for $f, g \in \mathscr{E}_+$ by

$$\langle f, K^{n\varepsilon} g\rangle_{\mathscr{H}} = \langle(\Theta f)g_{n\varepsilon}\rangle ,$$

and $g_{n\varepsilon}(U)$ denotes the translation of g by n units ε away from Π. In the continuum limit, $K \to e^{-H}$ gives the Hamiltonian and $K \leqslant I$ means that H is positive, $0 \leqslant H$.

4. Perturbation Theory Seems to Require Global Gauge Fixing

The first problem caused by gauge invariance is the "equivariance" problem. This problem illustrates that even in a lattice gauge theory it is useful to perform gauge fixing. This statement may at first appear surprising. However, let us consider a sequence of models with ε fixed. Take the limit in which the coupling constant $g \to 0$. We find that gauge fixing is required in this limit in order to obtain a perturbation theory based on functional integration. In fact perturbation theory is the standard method in constructive field theory to obtain insight into non-perturbative methods. Thus we are certainly interested in preserving compatibility with it.

The $g \to 0$ limit describes perturbation theory about the classical critical point of the action $A^{\varepsilon}(U)$ for which $A^{\varepsilon} = 0$. In other words $U(b) \equiv I$, which is the classical vacuum of physics. We parameterize the group variables U(b) above, by a Lie algebra field A(b), defined by

$$U(b) = \exp(ig\varepsilon A(b)) .$$

Then the Haar measure dU(b) can be written

$$dU = \sigma(g\varepsilon A)(g\varepsilon)^{(\dim G)} dA \tag{1.16}$$

where dA is Lebesgue measure on the Lie algebra. Also $\sigma(A)$ depends on G. For example, if $G = SU(2)$, then $\dim G = 3$, $\sigma(A) = (2\pi^2)^{-1}(\sin|A|/|A|)^2$ and

$$\sigma(g\varepsilon A) = \text{const} \left(\frac{\sin(|A|g\varepsilon)}{g\varepsilon|A|}\right)^2 = \text{const} \exp\left(-\frac{g^2\varepsilon^2}{6}|A|^2 + \cdots\right). \tag{1.17}$$

Expressing the integral (1.13) in terms of these variables,

$$\langle f \rangle = \frac{1}{Z} \int dA\, \sigma(g\varepsilon A) f(A) \exp\left(-\sum_{p\in T_\varepsilon} \frac{1}{2}|F|^2\varepsilon^d + \cdots\right) \tag{1.18}$$

where

$$F(p) = (\partial^\varepsilon A)(p) + \frac{1}{2} g \sum_{b<b'\in\partial p} i[A(b), A(b')] \tag{1.19}$$

and $b < b'$ denotes that b, b' are consecutive bonds in ∂p.

The range of the variable A in (1.16) is a compact subset of the Lie algebra. In the case of SU(2), this set is

$$\varepsilon g|A| \leqslant \pi . \tag{1.20}$$

However in the limit $\varepsilon g \to 0$, this domain fills $\mathbb{R}^3$.

Let us consider the $g = 0$ limit. Then the expectation (1.18) would have the form

$$\langle f \rangle = \lim_{g\to 0} \int f(A)\, d\mu_g(A) \tag{1.21}$$

where $d\mu_g$ is the probability measure $1/Z \exp(-g^{-2}A^\varepsilon(U))\, dU$ expressed in terms of A and $d\mu$ is formally a probability measure

$$\frac{1}{Z} \exp\left(-\frac{1}{2}\|\partial^\varepsilon A\|^2\right) dA \quad . \tag{1.22}$$

In other words, formally

$$d\mu_g \to d\mu \quad . \tag{1.23}$$

However $d\mu$ is not a probability measure! The normalizing factor $Z = \infty$, since the quadratic form $\|\partial^\varepsilon A\|^2$ vanishes on the space of gradients, which is the space of configurations A which are linearized gauge transformations. Thus to define a measure $d\mu$ we need to remove the space on which the quadratic form $\|\partial^\varepsilon A\|^2$ is degenerate. Hence we require gauge fixing.

Since we are studying a series expansion in g, and since we do this to obtain information about the limit $\varepsilon \to 0$, we have to investigate whether all gauge fixings are equally good from this point of view. In fact the Green's functions or covariances of the Gaussian distribution (1.22) have different singularities in different gauges. This affects the renormalizability of the perturbative expansion.

In order to study the ultraviolet problem, we wish to choose gauges for which the Green's functions have the best regularity properties. Examples of such gauges are Coulomb gauge, Landau gauges, etc. However, for these gauges, the inability to define a cross-section reemerges in all cases--independent of the volume or of the boundary conditions. Gribov and others showed that the cross-sections defined by the Coulomb and the Landau gauges cannot be chosen. In fact attempting to extend the section in Figure 1 results in intersection with certain orbits many times. It is even possible that it may have more singular behavior, such as intersecting the orbit on a Cantor set.

A more careful inspection of this argument shows that global gauge fixing is not really necessary for understanding perturbation theory. We have used the fact that in the limit $g \to 0$, the compact domain of integration defined by (1.20) expands to fill the entire space. The same is true if π in (1.20) is replaced by an arbitrary small number. Thus we recover the same perturbative expansion from the integral (1.18) restricted to an arbitrarily small and gauge invariant neighborhood $\mathfrak{A}_\delta$ of the critical orbit as illustrated in Figure 4. In $\mathfrak{A}_\delta$ the problem of choosing a section is a local problem rather than a global problem, and hence it poses no geometric or topological obstruction. Thus we can define the integral over $\mathfrak{A}_\delta/\mathfrak{G}$ by restricting the integral locally to the section defined by a gauge condition. It is convenient to choose a condition, such as Landau gauge defined by

$$\partial^* A = 0 \tag{1.24}$$

which does not explicitly depend on g. Thus the gauge condition has a natural $g \to 0$ limit. This limit defines a *global* gauge fixing for the abelian limiting action

$$\lim_{g \to 0} \frac{1}{g^2} A^\varepsilon(U) = \frac{1}{2} \|\partial^\varepsilon A\|^2 \tag{1.25}$$

and for the associated Gaussian measure (1.22).

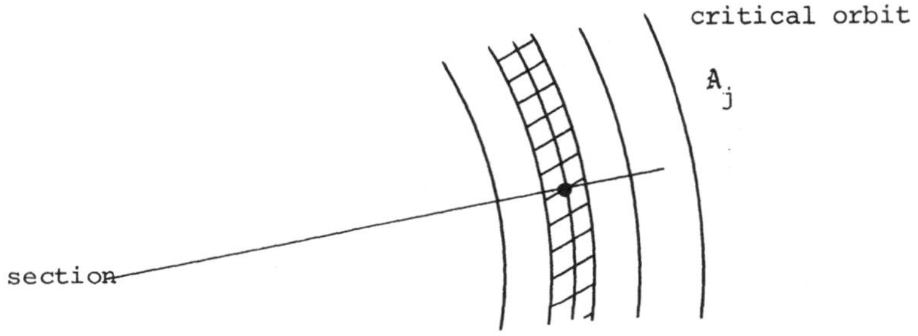

Fig. 4. Shaded orbits define $\mathfrak{A}_\delta$, namely those orbits which pass through the set $\varepsilon g|A - A_{cr}| < \delta$.

At the level of perturbation theory, this eliminates the problem; for perturbation theory the topological problem is not a problem at all. We now use this solution to guide us in our treatment of nonperturbative (i.e., exact) results in gauge field theory. We need to take advantage of perturbation theory insights into renormalization in order to understand the $\varepsilon \to 0$ limit. On the other hand we must avoid the problems of global gauge fixing.

5. The Solution: Phase Cell Localization and Local Gauge Fixing

Phase cell localization [6,2] provides a key to the ultraviolet problem and to the gauge fixing problem. We seek a partition of the degrees of freedom into subsets with two properties:

1) Each set is finite.

2) Disjoint sets are approximately independent.

This partition is called phase cell localization because each set is associated with a disjoint subset of phase space. The independence of disjoint sets of variables expresses the locality of interactions in space-time as well as the approximate independence in Fourier transform space. This picture motivates the decomposition of the function space integral corresponding to this decomposition into phase cells. Such a decomposition runs through all work in constructive quantum field theory.

Once we have prescribed the phase cells, we can consider the ultraviolet problem in a given cell. Since each cell has a small number of variables, independent of the lattice spacing ε, there is no ultraviolet problem in a cell. Thus the problem is reduced to studying the dependence between distinct cells. This problem will be studied using an inductive analysis.

We wish to implement this idea so that the phase cells are defined naturally in terms of momentum localization and space-time localization. In Figure 5 we illustrate such a decomposition schematically.

Likewise the gauge fixing problem in each cell poses no difficulty. This corresponds to the choice of a local section of $\mathcal{A}$. Furthermore, since the ultraviolet problem does not occur within cells, its problem does not affect the choice of gauge. The global gauge choice of Figure 1 is replaced by a local gauge choice of Figure 4.

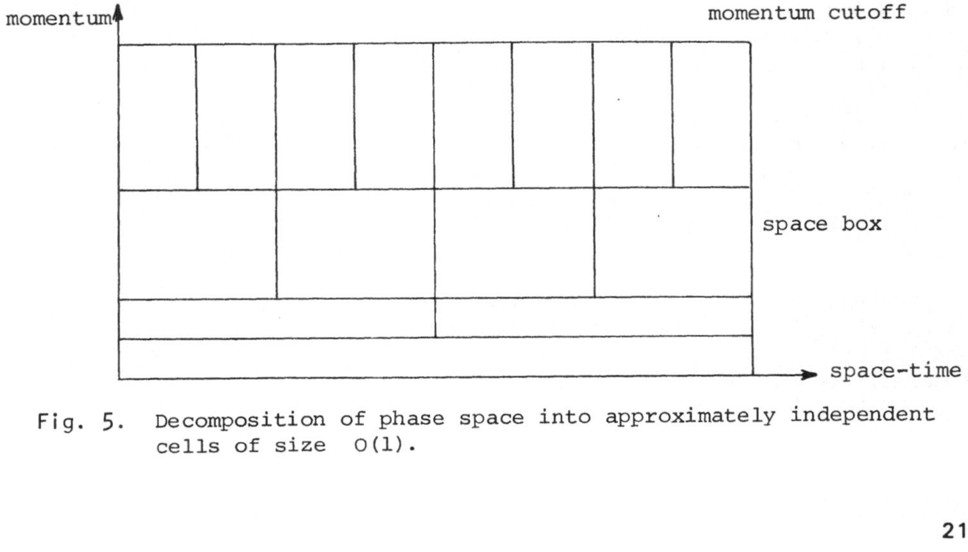

Fig. 5. Decomposition of phase space into approximately independent cells of size O(1).

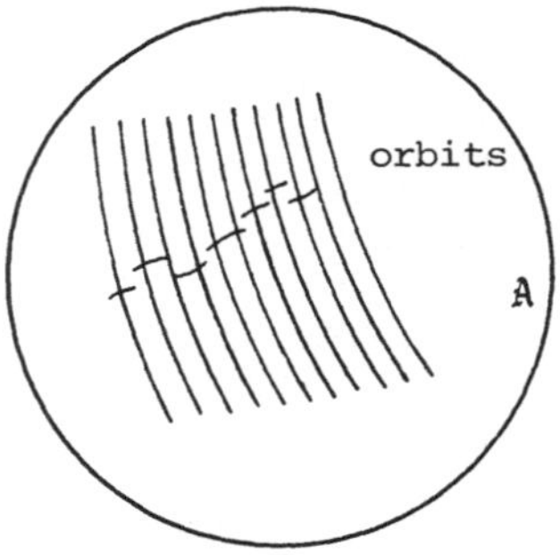

Fig. 6. Local gauge fixing within chosen phase cells.

We will implement this procedure in a very concrete way. We define a sequence of local gauges which capture the essential features of the gauge choices above. A local axial gauge gives rise to simple formulas and algebraic identities. A local Landau gauge is useful for studying regularity properties uniformly in ε, and for establishing the mathematical independence of distinct phase cells.

Quantitatively, this independence will reduce to exponential decay estimates for certain lattice Green's functions. The inductive procedure is based on the notion of an exponentially increasing length scale. Estimates on one scale are the hypotheses for establishing estimates on the next scale.

The definition of the phase cell decomposition for gauge theory is quite complicated, and it is the subject of all that follows. Here we describe the first step, the "block" construction and the definition of lattice averages.

On the lattice T_ε, consider a cover by cubes or "blocks" with side-length $L\varepsilon$ and L^d sites. The lattice of centers of these cubes is an $L\varepsilon$ lattice which we denote $T^1_{L\varepsilon}$. Define $B(y)$ as the block containing the $L\varepsilon$ site y. We wish to assign a gauge field on the T^1 lattice arising from a block average of the gauge field on the T_ε lattice.

Defining such an average has two aspects: (i) How to ensure gauge invariance, and (ii) How to average the group-valued fields. Question (i) has been the subject of many proposals, all of which involve considering contour variables which define parallel transport between the centers y, y' of two neighboring blocks. In particular both Kadanoff and Wilson have proposed several possible choices of appropriate variables [7]. For technical reasons related to establishing regularity we use a different set of contours following Balaban [8] which has also been used by Federbush [9]. Question (ii) must be answered in a way which reduces for small fields to averaging in the Lie algebra. Again there are many proposals which may have different advantages in different contexts. For example, both Wilson [7] and Federbush [9] have different averaging methods.

The contours $\{\Gamma_{yy'}\}$ which we use are built from three segments: a standard path Γ_{yx} from y to $x \in B(y)$; a straight line xx' parallel to yy'; and finally the inverse path to $\Gamma_{y'x'}$. The standard paths all lie on a maximal tree in $B(y)$, see Figure 7.

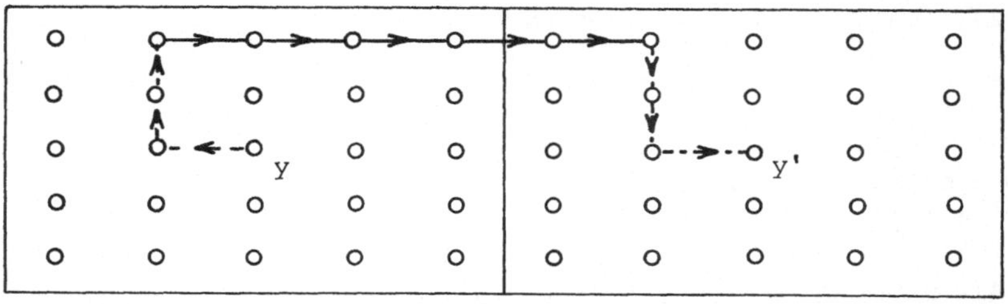

Fig. 7. A contour $\Gamma_{yy'}$ containing xx'. The set of contours $\{\Gamma_{yy'}\}$ is obtained by taking all sites $x \in B(y)$.

We define the average of $U(\Gamma_{yy'})$ by one of the following:

$$\bar{U}(yy') = \text{Proj}\left(\sum_{x \in B(y)} L^{-d} U(\Gamma_{yy'})\right) \quad , \quad \text{(Wilson average)} \tag{1.26}$$

where Proj X is uniquely defined for nonsingular X by a slightly modified polar decomposition of X. Write $X = Ue^{i\theta}H$ where $U \in SU(N)$, $\theta \in R$ and H is Hermitian. Proj $X = U$.

Define a real valued function d on the group element V by

$$d(V) = \sum_{x \in B(y)} \text{dist}^2(U(\Gamma_{yy'}),V) \quad , \quad \text{(Federbush definition)} \tag{1.27}$$

where dist(A,B) denotes the geodesic distance from A to B. Let $\bar{U}(y,y')$ be a V which minimizes $d(V)$. If the contour variables $U(\Gamma_{yy'})$ are close to each other, then V is uniquely defined.

Certain technical results were established using the following definition, but these proofs also extend to the other two definitions above:

$$\bar{U}(yy') = \exp\left[L^{-d} \sum_{x \in B(y)} \ln U(\Gamma_{yy'} \circ y'y)\right] U(yy') \quad . \tag{1.28}$$

This definition is motivated by a perturbative expansion of a solution V to the equation

$$\sum_{x \in B(y)} \ln[U(\Gamma_{yy'})V^{-1}] = 0 \quad . \tag{1.29}$$

All these definitions of averaging have the following three basic properties in common. These are the important ingredients of the proofs which follow:

(1) Gauge Covariance:

$$\overline{U^u} = (\bar{U})^u \quad . \tag{1.30}$$

(2) Small Field Approximation

If $U = \exp(iA)$ with A small, than

$$\bar{U} = \exp\left[i \sum_{x \in B(y)} L^{-d} A(\Gamma_{yy'}) + O(A^2) \right] \quad . \tag{1.31}$$

(3) $\bar{U}$ is an analytic function of U for regular U, as defined in Chapter 3.

The basic step in the phase cell expansion which we consider is implemented by a renormalization transformation. This is an integral transformation T defined [7] by

$$(T\rho)(V) = \int dU \; \delta(\bar{U}V^{-1})\rho(U) \quad . \tag{1.32}$$

Here δ denotes a product over bonds of the $L\varepsilon$ lattice, $T^1_{L\varepsilon}$, of Dirac measure on the group G. This transformation has the normalization property

$$\int dV(T\rho)(V) = \int dU\rho(U) \quad .$$

For gauge invariant functions ρ, we introduce local gauge fixing for the reasons explained above. We define local axial gauge in blocks of the lattice T_ε. In this gauge we set $U(b) = I$ on a maximal tree in each block. More precisely, we choose the tree $\cup\, \Gamma_{yx}$, the union of standard contours from y to $x \in B(y)$. Let

$$\delta_{Ax}(U) = \prod_{y \in T^1_{L\varepsilon}} \; \prod_{\substack{x \in B(y) \\ x \neq y}} \delta(U(\Gamma_{yx})) \tag{1.33}$$

denote the Dirac measure which fixes the local axial gauge. Thus for gauge invariant $\rho(U)$ we write

$$(T\rho)(V) = \int dU \; \delta(\bar{U}V^{-1})\delta_{Ax}(U)\rho(U) \quad . \tag{1.34}$$

We prove this identity by a simple application of the Faddeev-Popov procedure [10]. We have

$$\int du\, \delta_{Ax}(U^{u^{-1}}) = 1 \quad ,$$

where the integral is over the submanifold of gauge transformations u on the ε-lattice sites T_ε, which satisfy $u = I$ on the $L\varepsilon$-lattice $T^1_{L\varepsilon}$. Insert this identity under the integral (1.32), change the order of integration, and in the integral with respect to dU make the gauge transformation $U \to U^u$. This gauge transformation removes all dependence of the integrand on u, and yields the desired identity.

6. Steepest Descents

The renormalization transformation introduces the variables V and U. The variables approximately correspond to the first division into smaller momenta and larger momenta in the phase cells of Figure 5. (Here smaller momenta refers to the lower half of the figure.) In order to precisely define the decomposition into phase cells, we must not only focus on the region of localization, but also on the independence of the variables. Achieving independence suggests a slightly different decomposition, namely an expansion of $\mathcal{A}$ about its critical points. Such an expansion is a method of steepest descents. Let us explain this method in a general case.

Consider an integration over a domain M in a linear vector space. We are interested in the asymptotics $g \to 0$ of the integral

$$I(g) = \int_M dx\, e^{-\left[\frac{1}{g^2} f(x)+F(g,x)\right]} \tag{1.35}$$

where F is a regular function of g and $x \in M$. The method of steepest descent is to expand f about a critical point. For simplicity we assume that $f \restriction M$ has exactly one critical point x_0, and that this critical point is a minimum. Then with the scaled fluctuation variable $x - x_0 = gy$,

$$I(g) = g^{\dim M}\, e^{-\frac{1}{g^2} f(x_0)-F(g,x_0)} \int_M e^{-\frac{1}{2}\langle y,d^2 fy\rangle+H} dy \; . \tag{1.36}$$

Here $H(g,y)$ is a function obtained by taking remainders of the expansions. It is a regular function of g, and is $O(g)$ as $g \to 0$. Also $d^2 f$ is the Hessian of f.

The domain of integration M in (1.36) expressed in terms of the variable y is $g^{-1}(M-x_0)$. As $g \to 0$, this domain of integration tends to the whole vector space, in case x_0 is an interior point of M.

The asymptotic behavior of the integral in (1.36) over M is determined by standard perturbation theory. The series for the logarithm

$$\ell n \int_M \exp\left[-\frac{1}{2}\langle y,d^2 fy\rangle + H\right] dy = \sum_{n \geqslant 0} c_n g^n \tag{1.37}$$

is given by the cumulant expression, with the coefficients c_n given by Gaussian integrals.

The method described here cannot be applied directly in the case when we integrate over a manifold M rather than a vector space. This case arises in studying renormalization transformations. Each case requires the introduction of appropriate coordinates. For example, the integral (1.13) can be understood by introducing local coordinates for U given by Lie algebra variables $U = \exp(iB)$. Then $\mathcal{A}^{\varepsilon}(U) = A^{\varepsilon}(\exp iB)$ is minimized by $B = 0$, and $g^{-2}A^{\varepsilon}(I) = 0$. Thus the scaled variable $A = g^{-1}B$ can be introduced and the perturbative expansion studied in (1.21) is the same as that given by (1.36).

7. Infra-Red Behavior

The question of infra-red behavior is to determine how certain correlations behave asymptotically for large distances or sizes. For example, uniform rate of exponential decay of correlation functions establishes a mass gap in a model. We wish to know which gauge models have a mass gap. In the Higgs model this question can be studied perturbatively, as we do in more detail below. In other models (such as pure Yang-Mills theories) the question is more subtle. While it is believed that $d = 3,4$ pure Yang-Mills theories have a gap in the mass spectrum, no conclusive mathematical results have been established nor has a definitive physical picture emerged.

Another aspect of infra-red behavior is the rate of decay of loop variables $\langle \mathrm{Tr}\, U(\Gamma)\rangle$ where Γ is a closed loop. It is known in many cases that either

$$\langle \mathrm{Tr}\, U(\Gamma)\rangle \sim \exp[-O(1)\ \mathrm{Area}\ (\Gamma)] \tag{1.38}$$

or

$$\langle \mathrm{Tr}\, U(\Gamma)\rangle \sim \exp[-O(1)\ \mathrm{Perimeter}\ (\Gamma)] \tag{1.39}$$

where $O(1)$ is a constant depending on g, where Area (Γ) is the minimal surface bounded by Γ, and where Perimeter (Γ) is the length of Γ. Behavior (1.38) was established exactly using a cluster expansion by Osterwalder and Seiler for models with large coupling, $g \gg 1$. This followed similar conclusions by physicists using perturbative (strong coupling) expansions.

It is expected for $SU(n)$ models in dimension $d \leqslant 4$ and with arbitrary coupling constants that this area law holds. For dimension $d \geqslant 5$ one expects a phase transition: area law decay for large g and perimeter law decay for small g. There is no proof of either fact. See [11] for a survey of such questions.

In the abelian case, however, it is known that there is a phase transition for $d \geqslant 4$ [12]. This was proved by Guth and also by Fröhlich and Spencer. For the Villain approximation to the $d = 3$ model, an area law for all couplings (absence of a phase transition) was established by Göpfert and Mack. It is believed that the area law also holds for the $d = 3$ Wilson action, but it has not been established.

Let us now return to the abelian Higgs model as an illustration of constructive methods. The physics picture is easily understood by a perturbative analysis. In Sections 1.4-1.5 we argue that perturbation theory requires gauge fixing (which may be local). Once we have defined our model by such a procedure, we can then study one of two situations:

(I) We restrict attention to gauge invariant observables. This means we restrict attention to gauge invariant functions of (A,φ) and their vacuum expectation values. This is also the physically relevant question and the one we discuss at length below. As we observed in Section 1.3, gauge invariant expectations do not depend on our choice of gauge fixing. Thus the answers to physical questions--such as what is the mass gap, or what are the correlations--is independent of our choice of gauge.

(II) We study expectations of all functions of (A,φ), as discussed in the lectures of Wightman and of Strocchi. In this case all expectations of functions which are not gauge invariant depend on our initial

On the other hand, in low momentum cells we specify the unitary gauge. This exhibits the mass gap on length scales which are sufficiently large.

In fact, it is useful to use several gauges in each phase cell.

In order to implement such a scheme, and to interpret its consequence, we are tied to studying correlations of gauge invariant observables, i.e. problem I. Otherwise, the change of gauges necessary on each length scale would be impossible, and the gauge used to define the theory in the first place would appear extremely arbitrary.

Results [5,13] have been obtained on the Higgs effect and existence of a mass gap for unit lattice models by Osterwalder and Seiler, Nappi and Israel, Mack and Balaban, Brydges, Imbrie and Jaffe. Work on the ε-lattice with phase cell localization has been published [14] and is also in progress.

II. LATTICE MAXWELL THEORY

1. The Action

The expansion of the Wilson action in powers of the gauge potential A begins with a quadratic part. This is the lattice action S_M, where $S = S_M + O(A^3)$. Explicitly

$$S_M = \frac{1}{2}\langle \partial A, \partial A\rangle = \frac{1}{2}\sum_{p\in T_\varepsilon} \varepsilon^d |A(\partial p)|^2 \quad , \tag{2.1}$$

where $A(\partial p) = \Sigma_{b\in\partial p} A(b)\varepsilon^{-1}$ is the ε-lattice curl.

Choose an orthonormal basis in the Lie algebra $\mathfrak{G}$ of the gauge group G. Then the action (2.1) is the sum of the actions for each component of the $\mathfrak{G}$-valued field A. It is sufficient to consider one component of A, which takes real values. Furthermore it is convenient to allow A to take all real values. Now let S_M denote this action, which is the standard lattice approximation to the Maxwell action for continuum electromagnetism.

The properties of S_M play a central role in the analysis of S. For this reason it is instructive to first analyze S_M by itself. We now derive the effective actions and Green's functions obtained from renormalization transformations of S_M. We need all these Green's functions in the technical analysis of S. They also provide a good approximation to the short-distance behavior, and for some cases also to the long-distance behavior, of the Green's functions in models with interaction.

The expansion (2.1) was written for the ε-lattice. However, according to our strategy, we rescale S_M to the unit lattice and consider S_M on this lattice. With the canonical scaling, from the ε-lattice field A^ε to the unit lattice field A (defined on unit lattice bonds)

$$A^\varepsilon(\varepsilon b) = \varepsilon^{-(d-2)/2} A(b) , \qquad |b| = 1 \quad (\text{or } b\in T_1) .$$

The unit lattice action becomes

$$S_M(A) = \frac{1}{2} \langle \partial A, \partial A \rangle = \frac{1}{2} \sum_p (\partial A)^2 \quad , \tag{2.2}$$

where ∂A now denotes the unit lattice curl,

$$(\partial A)(p) = A(\partial p) = \sum_{b \in \partial p} A(b) \quad . \tag{2.3}$$

The Maxwell action inherits a gauge invariance from the Wilson action. In fact the gauge invariance is a linear approximation to the gauge invariance of the Wilson action. The gauge transformations are defined by

$$A \to A^{\lambda} = A - \partial\lambda \tag{2.4}$$

where λ is a real valued (scalar) function on the unit lattice and ∂ denotes the unit lattice gradient. Because of the gauge invariance, the functional $S_M(A)$ is a degenerate quadratic form. It is constant on orbits of the group of all gauge transformations (2.4). Since the A's take all real values, we must resolve this degeneracy before attempting to integrate functions such as $\exp(-S_M(A))$ with respect to product Lebesgue measure $\mathcal{D}A$.

2. Averaging

We define phase cell localization by local averages. Fixing a local average of a function approximately fixes the low momentum part of the Fourier transform of the function, while simultaneously localizing in space.

The block average of the unit lattice field A is defined as a linear transformation Q applied to A. In particular, if $c = (y,y')$ connects the corners of two nearest neighbor sites on the L-lattice, define QA on L-lattice bonds as follows: let (x,x') be a parallel translate of (y,y'), to some site $x \in B(y)$. Then

$$(QA)(c) = \sum_{x \in B(y)} L^{-(d+1)} A([xx']) \tag{2.5}$$

where

$$A([xx']) = \sum_{b \in (xx')} A(b) \quad . \tag{2.6}$$

What is the gauge covariance property of QA? Substituting $A^{\lambda} = A - \partial\lambda$, for A, we find

$$QA^{\lambda} = QA - Q\partial\lambda = QA - \partial^{L} Q\lambda \quad . \tag{2.7}$$

Here we use the scalar field average Q defined by

$$(Q\lambda)(y) = \sum_{x \in B(y)} L^{-d} \lambda(x) \quad . \tag{2.8}$$

Then

$$Q\partial = \partial^{L} Q \tag{2.9}$$

where ∂^L denotes the L-lattice difference operator.

Note that (2.7) ensures $QA^\lambda = QA$ whenever $Q\lambda = 0$ on each L-block, or more generally if $Q\lambda$ is globally constant. We can further summarize our discussion by the relation

$$QA^\lambda = (QA)^{Q\lambda} \quad . \tag{2.10}$$

3. Local Axial Gauge

The action S_M and the affine subspace defined by $QA = B$, i.e., fixing the average values of the gauge field, are still invariant with respect to the gauge transformations on the unit lattice which are described above.

Within each L-block there are L^d points and hence L^d gauge degrees of freedom. Fixing the average value $Q\lambda$ specifies a gauge degree of freedom for the block. The $L^d - 1$ remaining degrees of freedom are fixed by choosing the gauge $A = 0$ on bonds of a maximal tree in each block. In fact, a tree with L^d vertices has exactly $L^d - 1$ lines, so this counting degrees of freedom agrees with the construction of Section I.5.

We choose the tree in each block in the manner illustrated in Figure 8. All bonds within each block pointing along the x_d coordinate direction are fixed. All bonds in the block pointing along the x_{d-1} coordinate direction and which minimize x_d are fixed. All bonds which point in the x_{d-2} coordinate direction and which minimize x_d and x_{d-1} are fixed. Etc.

The bonds on which the gauge field A has been fixed to zero in local axial gauge can be regarded as belonging to some standard contours connecting points $x \in B(y)$ with the reference point y. We denote these contours by $\Gamma_{y,x}$, where $\Gamma_{y,x}$ is the line from x to y obtained by moving from x first in the d-th coordinate direction to the value y_d, then in the (d-1)-th coordinate direction to the value y_{d-1}, etc. The contour $\Gamma_{x,y}$ is defined as the contour $\Gamma_{y,x}$ taken with the opposite orientation.

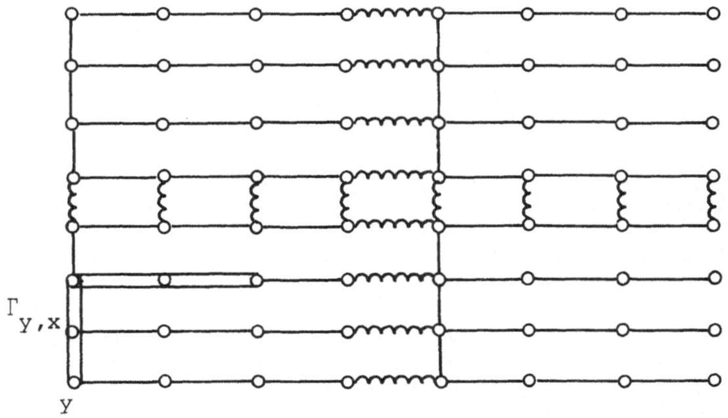

Fig. 8. Blocks with 4^d sites. The $4^d - 1$ bonds per block drawn in solid lines are fixed in the local axial gauge. The wavy bonds connect neighboring 4-blocks. The contour $\Gamma_{y,x}$ is denoted by double bonds.

In terms of the bonds $Ax(y)$, namely those bonds with endpoints in $B(y)$ which are fixed in our local axial gauge, define

$$\delta_{Ax}(A) = \prod_{y} \prod_{b \in Ax(y)} \delta(A(b)) \quad . \tag{2.11}$$

Then the axial gauge fixing can be represented by inserting the delta function $\delta_{Ax}(A)$ into integrals.

We are interested in the affine subspace of gauge configurations in the local axial gauge and with prescribed local averages. Let

$$\mathcal{H}_{Ax,B} = \{A:\ QA=B,\ A(b)=0 \quad \text{if} \quad b \in \cup_y Ax(y)\} \quad . \tag{2.12}$$

4. Renormalization Transformations

The Maxwell action has fields taking values in a noncompact domain. Therefore, as we have already remarked, it is necessary to introduce gauge fixing into the definition of the renormalization transformations. Define

$$(Te^{-S_M})(B) = \int dA\ \delta(QA-B)\delta_{Ax}(A)e^{-S_M(A)} \quad . \tag{2.13}$$

Here B is a vector field defined on L-lattice bonds. The transformation T can be extended to other functions, and we are especially interested in functions of the form $\psi \exp(-S_M)$ where ψ denotes a gauge invariant functional, such as those which arise in the study of correlations.

Finally, the integral (2.13) is Gaussian. But we have neglected to show that it is convergent. Since the integrand is positive, we can show that a translate of (2.13) is convergent. Choose A_0 to be any configuration in the affine subspace $\mathcal{H}_{Ax,B}$ defined by local axial gauge in $QA=B$. Write $A=A'+A_0$, so consider

$$\int dA'\ \delta(QA')\delta_{Ax}(A')\exp[-S_M(A') - S_M(A_0) - \langle \partial A', \partial A_0 \rangle] \quad . \tag{2.14}$$

The delta functions now restrict A' to the (linear) subspace $\mathcal{H}_{Ax,0}$ satisfying $QA'=0$ and the local axial gauge configuration. Thus the convergence of (2.14) is determined by whether $S_M(A') \geqslant 0$ has a nonzero minimum on $\mathcal{H}_{Ax,0}$.

PROPOSITION 2.1. $\partial \restriction \mathcal{H}_{Ax,0}$ *has no zero modes.*

COROLLARY 2.2. *A unique minimizing configuration* A_0 *exists for* $S_M(A)$ *on each affine subspace* $\mathcal{H}_{Ax,B}$.

Proof. A zero mode of $\partial \restriction \mathcal{H}_{Ax,0}$ is a nonzero $A \in \mathcal{H}_{Ax,0}$ for which $\partial A=0$. Suppose such an A exists. By the axial gauge condition, $A(b)=0$ on every bond b with both endpoints inside the same block. This can be established first for bonds $b \in \partial p$ for plaquettes in the (y_d, y_{d-1}) plane. Then proceed to $y_d=0$ and bonds in the (y_{d-1}, y_{d-2}) plane, followed by $y_d \neq 0$ in the same plane, etc. Hence we need only consider bonds connecting distinct blocks. For bonds b joining two particular blocks $B(y)$ and $B(y')$, the conditions $\partial A=0$ ensure that

$A(b) = \text{constant}$. However $(QA)(y,y') = 0$ ensures that the constant is zero. Thus A vanishes on all bonds, a contradiction.

The proposition ensures $S_M(A') > 0$, for $A \in \mathscr{H}_{Ax,0}$, $A \neq 0$. Since S_M is a quadratic form in a finite dimensional space, $S_M(A') \geq \lambda_0 \|A'\|^2$, where $\lambda_0 > 0$ is the minimum eigenvalue of the quadratic form $\partial^*\partial$. This ensures convergence of the integral (2.14). Since (2.14) equals (2.13), the latter integral also converges.

Now choose A_0 in (2.14) as the minimum of $S_M(A)$ on the affine space $\mathscr{H}_{Ax,B}$. For this choice of $A_0 = A_{cl}$, the term in the exponent in (2.14) which is linear in the variation A' must vanish. Hence

$$(Te^{-S_M})(B) = Ze^{-S_M(A_{cl})} = Z \exp\left(-\frac{1}{2}\langle\partial A_{cl}, \partial A_{cl}\rangle\right) , \tag{2.15}$$

where $Z = (Te^{-S_M})(B=0)$.

The minimizing configuration A_{cl} of S_M on $\mathscr{H}_{Ax,B}$ is obviously a linear function of B, and we denote the quadratic form $S_M(A_{cl})$ expressed as a function of B by $1/2\ \Delta_1$. In other words,

$$S_M(A_{cl}) = \frac{1}{2}\langle B, \Delta_1 B\rangle . \tag{2.16}$$

Thus

$$(Te^{-S_M})(B) = Z \exp\left(-\frac{1}{2}\langle B, \Delta_1 B\rangle\right) . \tag{2.17}$$

Note that (2.17) is gauge invariant under gauge transformation $\lambda(x)$ defined on the L-lattice. This can be easily seen from the fact that $(QA)(b)$, restricted to axial gauge configurations transforms as (2.10). But in axial gauge, $Q\lambda$ is the unspecified gauge degree of freedom in each block; $Q\lambda$ can be changed by unit lattice gauge transformation λ which is constant on each local L-block. Thus (2.10) defines a general L-block gauge transformation.

5. Back to the Unit Lattice

In order to iterate the renormalization transformation (2.13), we scale the field B from the L-lattice to the unit lattice. Therefore the renormalization transformation becomes a mapping on the space of functionals over the unit lattice. We denote the renormalization transformation on the unit lattice by T also, without danger of confusion.

We again use canonical scaling, which is correct since we expect the ultraviolet behavior of our gauge theory to be canonical. Let us denote the unit lattice field by A. Then

$$A(b) = L^{(d-2)/2} B(Lb) . \tag{2.18}$$

Let Δ_1 now denote the quadratic form in (2.16) rescaled to the unit lattice. Then

$$\begin{aligned}(Te^{-S_M})(A) &= Z \exp\left[-\frac{1}{2}\langle A, \Delta_1 A\rangle\right] \\ &= Z_1 \exp[-S_1(A)] \end{aligned} . \tag{2.19}$$

In (2.19) we define $Z_1 = Z$ and S_1 as the quadratic form given by Δ_1.

6. Iteration of the Renormalization Transformation

We can apply T to $\exp(-S_M)$ a number of times, yielding

$$(T^n e^{-S_M})(A) = Z_n e^{-\frac{1}{2}\langle A,\Delta_n A\rangle} = Z_n e^{-S_n(A)} \quad . \tag{2.20}$$

This is the result of iterating

$$(Te^{-S_n})(A) = Z^{(n+1)} e^{-S_{n+1}(A)} \quad , \tag{2.21}$$

where $Z_n = Z^{(1)} Z^{(2)} \cdots Z^{(n)}$, and $S_n(A) = 1/2 \langle A,\Delta_n A\rangle$.

It is actually necessary to verify that S_n exists! We have already checked the existence of S_1, but $S_1(A) = 1/2 \langle A,\Delta_1 A\rangle$ does not have the same general form as $S_M(A) = 1/2 \langle \partial A, \partial A\rangle$. Thus it is not clear that Proposition 2.1 and the argument above are sufficient to show the convergence of the integral defining $T \exp - S_1$, etc.

In order to continue iteratively, we note that S_1 (and correspondingly $S_2,\ldots,S_n,\ldots$) do resemble S_M. In fact

$$S_n(A) = \frac{1}{2}\langle A,\Delta_n A\rangle = \frac{1}{2}\langle \partial A,\sigma_n \partial A\rangle \quad . \tag{2.22}$$

Here σ_n is a strictly positive transformation. There exists a constant c such that

$$0 < cI \leqslant \sigma_n \quad . \tag{2.23}$$

The important property of c is that it is independent of n and of the number of lattice sites. This bound (2.21) ensures that each renormalization transformation of $S_M(A)$ is well-defined, and the estimates for the n-th iteration step do not depend on n. For a proof of (2.22-2.23), see [15].

The quadratic form σ_n and Δ_n have two other important properties: they are translation invariant, and their kernels have uniform exponential decay.

Because of the translation invariance of the basic quadratic form Δ_n, we can write a closed form expression for σ_n as a multiplication operator in Fourier transform space. We obtain this formula by computing the Gaussian integrals introduced above.

7. One Step Covariances

The calculation of the n-th renormalization step is associated with two important linear transformations: the covariance of a Gaussian integral and a transformation H_n which selects the minimum of the effective action S_{n-1}. Let us rewrite (2.21) as

$$(Te^{-S_n})(B) = \int dA\ \delta(QA - B)\delta_{Ax}(A)\exp\left[-\frac{1}{2}\langle A,\Delta_n A\rangle\right]. \tag{2.24}$$

This integral can be calculated as in the first step, translating to the minimum of $\langle A,\Delta_n A\rangle$ on the subspace $\mathcal{H}_{Ax,B}$. The inequality (2.23) ensures the absence of zero modes for Δ_n on $\mathcal{H}_{Ax,0}$, as in Proposition 2.1. Thus we have a Corollary 2.2 in this case too. This minimal configuration depends linearly on B and therefore determines the linear operator $H^{(n)}$ mapping B into the minimal configuration. Thus

$$A = A' + H^{(n)}B \tag{2.25}$$

which splits the quadratic form,

$$\langle A,\Delta_n A\rangle = \langle A',\Delta_n A'\rangle + \langle H^{(n)}B,\Delta_n H^{(n)}B\rangle \quad . \tag{2.26}$$

The second form on the right determines the new quadratic form Δ_{n+1} on the L-lattice.

The field A' in (2.25)-(2.26) is called the fluctuation field and the Gaussian integral with respect to A' defines the basic unit lattice covariance. We also need covariances with zero Dirichlet conditions outside a domain Λ.

DEFINITION 2.3. *The covariance* $C_\Lambda^{(n)}$ *is defined by*

$$\exp\left[\frac{1}{2}\langle J,C_\Lambda^{(n)}J\rangle\right] = Z^{(n)-1}\int \mathcal{D}A'_\Lambda\ \delta(QA')\delta_{Ax}(A')\exp\left[-\frac{1}{2}\langle A',\Delta_n A'\rangle + \langle A',J\rangle\right] \tag{2.27}$$

where $\mathcal{D}A'_\Lambda$ *denotes integration over variables localized in* Δ.

The most important property of $C_\Lambda^{(n)}$ is its approximate localization, expressed by the exponential decay of its kernel. This kernel, $C_\Lambda^{(n)}(b,b')$ is a function of two bond variables b and b', and it satisfies

$$|C_\Lambda^{(n)}(b,b')| \leqslant a\exp(-\delta|b-b'|) \quad , \tag{2.28}$$

where $|b-b'| = \mathrm{dist}(b,b')$ and where a and δ are positive constants, independent of n and Λ. The constants a and δ do depend on c in (2.23), on the exponential decay rate δ_0 for the kernel of σ_n and on the block size L.

8. Minimizers and Gauge Fixing

Aside from the unit lattice propagators, we also require n-step Green's functions which are associated with the n-fold composition of the renormalization transformation. The local axial gauges we have used for the unit lattice expressions are inappropriate for the analysis of n-step Green's functions. In fact the n-step Green's functions are naturally expressed on an L^{-n} lattice, rather than on a unit lattice, since they arise after n rescalings by L^{-1}. Thus understanding these Green's functions requires controlling their local regularity properties. Axial gauges, including the local axial gauges described here, have notoriously bad local regularity properties. Thus we need to reexpress the Green's functions in a better gauge.

Ultimately we are interested in gauge invariant expressions built from these Green's functions. Hence it may not be apparent that this change of gauge is necessary. However, the way that we establish estimates on the gauge independent quantities is to prove estimates for gauge dependent quantities in good gauges. We call a gauge which exhibits good local regularity properties an ultraviolet gauge.

The simplest ultraviolet gauges are modifications of the Feynman gauge or the Landau gauge. Let us discuss these gauges on the example of the n-fold iteration of the renormalization transformation applied to $\exp(-S_M)$. This composition is given by the integral

$$T^n\exp(-S_M)(B) = \text{const} \int dA\, \delta(B-Q_nA)\delta_{Ax,n}(A)\exp(-S_M) \quad , \tag{2.29}$$

where

$$\delta_{Ax,n}(A) = \prod_{j=0}^{n-1} \delta_{Ax}(Q_jA) \tag{2.30}$$

and Q_j is the j-fold composition of Q. The average Q_j is also given by the formula (2.5) applied on L^j-blocks, rather than on L-blocks. After the rescaling implicit in (2.29), the variables B are unit lattice fields while A is on the L^{-n}-lattice. Denote this "small" L^{-n}-lattice as the η-lattice.

While S_M is invariant under all η-lattice gauge transformations, the $\delta(B-Q_nA)$ factor is invariant under η-lattice gauge transformations λ which satisfy the restriction

$$Q_n\lambda = 0 \tag{2.31}$$

where Q_n now denotes an iterated average for the scalar quantity λ. In more detail

$$(Q_n\lambda)(y) = \eta^d \sum_{x\in B^n(y)} \lambda(x) \quad . \tag{2.32}$$

Recall that $B^n(y)$ denotes L^n blocks on the L^{-n}-lattice covering a unit cube with corner y. The stated invariance follows from the identity

$$Q_nA^\lambda - Q_nA = \partial Q_n\lambda \quad . \tag{2.33}$$

The modification mentioned above is connected with the restriction (2.31) on the gauge transformations. We use the Faddeev-Popov procedure and insert into the integrand (2.29) the identity

$$1 = \frac{\int d\lambda' \; \delta(Q_k'\lambda')\exp\left(-\frac{1}{2\alpha}\langle\partial^*A^{\lambda'},\partial^*A^{\lambda'}\rangle\right)}{\int d\lambda \; \delta(Q_k'\lambda)\exp\left(-\frac{1}{2\alpha}\langle\partial^*A^{\lambda},\partial^*A^{\lambda}\rangle\right)} \quad . \tag{2.34}$$

The restrictions on the gauge transformations given by the δ-functions above are chosen in such a way that all the expressions in the integral (2.29) with the exception of gauge fixing terms δ_{Ax}, are invariant. Change the order of integrations $\int dA \int d\lambda' \ldots$, and make the gauge transformation $-\lambda'$ in the integral $\int dA\ldots$. Change the order of integrations again and make a translation $\lambda \to \lambda + \lambda'$ in the integral $\int d\lambda \ldots$. We get

$$\begin{aligned}
(2.29) &= \text{const.} \int dA \; \delta(B-Q_kA)\left(\int d\lambda' \; \delta(Q_k'\lambda')\right. \\
&\quad \cdot \delta_{Ax}(Q_{k-1}A + \partial^{L^{-1}} Q_{k-1}'\lambda') \cdot \ldots \cdot \delta_{Ax}(A + \partial^{\eta}\lambda')) \\
&\quad \cdot \exp\left(-\frac{1}{2\alpha}\langle\partial^*A,\partial^*A\rangle\right)\left(\int d\lambda\delta(Q_k'\lambda)\exp\left(-\frac{1}{2\alpha}\langle\partial^*A^{\lambda},\partial^*A^{\lambda}\rangle\right)\right)^{-1}\exp[-S_M(A)] \\
&= \text{const.} \int dA \; \delta(B-Q_kA)\exp\left(-\frac{1}{2\alpha}\langle\partial^*A,\partial^*A\rangle\right) \\
&\quad \cdot \left(\int d\lambda \; \delta(Q_k'\lambda)\exp\left(-\frac{1}{2\alpha}\langle\partial^*A^{\lambda},\partial^*A^{\lambda}\rangle\right)\right)^{-1} \exp[-S_M(A)] \\
&= \text{const.} \int dA \; \delta(B-Q_nA)\exp\left[-\frac{1}{2\alpha}\langle\partial^*A,R_n\partial^*A\rangle\right]\exp[-S_M(A)] \quad .
\end{aligned} \tag{2.35}$$

The operator R is the orthogonal projection in the space $\ell_2(T_\eta)$ onto the subspace of configurations $\Delta\lambda$ which satisfy the restrictions $Q_n\lambda = 0$, see [15].

This is the modified Feynman gauge. We can obtain the modified Landau gauge by taking the limit $\alpha \to 0$.

We introduce the minimizer H_n for the action in Feynman gauge. By definition, the configuration H_nB solves the variational problem of minimizing the functional $S_M(A) + (2\alpha)^{-1}\langle\partial^*A,R_n\partial^*A\rangle$ under the restriction $Q_nA = B$. Let us denote by $\mathcal{H}_n(B)$ the affine subspace satisfying $Q_nA = B$. We assert that the minimization problem has a unique solution on $\mathcal{H}_n(B)$. This fact is a consequence of the strict positivity of the operator $\partial^*\partial + (2\alpha)^{-1}\partial R_n\partial^* + Q_n^*$, and will be discussed in the following section. It is a remarkable fact that H_nB is independent of α and therefore equals the minimizing configuration in Landau gauge.

It can be seen as follows: Let A_0 minimize $1/2\|\partial A\|^2 + (2\alpha)^{-1} \cdot \|R_n\partial^*A\|^2$ and assume the A_0 does not satisfy the Landau gauge condition $R_n\partial^*A = 0$. The second term in the functional is positive. We construct a gauge transformation λ satisfying $Q_n\lambda = 0$ and such that A_0^{λ} satisfies $R_n\partial^*A_0^{\lambda} = 0$. By gauge invariance, the first term in the functional is unchanged and the second term now equals zero. Hence the functional has a smaller value at the configuration A_0^{λ}, contrary to our assumption that A_0 is a minimum. To construct the gauge transformation λ we consider the equation $R_n \partial^* A^\lambda = R_n \partial^* A - R_n \Delta\lambda = R_n \partial^* A - \Delta\lambda = 0$. This equation $\Delta\lambda = R_n \partial^* A$ has a unique solution on the subspace $Q_n \lambda = 0$. Thus A_0 must satisfy the Landau gauge condition, and does not depend on α.

The minimizer H_n maps from configurations on the unit lattice to configurations on the $\eta = L^{-n}$-lattice. The kernel $H_n(b,c)$ has exponential decay (localization) for c a unit-lattice bond and b an η-lattice bond. In particular, there exist constants $a < \infty$, $\delta > 0$ such that

$$|H_n(b,c)|, \ |\nabla H_n(b,c)|, \ |\Delta H_n(b,c)| \leqslant a e^{-\delta|b-c|} \tag{2.36}$$

As a simple consequence of the bound on the minimizer, we derive the locality of the effective action. The formula

$$\Delta_n = H_n^* \partial^* \partial H_n \tag{2.37}$$

for the effective action exhibits the exponential decay of the kernel of Δ_n. This formula can be derived by making the translation

$$A = A' + H_n B \tag{2.38}$$

in the integral (2.35). This translation separates the dependence on B in this integral from the dependence on A'. It yields for (2.35)

$$\exp\left[-\frac{1}{2\alpha} \|R_n \partial^* H_n B\|^2 - \frac{1}{2} \|\partial H_n B\|^2\right] \quad . \tag{2.39}$$

Thus since $H_n B$ satisfies the Landau gauge condition, the first term in the exponent vanishes, and

$$\exp\left[-\frac{1}{2} \langle B, \Delta_n B\rangle\right] = \exp\left(-\frac{1}{2} \|\partial H_n B\|^2\right). \tag{2.40}$$

9. Green's Functions

The basic Green's functions which arise in this renormalization group method are gauge dependent. Three gauges are especially important: Feynman gauge, Landau gauge and axial gauge. In Feynman gauge the Green's function has the simplest form, namely

$$G_n = \left(\partial^*\partial + \frac{1}{\alpha} \partial R_n \partial^* + Q_n^* Q_n\right)^{-1} \quad . \tag{2.41}$$

Here R_n is the projection introduced in (2.35) and $\alpha^{-1} \partial R_n \partial^*$ provides the gauge fixing. The $Q_n^* Q_n$ arises from the averaging procedures. While these Green's functions do not appear in the discussion above, the properties of G_n are central for understanding the properties of minimizers H_n and of the covariance operators $C^{(n)}$.

In the axial and Landau gauges the Green's functions are defined by functional integral representations, since the gauge fixing is introduced by delta functions, as well as the averaging. In Landau gauge, call the Green's function $\mathfrak{G}_n$. Then

$$\exp\left(\frac{1}{2}\langle J,\mathfrak{G}_n J\rangle\right) = Z_n^{-1}\int dA\delta(Q_n A)\delta_{R_n}(R_n\partial^* A)\exp\left(-\frac{1}{2}\|\partial A\|^2 + \langle A,J\rangle\right). \quad (2.42)$$

The delta function δ_{R_n} denotes the delta function defined on the vector space Range R_n and concentrated at the origin.

The axial gauge Green's function $G_{n,Ax}$ is defined by replacing $\delta_{R_n}(R_n\partial^* A)$ in (2.42) with $\delta_{Ax,n}(A)$, in (2.29). This operator is related by gauge transformation to the operator in Landau gauge. For details, see the discussion in [14] for minimizers, which implies a formula for the Green's functions.

The Landau gauge Green's function can be expressed in terms of the Feynman gauge Green's function by the formula

$$\mathfrak{G}_n = G_n - \frac{1}{\alpha} G_n \partial R_n \partial^* G_n - G_n Q_n^*(Q_n G_n Q_n^*)^{-1} Q_n G_n \quad . \quad (2.43)$$

The properties of the minimizer in Feynman gauge follow from the representation

$$H_n = G_n Q_n^*(Q_n G_n Q_n^*)^{-1} \quad . \quad (2.44)$$

From these two representations, we see that we must investigate $Q_n G_n, Q_n^*$, and its inverse. Let us first describe G_n, which satisfies the bound

$$|\nabla^\alpha G_n(b,b')| \leqslant O(1)\,\frac{e^{-\delta|b-b'|}}{|b-b'|^{d-2+\alpha}} \quad (2.45)$$

where

$$\nabla^\alpha = \begin{cases} 1 & \alpha = 0 \\ \nabla & \alpha = 1 \\ \Delta & \alpha = 2 \end{cases}$$

denotes the appropriate derivatives of degree 0,1,2 for scalar functions defined on bonds. From this bound it follows that $Q_n G_n Q_n^*$ is a unit lattice operator whose kernel has exponential decay. This operator is translation invariant, and it satisfies the uniform bound

$$0 < c_1 \leqslant Q_n G_n Q_n^* \leqslant c_2 \quad . \quad (2.46)$$

The positive constants c_1, c_2 are independent of n and of the volume. The proof of this bound can be obtained from a closed form expression for Q_n and G_n in the momentum (Fourier transform) representation. This inequality and exponential decay of the kernel ensure exponential decay of the inverse kernel. This fact could be established directly using analyticity properties of the momentum space Green's functions. However, we present a more general method in Chapter IV.

As a result, we have a bound for the kernel of the inverse operator $(Q_nG_nQ_n^*)^{-1}$, namely for unit lattice bonds c,c',

$$|(Q_nG_nQ_n^*)^{-1}(c,c')| \leqslant O(1)\exp(-\delta|c-c'|) \quad . \tag{2.47}$$

The two inequalities (2.46, 2.47) ensure exponential decay of the kernel H_n as well as exponential decay for other Green's functions such as $\mathfrak{G}_n$ defined in (2.42).

10. Recurrence Relations

In this section we describe recurrence relations for Green's functions and minimizers which arise after n renormalization steps. The basic property is

$$H_{n+1} = H_n H^{(n)} + \text{gauge transformation} \tag{2.48}$$

where H_n is the n-step minimizer and $H^{(n)}$ the minimizer for going from step $n-1$ to step n. This identity is proved exactly as the proof of Equation (5.2.1) in [14] for minimizers in axial gauges. Then we use the relation (5.1.1) of the same paper connecting the minimizers in axial and Landau gauges by a gauge transformation.

The identity (2.48) implies a decomposition for variables A in integrals defining minimizers or Green's functions

$$A = A_0' + H^{(1)}A_1' + H^{(1)}H^{(2)}A_2' + H^{(1)}\cdots H^{(n-1)}A_{n-1}' + H^{(1)}\cdots H^{(n)}B \ , \tag{2.49}$$

where $B=0$ in the case of a Green's function. Using these relations, we write

$$A = A_0' + H_1A_1' + \cdots + H_{n-1}A_{n-1}' + H_nB + \partial\lambda \quad . \tag{2.50}$$

The fields A_j' are the fluctuation fields for the one-step renormalization transformations. They are independent random variables with $C^{(j)}$ the propagator for A_j'. This implies that the Green's function of (2.42) can be decomposed in the following way:

$$\mathfrak{G}_n = \sum_{j=0}^{n-1} H_jC^{(j)}H_j^* + \text{gauge transformation} \quad . \tag{2.51}$$

The recurrence relations (2.48, 2.51) provide a convenient tool to establish localization and exponential decay of Green's functions, as well as algebraic identities between them.

III. NON-ABELIAN GAUGE THEORY: ULTRAVIOLET PROBLEMS

As discussed in the introduction, we divide the ultraviolet stability analysis into two parts. The first part reduces to understanding the effective actions for small fields; this region is accessible by perturbation theory. The complementary analysis deals with large fields and relies on nonperturbative ideas. In this chapter we discuss the role of effective actions, which are basically perturbative in nature.

The strategy for establishing uniform stability bounds for nonabelian gauge theories differs from the corresponding proofs for scalar field theories. In the latter case, the action has the form

$$\mathcal{A} = \mathcal{A}_T + \mathcal{A}_V$$

where $\mathcal{A}_T$ is a "kinetic term" and $\mathcal{A}_V$ is formally bounded from below. Using this separation, the analysis generally proceeds perturbatively for small fields and using the fact that for large fields the potential term $\mathcal{A}_V$ is large (and positive).

For gauge theories, such a separation is not evident. Although the total action

$$\mathcal{A} = \mathrm{Tr}\, \Sigma \int |F|^2$$

is formally positive, and although $F = dA + A \wedge A$ yields a quartic potential of A, this potential has no useful positivity property by itself; the cross terms in F^2 are large and nonpositive.

The resolution of this problem is to make a decomposition of the form $\mathcal{A} = \mathcal{A}_T + \mathcal{A}_V$ only for small fields. In this case the interaction potential $\mathcal{A}_V$ is small. The analysis of the large field part of the integral must rely on positivity properties of the total action $\mathcal{A}$.

1. The First Renormalization Step

In the first renormalization step we need to evaluate an integral of the form

$$\int dU\ \delta(\bar{U}V^{-1})\,\delta_{Ax}(U)\chi\ \exp\left[-\frac{1}{g_0^2}\,A(U) - E\right] \quad . \tag{3.1}$$

The small field region we consider restricts the plaquette variables defined by U and V to a small neighborhood of the identity.

The goal of this section is to reduce the analysis of (3.1) to a type of integral which is frequently considered in statistical physics and quantum field theory. In particular, we reduce (3.1) to a multiple of an integral of the form

$$\int d\mu_C \chi\ e^{-V} \quad . \tag{3.2}$$

Here $d\mu_C$ is a Gaussian measure, and V yields a perturbation of the Gaussian, which is a small perturbation on the support of χ. We arrive at this representation in (3.17), as a result of the reasoning which follows.

Once (3.1) has been derived, there are two standard methods to analyze it further

(i) the cumulant expression,
(ii) the cluster expansion.

The cumulant expansion is especially useful for super-renormalizable theories, since after a finite number of expansion steps the ultraviolet renormalization cancellations no longer pose difficulty. The cluster expansion method is of more general applicability.

The general cumulant expansion gives a formula for the expectation

$$\langle e^{-V}\rangle = \exp\left[\sum_{n=1}^{N} \frac{1}{n!} \langle (-V)^n\rangle^T + O(|V|^{N+1})\,|\mathrm{vol}|\right] \tag{3.3}$$

where $|\mathrm{vol}|$ denotes the volume of the lattice. Note that $\langle V^n\rangle^T$ denotes the truncated expectation of V^n, and the expectation $\langle\,\rangle$ need not be Gaussian--the expectation may be with respect to

$$\frac{d\mu_C\, \chi}{\int d\mu_C\, \chi} \quad .$$

We wish to study the integral (3.1) by the method of steepest descents. Therefore we wish to know the critical points of the action $A(U)$ on the domain of integration restricted by the delta functions. It is a fact that $A(U)$ has exactly one critical point U_1, which minimizes $A(U)$. The minimizing configuration U_1 has two important properties:

(i) U_1 is an analytic function of V

(ii) U_1 is gauge covariant on the L-lattice corresponding to L-lattice transformations of V. Explicitly,

$$U_1(V^v) = U_1(V)^{v_0} \tag{3.4}$$

where v_0 is a unit lattice gauge transformation which is constant on L-blocks, namely

$$v_0(x) = v(y) \qquad \text{for} \quad x \in B(y) \quad . \tag{3.5}$$

We now expand U about U_1, defining

$$U = U'U_1 \quad . \tag{3.6}$$

We define

U_1 as the "background field,"

and

U' as the "fluctuation field."

The purpose of the characteristic function χ in (3.1) is to ensure that U is close to U_1, namely U' is close to the identity. The group variables U' are uniquely represented by Lie algebra variables A', where

$$U' = e^{iA'} \tag{3.7}$$

and where A' lies in a small neighborhood of zero (in the Lie algebra).

We expand the action $A(U)$ using Taylor's formula, as a series in A'. The lowest order terms yield

$$A(U'U_1) = A(U_1) + \langle A',J\rangle + \frac{1}{2}\langle A',\Delta A'\rangle + V_0(A') \quad , \tag{3.8}$$

where

$$V_0(A') = O(|A'|^3) \; , \qquad \text{for} \quad |A'| \to 0 \; . \tag{3.9}$$

The linear term J is in general nonzero even though U_1 is a critical point of $A(U)$. The reason for this is that the space of configurations U in the domain of integration specified by the delta functions in (3.1) is nonlinear. The first variation $\langle A',J\rangle$ vanishes only when A' belongs to the tangent space of this manifold. We determine the tangent space by looking more closely at the delta functions. Let

$$\bar{U}V^{-1} = e^{iQ(A')} \tag{3.10}$$

where

$$\bar{U}V^{-1} = (U'U_1)^{-}(\bar{U}_1)^{-1} \tag{3.11}$$

ensures that $Q(A') = O(|A'|)$ as $|A'| \to 0$. We expand $Q(A')$ to extract the linear term, LQA', so that

$$Q(A') = LQA' + C(A') \; , \qquad C(A') = O(|A'|^2) \; . \tag{3.12}$$

Remark that in (3.8)-(3.12), all the functions such as J, Δ, V_0, Q, C, etc. depend on U_1--even though this is not explicitly stated.

The axial gauge fixing expression $\delta_{Ax}(U)$ can be written in terms of the fluctuation field A' as $\delta_{Ax}(A')$. This is true because if $U(b) = 1$, for a bond b, then by (3.6) and the fact that U_1 satisfies the axial gauge conditions, we have that $U'(b) = I$. Hence $A'(b) = 0$.

The expansion (3.12) then shows that the tangent space to the manifold of integration is the set of A's for which

$$QA = 0 \quad ,$$

and such that A satisfies axial gauge. To assure the vanishing of the linear term $\langle A',J\rangle$ in the action, we parameterize the integration manifold by variables in the tangent space. Introduce the new variable A to linearize $Q(A')$. Choose $D(A)$ such that

$$D(A) = O(|A|^2)$$

and for

$$A' = A - D(A) \quad , \tag{3.13}$$

the function $Q(A')$ becomes

$$Q(A') = LQA \quad .$$

After this change of variables, the expansion (3.8) can be written

$$A(U'U_1) = A(U_1) - \langle D(A),J\rangle + \frac{1}{2}\langle A - D(A),\Delta(A - D(A))\rangle + V_0(A - D(A))$$

$$= A(U_1) + \frac{1}{2}\langle A,\Delta^{(0)}A\rangle + V(A) \tag{3.14}$$

where the quadratic term in $\langle D(A),J\rangle$ is included in the quadratic form $\Delta^{(0)}$ and where

$$V(A) = O(|A|^3) \; , \qquad |A| \to 0 \quad .$$

We can now rewrite the integral (3.1) in terms of the variable A. In fact, let us replace A by g_0A. Then

$$(3.1) = \exp\left[-\frac{1}{g_0^2}A(U_1) - E\right]\int dA\ \delta(QA)\delta_{Ax}(A)\chi\ \exp\left[-\frac{1}{2}\langle A,\Delta^{(0)}A\rangle\right.$$

$$\left. + \log\det\left(I - \left(\frac{\delta}{\delta A}D\right)(g_0A)\right) + \log\sigma(g_0A - D(g_0A)) - \frac{1}{g_0^2}V(g_0A)\right] . \tag{3.15}$$

Here σ arises as the density for the Haar measure dU with respect to dA' and the log det arises as the Jacobian of the change of variables $A' \to A$ in (3.13).

Let us define the Gaussian probability measure $d\mu_{C^{(0)}}$ with covariance $C^{(0)}$ by

$$d\mu_{C^{(0)}}(A) = Z^{(0)-1}\delta(QA)\delta_{Ax}(A)\exp\left(-\frac{1}{2}\langle A,\Delta^{(0)}A\rangle\right) dA \quad , \tag{3.16}$$

where $Z^{(0)}$ is the appropriate normalization. Then (3.1) can be written

$$(3.1) = \exp\left[-\frac{1}{g_0^2}A(U_1) - E + \log Z^{(0)}(U_1)\right]\int\chi\ \exp[-V^{(0)}]\ d\mu_{C^{(0)}} \tag{3.17}$$

where $V^{(0)}$ is the sum of the last three terms in the exponential occurring in (3.15). The covariance $C^{(0)}$ has a kernel $C^{(0)}(x,y)$ which is bounded and decays exponentially in x-y.

The function $D(A)$ in (3.13) is not uniquely determined by the property that it linearizes $Q(A')$ in (3.12). It is possible, moreover, to choose $D(A)$ that $\det(I-\delta D/\delta A)$ factorizes into a product over bonds of the L-lattice. Thus the logarithm of this determinant is local. The same choice shows that $\log\sigma$ is local, and hence $V^{(0)}$ is a local potential. (In later renormalization steps, the corresponding potential $V^{(k)}$ is not exactly local, but it is approximately local.)

The formula (3.17) reduces (3.1) to a class of integrals which can be studied by the cumulant expansion described above. The resulting integral has the general form $\exp(-\mathscr{E}^{(1)})$ where $\mathscr{E}^{(1)}$ is the contribution to the effective action arising in this renormalization step. The cumulant expansion gives a perturbative expression for $\mathscr{E}^{(1)}$ in powers of $V^{(0)}$. This sum can be reexpressed in powers of g_0, with a different remainder.

The total effective action after the first renormalization step is

$$A_1 = A_1(U_1) = -\frac{1}{g_1^2} A^{L^{-1}}(U_1) - E_1 + [\log Z^{(0)}(U_1) - \log Z^{(0)}(1)]$$

$$+ [\mathscr{E}^{(1)}(U_1) - \mathscr{E}^{(1)}(1)] \quad . \tag{3.18}$$

Here

$$\mathscr{E}^{(1)}(U_1) = \log\left[\int \chi \exp[-V^{(0)}]d\mu_{C^{(0)}}\right] \quad .$$

Here we have rewritten $g_0^{-2}A(U_1)$ as $g_1^{-2}A^{L^{-1}}(U_1)$ where A^{ε} is the rescaled action defined in (1.7). Thus

$$g_1^2 = g_0^2 L^{4-d}$$

is the coupling constant after the first renormalization step. Since $g_0^2 = g^2\varepsilon^{d-4}$, it follows that

$$g_1^2 = g^2(L\varepsilon)^{d-4} \quad , \tag{3.19}$$

and g_1 moves away from the fixed point. In (3.18),

$$E_1 = E - \log Z^{(0)}(1) - \mathscr{E}^{(1)}(1) \quad ,$$

where the two extra constants define the vacuum energy renormalization in the first step.

A major issue at this point concerns the gauge invariance of A_1. Because of the structure of (3.1), the action $A_1(U_1)$ is gauge invariant with respect to gauge transformations of the new variable V,

$$A_1(U_1(V)) = A_1(U_1(V^v)) \quad . \tag{3.20}$$

This gauge invariance (3.4) is in respect to gauge transformations v_0 which are constant on blocks. In fact A_1 has a much more general gauge transformation property which also follows from (3.1).

The action A_1 is defined on minimizing gauge configurations U_1. Using (3.18), we extend A_1 to all regular background fields U, where regularity means that the plaquette variables are sufficiently close to I. The decomposition into a background field U and a fluctuation field U' has the property that under gauge transformation of the background field, the fluctuation U' transforms under the adjoint action of the gauge group. This extended action is gauge invariant with respect to gauge transformations $U \to U^u$.

The only term in the action A_1 which is not given explicitly is the functional $\mathscr{E}^{(1)}$. This functional can be studied and understood using the cumulant expansion or the cluster expansion. The cumulant expansion, discussed in the beginning of this section, can be taken up to ninth order, with the remainder $O(\varepsilon^4)|T_1|$, which is small. The cluster expansion gives a representation of $\mathscr{E}^{(1)}$ as a sum of localized and exponentially small terms, of the form

$$\mathscr{E}^{(1)}(U_1) = \sum_X \mathscr{E}^{(1)}(X,U_1) \tag{3.21}$$

where $\mathscr{E}^{(1)}(X,U_1)$ depends on U_1 restricted to X, and

$$|\mathscr{E}^{(1)}(X,U_1)| \leqslant O(1)e^{-\kappa|X|} \quad .$$

The cumulant expansion also gives an expansion of the form (3.21), with an error $O(\varepsilon^4)|T_1|$.

The form of the effective action in (3.18) does not exhibit its convergence properties in closed form; these will be evident only after further rearrangement of the expansion--including renormalization. We discuss these issues only after n renormalization steps.

2. The Effection Action after n Renormalization Steps

We propose an outline for understanding the effective action $\exp[-A_n(g_n,U_n(V)]$ after n renormaliztion steps. Here V are the new field variables on the unit lattice obtained from n block constructions starting from the unit lattice, and n rescaling by L^{-1} to again obtain a unit lattice action. Let us denote the resulting lattice $T_1^{(n)}$.

The function $U_n(V)$ is the classical background configuration. It is defined as the solution to the minimization problem for the Wilson action, on the space of regular configurations $\mathfrak{U}_n(\varepsilon_0)$. This regularity is defined by the plaquette variables $\partial U(p) = U(\partial p)$ being (i) close to the identity and (ii) having small covariant derivative. Quantitatively

$$\mathfrak{U}_n(\delta) = \{U\colon |\partial U - I| < \delta\eta^2,\ |D_U^{\eta *}\partial U| < \delta\eta^2\}\ , \tag{3.22}$$

where $\eta = L^{-n}$ is the distance scale of the original lattice after n renormalization transformations. The minimization problem could be defined by

$$U \to A^{\eta}(U),\qquad U \text{ regular},\qquad \bar{U}^n = V\ ,\quad \text{Axial Gauge.} \tag{3.23}$$

Here $\bar{U}^n = V$ means that the n-th iterated block average of U (defined on the η-lattice) equals V (defined on the unit lattice). The axial gauge condition is defined by local axial gauges on n scales.

The space $U_n(\delta)$ of regular configurations is gauge invariant, and the Wilson action $A^\eta(U)$ is also gauge invariant. Furthermore, the averages transform covariantly under gauge transformations as

$$\overline{(U^u)}^n = (\bar{U}^n)^u \tag{3.24}$$

where the gauge transformation u on the left is defined on the η-lattice and on the right it is restricted to the unit lattice. Let $\mathfrak{G}_n$ denote the subgroup of gauge transformations u on the η-lattice which equal the identity on the unit lattice,

$$\mathfrak{G}_n = \{u\colon u|\text{unit lattice} = I\} \quad . \tag{3.25}$$

By (3.24), the condition $\bar{U}^n = V$ is invariant under gauge transformations in $\mathfrak{G}_n$.

Thus it is natural to drop the restriction to axial gauge in (3.23), and to consider the minimization problem as a problem on the space of gauge orbits under $\mathfrak{G}_n$. This is also important, for technical reasons, since we need to use Landau and other gauges, as well as axial gauge.

THEOREM. (Critical orbit for the Wilson action.) *Consider the map* $U \to A^\eta(U)$ *defined for* U *satisfying* $\bar{U}^n = V$ *and* $|\partial V - I| \leqslant \varepsilon \leqslant \delta$. *Let* $\delta > 0$ *be sufficiently small and* $B < \infty$ *sufficiently large. Then there exists a critical orbit* $U_n(V)$ *in* $\mathfrak{U}_n(B\varepsilon)$ *which minimizes* $A^\eta(U)$. *This orbit is unique in the larger space* $\mathfrak{U}_n(B\delta)$.

We define $U_n(V)$ as the background field after n renormalization steps. In the future we want to expand the function $U_n(V)$ about the point $V = V_0$. Write $V = V'V_0$, where V' is close to the identity. Thus $V' = \exp(iB)$ and

$$\begin{aligned} U_n(V) &= U_n(\exp(iB)V_0) \\ &= \exp[i\eta\mathscr{H}_n(B)]U_n(V_0) \quad \text{modulo a gauge transformation.} \end{aligned} \tag{3.26}$$

The gauge transformation is chosen to fix the Landau gauge for the function $\mathscr{H}_n(B)$.

The function $\mathscr{H}_n(B)$ is an analytic function in B. In fact the first term in this expansion is given by the minimizer of the linear problem,

$$\mathscr{H}_n(B) = H_n B + O(|B|^2) \quad . \tag{3.27}$$

The function $\mathscr{H}_n(B)$ is defined on the η-lattice, and is regular on that lattice. It satisfies the bounds

$$|\mathscr{H}_n(B)|, \qquad |\nabla^\eta_{U_n(V_0)} \mathscr{H}_n(B)| \leqslant C|B| \quad . \tag{3.28}$$

Here the constant C depends only on the dimension d, i.e. it is an absolute constant.

We now represent the effective action as

$$A_n(g_n, U_n(V)) = \frac{1}{g_n^2} A^{\eta}(U_n(V)) - \mathscr{E}_n(U_n(V)) \quad . \tag{3.29}$$

Here $1/(g_n^2)\, A^{\eta}(U_n(V))$ is the classical contribution to A_n, and $\mathscr{E}_n$ are the quantum corrections. The coupling constant g_n is defined by

$$g_n = g(L^n \varepsilon)^{\frac{4-d}{2}} \tag{3.30}$$

and is related to g_0 by n canonical scaling transformations.

The quantum corrections $\mathscr{E}_n(U_n)$ can be written

$$\mathscr{E}_n(U_n) = \sum_{j=1}^{n} \mathscr{E}^{(j)}(U_n) \tag{3.31}$$

where $\mathscr{E}^{(j)}$ arises from the j-th step. The term $\mathscr{E}^{(j)}$ can be expanded as

$$\mathscr{E}^{(j)}(U_n) = \sum_{x \in \mathscr{D}_j} \mathscr{E}^{(j)}(X, U_n) \quad , \tag{3.32}$$

where

$$|\mathscr{E}^{(j)}(X, U_n)| \leqslant O(1)e^{-\kappa|X|} \quad . \tag{3.33}$$

Here $O(1)$ is a constant independent of j. The domains X in which $\mathscr{E}^{(j)}(X,U_n)$ are localized are unions of $L^j\eta$-cubes and $|X|$ denotes the number of cubes. We assume also that the function $\mathscr{E}^{(j)}(X)$ can be extended to regular configurations U from the space $\mathfrak{U}_n(\delta)$, and that it is an analytic function on the space, satisfying the above bound. We have not specified yet the number δ in the above conditions. We take

$$\delta = g_n p(g_n) \ , \qquad p(g_n) = a(1 + \log g_n^{-1})^p \quad , \tag{3.34}$$

where a is a sufficiently large positive number, p is a positive integer > 4. We assume that g_n is small enough, so that the above theorems are valid for δ given by (3.34). This restriction on g_n is ensured by $(L^n\varepsilon)^{(4-d)/2}$ sufficiently small, i.e. by stopping the iteration of the renormalization transformation before reaching the unit lattice scale.

3. A Loop Expansion for the Effective Action

The bounds such as (3.33) for the effective action are not sufficiently strong to control vacuum energy and other divergences. In fact, summing (3.33) over X and J yields a bound of the form

$$|\mathcal{E}_n(U_n)| \leqslant O(1)|T_\eta|\eta^{-d} \quad , \tag{3.35}$$

which is characteristic of the vacuum energy divergence after rescaling to the unit lattice. In order to obtain a finite bound for $\mathcal{E}_n$, it is necessary to add the appropriate renormalization counterterms. In this case, we require the vacuum energy counterterm which is defined by $\mathcal{E}_n(1)$. Our goal is to prove that

$$|\mathcal{E}_n(U_n) - \mathcal{E}_n(1)| \leqslant O(1)|T_\eta| \quad , \tag{3.36}$$

with the constant $O(1)$ independent of n and of the torus $T_1^{(n)}$. This bound follows from corresponding bounds on the j-step effective actions. Let us introduce the logarithmic correction factor

$$q(\delta, L^j\eta) = \delta(1 + \log(L^j\eta)^{-1}) \quad . \tag{3.37}$$

In terms of this,

$$|\mathcal{E}^{(j)}(U_n) - \mathcal{E}^{(j)}(1)| \leqslant O(1)q(\delta, L^j\eta)^2(L^j\eta)^4|T_{L^{-j}}|$$

$$= O(1)q(\delta, L^j\eta)^2(L^j\eta)^{4-d}|T_\eta| \quad . \tag{3.38}$$

We derive this bound from bounds on expressions $\mathcal{E}^{(j)}(X,U_n)$ which are localized in the domain X. We require

PROPOSITION

$$|\mathcal{E}^{(j)}(X,U_n) - \mathcal{E}^{(j)}(X,I)| \leqslant O(1)q(\delta, L^j\eta)^2(L^j\eta)^4 e^{-1/2\,\kappa|X|} \quad . \tag{3.39}$$

Using this proposition we obtain the desired bound (3.37) by the standard cluster expansion estimate

$$\sum_{X\subset T_{L^{-j}}} e^{-1/2\,\kappa|X|} \leqslant O(1)|T_{L^{-j}}| \quad . \tag{3.40}$$

This proposition also means that in estimating $\mathcal{E}^{(j)}(X,U_n) - \mathcal{E}^{(j)}(X,I)$, it is possible to neglect contributions of the order of the right side. Thus, for example, we can neglect in the sum (3.32), those terms for which the localization domains X are large, namely

$$|X| \geqslant \frac{10}{\kappa}\, q(1, L^j\eta) \tag{3.41}$$

where κ occurs in (3.33). Such terms can be bounded by

$$|\mathscr{E}^{(j)}(X, U_n)| \leqslant O(1)\, e^{-1/2\, \kappa |X|} (L^j\eta)^5 \quad . \tag{3.42}$$

In fact, the power $(L^j\eta)^5$ can be improved to an arbitrary power $(L^j\eta)^N$ by taking the coefficient in (3.41) sufficiently large.

We only need to consider terms with localization size $|X| < 10\kappa^{-1} q(1, L^j\eta)$. We analyze terms in such localization domains by expanding them in loop variables of the fields. To construct this expansion, we apply the identity

$$U_n(V) = U_j(\bar{U}_n^j(V)) \quad . \tag{3.43}$$

This identity is a consequence of the following facts: The critical point $U_n(V)$ is defined on the space of U's satisfying $\bar{U}^n = V$. In a small neighborhood of $U_n(V)$, the configurations U which satisfy the condition $\bar{U}^j = \bar{U}_n^j(V)$ form a subspace of this neighborhood. These configurations also satisfy the condition $\bar{U}^n = V$. This ensures that $U_n(V)$ is also a critical point in the subspace, on which by definition the (unique) critical point is $U_j(\bar{U}_n^j(V))$. Let us define

$$V' = \bar{U}_n^j(V) \quad . \tag{3.44}$$

Then we conclude

$$\mathscr{E}^{(j)}(X, U_n(V)) = \mathscr{E}^{(j)}(X, U_j(V')) \quad . \tag{3.45}$$

PROPOSITION. *If* $|\partial U - I| < \delta L^{-2j}$, *then*

$$|\partial \bar{U}^j - I| \leqslant \delta + O(\delta^2) < 2\delta \quad . \tag{3.46}$$

As a consequence, the regularity condition (3.22) for configurations

$$|\partial U_n - I| < \delta\eta^2 \quad \text{ensures that} \quad |\partial V' - I| < 2\delta (L^j\eta)^2 \quad . \tag{3.47}$$

Consider X satisfying $|X| \leqslant 10\kappa^{-1} q(1, L^j\eta)$, and chose a cube $\square_1 \supset X$ of size $M_1 = [20\kappa^{-1} q(1, L^j\eta)]$. Take a second cube $\square_2$ of size $3M_1$ with the same center as $\square_1$. On $\square_2$ we introduce the axial gauge conditions for V', treating $\square_2$ as a single block. Let V'' denote the transformed configuration, and let y denote the center of $\square_2$. Thus

$$V''(x, x') = V'(\Gamma_{y,x} \cup \langle xx' \rangle \cup \Gamma_{yx'}^{-1}) \quad , \tag{3.48}$$

where $\langle xx' \rangle$ is a bond of the lattice $T^{(j)}$ contained in $\square_2$. The condition (3.47) ensures

$$|V''(x,x')-I| \leqslant 2\delta\ \mathrm{dist}(\langle x,x'\rangle,y)(L^j\eta)^2$$

$$\leqslant 60d\kappa^{-1}q(\delta,L^j\eta)(L^j\eta)^2 \quad . \tag{3.49}$$

The dependence of $\mathscr{E}^{(j)}(X,U_j(V'))$ on V' localized outside $\square_2$ is very weak, because of an exponential decay property of the function $U_j(V')$ generalizing the exponential decay properties of the minimizers H_j discussed in Chapter 2. We state

PROPOSITION. *Under the above hypotheses,*

$$\mathscr{E}^{(j)}(X,U_j(V')) = \mathscr{E}^{(j)}(X,U_j(V''|\square_2) + O((L^j\eta)^5 e^{-1/2\ \kappa|X|}) \ . \tag{3.50}$$

Define the Lie algebra variables B'' by

$$V'' = \exp(iB) \tag{3.51}$$

where B can be bounded by twice the right hand side of (3.49), so

$$|B| \leqslant O(1)q(\delta,L^j\eta)(L^j\eta)^2 \quad . \tag{3.52}$$

The function $\mathscr{E}^{(j)}(X,U_j(\exp(iB)))$ is analytic in B, and therefore has a Taylor expansion

$$\mathscr{E}^{(j)}(X,U_j(\exp iB)) = \mathscr{E}^{(j)}(X,U_j(I)) + \langle\frac{\delta}{\delta B}\mathscr{E}^{(j)}(X,U_j(I)),B\rangle$$

$$+\int_0^1 dt(1-t)\langle\frac{\delta^2}{\delta B^2}\mathscr{E}^{(j)}(X,U_j(\exp itB))B,B\rangle \ . \tag{3.53}$$

We have expanded to the second order because the second order remainder can be bounded by

$$O(1)\exp(-\kappa|X|)|B|^2 \leqslant O(1)q(\delta,L^j\eta)^2(L^j\eta)^4\exp(-\kappa|X|) \quad . \tag{3.54}$$

This bound, when summed over X, contributes to the right side of (3.38).

The first term in (3.53) contributes to the term we need to estimate, while the second term is zero, which is a special case of the general argument which follows:

Let $\mathscr{E}$ be a gauge invariant, differentiable function. For constant gauge transformations v,

$$\mathscr{E}(vVv^{-1}) = \mathscr{E}(V) \quad . \tag{3.55}$$

For $V = \exp(iB)$, we have

$$vVv^{-1} = \exp(iR(v)B) \quad , \tag{3.56}$$

where $R(v)B = vBv^{-1}$. This is the adjoint representation of the group $\mathfrak{G}$. We differentiate (3.55) to obtain

$$\langle R(v^{-1}) \frac{\delta}{\delta B} \mathscr{E}(1), B\rangle = \langle \frac{\delta}{\delta B} \mathscr{E}(1), B\rangle \tag{3.57}$$

for arbitrary B, and constant v. Hence

$$\frac{\delta}{\delta B} \mathscr{E}(1) = R(v^{-1}) \frac{\delta}{\delta B} \mathscr{E}(1) \quad . \tag{3.58}$$

Assuming semisimplicity, this ensures $(\delta/\delta B)\,\mathscr{E}(1) = 0$. Hence

$$\mathscr{E}^{(j)}(X, U_j(V'')) - \mathscr{E}^{(j)}(X, I) = O(1)q(\delta, L^j\eta)^2 (L^j\eta)^4 \exp(-\kappa|X|) \quad . \tag{3.59}$$

This bound, the bound (3.50) and the bound (3.42) yield (3.39).

4. Discussion of the Four-Dimensional Case

The bounds outlined above are borderline in four dimensions. In that case the power $(L^j\eta)^4$ needs to be replaced by $(L^j\eta)^{4+\varepsilon}$ in order to dominate logarithms and to give the additional convergence factor. This is possible if we improve the expansion (3.53), taking it to third order. The estimate on the error from the third order term is

$$O(1)\exp(-\kappa|X|)\,|B|^3 \leqslant O(1)q(\delta, L^j\eta)^3 (L^j\eta)^6 \exp(-\kappa|X|) \quad , \tag{3.60}$$

which is of the desired type.

Returning to (3.53), the first order term vanishes as before, and the constant term gives the vacuum energy renormalization. Thus the behavior of (3.53) is reduced to studying the second order term. This term demands an additional renormalization which is in fact a renormalization of the coupling constant--the only renormalization which is compatible with perturbative analysis.

According to this analysis, the coupling constant $g^2(\Lambda)$ which depends on the ultraviolet cutoff Λ satisfies the differential equation

$$\frac{dg^{-2}(\Lambda)}{d\,\ell n\,\Lambda} = \beta(g^2(\Lambda), \Lambda) \tag{3.61}$$

where β is a regular function which has a finite limit as $\Lambda \to \infty$. In the case of the lattice regularization, we have a discrete sequence of cutoffs $\Lambda_n = \varepsilon_n^{-1} = L^n$, and the differential equation is replaced by the difference

$$g_{n+1}^{-2} - g_n^{-2} = \beta_n(g_n^2) \tag{3.62}$$

where

$$g_n = g(L^n) \quad , \qquad \beta_n(g^2) = \beta(g^2, L^n) \quad .$$

The successive application of (3.62), starting with the term g_0^{-2}, contributes the term

$$\beta_{j-1}(g_{j-1}^2)A(U_n) \tag{3.63}$$

to the j-th effective action. Thus it is given by

$$\mathscr{E}^{(j)}(U_n) - \mathscr{E}^{(j)}(1) + \beta_{j-1}(g_{j-1}^2)A(U_n) \quad . \tag{3.64}$$

The main problem is to define the β function in an appropriate way. Such a definition follows from a detailed analysis of the quadratic terms in the expansion (3.53). In particular let ζ denote the coefficient of $1/2\|\partial B\|^2$ in the expansion of the quadratic part of the effective action,

$$\sum_X \frac{1}{2} < \frac{\delta^2}{\delta B^2} \mathscr{E}^{(j)}(X,U_j(1))B,B> = \zeta \frac{1}{2} \|\partial B\|^2 + O((L^j\eta)^{4+\alpha})|T_{L^{-j}}| \quad , \tag{3.65}$$

where $\alpha > 0$. The coefficient ζ in this expansion depends on the coupling constant. We define the function

$$\beta_{j-1} = \zeta \ , \tag{3.66}$$

and note that ζ is uniformly bounded.

We have an explicit formula for the coefficient ζ in terms of the effective action $\mathscr{E}^{(j)}$. In order to write it, we introduce the vacuum polarization tensor

$$\Pi_{\mu\nu}(x-y) = \frac{\delta^2}{\delta B_\mu(x)\delta B_\nu(y)} \mathscr{E}^{(j)}(U_j(1)) \quad . \tag{3.67}$$

In the momentum representation, ζ is given in terms of $\Pi_{\mu\nu}$ by

$$\zeta = -\frac{\partial^2}{\partial p_\mu \partial p_\nu} \tilde{\Pi}_{\mu\nu}(0) \quad , \qquad \text{for } \mu \neq \nu \quad , \tag{3.68}$$

and ζ does not depend on μ,ν.

Using β, we can compute a sequence (3.62) of running coupling constants g_n^2. What are the properties of β and g_n^2? The sequence $\{\beta_n(g^2)\}$ is uniformly bounded and convergent to a function $\beta(g^2)$, which is negative in a sufficiently small neighborhood of zero. This negativity ensures for n sufficiently large that

$$\beta'' \leqslant \beta_n(g^2) \leqslant \beta' < 0 \quad . \tag{3.69}$$

The equation (3.62), summed over n from k to $K = \log_L \varepsilon^{-1}$, yield

$$\frac{1}{g_K^2} - \frac{1}{g_k^2} = \sum_{n=k}^{K-1} \beta_n(g_n^2) \quad . \tag{3.70}$$

Denote the final coupling constant by

$$g_K^2 = g^2 \quad , \tag{3.71}$$

and use the bounds (3.69)-(3.70) to obtain

$$\frac{1}{g^2} - \beta' \log_L(L^k\varepsilon) \leqslant \frac{1}{g_k^2} \leqslant \frac{1}{g^2} - \beta'' \log_L(L^k\varepsilon) \quad . \tag{3.72}$$

This behavior of the running coupling constant is called asymptotic freedom. It ensures that the running coupling constant remain in a small neighborhood of zero so that a perturbative analysis is applicable at each step.

The properties of β the functions discussed above actually yield much more detailed asymptotics for the coupling constant. Let us consider, for example the first coupling constant. This is the "bare coupling constant" g_0 which we investigate as a function of the lattice spacing ε. Using (3.70), and the Taylor expansion for β,

$$\beta_n(g_n^2) = \beta_{n,0} + \beta_{n,2}g_n^2 + O(g_n^4) \quad . \tag{3.73}$$

Thus

$$\frac{1}{g_0^2} = \frac{1}{g_2} - \sum_{n=0}^{K-1} (\beta_{n,0} + \beta_{n+2,0}g_n^2 + O(g_n^4)) \quad . \tag{3.74}$$

The sequence $\beta_{n,0} \to \beta_0$, $\beta_{n,2} \to \beta_2$, where β_0, β_2 are the series coefficients for β. Hence the leading term in this expansion has the asymptotics

$$\sum_{n=0}^{K-1} \beta_{n,0} \sim \beta_0 K = \beta_0 \log_L \varepsilon^{-1} \quad . \tag{3.75}$$

The next term contributes $\beta_1 \log_L \log_L \varepsilon^{-1}$, where β_1 depends on β_0, β_2. The remaining terms contribute to $O(1)$. Thus

$$\frac{1}{g_0^2} = \frac{1}{g_2} - \beta_0 \log_L \varepsilon^{-1} - \beta_1 \log_L \log_L \varepsilon^{-1} + O(1) \quad . \tag{3.76}$$

This asymptotic behavior is typical for many models of quantum field theory and statistical physics having the property of asymptotic freedom. It has been rigorously established in certain models [16]. The above ideas are elaborated in [17].

5. The General Renormalization Step

In (3.29) we stated the general form of the effective action. Here we argue that the renormalization transformation actually preserve this form. In fact we prove a more general result for actions of the form

$$\text{Wilson action + Perturbation,} \tag{3.77}$$

showing that under the renormalization T this general form of the action is preserved. More specifically, we define a sequence of spaces $\{\mathbb{A}_n\}$, each $\mathbb{A}_n$ of the form (3.77), where $A_n \in \mathbb{A}_n$ and where

$$T\mathbb{A}_n \subset \mathbb{A}_{n+1} \quad . \tag{3.78}$$

Define $\mathbb{A}_n$ as follows: Consider the function

$$A_n(g,U) = \frac{1}{g^2} A^{\eta}(U) + \mathscr{E}_n(U) \quad . \tag{3.79}$$

Here $\eta = L^{-n}$, A^{η} denotes the Wilson action on the η-lattice and $U \in \mathfrak{U}_n(\delta)$, the space of regular configurations defined by (3.22). The function $\mathscr{E}_n$ has the properties

$$\mathscr{E}_n(U) = \sum_{j=1}^{n} \mathscr{E}^{(j)}(U) \quad . \tag{3.80}$$

The terms $\mathscr{E}^{(j)}$ in the sum (3.80) are associated with the scale L^{-j}, $j \leqslant n$. Therefore it is natural to consider $\mathscr{E}^{(j)}$ defined on the space $\mathfrak{U}_j(\delta) \supset \mathfrak{U}_n(\delta)$.

The functions $\mathscr{E}^{(j)}$ can be localized in domains X characteristic for the lattice $T_{L^{-j}}$. Define

$$\mathscr{D}_j = \{X:\ X \subset T_{L^{-j}},\ X \text{ connected union of M-cubes}\} \quad , \tag{3.81}$$

where M is a sufficiently large integer. We write

$$\mathscr{E}^{(j)}(U) = \sum_{X \in \mathscr{D}_j} \mathscr{E}^{(j)}(X,U) \quad , \tag{3.82}$$

where $\mathscr{E}^{(j)}(X,U)$ depends only on $U{\restriction}X$. Also we assume that the localized function $\mathscr{E}^{(j)}(X,U)$ is a regular and gauge invariant function on $\mathfrak{U}_j(\delta)$ satisfying the bounds

$$|\mathscr{E}^{(j)}(X,U)| \leqslant O(1)e^{-\kappa|X|} \quad . \tag{3.83}$$

Here $O(1)$ and κ are absolute constants, independent of j, X and U. The constant κ must also be sufficiently large. The number $|X|$ is the number of disjoint M-cubes contained in X. The properties (3.80)-(3.83) generalize (3.31)-(3.33).

Now we construct the renormalization transformation (3.78). There are three steps: integration, scaling, and extension. The first step is defined by the transformation T_1 defined on configurations on the η-lattice by

$$T_1(\exp(-A_n))(W) \equiv \int dV\ \delta(\bar{V}W^{-1})\delta_{Ax}(V)\chi\ \exp[-A_n(g,U_n(V))] \quad . \tag{3.84}$$

Here V is a unit lattice configuration (on the lattice $T_1^{(n)}$ of n-blocks), and W is an L-lattice configuration (on the lattice $T_L^{(n+1)}$ of (n+1) blocks). The configuration $U_n(V)$ is the critical configuration defined in III.2 by the variational problem (3.23). Here χ is a "small field" characteristic function. Also $\bar{V}$ is the block average defined in the introduction.

It is convenient to reexpress (3.84) on the unit lattice, and the second transformation T_2 does this scaling. The third transformation T_3 can only be explained after further analysis of (3.84). This analysis follows the lines of the analysis of the first renormalization step carried out in III.1. It begins with the saddle point method to expand (3.84). One must find the critical points of $A_n(g,U_n(V))$, restricted by $\bar{V}=W$, gauge fixing, and the characteristic function χ. The Theorem assures the existence of exactly one critical configuration $V^{(n)}(W)$ which satisfies

$$V^{(n)}(W) = \overline{U_{n+1}(W)}^{\,n} \quad . \tag{3.85}$$

We expand V around $V^{(n)}(W)$ and write

$$V = V'V^{(n)} \quad . \tag{3.86}$$

The characteristic function χ assures that V' is small so we can write

$$V' = \exp(igB) \quad . \tag{3.87}$$

The restriction of the characteristic function is $|B| < p(g)$.

The next step is to expand (3.84) in B. For the configuration $U_n(V) = U_n(\exp igBV^{(n)}(W))$ we use (3.26)-(3.27). We expand the Wilson action term using (3.8). The expressions in the delta function are linearized as in (3.13). After these transformations, we obtain

$$\begin{aligned}(3.84) = \text{const.}\ &\exp\left[-\frac{1}{g^2}A^{\eta}(U_{n+1}(W)) + \mathscr{E}_n(U_{n+1}(W))\right]\\ &\int dB\ \delta(QB)\delta_{Ax}(B)\chi\ \exp\Big[-\frac{1}{2}\langle B,\Delta^{(n)}B\rangle - \mathscr{V}^{(n)}(gB)\\ &+ \mathscr{E}_n(U_n(\exp igBV^{(n)}(W))) - \mathscr{E}_n(U_{n+1}(W))\Big] \quad .\end{aligned} \tag{3.88}$$

The quadratic terms $\Delta^{(n)}$ arise from taking the quadratic terms in the Wilson action, while the terms $\mathscr{V}^{(n)}$ arise from higher order terms and are small. By definition, the logarithm of the integral over B defines the new term $\mathscr{E}^{(n+1)}(U_{n+1}(W))$.

The integral (3.88) has the same general structure as (3.15) and (3.17). It is a Gaussian integral of a small interaction, which is localized. The localization of $\mathscr{E}_n$ follows from the inductive hypothesis. The localization of the potential $\mathscr{V}^{(n)}$ can be understood on the basis of the expansion (3.8) of the local Wilson action and the decay properties of the minimizer (3.27). With these locality properties in hand, we can represent (3.88), using either a cumulant expansion or a cluster expansion. This yields the localized representation (3.82) and the bounds (3.83) for the function $\mathscr{E}^{(n+1)}$.

Finally we note that $\mathscr{E}^{(n+1)}(U_{n+1}(W))$ is gauge invariant with respect to all gauge transformations of U_{n+1}. The discussion follows the discussion in the first section of gauge invariance after the first renormalization step.

The next transformation is the rescaling T_2 from the L-lattice to the unit lattice, or U_{n+1} from the η-lattice to the $L^{-1}\eta$-lattice. The only real operation in this scaling is a change in the Wilson action

$$\frac{1}{g^2} A^{\eta}(U_{n+1}) = \frac{1}{L^{4-d} g^2} A^{L^{-1}\eta}(U_{n+1}) \quad . \tag{3.89}$$

The last transformation T_3 replaces the critical configuration U_{n+1} by an arbitrary regular configuration in the space $\mathfrak{U}_{n+1}(\delta)$. This completes the construction of the renormalization map T. It also proves the inductive assumptions (3.29)-(3.33) for the effective actions.

6. Large Fields

In this section we discuss the separation of the domain of integration into large and small fields. The steepest descent analysis outlined above applies in the small field region. In the first step this decomposition into large and small fields is defined by the partition of unity

$$1 = \sum_{P^c} \prod_{p \in P} \chi(\{U : |U(\partial p) - 1| < \delta\}) \prod_{p \in P^c} \chi(U : |U(\partial p) - 1| \geqslant \delta\}) \ . \tag{3.90}$$

The large field region is determined by the set P^c. For example, we can take all M-cubes intersecting P^c, and their union defines Ω_1^c. For each plaquette $p \in P^c$ we use the bound

$$\exp\left[-\frac{1}{2g_0^2} |U(\partial p) - 1|^2\right] \leqslant \exp\left[-\frac{1}{2g_0^2} g_0^2 \, p^2(g_0)\right]$$

$$\leqslant \exp\left[-\frac{1}{2} p^2(g_0)\right] \leqslant O(g_0^N) \tag{3.91}$$

where N is arbitrarily large. For three dimensional theories, i.e. $d = 3$,

$$g_0 = g\varepsilon^{1/2}$$

so we also have a bound by an arbitrarily large power of ε. This power is important, including the bound on the sum over P^c in (3.38), where the factors $O(\varepsilon^{N/2})$ can be used to dominate the sum.

We have bounds on each term of (3.88). Considering only bounds from large field regions $p \in P^c$, we obtain

$$\sum_{P^c} O(\varepsilon^{N/2})^{|P^c|} \leqslant e^{O(\varepsilon^{(N/2-3)})|T_\varepsilon|} \quad . \tag{3.92}$$

So for $N/2 > 3$, we have a small contribution to the final ultraviolet stability bound.

Thus we should analyze the small field region, and this we do as in the sections above. The only technical modification is to restrict attention throughout to the small field region. Hence we introduce boundary conditions in estimates on the existence of critical configurations for the action, on the decay of Green's functions, etc. This poses only technical difficulties.

In successive renormalization steps we follow the same procedure: restrict the fields as in (3.90), estimate large field regions as in (3.91), and in the small field regions we expand as above. Renormalization yields small field estimates and the small factors in (3.91) control the sums over choices for the large field regions.

IV. DECAY OF CORRELATIONS

We relate the decay of the kernel $A(x,y)$ for a positive transformation A to the decay of its Green's function

$$G(x,y) = A^{-1}(x,y) \quad .$$

This generalizes the exponential decay of $(-\Delta + m^2)^{-1}(x,y)$, but allows for nonlocal A. The kernel $A(x,y)$ need not be supported on or near the diagonal, i.e. it may be nonlocal. The Green's function $G(x,y)$ need not be positive, as A may not satisfy the maximum principle.

We consider unit lattice operators on $\ell_2(Z^d)$, or more generally operators defined on $\ell_2(\Gamma)$ where Γ denotes lattice points, lattice bonds, lattice plaquettes, or so forth. Operators on the torus T^d may be estimated by summing our bounds. We make two fundamental assumptions about A, that of strict positivity and short range.

DEFINITION 1. *The operator* A *is strictly positive if for some* $m > 0$,

$$m^2 \leqslant A \quad . \tag{4.1}$$

DEFINITION 2. *A function* $a(i)$ *is a short range localizing function if*

(i) $\quad 0 < a(i)$, $\qquad$ (4.2)

(ii) $\quad a(i) \leqslant \text{const}(1 + |i|)^{-2(d+\delta)}$ $\qquad$ (4.3)

for some $\delta > 0$.

(iii) *Choose* δ *so* (ii) *holds. Let*

$$b(i) \equiv (1+|i|)^{d+\delta} a(i) \quad . \tag{4.4}$$

We assume that for some constant $K < \infty$

$$K^{-1} \leqslant \frac{b(i+j)}{b(i)} \leqslant K$$

for all i *and for* $|j| \leqslant 2d$.

(iv) *There are constants* $c < \infty$ *and* $\varepsilon > 0$ *such that for all* n, *the* n-*fold convolution* $b * \cdots * b$ *satisfies*

$$(b*b*\cdots *b)(i) \leqslant c^n b(\varepsilon i) \quad , \tag{4.5}$$

for some $\delta > 0$.

THEOREM 4.1. *Let* A *be strictly positive and let*

$$|A(x,y)| \leqslant a(x-y) \tag{4.6}$$

where $a(x)$ *is a short range localizing function. Then for* M *sufficiently large for* $|x-y| \geqslant M$, *and for* b *given by (4.4),*

$$|G(x,y)| \leqslant b((x-y)/M) \quad . \tag{4.7}$$

Example 1. An example of a short range localizing kernel with polynomial decay is

$$a(i) = (1+|i|)^{-2(d+\delta)} \tag{4.8}$$

for some $\delta > 0$. Clearly (i-iii) hold. But for $b(i) = (1+|i|)^{\delta} a(i) = (1+|i|)^{-2d-\delta}$, it follows that

$$(b*b)(i) \leqslant \text{const.}\, b(i) \quad ,$$

so (iv) holds too, with $\varepsilon = 1$. If the hypotheses of the theorem apply with this kernel, then $|G(x,y)| \leqslant b((x-y)/M)$ for $|x-y| \geqslant M$, and M sufficiently large.

Example 2. An example of a short range localizing kernel with exponential decay is

$$a(i) = \exp\left(-\zeta \sum_{k=1}^{d} |i_k|\right) \quad , \tag{4.9}$$

where $\zeta > 0$. In verifying (i-iv) we again see that (i-iii) are evident. Furthermore, since

$$|i| \leqslant \sum_{k=1}^{d} |i_k| \leqslant \text{const.}\,|i| \quad , \tag{4.10}$$

the verification of (iv) reduces to the case $d = 1$, which we now consider. Let $i \in Z$ and $\kappa < \zeta$. Define $c(i) = \exp(-\kappa|i|)$. Then

$$b(i) \leqslant \text{const.}\,c(i) = \text{const.}\,b(\kappa^2\zeta^{-1}i) \quad . \tag{4.11}$$

Thus we need only verify (iv) for $c(i)$. Consider the Fourier series

$$\tilde{c}(p) = \sum_{j=-\infty}^{\infty} \exp(-ipj-\kappa|j|)$$

$$= \sinh\kappa(\cosh\kappa-\cos p)^{-1} \quad . \tag{4.12}$$

The function $\cosh\kappa - \cos p$ is entire and does not vanish in the strip $|\mathrm{Im}\,p| \leqslant \kappa(1-\varepsilon)$ for any fixed $\varepsilon > 0$. Thus $\tilde{c}(p)$ is analytic and uniformly bounded in this strip. Let

$$c \equiv \sup_{|\mathrm{Im}\,p| \leqslant \kappa(1-\varepsilon)} |\tilde{c}(p)| \quad .$$

Note that $c(p)$ is periodic in $\mathrm{Re}\,p$ with period 2π.

Now we use the Cauchy integral theorem to dominate $c*\cdots*c$. By the periodicity of c in $\mathrm{Re}\,p$,

$$(c*c*c*\cdots c)(j) = \frac{1}{(2\pi)} \int_{-\pi}^{\pi} \tilde{c}(p)^n e^{ijp}\,dp$$

$$= \frac{1}{2\pi} \int_{-\pi+i\alpha}^{\pi+i\alpha} \tilde{c}(p)^n e^{ijp}\,dp \quad , \tag{4.13}$$

where $\alpha = (\mathrm{sgn}\,j)\kappa(1-\varepsilon)$, and $p + i\alpha$ lies inside the analyticity domain of the integrand. We estimate the integral by absolute values,

$$c*\cdots*c(j) \leqslant \frac{1}{2\pi} \int_{-\pi+i\alpha}^{\pi+i\alpha} |\tilde{c}(p)|^n |e^{ijp}|\,dp$$

$$\leqslant c^n e^{-|\alpha|j} = c^n c((1-\varepsilon)j) \quad , \tag{4.14}$$

which shows that $c(j)$ satisfies property (iv) as desired. Hence if the hypotheses of the theorem apply with $a(i)$ in (4.9), it follows that $|G(x,y)|$ has exponential decrease in $|x-y|$.

We now proceed to prove the theorem. We require two kinds of localizing functions. Let $\{\square_j\}$ denote a cover of $\mathbb{R}^d$ by 2M-cubes, where $\square_j$ is centered at the lattice site jM, $j \in Z^d$. Thus cubes $\square_i$, $\square_j$ overlap if $|i-j| \leq 2d^{1/2}$. A given cube $\square_i$ overlaps 5^d neighboring cubes. We also let $\square_j$ denote the projection of $\ell_2(Z^d)$ onto $\ell_2(Z^d \cap \square_j)$. The projections $\square_i$ and $\square_j$ are orthogonal for $|i-j| > 2d^{1/2}$.

We also use a smooth partition of unity $h_j^2(x)$ of R^d which is constructed as follows. Let $h(t)$, $t \in R$, be a positive C_0^∞ function such that

$$h(t) = \begin{cases} 1 & |t| \leq 1/3 \\ 0 & |t| \geq 2/3 \end{cases} \quad . \tag{4.15}$$

and

$$\sum_{j \in Z} h^2(t-j) = 1 \quad . \tag{4.16}$$

Assume furthermore that

$$\sup |h'(t)| \leq 10 \quad . \tag{4.17}$$

Let $j \in Z^d$

$$h_j(x) = \prod_{k=1}^{d} h\left(\frac{x_k}{M} - j_k\right) \quad . \tag{4.18}$$

Note that

$$h_j \square_j = h_j \quad , \tag{4.19}$$

and

$$\sum_{j \in Z^d} h_j^2(x) = 1 \quad . \tag{4.20}$$

Assuming (4.1), $m^2 \square_j \leq \square_j A \square_j$, so $\square_j A \square_j$ is invertible on $\ell_2(\square_j)$. Let

$$C_j = (\square_j A \square_j)^{-1} \upharpoonright \ell(\square_j) \quad , \tag{4.21}$$

and

$$C = \sum_j h_j C_j h_j \quad . \tag{4.22}$$

The operator C is an approximate inverse for A, and $C \to A^{-1}$ as $M \to \infty$. In fact, define R by $AC = I - R$.

PROPOSITION 4.2. *For* M *sufficiently large,* $\|R\| \leq O(M^{-\delta}) < 1$, so

$$G = A^{-1} + C(I-R)^{-1} = \sum_{n=0}^{\infty} CR^n \tag{4.23}$$

is a norm convergent series.

Proof. Expand the product AC as

$$AC = \sum_j Ah_j C_j H_j = \sum_j \Box_j Ah_j C_j h_j + \sum_j (1-\Box_j) Ah_j C_j h_j$$

$$= \sum_j h_j \Box_j A \Box_j C_j h_j + \sum_j \Box_j [A, h_j] \Box_j C_j h_j + \sum_j (1-\Box_j) Ah_j C_j h_j \quad .$$

Using (4.19)-(4.21), the first term on the right sums to I. Thus,

$$R = -\sum_j \Box_j [A,h_j] \Box_j C_j h_j + \sum_i (\Box_i - 1) Ah_i C_i h_i$$

$$= \sum_j \Box_j [h_j, A] \Box_j C_j h_j + \sum_{i,j} (\Box_i - 1) h_j^2 Ah_i C_i h_i \quad . \tag{4.24}$$

Since $(\Box_i - 1)h_j$ vanishes for $i = j$, we can write

$$R = \sum_{i,j} R_{ij} C_j h_j \tag{4.25}$$

with $R_{ii} = \Box_i [h_i, A] \Box_i$ and

$$R_{ij} = (\Box_j - 1) h_i^2 Ah_j \; , \qquad \text{for} \quad i \neq j \quad . \tag{4.26}$$

The transformations R_{ij} map $\ell_2(\Box_j) \to \ell_2(\Box_i)$. We formulate as three lemmas some intermediate bounds needed to complete the proof.

LEMMA 4.3. *Let* $T(x,y)$ *be the kernel of* T *on* $\ell_2(Z^d)$. *Then*

$$\|T\| \leqslant \left(\sup_x \sum_y |T(x,y)|\right)^{1/2} \left(\sup_x \sum_x |T(x,y)|\right)^{1/2} \quad .$$

Proof. For $f, g \in \ell_2(Z^d)$,

$$|\langle f, Tg\rangle| \leqslant \sum_{i,j} |f(i)T(i,j)g(j)|$$

$$= \sum_{i,j} (|f(i)|\,|T(i,j)|^{1/2})(|T(i,j)|^{1/2}|g(j)|) \quad ,$$

and the desired bound follows by the Schwarz inequality.

LEMMA 4.4. *Let* R *be given by (4.25). Then*

$$\|R\| \leq 2^d m^{-2} \left(\sup_i \sum_j \|R_{ij}\| \right)^{1/2} \left(\sup_j \sum_i \|R_{ij}\| \right)^{1/2} .$$

Proof. Note that $\Sigma_i \|\square_i f\|^2 = 2^d \|f\|^2$ and $\|C_j\| \leq m^{-2}$. Thus

$$|\langle f, Rg\rangle| \leq \sum_{i,j} \|\square_i f\| \|R_{ij} C_j h_j\| \|\square_j g\| .$$

The desired bound then follows by Lemma 3 and $\|R_{ij} C_j h_j\| \leq \|R_{ij}\| \|C_j\| \leq m^{-2} \|R_{ij}\|$.

LEMMA 4.5. *With the assumptions above, and* M *sufficiently large,*

$$\|R_{ij}\| \leq O(M^{-\delta}) b(M(i-j)/3) , \tag{4.27}$$

with δ, b of (4.4).

Proof. Consider first the case $i = j$. The kernel of R_{ii} is

$$R_{ii}(x,y) = A(x,y)(h_i(x) - h_i(y))\square_i(x)\square_i(y) . \tag{4.28}$$

Since h has a bounded derivative, using (4.18) gives

$$|h_i(x) - h_i(y)| \leq O(M^{-1})|x-y| .$$

But h_i is bounded by 1, so for any $\delta \in [0,1]$,

$$|h_i(x) - h_i(y)| \leq O(M^{-\delta})|x-y|^\delta .$$

It follows that

$$|R_{ii}(x,y)| \leq O(M^{-\delta})|x-y|^\delta |A(x,y)|$$
$$\leq O(M^{-\delta})|x-y|^\delta a(x-y) \leq O(M^{-\delta}) b(x-y) .$$

Hence by Lemma 4.3,

$$\|R_{ii}\| \leq O(M^{-\delta}) \sup_x \sum_y b(x-y) = O(M^{-\delta}) \sum_x b(x)$$
$$= O(M^{-\delta}). \tag{4.29}$$

For $i \neq j$, we infer from (4.26) that $R_{ij}(x,y)$ vanishes unless

$$x \in \operatorname{supp} h_i \cap (Z^d \backslash \Box_j) \qquad \text{and} \qquad y \in \operatorname{supp} h_j \quad .$$

On this set

$$|x-y| \geqslant \begin{cases} M/3 & |i-j| = 1 \\ M(|i-j|-4/3) & |i-j| > 1 \end{cases} \quad . \tag{4.30}$$

In particular,

$$|x-y| > \frac{1}{3} M|i-j| \tag{4.31}$$

in either case. We infer from (4.26) that, on its support, the kernel of R_{ij} satisfies

$$|R_{ij}(x,y)| \leqslant |A(x,y)| \leqslant a(x-y) = (1+|x-y|)^{-(d+\delta)} b(x-y) \quad . \tag{4.32}$$

Thus by (4.31) and Lemma 4.3,

$$\|R_{ij}\| \leqslant \left(\sup_x \sum_y |R_{ij}(x,y)|\right)^{1/2} \left(\sup_y \sum_x |R_{ij}(x,y)|\right)^{1/2}$$

$$\leqslant O(M^{-\delta}) \sup_{x,y} b(x-y) \leqslant O(M^{-\delta}) b(M(i-j)/3) \quad . \tag{4.33}$$

where the final sup over x,y is taken over $(x,y) \in R_{ij}(x,y)$. This completes the proof of the lemma.

We now return to Proposition 4.2. By Lemmas 4.4 and 4.5,

$$\|R\| \leqslant O(M^{-\delta}) 2^d m^{-2} \sup_i \sum_j b(M(i-j)/3) \quad .$$

By (4.3), and with fixed m, d,

$$\|R\| \leqslant O(M^{-\delta}) \quad ,$$

which establishes the proposition.

We now turn to the study of the Green's function $G(x,y) = (A^{-1})(x,y)$, using the series (4.23). This will yield the proof of Theorem 4.1. We can parameterize the terms in the expansion for $G(x,y)$ arising from (4.23) by a path $\omega = \{\omega_0, \omega_1, \ldots, \omega_{2n}\}$ where $\omega_j \in Z^d$ indexes a cube $\Box_{\omega_j}$ centered at $M\omega_j$. Here $x \in \Box_{\omega_0}$, $y \in \Box_{\omega_{2n}}$ and

$$G(x,y) = \sum_{\omega} \left(h_{\omega_0} C_{\omega_0} R_{\omega_1\omega_2} C_{\omega_2} h_{\omega_2} R_{\omega_3\omega_4} C_{\omega_4} h_{\omega_4} \right.$$

$$\left. \cdots R_{\omega_{2n-1}\omega_{2n}} C_{\omega_{2n}} h_{\omega_{2n}} \right)(x,y) \quad . \tag{4.34}$$

Because of the support properties of C,

$$|\omega_{2i} - \omega_{2i+1}| \leqslant 2d^{1/2}, \qquad i = 0,1,\ldots,n-1 \quad . \tag{4.35}$$

The sum (4.34) over walks ω is a version of the random walk representation for Green's functions. The kernel $G(x,y)$ can be estimated using $G(x,y) = \langle\delta_x, G\delta_y\rangle$ where $\|\delta_x\| = 1 = \|\delta_y\|$. Since $\|C_j\| \leqslant m^{-2}$ and $\|h_j\| = 1$, the series (4.33) yields

$$|G(x,y)| \leqslant \sum_{\omega} m^{-2n} \|R_{\omega_1\omega_2}\| \|R_{\omega_3\omega_4}\| \cdots \|R_{\omega_{2n-1}\omega_{2n}}\| + |C(x,y)| \quad , \tag{4.36}$$

where $|C(x,y)|$ arises from the $n=0$ contribution to (4.33). If $|x-y| > 2Md^{1/2}$, $C(x,y) = 0$. Alternatively $|C(x,y)| = |\langle\delta_x, C\delta_y\rangle| \leqslant \Sigma' \|C_j\| \leqslant 5^d m^{-2}$, where Σ' is restricted to the fewer than 5^d blocks $\square_j$ containing both x and y.

The other terms in (4.36) can be bounded using Lemma 4.5, as well as property (iii) of b, yielding

$$|G(x,y)| \leqslant |C(x,y)| + \sum_{n=1}^{\infty} O(M^{-\delta} 5^d m^{-2})^n (b_{M/3} * \cdots * b_{M/3})(\omega_0 - \omega_{2n}) \quad . \tag{4.37}$$

Here $b_{M/3}(i-j) = b(M(2-j)/3)$ appears in the bound the norm of R_{ij}. The sum over $\omega_0, \omega_2, \ldots, \omega_{2n}$ has been replaced by the sum over 5^{dn} terms with equal bounds which arise for fixed $\omega_1, \omega_3, \ldots, \omega_{2n-1}$. The sum over $\omega_1, \omega_3 \ldots, \omega_{2n-1}$ gives rise to the convolution of the $b_{M/3}$'s. Using property (iv) of the b's,

$$|G(x,y)| \leqslant |C(x,y)| + \sum_{n=1}^{\infty} O(M^{-\delta})^n b(\varepsilon M(\omega_0 - \omega_{2n})/3) \quad .$$

Choosing M sufficiently large so $O(M^{-\delta})^n$ is summable, we obtain for some new constant M_1,

$$|G(x,y)| \leqslant |C(x,y)| + b((x-y)/M_1) \tag{4.38}$$

which ensures the theorem.

REFERENCES

1. Osterwalder, K. and Schrader, R., "Axioms for Euclidean Green's functions, I, II", *Commun. Math. Phys.* 31, 83-112 (1973); *Commun. Math. Phys.* 42, 281-305 (1975).
   Fröhlich, J., Osterwalder, K. and Seiler, E., "On virtual representation of symmetric spaces and their analytic continuation", *Ann. Math.* 118, 461-489 (1983).

2. Glimm, J. and Jaffe, A., *Quantum Physics*, 2nd edition. New York: Springer-Verlag (1986). Collected Papers: Boston, Birkhäuser (1985).

3. Wilson, K.G., "Confinement of quarks", *Phys. Rev.* D10, 2445-2459 (1974).
   Seiler, E., "Gauge Theories as a Problem of Constructive Quantum Field Theory and Statistical Mechanics", *Lecture Notes in Physics*, Vol. 159. Berlin, Heidelberg, New York: Springer-Verlag (1982).

4. Singer, I.M., "Some remarks on the Gribov ambiguity", *Commun. Math. Phys.* 60, 7-12 (1978).

5. Osterwalder, K. and Seiler, E., "Gauge field theories on the lattice", *Ann. Phys.* 110, 440-471 (1978).

6. Glimm, J. and Jaffe, A., "Positivity of the ϕ_3^4 Hamiltonian", *Fort. Phys.* 21, 327-376 (1973).

7. Kandanoff, L., "The application of renormalization, group techniques to quarks and strings", *Rev. Mod. Phys.* 49, 267-296 (1977).
   Wilson, K.G., "Quantum chromodynamics on a lattice", in *Quantum Field Theory and Statistical Mechanics*, pp. 143-172 (1976); *Cargèse Lectures*, edited by M. Levy and P. Mitter, Plenum Press (1977).

8. Balaban, T., "Renormalization group methods in non-Abelian gauge theories", Harvard University preprint HUTMP B134 (1984).
   Balaban, T., "Averaging operations for lattice gauge theories", *Commun. Math. Phys.* 98, 17-52 (1985).

9. Federbush, P., "A phase cell approach to Yang-Mills theory III. Stability modified renormalization group transformation", University of Michigan preprint (1984).

10. Abers, E., and Lee, B.W., "Gauge theories", *Phys. Rept.* 9C, 1-141 (1973).
    Faddeev, L.D. and Slavnov, A.A., *Gauge Fields: Introduction to Quantum Theory*, Reading, MA: Benjamin-Cummings Publishing Co. (1980).

10. Cruetz, M., "Monte Carlo study of quantized SU(2) gauge theory", *Phys. Rev.* D21, 2308-2315 (1980).
    Cruetz, M., Jacobs, L. and Rebbi, C., "Monte Carlo study of Abelian lattice gauge theories", *Phys. Rev.* D20, 1915-1922 (1979).

12. Guth, A., "Existence proof of a nonconfining phase in four-dimensional U(1) lattice gauge theory", *Phys. Rev.* D21, 2291-2307 (1980).
    Fröhlich, J. and Spencer, T., "Massless phases and symmetry restoration in Abelian gauge theories and spin systems", *Commun. Math. Phys.* 83, 411-454 (1982).
    Göpfert, M. and Mack, G., "Proof of confinement of static quarks in three-dimensional U(1) lattice gauge theories for all values of the coupling constant", *Commun. Math. Phys.* 82, 545-606 (1982).

13. Israel, R.B. and Nappi, C.R., "Quark confinement in the two-dimensional lattice Higgs-Villain model", *Commun. Math. Phys.* **64**, 177-189 (1979). Mack, G., "Confinement of static quarks in two-dimensional lattice gauge theories", *Commun. Math. Phys.* **65**, 91-96 (1979). Balaban, T., Brydges, D., Imbrie, J. and Jaffe, A., "The mass gap for Higgs models on a unit lattice", *Ann. Phys.* **158**, 281-319 (1984).

14. Balaban, T., Imbrie, J. and Jaffe, A., "Renormalization of the Higgs Model: Minimizers, Propagators and the Stability of Mean Field Theory", *Commun. Math. Phys.* **97**, 299-330 (1985).

15. Balaban, T., "Propagators and renormalization transformations for lattice gauge theories, I", *Commun. Math. Phys.* **95**, 17-40 (1984).

16. Gawedzki, K. and Kupiainen, A., "Asymptotic freedom beyond perturbation theory" in *Critical Phenomena, Random Systems, Gauge Theories* (Les Houches 1984) Osterwalder, K. and Stora, R. (eds.); Amsterdam: North Holland, to be published (1985).

17. Balaban, T., "Renormalization group approach to lattice gauge field theories, I. Generation of effective actions in a small field approximation and a coupling constant renormalization", Harvard University preprint HUTMP B189 (1985).

PARTICLE STRUCTURE OF GAUGE THEORIES

Klaus Fredenhagen

Heisenberg fellow
II. Institut für Theoretische Physik
Universität Hamburg
D-2000 Hamburg 50

1. INTRODUCTION

Gauge theories are formulated in terms of gauge fields A_μ and matter fields ψ which are not directly connected with physical particles, although the use of notations like "quark" and "gluons" suggests such an interpretation. In fact, the structure of the set of particle states depends strongly on the dynamics, as you all know from the discussion on quark confinement, Higgs mechanism, charge screening and so on. A particular problem is the occurence of "charged" particles, i.e. particles which are separated from the vacuum by some superselection rule, the classical example being particles with half integer spin [1]. By the very definition of superselection rules there cannot exist an observable field which generates states of such a particle out of the vacuum.

In theories which have only global gauge symmetries this problem is often ignored. In these theories one has available besides the observable fields non observable fields which obey local commutation of anticommutation relations. These fields generate a set of superselection sectors which often contain all particle states. It may happen, however, that one has not sufficiently many fields at the beginning. As an example let me mention the Sine-Gordon theory; there the fields creating the soliton states from the vacuum are not contained in the original formulation of the theory (see e.g. [2]). It may also happen, that a non gauge invariant field does not create a new superselection sector out of the vacuum; this occurs in the case of spontaneous breakdown of gauge symmetry.

In gauge theories particles may exist which carry a charge related to the local gauge symmetry. Such a charge can be measured, according to Gauss' law, by the corresponding electric flux through an arbitrarily large surface surrounding the particle. There can never exist a local field creating such a particle from the vacuum, as may be seen by the following (standard) heuristic argument:

Let φ be a local field and Ω the vector representing the vacuum. The charge Q is the limit of the electric fluxes ϕ_R through a sphere with radius R around the origin. Then

$$Q\,\varphi(x)\Omega = \lim_{R\to\infty} \phi_R\, \varphi(x)\Omega = \lim_{R\to\infty} \varphi(x)\, \phi_R \Omega = \varphi(x) Q \Omega \tag{1.1}$$

hence if Ω is an eigenvector of Q $\varphi(x)\Omega$ is an eigenvector with the same eigenvalue.

Formally one may write down nonlocal fields creating charged particles. An example is the string field

$$\psi_{\mathcal{C}} = \psi(x)\, P\, e^{i\int_{\mathcal{C}} dy\, A(y)} \tag{1.2}$$

where $\mathcal{C}$ is a path from x to spacelike infinity and the symbol P denotes path ordering of the exponential. Another example is the electron field of quantum electrodynamics in the Coulomb gauge,

$$\psi_c(x) = \psi(x) \exp\{ie \int d^3\underline{y}\ \underline{A}(x^0,\underline{y}).(\underline{y}-\underline{x})|\underline{y}-\underline{x}|^{-3}\} \quad . \tag{1.3}$$

Unfortunately, it is very difficult to give a precise meaning to these nonlocal expressions. (cf. however [3]).

For avoiding nonlocal quantities one treats gauge theories usually in a formalism where the fundamental fields are local and act as operators in a vector space W which is equipped with an indefinite metric. There is a subspace V, containing the vacuum and being invariant under the application of gauge invariant operators (observables), on which the scalar product is nonnegative. The subspace V_o of V of vectors with length zero is also invariant under observables, hence there is a natural representation of observables by operators in the factor space V/V_o, whose completion is the space of physical states $\mathcal{H}_{phys}$.

A very difficult question is whether $\mathcal{H}_{phys}$ contains besides the vacuum sector of the theory also the charged sectors. This depends on the existence of certain elements in W which cannot be created by local fields out of the vacuum. One would like to derive their existence from a completeness property of W, but the absence of a natural notion of convergence obscures this possibility. The general structure of the indefinite metric formalism has been studied by Strocchi and Wightman [4] and later by Morchio and Strocchi [5]. I refer to the lectures of Prof. Strocchi for more details.

I want to start from a more general point of view. I consider the indefinite metric approach or the Euclidean functional integral approach as methods to compute the vacuum expectation values of gauge invariant quantities. I want to use only this information for a construction of the set of particle states. Actually, the explicit formulas for the observables in terms of the fundamental fields are never used. The only structure which is exploited is the association of regions $\mathcal{O}$ of Minkowski space to algebras $\mathfrak{A}(\mathcal{O})$ of Hilbert space operators; e.g.

$$\sum_{a,b} \bar{\psi}_a(x)\psi_b(y)\left(Pe^{i\int_{\mathcal{C}} A}\right)_{ab} \quad , \mathcal{C} \text{ path from x to y,} \tag{1.4}$$

is an observable which is localized in all regions $\mathcal{O}$ containing $\mathcal{C}$. For avoiding technical complications with domains of definition we restrict ourselves to bounded operators. For a quantum mechanical observable this can always be achieved by a suitable change of scale. This leads to the so-called algebraic framework of quantum field theory which has been proposed by Haag and Kastler [6].

2. THE ALGEBRAIC FRAMEWORK

According to Haag and Kastler [6], the basic object of a quantum field theory is an assignement of finitely extended space time regions $\mathcal{O}$ to operator algebras $\mathfrak{A}(\mathcal{O})$. Each algebra $\mathfrak{A}(\mathcal{O})$ is isomorphic to an algebra of bounded Hilbert space operators which is invariant under taking the adjoint (*-operation). $\mathfrak{A}(\mathcal{O})$ contains the unit operator and is closed with respect to the weak operator topology, i.e.

$$\lim (\Phi, A_\lambda \Psi) = (\Phi, A\Psi) \tag{2.1}$$

for all vectors Φ, Ψ and $A_\lambda \in \mathfrak{A}(\mathcal{O})$ for all λ imply $A \in \mathfrak{A}(\mathcal{O})$. Weakly closed *-invariant operator algebras have first been investigated by v. Neumann and are therefore called v. Neumann algebras. (For the mathematics of operator algebras see e.g. [7].)

The assignement $\mathcal{O} \rightarrow \mathfrak{A}(\mathcal{O})$ is called a local net. It has the following properties:

(1) Isotony: If $\mathcal{O}_1 \subset \mathcal{O}_2$ then $\mathfrak{A}(\mathcal{O}_1) \subset \mathfrak{A}(\mathcal{O}_2)$, and the unit operators of $\mathfrak{A}(\mathcal{O}_1)$ and $\mathfrak{A}(\mathcal{O}_2)$ coincide.

This property is obvious from the interpretation of $\mathfrak{A}(\mathcal{O})$ as well as from its construction. It enables us to consider the algebra of all local observables,

$$\mathfrak{A}_0 = \bigcup_{\mathcal{O}} \mathfrak{A}(\mathcal{O}) \quad . \tag{2.2}$$

Also $\mathfrak{A}_0$ can be considered as an operator algebra on some Hilbert space, e.g. the vacuum Hilbert space. Due to the existence of superselection sectors there are representations (*) of $\mathfrak{A}_0$ by operators in other Hilbert spaces which are not unitarily equaivalent to the identical representation in the vacuum Hilbert space. The weak operator topologies in inequivalent representations are different; the operator norm, however, and therefore also the closure of $\mathfrak{A}_0$ with respect to this norm

$$\mathfrak{A} = \overline{\mathfrak{A}_0}$$

are independent of the choice of the representation provided the representation is faithful (i.e. injective). $\mathfrak{A}$ is called the algebra of (quasi-local) observables (*-invariant normclosed algebras of Hilbert space operators are called C*-algebras). For more details see [6].

(2) Locality: If $\mathcal{O}_1$ is spacelike separated from $\mathcal{O}_2$ then, from Einstein causality, measurements in $\mathcal{O}_1$ and $\mathcal{O}_2$ cannot disturb each other, hence $[A,B] = 0$ for $A \in \mathfrak{A}(\mathcal{O}_1)$, $B \in \mathfrak{A}(\mathcal{O}_2)$.

(3) Covariance: Let $A \in \mathfrak{A}(\mathcal{O})$ be an observable and $L = (a, \Lambda)$ a Poincaré transformation in the identity component $P_+^\uparrow$ of the Poincaré group. There is a prescription assigning to A an observable $A_L \in \mathfrak{A}(L\mathcal{O})$. The mapping $\alpha_L : A \longrightarrow A_L$ is a symmetry transformation, i.e. it preserves all intrinsic properties of $\mathfrak{A}$, hence α_L is an automorphism of $\mathfrak{A}$:

(*) A representation π of a *-algebra $\mathfrak{A}$ is a linear mapping from $\mathfrak{A}$ into the algebra of bounded operators $B(\mathcal{H}_\pi)$ in some Hilbert space $\mathcal{H}_\pi$ such that
(i) $\pi(AB) = \pi(A)\pi(B)$
(ii) $\pi(A)^* = \pi(A^*)$

$$\begin{aligned} &\text{(i)} \quad \alpha_L(\lambda A) = \lambda \alpha_L(A) \\ &\text{(ii)} \quad \alpha_L(A+B) = \alpha_L(A) + \alpha_L(B) \\ &\text{(iii)} \quad \alpha_L(AB) = \alpha_L(A)\alpha_L(B) \\ &\text{(iv)} \quad \alpha_L(A^*) = \alpha_L(A)^* \end{aligned} \tag{2.4}$$

Moreover, if $L = L_1 L_2$ we have

$$\alpha_L = \alpha_{L_1} \alpha_{L_2} \tag{2.5}$$

hence $L \longrightarrow \alpha_L$ is a representation of $P_+^\uparrow$ by automorphisms of $\mathfrak{A}$ such that

$$\alpha_L(\mathfrak{A}(\mathcal{O})) = \mathfrak{A}(L\mathcal{O}) \tag{2.6}$$

(4) Stability: The systems we encounter in physics have in general certain stability properties. Whether this is merely, our inability to prepare unstable systems in reproducible experiments or whether it is a fundamental physical law, in any case it deeply influences the mathematical structure of the relevant models. Stability may be thought of as the existence of a state with "lowest energy". Unfortunately, the known ways of making precise the condition of stability need more technical input whose physical meaning is not fully clarified. I shall come back to this point later.

After having discussed the general properties of the set of observables we now have to consider the notion of a state. In the quantum mechanics of finitely many particles states are described by unit vectors in the Hilbert space of square integrable wave functions. In quantum field theory there is no a priori given Hilbert space of wave functions. The basic object is the algebra of observables $\mathfrak{A}$. If $\mathfrak{A}$ is realized by operators in some Hilbert space $\mathcal{H}$, each unit vector $\Psi \in \mathcal{H}$ describes a state. Let us consider the mapping

$$\mathfrak{A} \ni A \longrightarrow (\Psi, A\Psi) =: \omega_\Psi(A) \tag{2.7}$$

which associates to each observable A its expectation value in the state described by Ψ. The expectation values of all powers of A already fix the whole probability distribution of measured values of A since A is bounded. Hence we may identify a state by its expectation functional. This leads to the following definition:

Definition: A state on a C*-algebra $\mathfrak{A}$ (with unit) is a linear functional ω with

$$\begin{aligned} &\text{(i)} \quad \omega(A^*A) \geq 0 \quad , A \in \mathfrak{A} && \text{(positivity)} \\ &\text{(ii)} \quad \omega(1) = 1 && \text{(normalization)} \end{aligned} \tag{2.8}$$

Examples for states are the expectation functionals induced by unit vectors or density matrices in some Hilbert space representation of $\mathfrak{A}$. Actually every state is the expectation functional of some unit vector in a suitable representation of $\mathfrak{A}$:

Theorem 2.1 (GNS-construction) [7]
Let ω be a state on a C*-algebra $\mathfrak{A}$. Then there exists a Hilbert space $\mathcal{H}$, a representation π of $\mathfrak{A}$ by operators in $\mathcal{H}$ and a unit vector $\Omega \in \mathcal{H}$ such that

(i) $\{\pi(A)\Omega, A \in \mathfrak{A}\}$ is dense in $\mathcal{H}$

(ii) $(\Omega, \pi(A)\Omega) = \omega(A)$.

It is instructive to illustrate this theorem on the example of a state of the form $\omega(A) = \mathrm{Tr}\, \rho A$ with a density matrix ρ in a Hilbert space $\mathcal{H}_0$ where $\mathfrak{A} = B(\mathcal{H}_0)$ is the algebra of bounded operators in $\mathcal{H}_0$. Let $\mathcal{H}$ be the Hilbert space of Hilbert-Schmidt operators in $\mathcal{H}_0$ with the scalar product $(S,T) = \mathrm{Tr}\, S^*T$. Since the algebra of Hilbert-Schmidt operators is an ideal in $B(\mathcal{H}_0)$, $A \in B(\mathcal{H}_0)$ acts by left multiplication as an operator on $\mathcal{H}$,

$$\pi(A)T = AT \quad , A \in B(\mathcal{H}_0), T \in \mathcal{H} \quad . \tag{2.9}$$

π is a representation of $B(\mathcal{H}_0)$. The square root $\rho^{1/2}$ of the density matrix ρ is a Hilbert-Schmidt operator, hence an element of $\mathcal{H}$, and

$$\begin{aligned} (\rho^{1/2}, \pi(A)\rho^{1/2}) &= (\rho^{1/2}, A\rho^{1/2}) = \mathrm{Tr}\, \rho^{1/2} A \rho^{1/2} \\ &= \mathrm{Tr}\, \rho A = \omega(A) \end{aligned}$$

thus ω is a vector state in the representation π .

The advantage of the algebraic notion of a state is a larger flexibility in describing different physical situations. As an example let us look at a free theory of a charged scalar particle. We want to approximate charged states by chargeless states describing particle-antiparticle pairs by shifting the antiparticle "behind the moon". Let Φ_{xy} denote a unit vector describing a particle-antiparticle pair where the particle is near to the point x and the antiparticle near to y. In the limit y $\longrightarrow$ spacelike infinity the sequence $(\Phi_{x,y})_y$ does not converge strongly, and the weak limit is zero. Local measurements, however, are not influenced by the antiparticle at spacelike infinity, thus the expectation values of local observables converge, and the sequence of states $\omega_{xy}(A) = (\Phi_{xy}, A \; \Phi_{xy})$ converges pointwise to a functional on $\mathfrak{A}$ which is linear, positive and normalized and hence a state on $\mathfrak{A}$. The corresponding GNS construction gives a Hilbert space $\mathcal{H}$ and a representation π of $\mathfrak{A}$ in $\mathcal{H}$. It is not possible to identify $\mathcal{H}$ with the charge zero Hilbert space $\mathcal{H}_0$ by some unitary operator U such that

$$U A U^* = \pi(A) \tag{2.11}$$

(π is not unitarily equivalent to the identical representation on $\mathcal{H}_0$).

This follows (modulo some technicalities) from the fact that the local charge operators

$$Q_R = \int_{|\underline{x}|<R} d^3\underline{x} \;\; j_0(0,\underline{x}) \tag{2.12}$$

converge weakly to zero in $\mathcal{H}_0$ and to one in $\mathcal{H}$.

The set of all states of $\mathfrak{A}$ is very large, and it is difficult to determine the structure of the whole state space. For the purposes of particle physics, however, only those states must be considered which describe an incoming or outgoing configuration of finitely many particles. Such states should be vector states in a positive energy representation of $\mathfrak{A}$ [8].

<u>Definition</u>: A representation π of $\mathfrak{A}$ is called a positive energy representation if there is a unitary, strongly continuous representation U of the translation group in the representation space $\mathcal{H}_\pi$, implementing the translations on $\mathfrak{A}$,

$$U(x)\, \pi(A)\, U(x)^{-1} = \pi\alpha_x(A) \tag{2.13}$$

such that the generators $P = (P_0, \underline{P})$ of U,

$$U(x) = e^{ixP} \; , \quad xP = x^0 P_0 - \underline{x}.\underline{P} \; , \tag{2.14}$$

fulfil the relativistic spectrum condition

$$\mathrm{sp}\, P \subset \{ p \in \mathbb{R}^4 , p^2 \geq 0 , p_0 \geq 0 \} \equiv \overline{V_+} \; . \tag{2.15}$$

The existence of a faithful positive energy representation is a specific form of the stability requirement. Borchers [9] has shown that in a positive energy representation π it is always possible to modify U such that $\{U(x)\}$ is contained in the weak closure $\pi(\mathfrak{A})^{-}$ of the observable algebra $\pi(\mathfrak{A})$. This justifies the interpretation of P as energy-momentum. If U can be extended to a representation of the Poincaré group implementing the Poincaré transformations α_L, the energy-momentum spectrum is Lorentz invariant. However, as is well known, there are positive energy representations (e.g. coherent infrared representations of the free photon field [10, 11] and representations describing electrically charged states in quantum electrodynamics [12]) where such an extension is not possible. It is remarkable that nevertheless also in the general case there is a natural definition of the normalization of energy-momentum such that sp P is Lorentz invariant [13, 14]. We shall always use this definition of energy-momentum for positive energy representations.

3. CHARGED SINGLE PARTICLE STATES (*)

In gauge theories particles may occur which carry a gauge charge, i.e. a charge which is measurable in the spacelike complement $\mathcal{O}'$ of an arbitrarily large finitely extended region $\mathcal{O}$. Such particles can never be created by local fields. One may conjecture that there exist stringlike localized fields creating such particles; but it is difficult to guess properties of these nonlocal objects. Therefore we do not assume any a priori knowledge of the localization of particles. Instead of this we characterize particles by their energy-momentum properties. Let $H_m = \{p \in \mathbb{R}^4, p^2 = m^2, p_0 \geq 0\}$ be the single particle hyperboloid. We say that a positive energy representation π of $\mathfrak{A}$ contains charged single particle states if

$$H_m \subset \text{sp}\, P \subset H_m \cup \{p \in \mathbb{R}^4, p^2 \geq M^2\} \qquad (3.1)$$

for some $0 < m < M$.

The states in a representation may be partially classified by their charge quantum numbers. In the abstract framework considered here charge operators restricted to the representation π are simply those elements of the weak closure $\pi(\mathfrak{A})^{-}$ of $\pi(\mathfrak{A})$ which commute with each element in $\pi(\mathfrak{A})$ i.e. the elements of the center of $\pi(\mathfrak{A})^{-}$.

(*) The results in this section rely essentially on joint work with D. Buchholz [15, 16].

A positive energy representation π containing charged single particle states is called a single particle representation if all vector states in the representation space $\mathcal{H}_\pi$ have the same charge quantum numbers, i.e. the center of $\pi(\mathfrak{A})^-$ is trivial,

$$\pi(\mathfrak{A})^- \cap \pi(\mathfrak{A})' = \{\lambda 1, \lambda \in \mathbb{C}\} \quad . \tag{3.2}$$

Here for a subset $\mathcal{M}$ of the set $B(\mathcal{H})$ of all bounded operators in a Hilbert space $\mathcal{H}$ $\mathcal{M}'$ denotes the commutant of $\mathcal{M}$,

$$\mathcal{M}' = \{B \in \mathcal{B}(\mathcal{H}), [B,A] = 0 \quad \text{for all} \quad A \in \mathcal{M}\} \quad . \tag{3.3}$$

Representations with a trivial center are called factorial. Note that by (3.1) $0 \notin P$, hence there is no vacuum state in the representation π. By (3.2) a vacuum state in π would have the same charge quantum numbers as the single particle states, so the particle would be "chargeless".

Instead of requiring sharp charge quantum numbers, i.e. factoriality of the representation π , we could use the stronger assumption that π is irreducible; by Schur's Lemma, this means that the commutant of $\pi(\mathfrak{A})$ is trivial [7]. As a matter of fact, it turns out, under one additional assumption, the so called duality condition (3.27) discussed below, that each single particle representation is a multiple of an irreducible single particle representation [15], i.e. the representation space $\mathcal{H}_\pi$ of a single particle representation π can be written as a tensor product $\mathcal{H}_\pi = \mathcal{H}_1 \otimes \mathcal{H}_2$ of Hilbert spaces $\mathcal{H}_1$ and $\mathcal{H}_2$, and there is an irreducible single particle representation π_1 on $\mathcal{H}_1$ such that $\pi(A) = \pi_1(A) \otimes 1_{\mathcal{H}_2}$ for all $A \in \mathfrak{A}$.

The mentioned spectral properties of single particle states are typical for particles in a theory without massless particles. In the presence of massless particles more general situations occur; the single particle mass shell H_m will not be isolated from the rest of the spectrum, and it may also be that there is no discrete weight for H_m, as it is the case for infraparticles. The charged particles of quantum electrodynamics are probably examples for such a situation [17]. In these cases the spectral conditions admit many inequivalent representations of the observable algebra describing situations which are from the experimentalist's point of view indistinguishable. Buchholz [18] has proposed the concept of a charge class in which a lot of representations which differ only by some practically unobservable infrared cloud are kept together (cf. the lectures of Prof. Wightman and Prof. Strocchi). In spite of the considerable progress which has been achieved crucial problems, e.g. the scattering theory for infraparticles, are far from being solved.

If no massless particles are present, the particle definition in (3.1), (3.2) seems to be suitable. On a first sight one might believe that gauge symmetry requires the existence of massless particles corresponding to the gauge field. In fact, the associated term in the Lagrangian has no mass term. However, by a Higgs phenomenon or a similar effect it may happen that in the physical mass spectrum no massless particles exist. This is for instance expected in quantum chromodynamics.

Another possibility is , that the charged particles in gauge theories can only exist in the presence of massless particles. Swieca has investigated this question in a gauge theory with gauge group U(1) [19]. In such a theory the charge is the integral of the zeroth component of a conserved current which is the divergence of the field strength,

$$Q = \int d^3\underline{x} \; j_0(0,\underline{x}) \qquad , \; j_\mu = \partial^\nu F_{\mu\nu} \qquad . \tag{3.4}$$

j_μ and $F_{\mu\nu}$ are observable fields which should therefore obey local commutation relations,

$$[F_{\mu\nu}(x), F_{\rho\sigma}(y)] = 0 \qquad , \; (x-y)^2 < 0 \qquad . \tag{3.5}$$

For a scalar particle fulfilling the spectral condition (3.1) the form factor of $F_{\mu\nu}$ has the form

$$\langle q \,|\, F_{\mu\nu}(0) \,|\, \underline{p} \rangle = 2i(q_\mu p_\nu - q_\nu p_\mu) \frac{f((q-p)^2)}{(q-p)^2} \tag{3.6}$$

with $p_o = (\underline{p}^2 + m^2)^{1/2}$, $q_o = (\underline{q}^2 + m^2)^{1/2}$. f(0) is proportional to the charge of the particle. If the charge of the particle is nonzero, the form factor of $F_{\mu\nu}$ would be singular at zero momentum transfer. Swieca has given an argument that this singularity is incompatible with locality (3.5) and the spectral properties (3.1).

Swieca's argument may be sketched as follows. Let g be a strongly decreasing function with a Fourier transform $\tilde{g}$ such that $(m,\underline{0})$ + supp $\tilde{g}$ and $(m,\underline{0})$ - supp $\tilde{g}$ intersect the energy momentum spectrum only on the mass shell, and let $|\underline{p}\rangle$ denote the improper single particle state with spatial momentum $\underline{p}$ and the normalization

$$\langle \underline{p} | \underline{q} \rangle = 2\omega_{\underline{p}}\, \delta^3(\underline{p} - \underline{q}) \quad , \omega_{\underline{p}} = (\underline{p}^2 + m^2)^{1/2} \,. \tag{3.7}$$

Let $F_{oi}(g) = \int d^4x\, g(x)\, F_{oi}(x)$. Then $F_{oi}(g)\, |\underline{0}\rangle$ and $F_{oi}(g)^* \, |\underline{0}\rangle$ are single particle states. From the locality of F_{oi} and the strong decrease of g the expectation value of the commutator in the zero momentum state $|\underline{0}\rangle$,

$$\langle \underline{o} | [F_{oi}(\underline{x}), F_{oi}(g)] | \underline{o} \rangle \equiv h(\underline{x}) \tag{3.8}$$

is rapidly decreasing for large $\underline{x}$, and therefore the Fourier transform $\tilde{h}(\underline{p})$ is a smooth function. Since only single states contribute as intermediate states in (3.8) we find from (3.6)

$$\begin{aligned} \tilde{h}(\underline{p}) &= \{\langle \underline{o} | \tilde{F}_{oi}(\underline{p}) | -\underline{p} \rangle \langle -\underline{p} | F_{oi}(g) | \underline{o} \rangle \\ &\quad - \langle \underline{o} | F_{oi}(g) | \underline{p} \rangle \langle \underline{p} | \tilde{F}_{oi}(\underline{p}) | \underline{o} \rangle \} \\ &= -(2\pi)^3 8\omega_{\underline{p}}\, m^2 \,(f(t)/t)^2 \,[\tilde{g}((\omega_{\underline{p}} - m), \underline{p}) - \tilde{g}(-(\omega_{\underline{p}} - m), \underline{p})] \end{aligned} \tag{3.9}$$

where $t = (p - (m,\underline{0}))^2 = 2m(m - \omega_{\underline{p}})$ and $g((\omega_{\underline{p}} - m), \underline{p}) - g(-(\omega_{\underline{p}} - m), \underline{p}) = t\, G(\underline{p})$ with a smooth function G which does not vanish for $\underline{p} = 0$ for suitable g.

Eq. (3.9) implies that f(0) must vanish since otherwise $\tilde{h}$ would not be a smooth function. Thus the particle has zero charge.

Unfortunately, the argument of Swieca is not completely rigorous, due to the use of improper particle states with sharp momentum. It could be improved if it would be known that f is continuous at zero. One might think that this missing point in the proof is of minor importance. There is a widespread belief among theoretical physicists that "high precision" mathematical arguments have no physical meaning. On the contrary it is generally accepted that high precision measurements often lead to the detection of completely new phenomena. I would like to convince you that the search for a "high precision" argument might equally well provide completely new insights, and I think that Swieca's theorem on the absence of charged particles in a massive U(1) gauge theory is a good example.

Actually, the missing information on the continuity of f is of the same character as the information on the singularity of f(t)/t. Swieca's argument shows how one can proceed from the weaker property to the stronger.

In terms of position space both properties refer to the localization of the particle. Thus we may ask the more general question: how well are particles localized?

To find an answer we try to exhibit particle states which are as well localized as possible. The idea is the following one. Let $p \in H_m$ and let $\Phi \in \mathcal{H}_\pi$ with energy momentum spectrum $sp_P \Phi$ in a small neighbourhood of p. Let $A \in \mathfrak{A}(\mathcal{O})$. The ensemble described by $\pi(A)\Phi$ consists of particles in the original ensemble Φ which have been influenced by A, so they must have passed through $\mathcal{O}$, of particles which have not been influenced by A and keep therefore their momentum unchanged, and of components where additional particles have been created and whose momenta are therefore not on the mass shell H_m. Thus the components of $\pi(A)\Phi$ with momenta on the mass shell H_m but not in $sp_P\Phi$ should be localized near $\mathcal{O}$. To filter out only these components we choose a test function $f \in \mathcal{S}(\mathbb{R}^4)$ with $\operatorname{supp} \tilde{f} + sp_P \Phi \cap sp_P \Phi = \emptyset$, $\operatorname{supp} \tilde{f} + sp_P \Phi \cap sp\, P \subset H_m$ and set

$$B = \int d^4x \; f(x)\, \alpha_x(A) \qquad . \tag{3.10}$$

From the heuristic argument given before $\pi(B)\Phi$ is a candidate for a vector representing a well localized state.

There are different possibilities for a precise definition of the term "well localized". Here the appropriate notion of localization is concentration of the energy momentum content. I want to indicate the argument.

The argument relies essentially on a suitable utilization of almost local operators with a sharp momentum transfer which map single particle states onto single particle states. Here an operator B is called almost local, if there are observables $A_R \in \mathfrak{A}(\mathcal{O}_R)$, $\mathcal{O}_R = \{x \in \mathbb{R}^4, |x^0| + |\underline{x}| < R\}$ such that

$$\| B - A_R \| \, R^n \longrightarrow 0 \quad , \quad R \longrightarrow \infty \tag{3.11}$$

for all $n \in \mathbb{N}$. The operator B in (3.10) is an example for an almost local operator.

Now let B_1 and B_0 be almost local operators with momentum transfer such that $\pi(B_0)\Phi$ and $\pi(\alpha_x(B_1)\, B_0)\Phi$ are single particle states and $\pi\, \alpha_x(B_1)\Phi = 0$ for all $x \in \mathbb{R}^4$ (Fig. 3.1). Let $\tilde{B}_1(t, \underline{q})$ $= \int d^3\underline{x} \; e^{-i\underline{q}\underline{x}} \, \alpha_{(t,\underline{x})}(B_1)$. Then $\pi(\tilde{B}_1(t,\underline{q}))\Phi = 0$ and

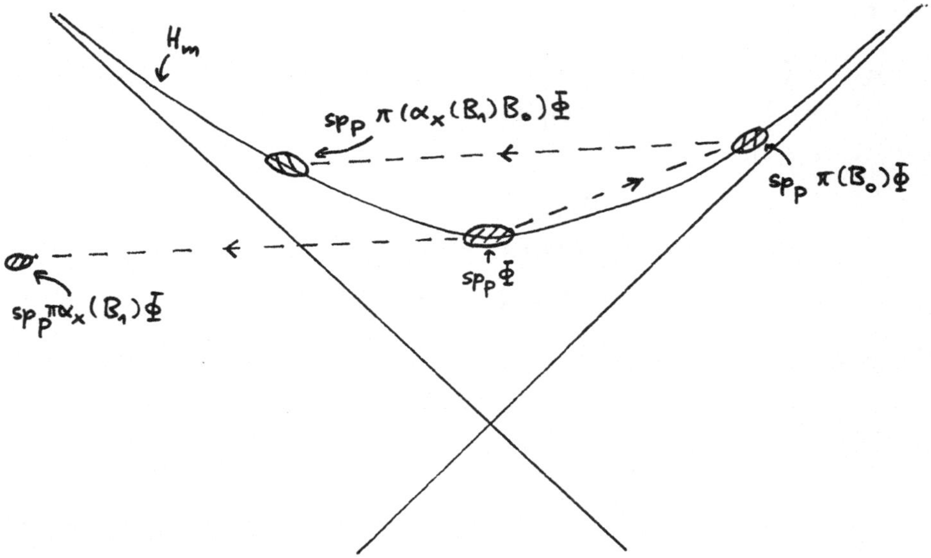

Figure 3.1: The choice of B_o and B_1

$$\begin{aligned}\pi([\tilde{B}_1(t,\underline{q}), B_0])\Phi &= \pi(\tilde{B}_1(t,\underline{q})B_0)\Phi \\ &= f(\underline{P})\,\pi(\tilde{B}_1(0,\underline{q})B_0)\,\Phi = f(\underline{P})\,\pi([\tilde{B}_1(0,\underline{q}), B_0])\Phi\,,\end{aligned} \tag{3.12}$$

$$f(\underline{P}) = \exp\{it[(\underline{P}^2+m^2)^{1/2} - ((\underline{P}+\underline{q})^2+m^2)^{1/2}]\}\,.$$

Now $B = [\tilde{B}_1(0,\underline{q}), B_0] = \int d^3\underline{x}\; e^{-i\underline{q}\underline{x}}\,[\alpha_{(0,\underline{x})}(B_1), B_0]$ and $B' = [\tilde{B}_1(t,\underline{q}), B_0]$ are almost local operators. Thus we have found a relation

$$\pi(B')\Phi = f(\underline{P})\,\pi(B)\Phi \tag{3.13}$$

with almost local operators B and B'. If B would be invertible, the measurement of the function $f(\underline{P})$ of the momentum operator in the state $\pi(B)\Phi$ could be replaced by the almost local operator $\pi(B'B^{-1})$, so in a sense, the momentum content of the particle in the state $\pi(B)\Phi$ is essentially localized in a finite region.

By using multiple commutators and by smearing over the time variables one can establish the relation (3.13) for an arbitrary smooth function f such that the almost local operator on the right hand side does not depend on f. It is not clear whether one can choose an invertible $B^{(*)}$. The precise theorem which can be derived is the following one:

Theorem 3.1 [14]. Let $p \in H_m$ and let Δ be a neighbourhood of p. There exists some $\Phi \in \mathcal{H}_\pi$ with $sp_P \Phi \subset \Delta$ and almost local operators B, B_μ, $\mu = 0, \ldots, 3$ such that

(i) $\pi(B)\Phi \neq 0$

(ii) $P_\mu \pi(B)\Phi = \pi(B_\mu)\Phi$

(iii) $\pi(B_\mu^* B)\Phi = \pi(B^* B_\mu)\Phi$

Now let ω be the state on $\mathfrak{A}$ induced by a vector $\pi(B)\Phi$, $\pi(B)$, Φ chosen according to Thm 3.1 with $\|\pi(B)\Phi\| = 1$. Then for $A \in \mathfrak{A}(G)$

$$\begin{aligned} \partial_\mu \omega\alpha_x(A) &= i(\pi(B)\Phi, [P_\mu, \alpha_x(A)]\pi(B)\Phi) \\ &= i\{(\pi(B_\mu)\Phi, \pi\alpha_x(A)\pi(B)\Phi) - (\pi(B)\Phi, \pi\alpha_x(A)\pi(B_\mu)\Phi)\} \\ &= i\{(\Phi, \pi[B_\mu^*, \alpha_x(A)]\pi(B)\Phi) - (\Phi, \pi[B^*, \alpha_x(A)]\pi(B_\mu)\Phi) \\ &\quad + (\Phi, \pi\alpha_x(A)\pi(B_\mu^* B - B^* B_\mu)\Phi)\} \end{aligned} \tag{3.14}$$

where the last term disappears because of Thm 3.1 (iii). Hence

$$|\partial_\mu \omega\alpha_x(A)| \leq \text{const}\{\|[B_\mu^*, \alpha_x(A)]\| + \|[B^*, \alpha_x(A)]\|\} \; . \tag{3.15}$$

Thus the derivative of $x \longrightarrow \omega\alpha_x(A)$ decreases rapidly in spacelike directions. $x \longrightarrow \omega_x(A)$ will tend therefore to a constant for x tending to spacelike infinity. This constant is independent of the direction in more than two dimensions. In two space time directions there may be different limits for x tending to the right and tending to the left; the particle is then called a soliton [16]. In any case the limit ω_0 is a translation invariant state which can be interpreted as the vacuum. The corresponding GNS representation π_0 is a positive energy representation with a unique (up to a phase) translation invariant unit vector Ω and a mass gap:

$$sp P \subset \{0\} \cup \{p \in \mathbb{R}^4, p^2 \geq (M-m)^2, p_0 \geq 0\} \tag{3.16}$$

We have found the following theorem:

Theorem 3.2. For all $A \in \mathfrak{A}_0$, $\omega\alpha_x(A) - \omega_0(A)$ is rapidly decreasing in spacelike direction $(|\underline{x}| - |x^0| \longrightarrow \infty)$.

(*) Note that no assumption on the multiplicity of spP on H_m. i.e. on the number of components of single particle wave functions has been made.

This result does not depend on the use of bounded operators. It holds equally well in the framework of Wightman fields. Specializing to the case of a U(1) gauge theory, it means that the expectation values of the electric field decrease rapidly in spatial direction, hence the electric charge of the state ω is zero. Since all particle states in the representation π have the same charge, this proves Swieca's theorem [16].

In the Haag-Kastler framework a much stronger result holds. From (3.15) one concludes that

$$|\partial_\mu \omega\alpha_x(A)| \leq \|A\| \, h(R) \qquad (3.17)$$

with a rapidly decreasing function h, where R is chosen such that $\alpha_x(A)$ commutes with $\mathfrak{A}(\mathcal{O}_R)$. This shows that $\omega\alpha_x$ converges to ω_0 uniformly on large subalgebras of $\mathfrak{A}$ corresponding to certain unbounded regions. Let G be a region containing a path x(s) to spacelike infinity with uniformly bounded tangent vectors $\dot{x}^\mu(s)$ such that for some $\varepsilon > 0$

$$\mathcal{O}_{s^\varepsilon} + x(s) \subset G \quad . \qquad (3.18)$$

Roughly speaking, G may be though of as a string which fattens.

Let $A \in \mathfrak{A}^c(G) = \{A \in \mathfrak{A}, [A,B] = 0$ for all $B \in \mathfrak{A}(\mathcal{O}), \mathcal{O} \subset G\}$.
Then

$$|(\omega\alpha_{-x(s)} - \omega_0)(A)| = |\int_s^\infty ds' \, \dot{x}^\mu(s') \, \partial_\mu(\omega\alpha_{-x(s')}(A))|$$
$$\leq \|A\| \, H(s) \qquad (3.19)$$

with the strongly decreasing function $H(s) = \sup|\dot{x}^\mu(s')| \int_s^\infty h(s'^\varepsilon) ds'$.
Thus the convergence $\omega\alpha_{-x(s)} \longrightarrow \omega_0$ is uniform on $\mathfrak{A}^c(G)$. Therefore ω_0 can be extended to a normal (*) state ω_G on $\pi(\mathfrak{A}^c(G))^-$.
Thus there is a density matrix ρ_G in $\mathcal{H}_\pi$ such that

$$\omega_0(A) = \mathrm{Tr}\, \rho_G \, \pi(A) \quad , \quad A \in \mathfrak{A}^c(G) \quad . \qquad (3.20)$$

(*) A state is called normal if it is weakly continuous on uniformly bounded subsets.

Using more advanced methods of the theory of v. Neumann algebras (Araki's theory of natural cones [7]) one can even find a unit vector $\Psi_G \in \mathcal{H}_\pi$ with

$$\begin{aligned} &(i) \quad (\Psi_G, \pi(A)\Psi_G) = \omega_0(A) \quad , A \in \mathfrak{A}^c(G) \\ &(ii) \quad \| \Psi_G - \pi(B)\Phi \|^2 \leq \| (\omega_0 - \omega) \upharpoonright \mathfrak{A}^c(G) \| \\ &(iii) \quad \overline{\pi(\mathfrak{A}^c(G))\Psi_G} = \overline{\pi(\mathfrak{A})\pi(B)\Phi} \quad . \end{aligned} \tag{3.21}$$

Hence there is a vector state in the single particle representation π which looks like the vacuum in the spacelike complement of G. Moreover, this state differs from ω on the whole algebra $\mathfrak{A}$ not more than on the algebra $\mathfrak{A}^c(G)$. If we further assume that $\pi(B)\Phi$ is cyclic for $\pi(\mathfrak{A})$, i.e. $\pi(\mathfrak{A})\pi(B)\Phi$ is dense in $\mathcal{H}_\pi$, we can use Ψ_G for a definition of a unitary charge generating operator. Let V_G be an operator from $\mathcal{H}_0$ to $\mathcal{H}_\pi$ with

$$V_G \pi_0(A)\Omega = \pi(A)\Psi_G \quad , A \in \mathfrak{A}^c(G) \quad . \tag{3.22}$$

$\pi_0(\mathfrak{A}^c(G))\Omega$ is dense in $\mathcal{H}_0$ according to the Reeh-Schlieder Theorem [8], $\pi(\mathfrak{A}^c(G))\Psi_G$ is dense in $\mathcal{H}_\pi$ according to (3.21) and the assumed cyclicity of $\pi(B)\Phi$ for $\pi(\mathfrak{A})$. Moreover

$$\begin{aligned} \| V_G \pi_0(A)\Omega \|^2 &= \| \pi(A)\Psi_G \|^2 = (\Psi_G, \pi(A^*A)\Psi_G) \\ &= \omega_0(A^*A) = (\Omega, \pi_0(A^*A)\Omega) = \| \pi_0(A)\Omega \|^2 \quad , \end{aligned} \tag{3.23}$$

hence V_G is a unitary operator from $\mathcal{H}_0$ onto $\mathcal{H}_\pi$. V_G intertwines the representations π_0 and π , restricted to $\mathfrak{A}^c(G)$,

$$V_G \pi_0(A) = \pi(A) V_G \quad , A \in \mathfrak{A}^c(G) \quad . \tag{3.24}$$

V_G may be interpreted as an operator which generates a charge within the region G.

V_G has similar localization properties as the formal operators $\psi_{\mathcal{C}}$ of (1.2), restricted to the vacuum Hilbert space $\mathcal{H}_0$, and may be considered as a mathematically rigorous version of such an operator. If local fields exist which create the particle states from the vacuum, one can even find unitary operators $V_\mathcal{O}$ associated to bounded regions $\mathcal{O}$ such that

$$V_G \pi_0(A) = \pi(A) V_G \quad , A \in \mathfrak{A}(\mathcal{O}') \tag{3.25}$$

where $\mathcal{O}'$ is the spacelike complement of $\mathcal{O}$ and $\mathfrak{A}(\mathcal{O}')$ is the C*-algebra generated by the algebras $\mathfrak{A}(\mathcal{O}_1)$ with $\mathcal{O}_1 \subset \mathcal{O}'$.

Superselection sectors corresponding to representations π fulfilling relation (3.25) are called locally generated. The structure of locally generated superselection sectors has been analyzed in general by Doplicher, Haag and Roberts (DHR) [20]. Their analysis extends and partially corrects an earlier analysis of Borchers [21]. Using the "duality assumption"

$$\pi_0(\mathfrak{A}(\mathcal{O}'))' = \pi_0(\mathfrak{A}(\mathcal{O})) \tag{3.26}$$

they showed that there is a composition law of sectors, corresponding to the idea that charges can be added. Furthermore, they prove that (in more than 2 space time dimensions) there is an intrinsic notion of statistics leaving only the possibilities of (para-) Bose, (para-) Fermi and infinite statistics. The pathological case of infinite statistics has been ruled out for single particle representations by an application of Thm 3.1 [22, 15]. In the case of finite statistics, Doplicher, Haag and Roberts derived the existence of antiparticles and of multiparticle scattering states. Very recently [23], Doplicher and Roberts showed that there is always a compact group (the "global gauge group") whose irreducible representations label the locally generated superselection sectors, and they construct a C*-algebra $\mathcal{F} \supset \mathfrak{A}$ (the "field algebra") on which the gauge group acts by automorphisms such that $\mathfrak{A}$ is the invariant part of $\mathcal{F}$, and which contains enough local charged fields to create all locally generated superselection sectors out of the vacuum.

By analogous methods one can perform a similar analysis for representations fulfilling relation (3.24), and one finds stringlike localized field operators and (in more than 3 space time dimensions) a full set of particle states, including antiparticle states and all incoming and outgoing multiparticle scattering states [15]. Instead of the duality assumption (3.26) one uses in this analysis the assumption

$$\pi_0(\mathfrak{A}^c(G))^- = \pi_0(\mathfrak{A}(G))' \tag{3.27}$$

which can be proven for a sufficiently large set of stringlike regions in the Wightman framework of field theory [24].

One may ask whether a more clever analysis of single particle states in massive theories will always lead to the DHR type of localization. This would mean that gauge charges can never occur in massive theories and would be a very strong generalization of Swieca's Theorem. Such a possibility cannot be ruled out in the moment; the fact, however, that the weaker property (3.24) already leads to the usual structure in the set of particle states shows that there is no intrinsic inconsistency in the stringlike localization. There are, on the other hand, severe dynamical restrictions imposed by this kind of localization. These will be discussed in the next section.

4. DYNAMICAL IMPLICATIONS OF THE EXISTENCE OF GAUGE CHARGES

We have derived localization properties of single particle states which show that there always exist field operators generating the particle states from the vacuum which are localized in a stringlike region G. We shall now concentrate on the case where this localization cannot be improved to a finite localization of the DHR-type.

In such a case there are no nonzero operators $V_{\mathcal{O}}$ fulfilling relation (3.21). The commutant of $\pi_0 \oplus \pi(\mathfrak{A}(\mathcal{O}'))$,

$$\pi_0 \oplus \pi(\mathfrak{A}(\mathcal{O}'))' = \{ B \in \mathcal{B}(\mathcal{H}_0 \oplus \mathcal{H}_\pi), [B, \pi_0 \oplus \pi(A)] = 0 \ \forall A \in \mathfrak{A}(\mathcal{O}')\} \quad (4.1)$$

consists therefore only of operators $C_0 \oplus C$, $C_0 \in \pi_0(\mathfrak{A}(\mathcal{O}'))'$ and $C \in \pi(\mathfrak{A}(\mathcal{O}'))'$. The bicommutant, i.e. the commutant of the commutant, is then

$$\pi_0 \oplus \pi(\mathfrak{A}(\mathcal{O}'))'' = \pi_0(\mathfrak{A}(\mathcal{O}'))'' \oplus \pi(\mathfrak{A}(\mathcal{O}'))'' \quad . \quad (4.2)$$

Since the bicommutant coincides with the weak closure (v. Neumann's bicommutant theorem) we find that the "charge operator"

$$Q = 1 \oplus -1 \quad (4.3)$$

is contained in the weak closure of $\pi_0 \oplus \pi(\mathfrak{A}(\mathcal{O}'))$ for all $\mathcal{O}$. This may be considered as an abstract version of Gauss' law.

Now from the localization properties of charged states derived in Sect.3 Q cannot be a sum of operators which are localized in double cones in $\mathcal{O}'$. Thus Q cannot be the sum of partial electric fluses as it is the case for

U(1) gauge theories. This might be interpreted as a generalized Swieca theorem:

In massive gauge theories only multiplicative charges can occur.

If we think of the total charge operator Q to be the product of the electric fluxes E_1 and E_2 through two opposite halfspheres, $Q = E_1E_2$, we have

$$|\omega(E_i) - \omega_0(E_i)| \approx 0 \quad , \; i = 1,2 \tag{4.4}$$

and

$$\omega_0(E_1E_2) = 1 \quad , \quad \omega(E_1E_2) = -1 \quad . \tag{4.5}$$

Now (4.4) and (4.5) are only compatible if the fluxes E_1 and E_2 are strongly correlated already in the vacuum. Hence the nonvanishing of the correlation

$$\omega_0(E_1E_2) - \omega_0(E_1)\,\omega_0(E_2) \tag{4.6}$$

is a necessary condition for the existence of a charged particle corresponding to the charge E_1E_2. We shall see that (4.6) will be the basis for a confinement criterion.

Let us investigate the dependence of the charge generator V_G on the stringlike region G. Let G_1 be another stringlike region. Then from (3.18) (ii)

$$\|V_G\Omega - V_{G_1}\Omega\|^2 \leq 2\,\{\|(\omega-\omega_0)\upharpoonright \mathfrak{A}^c(G)\| + \|(\omega-\omega_0)\upharpoonright \mathfrak{A}^c(G_1)\|\} \tag{4.7}$$

Now from (3.16), if $G \cap G_1 \supset G_R$ we have

$$\|V_G\Omega - V_{G_1}\Omega\|^2 \leq 4\,H(R) \tag{4.8}$$

where H is the rapidly decreasing function in (3.19), hence the asymptotic direction of the string is not visible. If we interpret $V_{G_1}^{-1}\,V_G$ as an operator which shifts a charge within G to infinity and brings it back in G_1 we see that such a charge transfer on a closed loop has an expectation value near to 1,

$$|(\Omega, V_{G_1}^{-1} V_G\Omega) - 1| \leq \|(V_G - V_{G_1})\Omega\| \leq 2\,H(R)^{1/2} \quad . \tag{4.9}$$

If, on the other hand, $G_1 \subset G_R'$, the states induced by $V_{G_1}\Omega$ will converge weakly to ω_0 in the limit $R \to \infty$; one can show that this implies that the scalar product with $V_G \Omega$ tends to zero,

$$(V_{G_1}\Omega, V_G \Omega) \longrightarrow 0 \quad , R \longrightarrow \infty \quad . \tag{4.10}$$

If $G_1 = G \cap \mathcal{O}_R'$, $V_{G_1}^{-1} V_G$ may be interpreted as a charge transfer inside of G into the spacelike complement of $\mathcal{O}_R$. The vacuum expectation values of such quantities are small,

$$(\Omega, V_{G_1}^{-1} V_G \Omega) \longrightarrow 0 \quad , R \longrightarrow \infty \tag{4.11}$$

In Section 6 we shall see that the comparison of the behaviour of charge transfers on open strings (4.11) with that on closed loops (4.9) leads to another confinement criterion.

5. THE $\mathbb{Z}_2$ HIGGS MODEL

We now want to confront the results of the general analysis with the structure of a lattice gauge theory; as a simple example we take the $\mathbb{Z}_2$ Higgs model which is a gauge theory with gauge group $\mathbb{Z}_2$ coupled to a $\mathbb{Z}_2$ valued Higgs field. This model has first been introduced by Wegner [25]. The present analysis relies mainly on joint work with M. Marcu [26].

Let us first look at the associated classical statistical mechanical system, i.e. the Euclidean theory from the point of view of quantum field theory. On a hypercubic lattice $\mathbb{Z}^{d+1}$, $d \geq 2$ we have a gauge field $\tau(b) = \pm 1$, which is defined on the lattice bonds b of $\mathbb{Z}^{d+1}$, and a Higgs field $\sigma(x)$ which is defined on the sites $x \in \mathbb{Z}^{d+1}$. The Hamilton function (the Euclidean action) is

$$\mathcal{H}(\tau, \sigma) = \beta_g \sum_p \delta\tau(p) + \beta_h \sum_b \tau(b)\, \delta\sigma(b) \tag{5.1}$$

where $\beta_g, \beta_h > 0$ are coupling constants, p runs over the plaquettes and b over the bonds of the lattice and δ denotes the "exterior derivative"

$$\delta\tau(p) = \prod_{b \in \partial p} \tau(b) \quad , \quad \delta\sigma(b) = \prod_{x \in \partial b} \sigma(x) \quad . \tag{5.2}$$

The phase diagrams of the model is expected to have the form shown in Fig. 5.1. It consists out of two phases, the screening/confinement phase (I) and the free charge phase (II) [27]. In both phases there exist subregions (the shaded regions in Fig. 1) where convergent expansions are known (see [28] for the screening/confinement phase and [29] for the free charge phase).

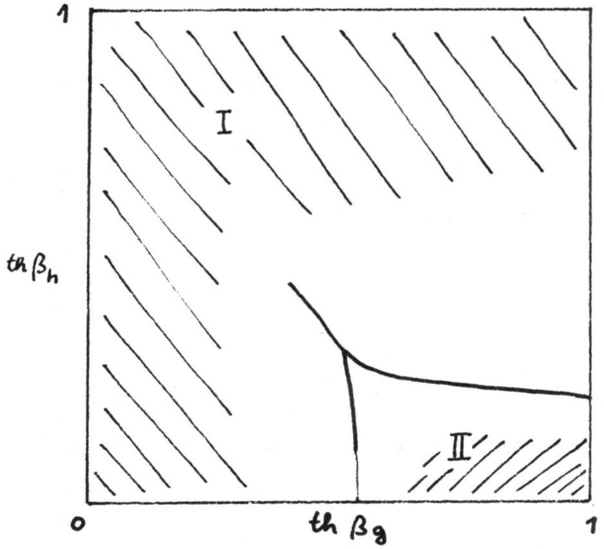

Figure 5.1: The phase diagram of the $\mathbb{Z}_2$ Higgs model (conjectured)

The corresponding quantum system in the temporal gauge is defined on a lattice $\mathbb{Z}^d$ which represents the space. The time is continuous in the quantum system. On each lattice bond $\underline{b}$ there are Pauli matrices $\tau_3(\underline{b})$ and $\tau_1(\underline{b})$ representing the gauge field and the electric field, respectively. On each lattice point $\underline{x}$ there are Pauli matrices $\sigma_3(\underline{x})$ and $\sigma_1(\underline{x})$, representing the Higgs field and its canonical momentum, resp. . One has the "canonical commutation relations"

$$\begin{aligned} &\sigma_i(\underline{x})^2 = \tau_i(\underline{b})^2 = 1 \ , \quad \sigma_i(\underline{x})\,\tau_j(\underline{b}) = \tau_j(\underline{b})\,\sigma_i(\underline{x}) \ , \\ &\sigma_i(\underline{x})^* = \sigma_i(\underline{x}) \ , \quad \tau_j(\underline{b})^* = \tau_j(\underline{b}) \quad , \ i,j = 1,3 \quad , \\ &\sigma_3(\underline{x})\,\sigma_1(\underline{x}) = -\,\sigma_1(\underline{x})\,\sigma_3(\underline{x}) \quad , \ \tau_3(\underline{b})\,\tau_1(\underline{b}) = -\,\tau_1(\underline{b})\,\tau_3(\underline{b}) \ . \end{aligned} \tag{5.3}$$

Fields at different points or bonds commute.

Let $\mathfrak{F}$ be the *-algebra which is generated by these fields. Gauge transformations on $\mathfrak{F}$ are implemented by the operators

$$q(\underline{x}) = \sigma_1(\underline{x})\ \delta^*\tau_1(\underline{x}) \tag{5.4}$$

where $\delta^*\tau_1(\underline{x}) = \prod_{\underline{b}:\underline{x}\in\partial\underline{b}} \tau_1(\underline{b})$ denotes the "divergence" of τ_1. $q(\underline{x})$ can be interpreted as the external charge at the point $\underline{x}$. In a U(1) theory in the continuum (5.4) corresponds to the difference between the charge density and the divergence of the electric field.

Observables A are defined to be gauge invariant elements of $\mathcal{F}$, i.e.

$$A\, q(\underline{x}) = q(\underline{x})\, A \quad , \underline{x} \in \mathbb{Z}^d \quad . \tag{5.5}$$

Let $\mathfrak{A}$ denote the algebra of observables. $\mathfrak{A}$ has a nontrivial center which is generated by the external charges. If in a representation π of $\mathfrak{A}$ Gauss' law holds,

$$\pi(\sigma_1(\underline{x})) = \pi(\delta^* \tau_1(\underline{x})) \quad , \underline{x} \in \mathbb{Z}^d \quad , \tag{5.6}$$

one has $\pi(q(\underline{x})) = 1$ for all external charges. Let I denote the ideal in $\mathfrak{A}$ which is generated by the operators $q(\underline{x}) - 1$. For representations fulfilling Gauss' law the relevant algebra of observables is

$$\mathcal{B} = \mathfrak{A}/I \quad . \tag{5.7}$$

$\mathcal{B}$ is generated by the rest classes modulo I of $\tau_1(\underline{b})$ and $\tau_3(\underline{b})\,\delta\,\sigma_3(\underline{b})$ which may be called $u_1(\underline{b})$ and $u_3(\underline{b})$, respectively. $u_1(\underline{b})$ and $u_3(\underline{b})$ have the algebraic properties of Pauli matrices (5.3). For a set $\underline{L}$ of bonds let $\mathcal{B}(\underline{L})$ denote the algebra generated by $u_i(\underline{b})$, i = 1, 3, $\underline{b} \in \underline{L}$.

It is interesting to note that the quantum system described by $\mathcal{B}$ has no locally generated charges. This fact holds independently of the dynamics. To see this we first observe that the relative commutant $\mathcal{B}(\underline{L})^c$ of $\mathcal{B}(\underline{L})$,

$$\mathcal{B}(\underline{L})^c = \{B \in \mathcal{B} \mid [B, B_1] = 0 \quad \forall\, B_1 \in \mathcal{B}(\underline{L})\} \quad , \tag{5.8}$$

coincides with the algebra $\mathcal{B}(\underline{L}^c)$ where $\underline{L}^c$ denotes the complement of $\underline{L}$ in the set of all lattice bonds. This property would be absent for instance in d = 2 dimensions for the algebra generated by $u_1(\underline{b})$ and $\delta u_3(\underline{p})$. There $\prod_{b \in M} u_3(\underline{b})$ for some closed curve $\underline{M}$ commutes with all operators $u_1(\underline{b}')$ with $\underline{b}' \notin M$ but is a product of all plaquette operators $\delta u_3(\underline{p})$ where $\underline{p}$ is surrounded by $\underline{M}$.

The next step is more abstract. Let π_0 and π be representations of $\mathcal{B}$ which are disjoint, i.e. the "charge operator" Q = 1 $\oplus$ − 1 is contained in the weak closure of $\pi_0 \oplus \pi(\mathcal{B})$. Thus there is a sequence $A_n \in \mathcal{B}$ with $\|A_n\| \leq 1$ such that $\pi_0(A_n)$ converges weakly to 1 and $\pi(A_n)$

(*) Each element of the weak closure of a *-algebra of bounded operators on a separable Hilbert space is the weak limit of a bounded sequence according to Kaplanski's density theorem (see e.g. [7]).

to - 1 (*). Let $\underline{L}$ be a finite set of bonds and let $G_{\underline{L}}$ denote the (finite) group which is generated by $u_1(\underline{b})$ and $u_3(\underline{b})$, $\underline{b} \in \underline{L}$. Let $m_{\underline{L}}$ denote the "conditional expectation"

$$m_{\underline{L}}(B) = |G_{\underline{L}}|^{-1} \sum_{U \in G_{\underline{L}}} U B U^{-1} \tag{5.9}$$

$m_{\underline{L}}(B)$ commutes with $G_{\underline{L}}$ and therefore also with $\mathcal{B}(\underline{L})$. Thus $m_{\underline{L}}(A_n) \in \mathcal{B}(\underline{L}^c)$ and

$$\pi_0(m_{\underline{L}}(A_n)) \longrightarrow 1 \quad , \quad \pi(m_{\underline{L}}(A_n)) \longrightarrow -1 \tag{5.10}$$

since $m_{\underline{L}}$ is weakly continuous. Thus

$$Q \in \pi_0 \oplus \pi\,(\mathcal{B}(\underline{L}^c))^- \tag{5.11}$$

which is the abstract version of Gauss' law discussed in Section 4.

Whereas there are no locally generated superselection sectors there is an uncountable number of mutually disjoint representations. We are interested in the question whether there are, besides the vacuum, other positive energy representations of $\mathcal{B}$. This question cannot be answered in the kinematical framework described above, instead we have to introduce a dynamics, and the answer will strongly depend on the dynamics.

A convenient way of introducing a dynamics in a lattice model is the Euclidean method. There the local Hamiltonians $H_{\underline{\Lambda}}$ (*) are defined implicitely as $(-\ln T_{\underline{\Lambda}})$ where the local transfer matrices $T_{\underline{\Lambda}}$ are positive, invertible operators in $\mathcal{F}$. For the gauge invariant Ising model with the Hamilton function (5.1) the transfer matrix is (in the temporal gauge)

$$\begin{aligned} T_{\underline{\Lambda}} &= e^{\frac{1}{2}A_{\underline{\Lambda}}}\, e^{B_{\underline{\Lambda}}}\, e^{\frac{1}{2}A_{\underline{\Lambda}}} , \\ A_{\underline{\Lambda}} &= \beta_h \sum_{\underline{b} \subset \underline{\Lambda}} \delta \sigma_3(\underline{b}) \tau_3(\underline{b}) + \beta_g \sum_{\underline{p} \subset \underline{\Lambda}} \delta \tau_3(\underline{p}) , \\ B_{\underline{\Lambda}} &= \beta_h^* \sum_{\underline{x} \in \underline{\Lambda}} \sigma_1(\underline{x}) + \beta_g^* \sum_{\underline{b} \subset \underline{\Lambda}} \tau_1(\underline{b}) , \quad \beta^* = -\tfrac{1}{2} \ln \operatorname{th} \beta . \end{aligned} \tag{5.12}$$

The local time evolution on $\mathcal{F}$ is defined by

$$\alpha_z^{\underline{\Lambda}}(A) = e^{izH_{\underline{\Lambda}}} A\, e^{-izH_{\underline{\Lambda}}} , \quad z \in \mathbb{C} \tag{5.13}$$

(*) Here and in the following $\underline{\Lambda}$ will denote a box.

For integer imaginary values of z the locality properties of $T_{\underline{\Lambda}}$ imply that $\alpha_z^{\underline{\Lambda}}(A)$ becomes independent of $\underline{\Lambda}$ for $\underline{\Lambda}$ sufficiently large. Thus

$$\alpha_{in}(A) = \lim_{\underline{\Lambda} \uparrow \mathbb{Z}^d} \alpha_{in}^{\underline{\Lambda}}(A) \tag{5.14}$$

exists for all $n \in \mathbb{Z}$. α_i is an algebraic automorphism of $\mathfrak{F}$ which is not compatible with the *-operation,

$$\alpha_i(A)^* = \alpha_{-i}(A^*) \tag{5.15}$$

and $\alpha_{in} = (\alpha_i)^n$, $n \in \mathbb{Z}$.

It is not known whether $\alpha_z^{\underline{\Lambda}}$ converges for other values of z (in some sense). Under certain conditions one can construct α_t for real t from α_i (essentially by using the "symmetric resolvent" $(e^{\lambda}\alpha_i + e^{-\lambda}\alpha_{-i})^{-1}$, $\lambda \in \mathbb{R}$; α_i is then called the analytic generator of α_t [30]). It seems, however, that these conditions are not satisfied in the present model.

Instead of defining the real time evolution directly we study the problem in a "vacuum" representation of $\mathfrak{B}$. Unfortunately, also the term "vacuum" = "ground state" is usually defined by means of the time evolution. It turns out, however, that one can characterize ground state also in terms of α_i.

Definition: A state ω_0 on $\mathfrak{F}$ is called a ground state with respect to α_i if

$$0 \leq \omega_0(A^* \alpha_i(A)) \leq \omega_0(A^*A) \tag{5.16}$$

for all $A \in \mathfrak{F}$.

A simple consequence of the definition is the α_i-invariance of ω_0 . In fact

$$\langle B, A \rangle = \omega_0(B^* \alpha_i(A)) \tag{5.17}$$

is a positive sesquilinear form on $\mathfrak{F}$ and therefore hermitean,

$$\overline{\langle B, A \rangle} = \langle A, B \rangle \quad . \tag{5.18}$$

Thus, for B = 1

$$\omega_0(\alpha_i(A)) = \langle 1, A \rangle = \overline{\langle A, 1 \rangle} = \overline{\omega_0(A^*)} = \omega_0(A) \quad . \tag{5.19}$$

Let $(\pi_0, \mathcal{H}_0, \Omega)$ denote the GNS representation of $\mathcal{F}$ induced by ω_0, i.e.

$$\begin{aligned} \mathcal{H}_0 &= \overline{\pi_0(\mathcal{F})\Omega} \quad , \\ (\Omega, \pi_0(A)\Omega) &= \omega_0(A) \quad , \quad A \in \mathcal{F} \quad . \end{aligned} \tag{5.20}$$

In $\mathcal{H}_0$ one can define the global transfer matrix T_0 by

$$T_0\, \pi_0(A)\Omega = \pi_0(\alpha_i(A))\Omega \quad , A \in \mathcal{F} \, . \tag{5.21}$$

We have

$$(\pi_0(A)\Omega, T_0\, \pi_0(A)\Omega) = \omega_0(A^* \alpha_i(A)) \tag{5.22}$$

hence from the definition of a ground state

$$0 \leq T_0 \leq 1 \quad . \tag{5.23}$$

Moreover, T_0 has a densely defined inverse,

$$T_0^{-1} \pi_0(A)\Omega = \pi_0 \alpha_{-i}(A)\Omega \quad , \quad A \in \mathcal{F} \, . \tag{5.24}$$

Thus we can define the global Hamiltonian by

$$H_0 = - \ln T_0 \tag{5.25}$$

and the real time translations by

$$\hat{\alpha}_t(\pi_0(A)) = e^{itH_0}\, \pi_0(A)\, e^{-itH_0} \quad . \tag{5.26}$$

If we insert t = in in (5.26) we find

$$\hat{\alpha}_{in} \circ \pi_0 = \pi_0 \circ \alpha_{in} \tag{5.27}$$

thus (5.26) is consistent with (5.14).

It is an open problem whether the time translations $\hat{\alpha}_t$ leave the norm closure of $\pi_0(\mathcal{F})$ or at least its weak closure invariant. We therefore have also to consider the probably larger algebra $\hat{\mathcal{F}}$ which is generated by $\hat{\alpha}_t \pi_\circ(\mathcal{F})$, $t \in \mathbb{R}$, and the analogously defined algebras $\hat{\mathfrak{A}}$ and $\hat{\mathfrak{B}}$.

If one introduces the dynamics by the Hamiltonian method, i.e. by choosing a local expression for $H_{\underline{\Lambda}}$, e.g.

$$H_{\underline{\Lambda}} = A_{\underline{\Lambda}} + B_{\underline{\Lambda}} \tag{5.28}$$

one can construct the time evolution directly in the algebra. Moreover, this time evolution fulfils a relativistic causality condition (with maximal signal velocity) up to exponential tails [7]. Unfortunately, in this framework it is much more difficult to construct the ground state than in the Euclidean case. Therefore we prefer the Euclidean method.

A ground state of $\mathcal{F}$ with respect to α_i can be defined in terms of a Gibbs state of the Euclidean model. Let $\langle\ \rangle$ denote a Gibbs state of the gauge invariant Ising model in the temporal gauge and assume that $\langle\ \rangle$ fulfils Osterwalder-Schrader positivity (reflection positivity) for hyperplanes containing a lattice hyperplane or lying half between two neighbouring lattice hyperplanes. Define a state $\omega_\circ$ on $\mathcal{F}$ by

$$\omega_\circ\Big(\prod_n \alpha_{i^n}\Big(\prod_{\underline{x}\in \underline{M}_n} \sigma_3(\underline{x}) \prod_{\underline{b}\in \underline{L}_n} \tau_3(\underline{b})\Big)\Big) = \Big\langle \prod_{x\in M} \sigma(x) \prod_{b\in L} \tau(b)\Big\rangle . \tag{5.29}$$

where $M = \bigcup_n \{n\}\times \underline{M}_n$, $L = \bigcup_n \{n\}\times \underline{L}_n$, M_n, L_n being finite sets of sites and bonds in $\mathbb{Z}^d$, respectively. Then $\omega_\circ$ is a ground state of $\mathcal{F}$ with respect to α_i [31].

A Gibbs state with the properties mentioned above may be obtained as the limit of local Gibbs states with free boundary conditions. This limit always exists as a consequence of Griffith inequalities [32].

6. CHARGED STATES OF THE $\mathbb{Z}_2$ THEORY

We now want to find charged states of the model. By definition, a charged state is a state which cannot be represented by a vector in the vacuum Hilbert space and which has finite energy. The latter property means more precisely that in the GNS representation π induced by this state there is a positive bounded operator T - the transfer matrix in the representation π - such that

$$T\pi(A) = \pi\alpha_i(A)T \quad , \ A \in \mathcal{F} \quad . \tag{6.1}$$

The idea for the construction of a charged state is simple. One creates a charge at some point $\underline{x}$ together with a compensating charge and transports the compensating charge to infinity. In a gauge theory the charges are connected by electric flux lines, so one has to arrange these flux lines in such a way that the limit state has finite energy.

A convenient choice of the flux lines is obtained in the following way (*). Let $\underline{x}_r = (2r, 0, \ldots, 0) \in \mathbb{Z}^d$, $r \in \mathbb{N}$, and Let $\underline{L}_r$ be the path along the 1-axis from the origin to $\underline{x}_r$. Let

$$\Phi_r = \sigma_3(\underline{0})\, \sigma_3(\underline{x}_r)\, T_0^r\, \tau_3(\underline{L}_r)\, \Omega \tag{6.2}$$

with $\tau_3(\underline{L}_r) = \prod_{\underline{b} \in \underline{L}_r} \tau_3(\underline{b})$.

The application of the r-th power of the transfer matrix to the string state vector $\tau_3(\underline{L}_r)\Omega$ suppresses its high energy components. Now consider the state

$$\omega_r(A) = \frac{(\Phi_r, A\, \Phi_r)}{\|\Phi_r\|^2}, \quad A \in \mathcal{F} \quad . \tag{6.3}$$

ω_r is interpreted as a state where the charges have been separated by a distance 2r such that the energy remains bounded, independently of r. In fact, for $n \in \mathbb{N}$

$$\begin{aligned} \omega_r(e^{nH_0}) &\leq \text{const} \frac{\| T_0^{r-n}\, \tau_3(\underline{L}_r)\Omega \|}{\| T_0^r\, \tau_3(\underline{L}_r)\Omega \|} \\ &= \text{const} \frac{\langle \tau(M_{2r,2(r-n)}) \rangle^{1/2}}{\langle \tau(M_{2r,2r}) \rangle^{1/2}} \end{aligned} \tag{6.4}$$

where $M_{k,l}$ denotes the rectangular loop in the (0-1)-plane in $\mathbb{Z}^{d+1}$ with side k in the 0-direction and l in the 1-direction. The perimeters of the loops in the numerator and the denominator differ by 4n, so it is plausible that the perimeter law for the Wilson loop which is known to hold in the whole region $\beta_h > 0$ because of Griffiths inequalities implies that $\omega_r(e^{nH_0})$ is bounded uniformly in r. (For a proof see [26] .)

(*) Another method which also leads to the construction of charged states has been invented by Szlachányi [33].

Using the convergent cluster expansions one can show that the sequence ω_r converges to a state ω in the free charge phase as well as in the screening/confinement phase. Let us first look at the free charge phase. The local charge operator is

$$Q_{\underline{\Lambda}} = \prod_{\underline{b} \in \partial^* \underline{\Lambda}} \tau_1(\underline{b}) \tag{6.5}$$

where $\partial^* \underline{\Lambda}$ is the set of bonds with exactly one endpoint in $\underline{\Lambda}$. We find

$$\frac{\omega(Q_{\underline{\Lambda}})}{\omega_0(Q_{\underline{\Lambda}})} \longrightarrow -1 \quad , \quad \underline{\Lambda} \nearrow \mathbb{Z}^d \tag{6.6}$$

whereas for each local excitation of ω_0,

$$\omega_F(A) = \frac{\omega_0(F^* A F)}{\omega_0(F^* F)} \quad , \; F \in \mathcal{F} \quad , \tag{6.7}$$

one has

$$\frac{\omega_F(Q_{\underline{\Lambda}})}{\omega_0(Q_{\underline{\Lambda}})} \longrightarrow 1 \quad . \tag{6.8}$$

This supports the interpretation of ω as a charged state. A second indication that ω is not in the vacuum sector comes from the weak convergence of $\Phi_\tau \|\Phi_\tau\|^{-1}$ to zero. This can be shown for all (β_g, β_h) such that the pure gauge theory with coupling β_g satisfies the perimeter law and such that β_h is sufficiently small (dependent on the parameter in the perimeter law). The third property of the free charge phase which indicates the presence of charges is the existence of large vacuum fluctuations between electrical fluxes as discussed in Sect. 4. Let $\underline{\Lambda}_R$ be a cube with side length R and let S_ℓ and S_r denote the left and right half of the boundary $\partial^* \underline{\Lambda}_R$ of $\underline{\Lambda}_R$. Let

$$E_\ell = \prod_{\underline{b} \in S_\ell} \tau_1(b) \tag{6.9}$$

be the electric flux through S_ℓ and E_r the electric flux through S_r. Then $Q_{\Lambda_R} = E_\ell E_r$ and

$$\frac{\omega_0(E_e)\,\omega_0(E_r)}{\omega_0(Q_{\underline{\Lambda}_R})} \sim e^{-\alpha|\partial^* \underline{\Lambda}_R|} , \tag{6.10}$$

$$|\partial^* \underline{\Lambda}_R| \sim R^{d-1} .$$

Let us compare these results with the corresponding properties in the screening/confinement phase. There one finds a vector Φ in the vacuum Hilbert space $\mathcal{H}_0$ such that

$$(\Phi, A\Phi) = \omega(A) . \tag{6.11}$$

Φ is obtained as the limit of $\sigma_3(\underline{\varrho})\, T^r \sigma_3(\underline{\varrho})\Omega \; \| T^r \sigma_3(\underline{\varrho})\Omega\|^{-1}$ for $r \to \infty$. Thus ω is certainly not a charged state. For the charge operator one finds

$$\frac{\omega(Q_{\underline{\Lambda}})}{\omega_0(Q_{\underline{\Lambda}})} \longrightarrow 1 , \quad \underline{\Lambda} \uparrow \mathbb{Z}^d . \tag{6.12}$$

The sequence $\Phi_r \|\Phi_r\|^{-1}$ converges weakly to $(\Omega,\Phi)\Phi$ with $(\Omega,\Phi) \neq 0$. Especially

$$\frac{(\Omega,\Phi_r)}{\|\Phi_r\|} \longrightarrow (\Omega,\Phi)^2 . \tag{6.13}$$

Last not least, the correlations of electric fluses are much weaker; one finds

$$\frac{\omega_0(E_e)\,\omega_0(E_r)}{\omega_0(Q_{\underline{\Lambda}_R})} \sim e^{-\alpha|\partial^* S_r|} , \tag{6.14}$$

$$|\partial^* S_r| \sim R^{d-2} .$$

If we formulate these results in the framework of the Euclidean theory we find three order parameters which seem to be suitable for the distinction of the free charge phase from the screening/confinement phase. The first one is the expectation value of the charge operator in the state ω :

$$\rho_1 = \lim_{R\to\infty} \frac{\langle \overset{R\;\;R}{\square\!\text{[dashed loop]}} \rangle}{\langle \underset{R}{\square} \rangle \langle \underset{R}{\text{[dashed loop]}} \rangle} \tag{6.15}$$

where $\square R$ means a square like Wilson loop with side length R and $\lozenge R$ means a dual loop (a loop in the dual theory in d+1 = 3 dimensions and a closed surface in the dual theory in d+1 = 4 dimensions) with side length R. We have ϱ_1 = 1 in the screening/confinement region and $\varrho_1 = -1$ in the free charge phase.

The second one measures the overlap of the vacuum with the approximate charged state $\Phi_r \|\Phi_r\|^{-1}$ in the limit $r \to \infty$:

$$\varrho_2 = \lim_{r\to\infty} \frac{\langle \overset{2r}{\sqcap} \rangle}{\langle \underset{2r}{\square} \rangle^{1/2}} \tag{6.16}$$

We have $\varrho_2 > 0$ in the screening/confinement phase and $\varrho_2 = 0$ in the free charge phase.

The third one is sensitive to the correlations of electric fluxes:

$$\varrho_3(R) = \frac{\langle R\lozenge \rangle \langle \lozenge R \rangle}{\langle \lozenge R \rangle} \tag{6.17}$$

If behaves like $e^{-\text{const}\, R^{d-1}}$ in the free charge phase and like $e^{-\text{const}\, R^{d-2}}$ in the screening/confinement phase. Note that in d+1 = 3 dimensions where the theory is selfdual, $\varrho_3(\infty)$ is the dual of ϱ_2.

All these order parameters may be used as confinement criteria in gauge theories with matter fields. For ϱ_2 this has been proposed in some detail in [34].

The order parameter ϱ_2 has also been tested in Monte Carlo simulations [35]. It shows the expected behaviour beyond the region of convergence of cluster expansions. There is, however, a region in the screening/confinement phase where the results are not yet conclusive. This on the first sight unpleasant fact has an interesting explanation. It is connected probably with the following behaviour of $\varrho_2(r)$ for finite r. For small r $\varrho_2(r)$ decreases in a similar way as in the free charge phase. Then, at a certain r it starts to increase again up to some finite value. This turning point r_f may be interpreted as the distance where fragmentation of the string sets in. It coincides with the transition from the area law to the perimeter law for the Wilson loop. Rough estimates indicate that r_f is very large in this region, thus one cannot see the asymptotic value of ϱ_2 on a relatively small lattice (22^{d+1} lattice points).

There have been several other attempts to find an order parameter which distinguishes the free charge phase from the screening/confinement phase [36, 37, 38]. In general they do not reproduce the known phase diagram; most of them indicate an artificial transition between the screening and the confinement region. There is one order parameter proposed by Bricmont and Fröhlich [36] which looks very similar as the order parameter ρ_2. Bricmont and Fröhlich argue that the expectation value of a straight string

$$a(r) = \langle \overset{r}{\longrightarrow} \rangle \tag{6.18}$$

behaves like

$$a(r) \sim e^{-\mathrm{const}\, r} \tag{6.19}$$

in the screening/confinement phase and like

$$a(r) \sim r^{-d/2} e^{-\mathrm{const}\, r} \tag{6.20}$$

in the free charge phase. As a test which behaviour is present they propose to look whether the limit

$$\rho_{BF} = \lim_{r \to \infty} \frac{a(r)^2}{a(2r)} \tag{6.21}$$

vanishes.

In the language of the quantum model, ρ_{BF} is

$$\rho_{BF} = \lim_{r \to \infty} \frac{(\Omega, \sigma(\underline{0}) T_0^{r} \sigma(\underline{0}) \Omega)^2}{\| T_0^{r} \sigma(\underline{0}) \Omega \|^2} = (\Omega, \Phi)^2 \tag{6.22}$$

with Φ from (6.11). Hence in the screening/confinement region from (6.13) ρ_{BF} coincides with ρ_2. In the free charge region, however, this seems unlikely. Namely, ρ_{BF} does not vanish if the highest spectral value of the transfer matrix in the sector with external charge at the origin is an eigenvalue [26]. The corresponding eigenstate may be considered as a bound state of a dynamical charge with the external charge, i.e. it is the "hydrogen-atom" of this model. The existence of such bound states does not exclude in general the existence of isolated charged particles, hence the transition indicated by ρ_{BF} probably does not coincide with the transition from the free charge phase to the screening/confinement phase. It would be very interesting to verify this conjecture. Some work in this direction has been done by Bricmont and Fröhlich [39].

7. PARTICLE STRUCTURE IN THE CHARGED SECTOR

We now want to investigate the particle structure of the present model more closely. Isolated particle shells in the joint spectrum of the transfer matrix and the translation operators have been found in the vacuum sector of several models. Schor proved their existence in strongly coupled pure gauge theories [40]. By his methods actually a rich class of stable particles was found [41, 42, 43, 44]. Another method has been developed by Bricmont and Fröhlich. They compute power corrections to the exponential decay of 2-point functions and derive the existence of particles from there [39].

A necessary condition for a corresponding proof in the charged sector is the construction of a transfer matrix and of translation operators in the charged representation. Let $(\mathcal{H}, \pi, \Phi)$ be the GNS representation induced by ω (Thm. 2.1). Let $\Phi_{\underline{o}} = \pi(\sigma_3(\underline{o}))\Phi$ and let

$$\omega_{\underline{o}}(A) = (\Phi_{\underline{o}}, \pi(A)\Phi_{\underline{o}}) \quad , A \in \mathcal{F} \quad . \tag{7.1}$$

$\omega_{\underline{o}}$ is a state with an external charge at the origin. Moreover, $\omega_{\underline{o}}$ is invariant under α_i. The transfer matrix T in $\mathcal{H}$ is now defined by

$$T\pi(A)\Phi_{\underline{o}} = \pi\alpha_i(A)\Phi_{\underline{o}} \quad , A \in \mathcal{F} \quad . \tag{7.2}$$

T satisfies the relation

$$T\pi(A) = \pi\alpha_i(A)T \quad , A \in \mathcal{F} \tag{7.3}$$

and has the densely defined inverse

$$T^{-1}\pi(A)\Phi_{\underline{o}} = \pi\alpha_{-i}(A)\Phi_{\underline{o}} \quad , \quad A \in \mathcal{F} \, . \tag{7.4}$$

Moreover

$$0 \leq T \leq e^{\alpha} \tag{7.5}$$

where α is the parameter occuring in the perimeter law of the Wilson loop [26].

The lattice translations $\underline{x}$ act as automorphisms $\alpha_{\underline{x}}$ of the algebra $\mathcal{F}$,

$$\begin{aligned} \alpha_{\underline{x}}\,\sigma_i(\underline{y}) &= \sigma_i(\underline{y}+\underline{x}) \quad , \; i = 1,3 \\ \alpha_{\underline{x}}\,\tau_i(\underline{b}) &= \tau_i(\underline{b}+\underline{x}) \quad , \; i = 1,3 \quad . \end{aligned} \tag{7.6}$$

Let $\omega_{\underline{x}} = \omega_0 \circ \alpha_{\underline{x}}$. We have the following theorem:

Theorem 7.1 [26] There is a unique vector $\Phi_{\underline{x}} \in \mathcal{H}$ such that

$$\text{(i)} \quad T\Phi_{\underline{x}} = \Phi_{\underline{x}}$$

$$\text{(ii)} \quad (\Phi_{\underline{x}}, \pi(A)\Phi_{\underline{x}}) = \omega_{\underline{x}}(A)$$

$$\text{(iii)} \quad (\pi(\sigma_3(\underline{x}))\Phi, \Phi_{\underline{x}}) > 0$$

Now the translation operators can be defined by

$$U(\underline{x})\,\pi(A)\Phi_0 = \pi\alpha_{\underline{x}}(A)\Phi_0 \quad , \quad A \in \mathcal{F} \quad . \tag{7.7}$$

They have the following properties:

Theorem 7.2 [26]

$$\text{(i)} \quad U(\underline{x})U(\underline{y}) = U(\underline{x}+\underline{y})$$

$$\text{(ii)} \quad U(\underline{x})\,\pi(A)\,U(-\underline{x}) = \pi\alpha_{\underline{x}}(A)$$

$$\text{(iii)} \quad [U(\underline{x}), T] = 0$$

$$\text{(iv)} \quad (\Psi, U(\underline{x})\Psi) \underset{|\underline{x}|\to\infty}{\longrightarrow} 0 \quad \text{for all } \Psi \in \mathcal{H}$$

Thm 7.2 (iv) shows that the charged representation is really different from the vacuum representation. Namely, for the translation operators $U_0(\underline{x})$ in the vacuum Hilbert space $\mathcal{H}_0$, defined by

$$U_0(\underline{x})\,\pi_0(A)\Omega = \pi_0\alpha_{\underline{x}}(A)\Omega \tag{7.8}$$

one has instead of Thm 7.2 (iv)

$$(\Psi, U_0(\underline{x})\Psi) \underset{|\underline{x}|\to\infty}{\longrightarrow} |(\Omega, \Psi)|^2 \tag{7.9}$$

for all $\Psi \in \mathcal{H}_0$. (For a more precise discussion see [26, Sect. 7]).

There are many open questions. The first one concerns the existence of an isolated mass shell in the joint spectrum of T and $U(\underline{x})$. It is conceivable that there are methods similar to those used by Schor or by Bricmont and Fröhlich by which one can show the existence of charged particles. Provided these single particle states exist one would like to construct multiparticle scattering states, i.e. to develop a lattice version of the Haag-Ruelle scattering theory [45, 46]. Here the lack of locality of the real time translation in the Euclidean lattice theory will cause some problems, and it may be easier to work in the Hamiltonian formalism (but there one would have to show first the existence of charged states).

The next question is whether these particles will have a well defined statistics, whether antiparticles exist and whether there is a global gauge group labeling the charge sectors. As mentioned in Section 3 these questions have a positive answer in the general framework of quantum field theory in continuous space time.

A very important question is whether the continuum limit exists and whether the charge structure survives in this limit. In this respect it is interesting that the $\mathbb{Z}_2$ Higg model in d+1 = 4 dimensions seems to have a second order phase transition between the free charge and the screening phase [47]. The existence of a second order phase transition is a necessary condition for the existence of a continuum limit.

There are many other lattice models where similar questions could be investigated. Some work has been done on the U(1) Higgs model. Barata and Wreszinski have shown that in some part of the screening/confinement region the expectation value of the charge operator in the state ω_r defined in Eq. (6.3) vanishes in the limit of large r [48, 49]. There is also some recent work by Brydges and Seiler [50] and of Kennedy and King [51] on the noncompact U(1) Higgs model which has been mentioned in the lecture of Prof. Wightman.

REFERENCES

1. G. C. Wick, E. P. Wigner, A. S. Wightman: Intrinsic parity of elementary particles. Phys. Rev. 88. 101 (1952)
2. J. Fröhlich: New superselction sectors ("soliton states") in two dimensional Bose quantum field models. Commun. Math. Phys. 47, 269 (1976)
3. O. Steinmann: Perturbative QED in terms of gauge invariant fields. Ann. Phys. (NY) 157, 232 (1984)
4. F. Strocchi, A. S. Wightman: Proof of the charge superselection rule in local relativistic quantum field theory. Journ. Math. Phys. 15, 2198 (1974)
5. G. Morchio, F. Strocchi: Infrared singularities, vacuum structure and pure phases in local quantum field theory. Ann. Inst. Henri Poin. XXXIII, 251 (1980)
6. R. Haag, D. Kastler: An algebraic approach to quantum field theory. Journ. Math. Phys. 5, 848 (1964)
7. O. Bratteli, D. Robinson: Operator algebras and statistical mechanics I, II, Berlin, Heidelberg, New York: Springer 1981
8. H.-J. Borchers: On the vacuum state in quantum field theory II. Commun. Math. Phys. 1, 57 (1965)
9. H.-J. Borchers: Energy and momentum as observables in quantum field theory. Commun. Math. Phys. 2, 49 (1966)
10. G. Roepstorff: Coherent photon states and spectral condition. Commun. Math. Phys. 19, 301 (1970)

11. A. Stoffel: Eigenschaften von Infrarotdarstellungen. Diplomarbeit Hamburg 1973 (unpublished)

12. J. Fröhlich, G. Morchio, F. Strocchi: Infrared problem and spontaneous breaking of the Lorentz group in QED. Phys. Lett. 89B, 61 (1979)

13. H.-J. Borchers, D. Buchholz: The energy momentum spectrum in local field theories with broken Lorentz symmetry. Commun. Math. Phys. 97, 169 (1985)

14. H.-J. Borchers: Locality and covariance of the spectrum. Bielefeld University preprint (1984)

15. D. Buchholz, K. Fredenhagen: Locality and the structure of particle states. Commun. Math. Phys. 84, 1 (1982)

16. D. Buchholz, K. Fredenhagen: Charge screening and mass spectrum in Abelian gauge theories: Nucl. Phys. B154, 226 (1979)

17. B. Schroer: Infrateilchen in der Quantenfeldtheorie. Fortschr. Physik 11, 1 (1963)

18. D. Buchholz: The physical state space of quantum electrodynamics. Commun. Math. Phys. 85, 49 (1982)

19. J. A. Swieca: Charge screening and mass spectrum. Phys. Rev. D13, 312 (1976)

20. S. Doplicher, R. Haag, J. E. Roberts: Local observables and particle statistics. Commun. Math. Phys. 23, 199 (1971) and 35, 49 (1974)

21. H.-J. Borchers: Local rings and the connection of spin with statistics. Commun. Math. Phys. 1, 281 (1965)

22. K. Fredenhagen: On the existence of antiparticles. Commun. Math. Phys. 79, 141 (1981)

23. S. Doplicher, J. E. Roberts: Compact Lie groups associated with endomorphisms of C*-algebras. Bielefeld University preprint (1984)

24. J. J. Bisognano, E. H. Wichmann: On the duality condition for a hermitean scalar field. Journ. Math. Phys. 16, 985 (1975)

25. F. Wegner: Duality in generalized Ising models and phase transitions without local order parameters. Journ. Math. Phys. 12, 2259 (1971)

26. K. Fredenhagen, M. Marcu: Charged states in $_2$ gauge theories. Commun. Math. Phys. 92, 81 (1983)

27. E. Fradkin, S. Shenker: Phase diagrams of lattice gauge theories with Higgs fields. Phys. Rev. D19, 3682 (1979)

28. K. Osterwalder, E. Seiler: Gauge field theories on a lattice. Ann. Phys. 110, 440 (1978)

29. R. Marra, S. Miracle-Sole: On the statistical mechanics of the gauge invariant Ising model. Commun. Math. Phys. 67, 233 (1979)

30. I. Cioranescu, L. Zsido: Analytic generators for one-parameter groups. Tôhoku Math. J. 28, 327 (1976)

31. K. Fredenhagen: On the existence of the real time evolution in Euclidean lattice gauge theories (to appear in Commun. Math. Phys.)

32. D. Ruelle: Statistical mechanics. Reading MA: Benjamin 1969

33. K. Szlachányi: Non-local charged fields and the confinement problem. Phys. Lett. 147B, 335 (1984)

34. K. Fredenhagen, M. Marcu: A confinement criterion for QCD with dynamical quarks. DESY 85-008 (preprint)

35. Th. Filk, K. Fredenhagen, M. Marcu: Order parameters for lattice gauge theories with matter fields. (to be published)

36. J. Bricmont, J. Fröhlich: An order parameter distinguishing between different phases of lattice gauge theories with matter fields. Phys. Lett. 122B, 73 (1983)

37. G. Mack, H. Meyer: A disorder parameter that tests for confinement in gauge theories with quark fields. Nucl. Phys. 200B, 249 (1982);
H. Meyer-Ortmanns: The vortex free energy in the screening phase of the Z(2) Higgs model. Nucl. Phys. B225 FS 11 , 115 (1984);
H. Meyer-Ortmanns: Unexpected behavior of an order parameter for lattice gauge theories with matter fields. Nucl. Phys. B230 FS 10 , 31 (1984)

38. J. Bricmont, J. Fröhlich: Defect free energies in lattice gauge theories with matter fields. Nucl. Phys. B230 FS 10 , 407 (1984)

39. J. Bricmont, J. Fröhlich: Statistical mechanical methods in particle structure analysis of lattice field theories. Nucl. Phys. B251 FS 13 , 517 (1985) and Commun. Math. Phys. 98, 553 (1985)

40. R. S. Schor: Existence of glueballs in strongly coupled lattice-gauge theories. Nucl. Phys. B222 FS 8 , 71 (1983)

41. R. S. Schor: Excited glueball states on a lattice from first principles. Phys. Lett. 132B, 161 (1983)

42. R. S. Schor: Glueball spectroscopy in strongly coupled lattice gauge theories. Commun. Math. Phys. 92, 369 (1984)

43. M. O'Caroll, G. Braga, R. S. Schor: Glueball mass spectrum and mass splitting in 2+1 strongly coupled lattice gauge theories. Commun. Math. Phys. 97, 429 (1985)

44. M. O'Caroll, G. Braga: Analyticity properties and a convergent expansion for the glueball mass and dispersion curve of strongly coupled Euclidean 2+1 lattice gauge theories. Journ. Math. Phys. 25, 2741 (1984)

45. R. Haag: Quantum field theories with composite particles and asymptotic behaviour. Phys. Rev. 112, 669 (1958)

46. D. Ruelle: On the asymptotic condition in quantum field theory. Helv. Phys. Acta 35, 147 (1962)

47. Th. Filk, K. Fredenhagen, M. Marcu: A 2nd order phase transition in the 4 dimensional $_2$ Higgs theory. (to be published)

48. J. C. A. Barata: Sobre a relacao entre ausência de carga e "gap" de massa em teorias de gauge na rede. Thesis Sao Paulo 1985 (unpublished)

49. J. C. A. Barata, W. F. Wreszinski: Absence of charged states in the U(1) Higgs lattice gauge theory. (to appear in Commun. Math. Phys.)

50. D. C. Brydges, E. Seiler: Absence of screening in certain lattice gauge and plasma models. Princeton University preprint 1985

51. T. Kennedy, C. King: Symmetry breaking in the lattice Abelian Higgs model. Princeton University preprint 1985

INFRARED PROBLEM, HIGGS PHENOMENON AND LONG RANGE INTERACTIONS

G. Morchio*† and F. Strocchi**

*Dipartimento di Fisica dell'Università, Pisa, Italy
**International School for Advanced Studies, Trieste, Italy
and International Centre for Theoretical Physics, Trieste

I. A NON-PERTURBATIVE APPROACH TO THE INFRARED PROBLEM IN QED

1.1 *Conventional wisdom and questions of principle*

The infrared problem in QED is associated to the zero mass of the photon; at the level of (perturbative) Green's functions the removal of the infrared cutoff (e.g. a fictitious photon mass μ) is completely under control even from a rigorous point of view[1]. The essential part of the problem occurs when one tries to remove the infrared cutoff in scattering amplitudes.

The standard perturbative approach[2] is that to compute transition probabilities, to a given perturbative order n, one has to sum over all processes, in which soft photons with energy smaller than ΔE (the resolution of the experimental apparatus) are present in the initial or in the final state. With this prescription the removal of the infrared cutoff yields (infrared) finite ΔE dependent transition probabilities (to each perturbative order). Natural questions arise: i) is there an underlying theory independent of ΔE or should one take the point of view that each experimental apparatus defines a theory? Can one find ΔE independent amplitudes? ii) one cannot control the ΔE dependence of the experimental cross sections since the effective expansion parameter is $\alpha \ln(\Delta E/m)$ and one needs a leading order summation prescription (exponentiation?), i.e. information on a ΔE independent theory, iii) ΔE is not a removable cutoff, since transition probabilities $P^{\Delta E}$ (summed over all orders) are expected[3] to be $O(\Delta E)$ as $\Delta E \to 0$ and it is therefore impossible to extrapolate.

An improvement and significant step towards a ΔE independent theory was provided by Chung[4] and Kibble[5]. The idea is that, instead of summing over soft photon processes, one computes transition amplitudes (in terms of Feynman diagrams) between non-Fock coherent states characterized by coherent functions, with singular behaviour as $p/p\cdot k$ in the infrared region $k \to 0$, and otherwise related to the preparation of the in/out states. The removal of

† Supported in part by INFN, Sezione di Pisa.

the infrared cutoff then yields finite S-matrix elements between such states. The correct definition of the asymptotic states takes care of the dependence on the experimental resolution of the previous approach (in a non-Fock coherent state it is impossible to count the number of (soft) photons!). Chung's analysis is based on the (perturbative) factorization of the infrared singularities in transition amplitudes[6]: the infrared divergent part due to soft photon emission (or absorption) in S-matrix elements factorizes in the form

$$\exp \quad e\, a^*\left(\frac{p^{out}}{p^{out}\cdot k} - \frac{p^{in}}{p^{in}\cdot k}\right) \quad ,$$

it depends only on the initial and final momenta and it is essentially the same as in the Bloch-Nordsieck model[7]. In a certain sense, the coherent states provide a non-perturbative description of the "soft photon cloud" attached to a charged state; transition amplitudes between coherent states involve an infinite sum of Feynman diagrams. Chung's result can be briefly summarized as a perturbative expansion based on the Bloch-Nordsieck (BN) ansatz on the asymptotic states.

Clearly a non-perturbative justification of the BN ansatz is needed from a general point of view and in particular the following questions arise: i) are the asymptotic charged states coherent states for A_μ (interaction picture)? ii) which set of coherent states are relevant; does one need a non-separable Hilbert space as advocated by Kibble[5]? iii) since different (charged) momenta define disjoint coherent representations with infrared BN factors[8], should one conclude that the momentum of the charged particles is superselected? iv) are there non-perturbative features with respect to Chung's analysis? Can one get a non-perturbative control on the general structure?

Rigorous information can be obtained from non-relativistic models like the BN model[7], which describes the electromagnetic field interacting with a classical current, or the Blanchard-Pauli-Fierz (BPF) model[9] which describes a Schroedinger charged particle interacting with the quantized e.m. field and an external potential V(x),

$$H_{BPF} = \frac{1}{2m}(\vec{p} + e\vec{A}(\vec{x}))^2 - eA_o(\vec{x}) + V(\vec{x}) + H_o^{e.m.} \tag{1.1}$$

The latter is completely under control in the dipole approximation ($-eA_o(\vec{x}) \to -eA_o(0)$, $(\vec{p}+e\vec{A}(\vec{x}))^2 \to \vec{p}^2 + 2e\vec{A}(0)\vec{p}$). In both models charged states define coherent representations for A_μ^{as} (as=in/out). There is however an arbitrariness in the choice of the representation of $A_\mu(\vec{x},t=0)$, which is crucial for the non-Fock character of the above coherent representations. (The ambiguity is inevitable in the Coulomb gauge and it is resolved by locality in local gauges, see below). Both models exhibit (non-perturbative) exponentiation of the infrared singularities. In the BN model one is led to consider an uncountable set of inequivalent scattering representations, incoherently labelled by the particle momenta, and furthermore depending on the arbitrariness mentioned above; this has lead Kibble to advocate a non-separable space, in general. In the BPF model one has a non-separable space if all the Lienardt-Wiechert potential are allowed. They are labelled by the velocity of the frame in which the LW potential reduces to the Coulomb one; such velocity parameter is superselected whereas the particle momentum $\vec{p}$ is not. In both models a separable space is obtained if a local formulation is used.

The lesson from "soluble" models (especially BPF) has been extrapolated to quantum electrodynamics by Kulish and Fadde'ev[9]. Their analysis is based on Dollard's idea of asymptotic dynamics[10] which governs the time evolution of asymptotic states and is (in general) different from the free Hamiltonian (H_o) dynamics

$$H_{as}(t) = H_o + V_{as}(t) ,$$

where

$$V_{as}(t) = (2\pi)^{-3/2}\int \frac{d^3k}{\sqrt{2k_o}} [a_\mu^+(-\vec{k}) + a_\mu(\vec{k})] J^\mu_{as}(\vec{k},t),$$

$$J^\mu_{as}(\vec{k},t) = - e\int \frac{d^3p}{p_o} p^\mu \rho(\vec{p}) \exp[i \frac{\vec{p}\cdot\vec{k}}{p_o} t]$$

and

$$\rho(\vec{p}) = \sum_n [b_n^+(\vec{p})b_n(\vec{p}) - d_n^+(\vec{p})d_n(\vec{p})]$$

is the charge density operator of asymptotic negative (b) and positive (d) charges. On a state $\Psi(p,q)$ of charged particles with momenta p (and charge $-e$) and q ($+e$), J^μ_{as} reduces to the classical (BN type) currents of point charges with momenta p and q. The so-obtained asymptotic dynamics is similar to that of the BPF Hamiltonian:

$$U_{as}(t) = \exp [-iH_o t] \exp[R(t)] \exp [i\Phi(t)] ,$$

where

$$R(t) = e(2\pi)^{-3/2}\int \frac{d^3p d^3k}{\sqrt{2k_o}} \rho(\vec{p}) \frac{p^\mu}{p\cdot k} [a_\mu^+(k) \exp(i \frac{k\cdot p}{p_o} t) - h.c.]$$

and

$$\Phi(t) = \frac{e^2}{8\pi}\int d^3p d^3q\, p\cdot q\, ((p\cdot q)^2 - m^4)^{-1/2} :\rho(p)\rho(q):$$

is the generalization of the Coulomb phase

$$\phi(t) = (em/p)\ \mathrm{sign}\,t\ \ln(|t|/t_o)$$

in potential scattering.

The S-operator is then defined by

$$S(t_1,t_2) = U^\dagger_{as}(t_1)\, e^{-iH(t_1-t_2)}\, U_{as}(t_2),$$

(in contrast with the Dyson S-operator defined by taking $H_{as} = H_o$). S (formally) acts on a Hilbert space

$$\mathcal{H} = \mathcal{H}_{\text{charges}} \times \mathcal{H}_{\text{ph}} ,$$

where $\mathcal{H}_{ph}$ is a non-separable (Von Neumann infinite tensor product) space describing the photon states. Since R(t) is infrared singular as operator on the photon Fock space $\mathcal{H}_F$, exp[-R(t)] is not a well-defined (unitary) operator in $\mathcal{H}_F$; it can be defined on $\mathcal{H}$ and

$$\mathcal{H}_{as} \equiv \exp[-R(t)]\ \mathcal{H}_F \neq \mathcal{H}_F .$$

Asymptotic limits (are then expected to) exist in $\mathcal{H}$, but not in the space $\mathcal{H}_{ch} \times \mathcal{H}_F$, in which the theory is defined (in the interaction picture) for finite times.

The Kulish-Fadde'ev (KF) analysis is not free of conceptual questions and problems:

i) do the asymptotic limits of the fields in the Heisenberg picture exist in the Hilbert space in which the theory is defined for finite times (Wightman space)?

ii) in KF analysis the photon transversality condition (Gupta-Bleuler condition) is satisfied by an ad hoc introduction of an external four vector $c_\mu(\vec{k})$ and this raises non-trivial questions about the Lorentz invariance;

iii) a critical analysis of the problem led Zwanziger[11] to conclude that it is actually impossible to satisfy the (covariant) Bloch-Nordsieck ansatz for the charged states and the Gupta-Bleuler (GB) condition: the BN ansatz at the basis of Chung's analysis for a charged state Ψ_p of momentum p reads

$$a_\mu^{as}(k)\ \Psi_p = (e \frac{p_\mu}{p\cdot k}\ \rho(p\cdot k) + k_\mu g(p\cdot k))\ \Psi_p , \tag{1.2}$$

where a_μ^{as} is the destruction operator for asymptotic photons, $\rho(p,k) \to 1$ as $k \to 0$ and g is a gauge function, so that

$$\lim_{k\to 0} k^\mu a_\mu^{as}(k)\ \Psi_p = Q\Psi_p = e\Psi_p \neq 0 , \tag{1.3}$$

in contrast with the GB condition (Zwanziger's paradox).

Zwanziger's analysis indicates that despite the folklore wisdom, the infrared problem is plagued by unexpected conceptual difficulties and that it might be impossible to reconcile the Lorentz covariance, the Bloch-Norsieck ansatz and the Gupta-Bleuler condition. More generally the above discussion should indicate a confused situation at the level of general principles and the need for a non-perturbative understanding of the infrared problem. This is also of interest for non-abelian gauge theories and it is in any case a prerequisite, since it is a prototype of non-perturbative infrared phenomena occurring at scales of order $m^{-1} e^{1/\alpha}$. In QED this is the characteristic scale of soft photon emission and since such scale is very large it leads to small corrections in ordinary experiments, but it is not so in QCD since $\alpha_{QCD}(1\ \text{Gev}) \simeq 1$ and the corresponding infrared scale is of order of Λ^{-1}_{QCD}.

1.2 Basic problems. Gauss' law and construction of charged states. Asymptotic e.m. field algebra

The above difficulties stimulate an approach to the infrared problem from first principles. The basic problem is the construction of the asymptotic fields and in particular

a) the existence of the asymptotic limits for the photon field;

b) the characterization of the representations of A_μ^{as}, especially those corresponding to charged states;

c) the existence and construction of charged asymptotic fields and the strictly related infraparticle problem.

The substantial part of the infrared problem is b); it is much more difficult than in a Yukawa theory with massless bosons because of the intrinsic non locality of the charged fields due to the local Gauss' law[12,13] and of the charge superselection rule[14]. More generally this is related to the problem of the construction of charged representations of the local observable algebra; the difficulty is that such representations cannot be obtained by local operations (more generally by local morphisms) on the observable algebra in contrast to the case in which there is no local Gauss' law[15]. The latter amounts to a non-local constraint since it requires a non-trivial (electric) flux at spacelike infinity and this cannot be obtained by local operations. (For an algebraic approach to this problem see Ref. 16 and Fredenhagen lectures in this Proceedings).

The existence and construction of asymptotic limits has been one of the remarkable achievements of general quantum field theory for the case of massive particles (Haag-Ruelle scattering theory)[17]. The crucial role played by the mass gap condition (and the strong cluster property) prevented a simple extension to the massless case. This problem was solved by Buchholz[18] by exploiting the Huyghens' principle and locality, under the following general assumptions

A_1) There is a local C*-algebra $\mathcal{A}$

$$\mathcal{A} = \bigcup_{\mathcal{O}} \mathcal{A}(\mathcal{O})$$

where $\mathcal{O}$ denote a bounded region of space time (double cones) and $\mathcal{A}(\mathcal{O})$ is the associated local field algebra ($\mathcal{A}(\mathcal{O})$ can be thought as generated by bounded functions of Wightman fields localized in $\mathcal{O}$).

A_2) a representation of $\mathcal{A}$ in a separable Hilbert space H, in which the space time translations are unitarily represented

$$U(a)\,\mathcal{A}(\mathcal{O})\ \ U(a)^{-1} = \mathcal{A}(\mathcal{O}+a)$$

and satisfy the relativistic spectrum condition ($H^2 - \vec{P}^2 \geq 0$). The vacuum state Ψ_o is translationally invariant and there are massless one-particle states $\overset{o}{\Phi}$ ($\Phi \perp \Psi_o$)

$$(H - |\vec{P}|)\, \Phi = 0 .$$

H_1 will denote the space of such states and P_1 the corresponding projector.

A_3) the vacuum state Ψ_o is cyclic with respect to $\mathcal{A}$

$$H = \overline{\{\mathcal{A}\Psi_o\}} .$$

One then has

THEOREM 1.1 (Buchholz) Given $A \in \mathcal{A}(O)$, then

a_1) one can construct an asymptotic operator A^{out} such that

$$A^{out}\Psi_o = P_1 A \Psi_o \in H_1$$

i.e. A^{out} creates one particle massless (asymptotic) states.

a_2) $A^{out}(x)$ satisfies free field equations: $\Box A^{out}(x) = 0$.

a_3) A^{out} has localization properties

$$[A^{out}, B] = 0, \quad \forall\, B \in O_1 \subset O_+ ,$$

where O_+ denotes the forward lightcone with apex at the top of the double cone O.

a_4) A^{out} satisfies local commutativity and has a c-number commutator: given $A_i \in \mathcal{A}(O_i)$, $i = 1, 2$,

$$[A_1^{out}, A_2^{out}] = \langle \Psi_o, A_1 P_1 A_2 \Psi_o \rangle - \langle \Psi_o, A_2 P_1 A_1 \Psi_o \rangle$$

and if O_1 is spacelike with respect to O_2

$$[A_1^{out}, A_2^{out}] = 0$$

a_5) if $A = A^*$, then A^{out} is self adjoint.

Buchholz's theory provides a solution of problem a), i.e. the construction of the asymptotic photon fields[18,19]. By using Buchholz' result it is possible to discuss[19,20] the infrared problem from first principles and in particular to obtain[20,21]:

1) a characterization of the charged states in terms of the asymptotic e.m. field algebra;

2) a solution of the infraparticle problem;

3) a clarification of the relation between the non-Fock coherent property of the charged states and the breaking of the Lorentz group.

1.3 *Charged scattering representations. Asymptotic dynamics*

To characterize charged scattering states we shall start from general physical considerations as a guide to the mathematical formulation. The main point is that the characterization of a charged state involves not only the measurement of (charged) particle observables like momentum $\vec{p}$, spin $\vec{s}$ etc., but also of its associated asymptotic electromagnetic radiation. Thus a charged state Ψ defines its associated e.m. radiation, more generally a representation of the asymptotic e.m. algebra $\mathcal{A}(F^{as}_{\mu\nu})$: $< \Psi, \mathcal{A}(F^{as}_{\mu\nu})\Psi >$. This is a peculiar basic difference with respect to the massive case where, by suitable operations (like counters with sufficiently sharp energy resolution) one can always simplify the description by reducing to states without massive "photons", i.e. to Fock states for the "photon" asymptotic algebra. In the massless (photon) case it is in general impossible to eliminate the asymptotic e.m. radiation (on the semiclassical basis the number of asymptotic photons is actually infinite in a charged state!). In the standard perturbative approach discussed above, one adopts a non-sharp description of charged states, by not specifying all the associated asymptotic e.m. radiation with energy smaller than ΔE (see however the above-mentioned problems). A more natural and satisfactory choice[20] is to characterize a charged state as a representation of $\mathcal{A}^{as} \equiv \mathcal{A}(F^{as}_{\mu\nu})$, i.e. as a state described in terms of charged particle observables and of properties of the asymptotic e.m. radiation, as for example the energy spectrum, but not (in general) the photon number. Since only states with finite energy can be physically realized we will further require that the space time translations are unitarily implemented and satisfy the relativistic spectrum condition in the representations of $\mathcal{A}^{as}$ provided by a (physical) charged state (scattering representations). As we shall see, this approach will provide a clear identification of the asymptotic dynamics and a solution of the infraparticle problem.

More precisely our discussion will be based on the following mathematical structures[20,21]

I. The asymptotic limits $F^{as}_{\mu\nu}$, (as=in/out), of the electromagnetic fields exist as free fields

$$\partial^{\mu}F^{as}_{\mu\nu} = 0, \quad \partial^{\mu}{*F}^{as}_{\mu\nu} = 0 , \tag{1.4}$$

(on a dense domain of the physical Hilbert space H_{phys}) and they satisfy free fields commutation relations (CCR).

II. The space-time translations

$$F^{as}_{\mu\nu}(x) \rightarrow F^{as}_{\mu\nu}(x+a) \tag{1.5}$$

are described by unitary operators U(a) on H_{phys} and the corresponding generators $(H,\vec{P})$ satisfy the relativistic spectrum condition.

To avoid technical domain problems and to get a simple algebraic structure, it is convenient to construct bounded operators or observables associated to $F^{as}_{\mu\nu}$, i.e. the Weyl algebra $\mathcal{O}^{as}$ of the asymptotic e.m. field

$$\mathcal{O}^{as} = \{W(f) = \exp i\, F^{as}_{\mu\nu}(f^{\mu\nu}),\ f_{\mu\nu} \in \mathcal{S}_{real}(R^4)\} . \tag{1.6}$$

The following is a sufficient condition[22,20] for the construction of the Weyl algebra $\mathcal{O}^{as}$: the (Wightman) fields $F^{as}_{\mu\nu}$ satisfy the following inequality in the sense of quadratic forms

$$\pm F^{as}_{\mu\nu}(f) \leq |f|\ (H+1) , \tag{1.7}$$

where $f_{\mu\nu} \in \mathcal{S}_{real}(R^4)$ and $|f|$ is a Schwartz seminorm. This easily follows from sect. 7 of the first Ref. 22. Eq. (1.7) also implies $\pm F^{as}_{\mu\nu}(f) \leq H+\Delta(f_{\mu\nu})$, Δ being a c-number. (The latter inequality is enough for the existence of the Weyl operators of $F^{as}(f)$, by Theor. 1_M, M=2 of the first Ref. 22). In the following we shall assume that the asymptotic fields F^{as} satisfy condition (1.7). The Poincaré group is assumed to define automorphisms $\alpha_{\Lambda,a}$ of the observable algebra $\mathcal{O}^{as}$.

We shall now derive general properties of the representations of $\mathcal{O}^{as}$, in which the space time translations are unitarily implemented and the relativistic spectrum condition is satisfied (scattering representations).

THEOREM 1.2 Let π be a scattering representation of $\mathcal{O}^{as}$, then π is "locally Fock in momentum space", i.e. the restrictions of π to the algebra $\mathcal{O}^{as}_{\rho}$

$$\mathcal{O}^{as}_{\rho} \equiv \{W(f) = \exp i\, F^{as}_{\mu\nu}(f^{\mu\nu}),\ \text{with}$$

$$\text{supp}\ \tilde{f}_{\mu\nu}(p) \cap \{p:\ p_0 < \rho\} = \emptyset\} \tag{1.8}$$

is quasi equivalent (i.e. equivalent apart from multiplicities) to the Fock representation, $\Psi\rho$.

The proof follows the ideas of Ref. 23: one starts from a vector Ψ_1 with bounded energy spectrum and one applies to it the photon destruction operator $a_\lambda(f)$, supp $f \subset \{|\vec{k}| \geq \rho\}$. By eq. (1.7), any product of the a's can be applied to Ψ_1 with the result of successively lowering the upper bound of the energy spectrum by ρ, (each time one takes $a_\lambda(g)$ such that $a_\lambda(g)\Psi_n \neq 0$). Since $H \geq 0$, this is only possible a finite number of times and the last time gives a Fock vacuum.

THEOREM 1.3[20] In a scattering representation π of $\mathcal{O}^{as}$ (as=in/out) one can decompose

$$H = H^{ph} + H^{c},\quad \vec{P} = \vec{P}^{ph} + \vec{P}^{c} \tag{1.9}$$

where $(H^{ph}, \vec{P}^{ph})$ are selfadjoint operators affiliated to $\pi(\mathcal{A}^{as})''$ (i.e. they are "functions" of $F^{as}_{\mu\nu}$) with

$$\text{spectrum } (H^{ph}, \vec{P}^{ph}) = \bar{V}_+ \,, \quad H^{ph} \le H, \tag{1.10}$$

and $(H^c, \vec{P}^c)$ are selfadjoint operators affiliated to $\pi(\mathcal{A}^{as})'$, (i.e. the commute with $\mathcal{A}^{as}$), and satisfy

$$\text{spectrum } (H^c, \vec{P}^c) \subseteq \bar{V}_+ \tag{1.11}$$

(the above decomposition (1.10) will in general depend on as=in/out, but the subscript has been omitted for simplicity).

Proof Since by Theor. 1.2 the restriction of π to $\mathcal{A}^{as}_\rho$ is quasi equivalent to the Fock representation π_F one has an explicit expression for the space time translations on $\pi(\mathcal{A}^{as}_\rho)$: $\exp i\,(t\,H_\rho - \vec{a}\cdot\vec{P}_\rho)$, where

$$H_\rho = \sum_{\lambda=1,2} \int_{k>\rho} d^3k\, |\vec{k}|\, a^*_\lambda(\vec{k})\, a_\lambda(\vec{k}) \,, \tag{1.12}$$

$$\vec{P}_\rho = \sum_{\lambda=1,2} \int_{k>\rho} d^3k\, \vec{k}\, a^*_\lambda(\vec{k})\, a_\lambda(\vec{k}) \,, \tag{1.13}$$

(as weak integrals over the dense domain D_ρ of states having a finite number of photons with energy $> \rho$; then H_ρ, $\vec{P}_\rho$ are essentially selfadjoint (e.s.a.) on D_ρ, affiliated to $\overline{\pi(\mathcal{A}^{as}_\rho)} = \pi(\mathcal{A}^{as}_\rho)''$, the bar denoting the weak closure).

Clearly, since $\exp i\,Ht$ and $\exp i\,tH_\rho$ define the same automorphism on $\overline{\pi(\mathcal{A}^{as}_\rho)}$, to which $\exp i\,H_\rho t$ belongs,

$$[\exp i\,tH, \exp i\,sH_\rho] = 0 \tag{1.14}$$

on $\mathcal{H}_\pi$ (the Hilbert space of the representation π). We then define

$$V_\rho(t) \equiv \exp i\,tH \,\exp(-i\,tH_\rho) \,, \tag{1.15}$$

which is a unitary group in $\mathcal{H}_\pi$ affiliated to $\overline{\pi(\mathcal{A}^{as}_\rho)}'$. Since by eq. (1.14) H, H_ρ have a common spectral resolution and therefore a common set of analytic vectors, the generator H^c_ρ of $V_\rho(t)$ is e.s.a. on $D_H \cap D_{H_\rho}$ and

$$H^c_\rho = H - H_\rho \tag{1.16}$$

on that domain. Clearly H^c_ρ is affiliated to $\overline{\pi(\mathcal{A}^{as}_\rho)}'$.

Furthermore since quasi equivalence implies isomorphism of the corresponding Von Neumann algebras (Ref. 24, Sect. 5.3) and the Fock representation defines a type I_∞ factor (i.e. a Von Neumann algebra isomorphic to the algebra of all bounded operators in an infinite-dimensional Hilberts space), by Theor. 1.2 $\pi(\mathcal{A}^{as}_\rho)$ is a factor of type I_∞ and therefore (see e.g. Ref. 24, App. A, A.36) $\mathcal{H}_\pi$ can be written as the tensor product of a space $\mathcal{H}_\rho$ in which $\overline{\pi(\mathcal{A}^{as}_\rho)}$ acts and a space $\mathcal{H}^c_\rho$ in which $\overline{\pi(\mathcal{A}^{as}_\rho)}'$ acts.

The equation $H = H_\rho + H^c_\rho$ then implies: i) $H^c_\rho \geq 0$, since $H \geq 0$ and spectrum H_ρ goes down to zero in π_F; and therefore ii) $H_\rho \leq H$. The same argument applied to $\tilde{H} \equiv H + \vec{\lambda}\cdot\vec{P}$, $|\vec{\lambda}| \leq 1$, implies that the relativistic spectrum condition holds for both $(H_\rho, \vec{P}_\rho)$ and $(H^c_\rho, \vec{P}^c_\rho)$.

We have now to discuss the limit $\rho \to 0$ of the above decomposition (1.16); we start by proving that

$$\operatorname*{s-lim}_{\rho\to 0} \exp i\, tH_\rho = U(t) \tag{1.17}$$

exists and defines a unitary operator $U(t)$ whose generator yields H^{ph}. In fact, $H_\rho \leq H$, $[H, H_\rho] = 0$ imply $H^2_\rho \leq H^2$ and $D_H \subset D_{H_\rho}$; furthermore $[H_\rho, H_{\rho'}] = 0$ and, for $\rho > \rho'$, $0 < H_{\rho'} - H_\rho < H$, which implies $(H_{\rho'} - H_\rho)^2 \leq H^2$. Then, for any $\Psi \in D_H$

$$||(e^{itH_\rho} - e^{itH_{\rho'}})\Psi||^2 = -\int_0^t ds \frac{d}{ds} (<\Psi, e^{-is(H_\rho - H_{\rho'})}\Psi> + <\rho \leftrightarrow \rho'>)$$

$$= i\int_0^t ds <\Psi, (H_\rho - H_{\rho'})\, e^{-is(H_\rho - H_{\rho'})}\Psi> + <\rho \leftrightarrow \rho'>$$

$$\leq 2|t| <\Psi, (H_{\rho'} - H_\rho)\Psi>. \tag{1.18}$$

Since $h_\rho(\Psi) \equiv <\Psi, H_\rho\Psi>$ is a bounded ($H_\rho < H$) decreasing ($H_{\rho'} < H_\rho$ for $\rho' > \rho$) function of ρ, it has a limit when $\rho \to 0$ and the r.h.s. of eq. (1.18) vanishes, when $\rho, \rho' \to 0$. By a similar argument, one proves that the limit $U(t)$ is strongly continuous in t and that $g_\rho(t) \equiv <\Psi, \exp iH_\rho t\Psi>$ converge, as $\rho \to 0$, as distributions in t (bounded by $||\Psi||^2$). Thus, H^{ph} is well defined and

$$H^{ph} \geq 0 \quad , \tag{1.19}$$

(since the Fourier transforms of g_ρ have positive supports, the support of the limit is positive). Similarly, one proves that s-lim $V_\rho(t)$ exists as a strongly continuous unitary group $V(t)$ and the generator is positive

$$H^c \geq 0 \quad . \tag{1.20}$$

Finally

$$\exp iHt = U_\rho V_\rho \overset{s}{\to} U(t)\, V(t)$$

and $[U(t), V(t)] = 0$, so that on D_H

$$H = H^{ph} + H^c . \tag{1.21}$$

In a similar way one constructs $H^{ph} + \vec{\lambda}\cdot\vec{P}^{ph}$, and $H^c + \vec{\lambda}\cdot\vec{P}^c$, $|\vec{\lambda}| \leq 1$; the spectral conditions follow from those of the infrared cutoffed operators (H_ρ, P_ρ), as for $\vec{\lambda} = 0$.

Remark. The above theorem provides an identification of the asymptotic dynamics of the photons $(H^{ph}, \vec{P}^{ph})$ and of the charges $(H^c, \vec{P}^c)$ and therefore a solution of one of the basic problems of scattering theory (as clarified by

Dollard[10] infrared divergences in Coulomb scattering are due to a wrong identification of the asymptotic dynamics).

The above theorem also shows that the scattering representations of $\mathfrak{A}^{as}$ are labelled by the spectrum of P^c and one has a clear separation of the charge quantum numbers and the asymptotic photon description*. In particular, one may discuss the properties of the scattering representations of $\mathfrak{A}^{as}$ in terms of the charged particle quantum numbers.

Before closing this paragraph, we mention a characterization of a class of scattering representations, whose relevance for the infrared problem in QED has been discussed in the previous paragraphs and it will be strongly supported by the analysis in the local formulation of QED in Sects. 2.2, 2.3.

<u>THEOREM 1.4</u>[20] Let π be an irreducible† scattering representation and θ a cyclic vector of π in the domain of H with the property that

III) the difference between the quantum and the classical energy density of the e.m. radiation associated to θ goes to zero when $k \to 0$, as k^{ε}, $\varepsilon > 0$, after angle integration (<u>classical correspondence "principle"</u>).

Then, π is a generalized <u>coherent state representation</u>

$$\pi(F^{as}_{\mu\nu}) = \pi_F(F^{as}_{\mu\nu}) + \langle \theta , F^{as}_{\mu\nu}\, \theta \rangle \tag{1.22}$$

and therefore it is completely characterized by the <u>classical free field</u>

$$f^{\theta}_{\mu\nu}(x) \equiv \langle \theta, F^{as}_{\mu\nu}(x)\, \theta \rangle . \tag{1.23}$$

It is convenient to extract the $\delta(k^2)$ in the Fourier transform $\tilde{f}^{\theta}_{\mu\nu}(k)$ of $f^{\theta}_{\mu\nu}$

$$\tilde{f}^{\theta}_{\mu\nu}(k) = \hat{f}^{\theta}_{\mu\nu}(k)\delta(k^2). \tag{1.24}$$

Then, the condition of finite energy implies[8]

$$\hat{f}^{\theta}_{\mu\nu}(k) \in L^2(d^3k/|\vec{k}|^2) . \tag{1.25}$$

Two representations π_1, π_2 are equivalent iff

$$\hat{f}^{(1)}_{\mu\nu}(k) - \hat{f}^{(2)}_{\mu\nu}(k) \in L^2(d^3k/|\vec{k}|^3) . \tag{1.26}$$

---

* The strategy of extracting the associated asymptotic e.m. radiation in the description of a charged state has been taken by Stapp as the basis of his analysis of the infrared problem[25].

† For a reducible π, if (almost) all its (factor) subrepresentations π_i (obtained by central decomposition) contain a cyclic vector θ_i satisfying III), then (almost) all π_i are coherent state representations[20].

1.4 *A solution of the infraparticle problem*

As we shall see, the above framework also allows a solution of the so-called infraparticle problem. This consists in the dissolution of the singular spectrum of P^2: as a consequence of the interaction of the charged (massive) fields with the massless photon field, their two point function does not longer have a sharp one particle contribution or a $\delta(p^2+m^2)$ in its Fourier transform. The evidence comes from the (2-dimensional) Schroer model[26] described by the field equations

$$\Box \phi(x) = 0 ,$$

$$(\gamma^\mu \partial_\mu + M)\, \psi(x) = ig\!:\partial_\mu \phi(x)\, \gamma^\mu \psi(x)\!:$$

$$\equiv ig \lim_{y\to x} [\gamma^\mu \partial_\mu \phi(x)\psi(y) - g\, \gamma^\mu \partial_\mu \Delta^+(x-y)\psi(y)]. \tag{1.27}$$

(The model tries to mimic the QED case with the correspondence $A_\mu \to \partial_\mu \phi$). For the two point function of the charged field one has

$$< \psi(x)\bar{\psi}(y) >_o = -i \int_{-\infty}^{\infty} \rho(\mu)\, S^+(x-y,|\mu|)\, d\mu , \tag{1.28}$$

where the spectral measure $\rho(\mu)$ no longer has a $\delta(\mu-M)$ contribution, but rather behaves like

$$\Gamma(g^2/2\pi)^{-1}\, \theta(\mu-M)\, (\mu-M)^{(g^2/2\pi-1)} \underset{g^2\to 0}{\to} \delta(\mu-M) , \tag{1.29}$$

in the neighbourhood of M. Correspondingly, the fermion propagator $S_F(p)$ behaves like $(p^2-M^2)^{g^2/2\pi-1}$, rather than having a pole. In conclusion, there is no sharp hyperboloid for $g^2 \neq 0$, no eigenstate of P^2 and no one particle state with definite mass (infraparticle phenomenon).

The same evidence comes from the Bloch-Norsieck model with Dirac equation (in four dimenions[27,28])

$$\Box A_\mu = - e\, \bar{\psi}\, u_\mu\, \psi, \tag{1.30}$$

$$(\gamma^\mu \partial_\mu + M)\psi(x) = ie\, A_\mu(x)\gamma^\mu\, \psi(x) ,$$

where u_μ is a fixed four vector. The fermion propagator is

$$G(p) = \left|1 - \frac{u\cdot p}{m}\right|^{\beta} \frac{1}{m-u\cdot p} , \tag{1.31}$$

where

$$\beta = - \frac{\alpha}{2\pi}\, (3 - d_\ell) \tag{1.32}$$

and d_ℓ is the gauge parameter occurring in the photon propagator

$$D_{\mu\nu}(k) = - (g_{\mu\nu} - \frac{k_\mu k_\nu}{k^2})\, \frac{1}{k^2} - \frac{d_\ell}{k^2}\, \frac{k_\mu k_\nu}{k^2} .$$

Finally there are indications from QED calculations[5,11,2] that the fermion two point function behaviour near the mass shell is given by

$$\frac{\not p+m}{\Gamma(1-\beta)} \; \frac{1}{2m^2} \, p_\mu \frac{\partial}{\partial p_\mu} \left[\frac{\theta(p_0)\theta(p^2-m^2)}{(p^2-m^2)^\beta}\right] . \tag{1.33}$$

The "mass-shell" is somewhat smeared out and there is no sharp relation between the energy and momentum of the electron, which can be justified as a modification due to the soft photon emission or absorption. (Similarly the pole in the propagator is turned into a branch point). The phenomenon seems to disappear in the Yennie gauge[6] where $\beta = 0$, but unfortunately this gauge is plagued by dipole ghosts and the identification of the physical spectrum is not obvious. That the phenomenon is a real physical phenomenon can be seen in the Coulomb gauge[5], where no unphysical degree of freedom appears and the mass shell singularity of the propagator is that corresponding to

$$\beta = \frac{\alpha}{\pi}\left[\frac{1}{v}\ln\frac{1+v}{1-v} - 2\right] , \quad v \equiv |\vec{p}/p_o| . \tag{1.34}$$

The infraparticle phenomenon shows that the infrared problem does not only involve the construction of the asymptotic limit of the fields describing massless particles[18,19], but, even more seriously, the asymptotic limit of the charged fields, since one of the basic assumptions of the Haag-Ruelle (or LSZ) theory, namely the existence of charged one-particle states with definite mass, is not valid.

The above approach provides a strategy for solving the infraparticle problem. The idea is that the charged particle mass is related to the asymptotic dynamics of the charges and therefore it is the <u>spectrum</u> of P^c which is relevant. Charged one particle states then correspond to sharp (isolated) hyperboloids in the spectrum $\sigma(P^c)$

$$(P^c)^2 = m^2 .$$

Charged particle singularities can then be seen, <u>once</u> the energy of the asymptotic e.m. radiation has been subtracted out. This provides a clear cut definition of charged particle mass and it relies on the characterization of charged states in terms of representations of the asymptotic e.m. field algebra, rather than of the interacting gauge invariant algebra. This makes possible an approach to scattering theory along the following lines[20]. Let $\Psi_{(m)}$ be a charged one-particle state corresponding to $(P^c_{as})^2 = m^2$ and ψ a field operator which interpolates between the vacuum and $\Psi_{(m)}$

$$< \Psi_{(m)}, \psi \, \Psi_o > \neq 0 .$$

One defines

$$\psi(x) = \exp i\,(t\,H^c_{as} - \vec{x}\cdot\vec{P}^c_{as})\psi \exp[-i(t\,H^c_{as} - \vec{x}\cdot\vec{P}^c_{as})] ,$$

$$\psi_f(t) = (2\pi)^{-3/2}\int d^4p\; \tilde\psi(\vec{p},p_o)\; e^{-it\sqrt{\vec{p}^2+m^2}}\; \tilde f(\vec{p},p_o) ,$$

$$\text{supp}\,\tilde f \cap \{p^2 = m^2\} \neq \emptyset .$$

Then the asymptotic limit, (as=out),

$$s - \lim_{t\to+\infty} e^{iHt} e^{-it H^{ph}_{out}} \psi_f(t)\Psi_o$$

exists and defines a vector $\Psi_{out}(f) \in H^{(1)}$, the charged one-particle subspace ($(p^c)^2 = m^2$). Further work is needed for the asymptotic limit of a product of charged fields.

It is instructive to work out the above ideas in the Schroer model, the Bloch-Nordsieck model and in the BPF model[30].

II. BREAKING OF THE LORENTZ GROUP: GENERAL PROPERTIES OF CHARGED STATES

2.1 *Breaking of the Lorentz group and its physical meaning*

It is now possible to clarify the relation between non-Fock coherent representations and the Lorentz invariance, an issue on which the different approaches to the infrared problem are in contrast. Kibble[5] uses a non-separable (infinite tensor product) space, in which all coherent representations are allowed and in this big space one has Lorentz covariance. Kulish and Fadde'ev[9] use a separable space and have Lorentz covariance in the Gupta-Bleuler space H, which contains unphysical states, however the fulfilment of the Gauss' law constraint is obtained via the introduction of an external vector c_μ which seems to destroy the Lorentz invariance. Zwanziger[11] introduces an external classical current to satisfy the Gupta-Bleuler (GB) condition.

In the framework discussed in the previous sections, we consider the one charged particle space $H^{(1)}$, corresponding to $(p^c)^2 = m^2$. $H^{(1)}$ can be decomposed as a direct integral over the spectrum of P^c

$$H^{(1)} = \int_{p^2=m^2} d\mu(p)\, H^{(1)}_p \tag{2.1}$$

Guided by the physical requirement that besides momentum one particle charged states are labelled by a countable set of quantum numbers we assume that $H^{(1)}_p$ is a separable space, $\forall p$.

<u>Lemma 1</u> If the Lorentz group is implementable in $H^{(1)}$ then

$$U(\Lambda):\ H^{(1)}_p \to H^{(1)}_{\Lambda p}. \tag{2.2}$$

<u>Proof of Lemma 1</u> Let P_μ denote the generator of the space time translations, then by Poincaré group law

$$U(\Lambda)\, P_\mu\, U(\Lambda)^{-1} = \Lambda^\nu_\mu\, P_\nu .$$

By construction, P^{ph} is covariant under $U(\Lambda)$ and therefore so is also $P^c = P - P^{ph}$ and (2.2) follows.

In particular the little group L_p of p leaves H^1_p invariant and we shall restrict our attention to representations U(Λ) such that L_p is represented by weakly measurable operators in $H^{(1)}_p$, i.e. the matrix elements of L_p in H^1_p are measurable functions of the Lorentz parameters. This technical condition can be hardly dispensed with for a reasonable physical interpretation of the Lorentz covariance.

THEOREM 2.1[21] (Breaking of the Lorentz group). If $H^{(1)}$ contains (in the decomposition (2.1)) a non-Fock coherent representation $H^{(1)}_{p,i}$, then the Lorentz group is not implementable in $H^{(1)}$.

Lemma 2 Let $f_{\mu\nu}(p,)$, $k^2 = 0$, $p^2 = m^2$, denote a family of functions characterizing coherent states (eq. (1.23)(1.25)), then if $f_{\mu\nu}(p,k)$ is Lorentz covariant, namely

$$f_{\mu\nu}(p,k) = \Lambda^{\rho}_{\mu}\Lambda^{\sigma}_{\nu} f_{\rho\sigma}(\Lambda^{-1}p, \Lambda^{-1}k), \tag{2.3}$$

one has $f_{\mu\nu}(p,k) = 0$.

Proof of Lemma 2 Eq. (2.3) implies that $f_{\mu\nu}$ can be written in the following form

$$f_{\mu\nu}(p,k) = (p_\mu k_\nu - k_\mu p_\nu)\, g(p\cdot k) + \varepsilon_{\mu\nu\rho\sigma}\, p^\rho k^\sigma h(p\cdot k), \tag{2.4}$$

Then, the first free Maxwell equations (1.4) imply

$$k_\nu\; k\cdot p\; g(k\cdot p) = 0$$

and, since by the finite energy condition $f_{\mu\nu} \in L^2(d^3k/|k|^2)$, one has $g = 0$. Similarly the second Maxwell equations (1.4) yield $h = 0$.

Lemma 3 If $f_{\mu\nu}(\bar{p},k)$ is covariant under the little group $L_{\bar{p}}$ of $\bar{p}$, then the equation

$$f_{\mu\nu}(\Lambda\bar{p}, \Lambda k) = \Lambda^{\rho}_{\mu}\, \Lambda^{\sigma}_{\nu} f_{\rho\sigma}(\bar{p},k) \tag{2.5}$$

defines a function covariant under the Lorentz group.

Proof of Lemma 3 One easily checks that if p can be reached from $\bar{p}$ by Λ_1 and Λ_2, the definitions of $f_{\mu\nu}(p,k)$ via eq. (2.5) coincide, i.e. eq. (2.5) provides a good definition. The Lorentz covariance is immediate. (Lemma 3 also provides a simple proof of eq. (2.4); for example take $p = \bar{p} = (m,\vec{0})$, then by rotation covariance one has $f_{oi}(\bar{p},k) = k_i\, f(|\vec{k}|)$. Similarly one proceeds for $f_{ij}(\bar{p},k)$).

Lemma 4 Let the Lorentz group be implementable in $H^{(1)}$ and $\theta_{\bar{p}} \in H^{(1)}_{\bar{p}}$ a state with coherent function $f_{\mu\nu}(\bar{p},k)$, then the representation defined by $\theta_{\bar{p}}$ contains a vector with coherent function covariant under $L_{\bar{p}}$.

Proof of Lemma 4 Average $f_{\mu\nu}(\bar{p},k)$ over $L_{\bar{p}}$ (V ≡ Volume of $L_{\bar{p}}$)

$$g_{\mu\nu}(\bar{p},k) \equiv f_{\mu\nu}(\bar{p},k) + V^{-1}\int dg\ [-f_{\mu\nu}(\bar{p},k) + (\Lambda(g)\Lambda(g)f)_{\mu\nu}(\bar{p},\Lambda(g)^{-1}k)] \tag{2.6}$$

We have to show that i) the second term on the r.h.s. of eq. (2.6) is well defined and belongs to $L^2(d^3k/k^3)$, and ii) $g_{\mu\nu}(\bar{p},k) = (\Lambda\Lambda g)_{\mu\nu}(\bar{p},\Lambda^{-1}k)$. Given i), ii) follows by simple computation. For i) we show that the integrand on the r.h.s. is strongly continuous in $g \in L_{\bar{p}}$, in $L^2(d^3k/k^3)$. In fact, $U(\Lambda)$, $\Lambda \in L_{\bar{p}}$, leaves $H_{\bar{p}}^{(1)}$ invariant by Lemma 1, is strongly continuous (since it is weakly measurable and $H_p^{(1)}$ is separable) and then $U(\Lambda)\theta_p$ is norm continuous; then the corresponding states are continuous in the norm $\|\ \|_{\mathcal{A}}$ of the states on $\mathcal{A}^{as}$

$$\|\phi_1 - \phi_2\|_{\mathcal{A}} = \sup_{\|A\|=1} |\phi_1(A) - \phi_2(A)| = \sup_{\|A\|=1} |\langle\Psi_1, A\Psi_1\rangle - \langle\Psi_2, A\Psi_2\rangle|$$

$$\leq 2\ \|\Psi_1 - \Psi_2\|\ .$$

Now, given ϕ_1, ϕ_2 with coherent functions f_1, f_2 a simple computation gives

$$\|\phi_1 - \phi_2\|_{\mathcal{A}} \geq C\ \|f_1 - f_2\|_{L^2}$$

Then $U(\Lambda)$ defines a continuous mapping of the corresponding coherent functions.

The integral on the r.h.s. of eq. (2.6) is over a compact set (and the integrand is norm continuous); then it exists as a Riemann integral in $L^2(d^3k/k^3)$ and it defines a function of $L^2(d^3k/k^3)$. Hence $g_{\mu\nu}(\bar{p},k)$ belongs to the equivalence class of $f_{\mu\nu}(\bar{p},k)$.

Proof of Theor. 2.1 If the Lorentz group is implementable and $\theta_p \in H_p^{(1)}$ is a coherent state, then by Lemma 4 the representation defined by θ_p contains a state with coherent function covariant under L_p, which then defines by Lemma 3 a Lorentz covariant function, which vanishes by Lemma 2. Hence θ_p defines a Fock representation.

Remark 1 The transformation properties of (generalized) coherent states have previously been considered by Roepstorff[8] who proved the non-implementability of the Lorentz group in any non-Fock irreducible (i.e. fixed p) coherent representation. The above Theorem extends the proof to the whole one-particle charged sector $H^{(1)}$, which contains all the momenta[21].

Remark 2 The evidence for the non-Fock character of the charged coherent states comes from simple infrared models, rigorous non-relativistic models, semiclassical limit etc. (all yield the BN singular behaviour of the type $p/p\cdot k$) and as discussed in Sect. 1.1. all the conventional wisdom relies on this property. The above theorem abstracts general structure properties with respect to the explicit computation based on the standard BN coherent functions[20].

The general relevant point is that to each charged state one associates a free e.m. radiation i.e. a free field coherent function. The Lienard-

Wiechert (LW) potential describes a covariant but not free e.m. field. The LW potentials are the Lorentz transform of the Coulomb or Gauss field, are labelled by the frame velocity $\vec{v}$ and yield a non-trivial electric flux at spacelike infinity in any direction $\vec{n}$, $\Phi(\vec{n})$:

$$k^i f_{oi}^{LW}(\vec{v}=0,k) = k^i \delta(k_o)\, k_i/|\vec{k}|^2 = \delta(k_o),$$

$$n^i f_{oi}^{LW} = \delta(k_o)\, \frac{\vec{n}\cdot\vec{k}}{|\vec{k}|^2} \quad .$$

The flux $\Phi(\vec{n})$ is different in different Lorentz frames and it defines a superselection rule since it commutes with the algebra of local observables. Thus, one has an uncountable number of inequivalent representations, labelled by $\Phi(\vec{n})$ or equivalently by $\vec{v}$, $H_{\vec{v}}$. One may then argue[19] that the Lorentz group is broken in the sense that it does not leave stable each irreducible representation of the algebra of local observables (stability would be obtained by considering reducible representations $\int^{\oplus} dv\, H_{\vec{v}}$). It should be stressed that our p^c is very different from the LW velocity parameter $\vec{v}$, which is superselected, whereas p^c is not, as we shall see below.

<u>Remark 3</u> The breaking of the Lorentz group can be also understood on the basis of semiclassical considerations. In fact, at the semiclassical limit the e.m. radiation associated to an electron which is accelerated from $\vec{p}=0$ to $\vec{p}$ is of the form

$$F_{oi}(k) \sim q\, \frac{k_o p_i}{p\cdot k}\, (\delta_{ij} - \frac{k_i k_j}{|\vec{k}|^2})\, \delta(k^2) \qquad (2.7)$$

whereas that associated to an electron boosted from $\vec{p}=0$ to $\vec{p}$ through a Lorentz transformation vanishes.

On this basis one can also see that there is no Lorentz breaking for a state describing a pair (+q,−q) of charged particles with momenta p_1, p_2. The associated e.m. radiation is of the form

$$F_{\mu\nu}(k) \sim q\, [k_\mu (\frac{p_1}{p_1\cdot k} - \frac{p_2}{p_2\cdot k})_\nu - (\mu\leftrightarrow\nu)]\, \delta(k^2)\, \varepsilon(k)$$

and it is a <u>covariant</u> function of the <u>two</u> momenta. The separation of this radiation into two functions involving only one momentum (p_1 and p_2 respectively) is incompatible with the Maxwell equations and the Lorentz covariance. (The naive Lorentz covariant separation $q[(k_\mu p_\nu - k_\nu p_\mu)/p\cdot k]\delta(k^2)\varepsilon(k)$ violates the Maxwell equations and corresponds to Zwanziger introduction of a classical external current). It is clear that the non-covariance is related to the tensor character of the Maxwell equations.

The above argument also shows that the breaking of the Lorentz group arises whenever the description of a charged one particle state does not involve the oppositely charged particle (−q), from which it has been separated in the preparation process. The <u>removal of charge −q "behind the moon"</u>, in the sense that one neglects its associated radiation, necessarily leads to a breaking of the Lorentz group.

Remark 4 The large distance behaviour of the electric field cannot be changed by a scattering or in general by a localized process, since by causality spacelike separated regions cannot influence each other; it is however changed by a Lorentz transformation which is a non-local operation (and in fact affects the electric flux $\Phi(\vec{n})$). Now, given two (one particle) charged states with different momenta p_1, p_2 (defining representations of the algebra $\mathcal{A}$ of local observables with non-trivial Gauss flux at spacelike infinity), they cannot be related by a Lorentz transformation since the latter changes the electric flux at infinity, whereas if the momentum p^c is not superselected, there are local observable operators (that is physically realizable operations), which lead from $|p_1\rangle$ to $|p_2\rangle$ and by causality cannot change the electric flux at infinity. Thus the breaking of the Lorentz group is necessary to avoid the superselection of the charge momentum p^c.

Remark 5 A final remark about the possible physical consequences of the Lorentz breaking may be useful. To this purpose consider the measurement of the e.m. radiation at large distances associated to an electron of momentum p coming out of an "accelerator". Changing the electron momentum $p \to p'$ either by raising the accelerator potential or by Lorentz boosting the entire system gives rise to electron states of momentum p' with different associated electromagnetic radiation, i.e. the description of a charged state in terms of one particle observables and its associated e.m. radiation, cannot be Lorentz covariant. To detect a large number of soft photons one must go to large distances so that the effects are small for one electron. One expects however sizeable effects, due to the non-Fock coherent character of the state, when macroscopic currents ($\sim 10^{20}$ electrons/sec) are present. Large effects would also be expected for a single particle if $\alpha \sim 10^{-1}$ (rather than 1/137), as in the QCD case.

2.2 *Infrared problem and local formulation of QED. Bloch-Nordsieck infrared singularities*

The previous analysis raises some problem with respect to the conventional local, covariant (renormalizable) formulation of QED. In particular: i) there is an apparent conflict between the Lorentz covariance of the Wightman functions, in the local GB formulation (at each perturbative order), and the Lorentz breaking in the physical charged sectors, ii) the characterization of the charged states in terms of non-convariant coherent factors leads to the Lorentz breaking, whereas covariant coherent factors, which look natural in the GB formulation, lead to serious difficulties with the GB condition (Zwanziger's paradox). It is then worthwhile to approach the problem from first principles in order to get: 1) a non-perturbative justification of the BN ansatz, 2) the compatibility between the GB condition and the non-Fock infrared behaviour of the charged states, 3) a non-perturbative construction of charged states in the GB formulation. Our analysis[31] will be based on the following general structure:

I. A local field algebra $\mathcal{F}$ generated by the Wightman fields A_μ, ψ, $\bar{\psi}$ and a cyclic vacuum Ψ_o.

II. A gauge group G acting on $\mathcal{F}$, leaving the observable algebra $\mathcal{A}$, invariant and with the property that the gauge transformations of the first kind are generated by the electric current j_μ (the source of A_μ):

$$\Box A_\mu = j_\mu , \qquad \partial^\mu j_\mu = 0 \quad , \tag{2.8}$$

$$\lim_{R\to\infty} [Q_R^{el}, \psi] = q\,\psi \quad , \qquad \lim_{R\to\infty} [Q_R^{el}, A_\mu] = 0 \quad , \tag{2.9}$$

$$Q_R^{el} \equiv j_o(f_R\alpha) = \int d^4x f_R(\vec{x})\alpha(x_o) \ ,$$

$$(f_R(\vec{x}) = f(|\vec{x}|/R), \quad f(x) = 1, \ |x| < 1, \ f(x) = 0, \ |x| > 1+\varepsilon,$$

$$f \in D(R^3), \ \alpha \in D(R^1), \quad \alpha(x_o) = \alpha(-x_o), \quad \int \alpha(x_o)dx_o = 1) \ .$$

The gauge transformations of the second kind are generated by

$$Q^\Lambda \equiv \int d^3x \ \Lambda(\vec{x},x_o) \overleftrightarrow{\partial_o} \partial^\sigma A_\sigma(\vec{x},x_o) \ ,$$

with $\Box\Lambda = 0$, $\Lambda(\vec{x},x_o) \in S(R^3)$:

$$\begin{aligned} [Q^\Lambda, \psi(x)] &= q\ \Lambda(x)\psi(x) \ , \\ [Q^\Lambda, A_\mu(x)] &= \partial_\mu \Lambda(x) \ . \end{aligned} \tag{2.10}$$

Proposition 1 Eq. (2.8) implies that ∂A is a free field and eq. (2.10) yields

$$[\partial A^-(x), \psi(y)] = q\ \Delta^-(x-y)\psi(y) \tag{2.11}$$

(q is the renormalized charge).

Proof It follows from the free propagation of Λ and from $\Delta = \Delta^+ + \Delta^-$ being the Green's function for the wave equation.

Proposition 2 Eq. (2.11) implies that for any local states Φ, Ψ ($\Phi=\psi(g)\Psi_o$, etc.)

$$\lim_{R\to\infty} <\Phi,\ \partial_o\partial A(f_R\alpha)\Psi> = qf(0)\ <\Phi,\Psi> \ . \tag{2.12}$$

Proof The Gupta-Bleuler condition for the vacuum, $\partial A^-\Psi_o = 0$, (also implied by the spectral condition) gives

$$<\Phi,\ \partial_o\partial A(x)\Psi> = <[\partial_o\partial A^-(x), \psi(g)]\Psi_o, \Psi> + <\Phi, [\partial_o\partial A^-(x),\ \psi(f)]\Psi_o>$$

$$= \int d^4y\ q\ \{<\psi(\Delta^-(x-y)g(y))\Psi_o, \Psi> + <\Phi, \psi(\Delta^-(x-y)f(y))\Psi_o>.$$

Then eq. (2.12) easily follows.

We now exploit Prop. 2 under the following assumption

III The asymptotic limits A_μ^{as}, as = in/out, exist as free fields satisfying the CCR's and covariant under the Poincaré transformations (on a dense domain stable under the time translations U(t) and the Hamiltonian).

Proposition 3[29] If III holds, then

$$\partial A = (\partial A)^{as} = \partial^\mu A_\mu^{as} .$$

Proposition 4 In the above framework, $\forall\Phi$, Ψ, local states with

$$< \Phi, A_\mu^{as}(k)\Psi > \equiv - F_\mu(k)\delta(k^2)\,\varepsilon(k) , \tag{2.13}$$

we have

$$\lim_{R\to\infty} \tfrac{1}{2}[(k^+/R)^\mu F_\mu(k^+/R) + (k^-/R)^\mu F_\mu(k^-/R)] =$$
$$= (2\pi)^{-3/2} q < \Phi,\Psi> , \quad (k^\pm)^\mu \equiv (\pm|\vec{k}|, \vec{k}) , \tag{2.14}$$

where the limit has to be understood in the distributional sense. Eq. (2.14) implies

$$F_\mu(k) \notin L^2(d^3k/|\vec{k}|) .$$

Similarly, putting

$$< \Phi, F_{\mu\nu}^{as}\Psi > \equiv iF_{\mu\nu}(k)\delta(k^2)\,\varepsilon(k) ,$$

$$\lim_{R\to\infty} \tfrac{1}{2}\, k^i/k_o[F_{oi}(k^+/R) + F_{oi}(k^-/R)] = (2\pi)^{-3/2} q <\Phi,\Psi> ,$$

and

$$F_{oi}(k) \notin L^2(d^3k/|\vec{k}|^3) ,$$

(Bloch-Nordsieck infrared singularities).

Proof By Prop. 3 and Prop. 2

$$\lim_R <\partial_o \partial A(f_R\alpha)> = \lim_R (2\pi)^3 \tfrac{1}{2} \int d^3k\, \tilde{f}(\vec{k})\, \alpha(\frac{|\vec{k}|}{R})$$

$$[(k^+/R)\cdot F(k^+/R) + (k^-/R)\cdot F(k^-/R)] = (2\pi)^{-3/2} q \int d^3k\, \tilde{f}(k) <\Phi,\Psi>$$

Proposition 5 Let Ψ_u be a local state covariant in u_μ

$$U(\Lambda)\,\Psi_u = \Psi_{\Lambda u} ,$$

then the corresponding function, eq. (2.13), has the form

$$F_\mu(k,u) = q\,\frac{u_\mu}{u\cdot k}\,\rho(k) + \text{gauge terms},$$

with $\frac{1}{2}(\rho(k^+) + \rho(k^-)) \to (2\pi)^{-3/2}$, as $\vec{k} \to 0$, (<u>Bloch-Nordsieck covariant factors</u>).

The above states Ψ_u can be easily constructed. E.g., in the spinless case, choose f as a rotationally invariant test function and define $\Psi_{\bar{u}=(1,0,0,0)} \equiv \psi(f)\Psi_o$. Then $\Psi_{u=\Lambda\bar{u}} = U(\Lambda)\psi(f)\Psi_o = U(\Lambda)\Psi_{\bar{u}}$.

Alternatively, assume the decomposition $P = P^{ph} + P^c$, with

$$P^{ph}_\mu = \int d^3k\, k_\mu\, a^+_\nu(k)\, a^\nu(k).$$

Decompose the local states over the spectrum of P^c

$$\Psi_{(m)} = \int_{p^2=m^2} d^3p\, \Psi_p \qquad .$$

Then, the functions corresponding to Ψ_p have the form

$$F_\mu(k,p) = q\, \frac{p_\mu}{p\cdot k}\, \rho(k) + \text{gauge terms}$$

with $\frac{1}{2}(\rho(k^+) + \rho(k^-)) \to (2\pi)^{-3/2}$, as $\vec{k} \to 0$ [31].

In conclusion, we have obtained a relation between locality and the infrared behaviour of the charged states; in particular, for local states we got covariant BN-type expectation values. One should not be surprised that non-trivial information is obtained by locality. This can be clearly seen in the soluble BN, BPF models, where

$$A^{in/out}_\mu(\vec{k},t) = A^{free}_\mu(\vec{k},t) + F^{in/out}_\mu(\vec{k},t) ,$$

with

$$F^{in/out}_\mu(k) \sim q(p^{in/out}_\mu/p\cdot k)\delta(k^2)\varepsilon(k) \qquad \text{(BN)}$$

$$\sim q(p_\mu/\omega)\delta(k^2)\varepsilon(k) \qquad \text{(BPF)}$$

$$A^{free}_\mu(\vec{k},0) = A_\mu(\vec{k},0),\quad \dot{A}^{free}_\mu(\vec{k},0) = \dot{A}_\mu(\vec{k},0).$$

The representation of $A^{in/out}_\mu$ depends on the choice of the representation for $A_\mu(\vec{k},t)$ at finite times and in general (as remarked in Sect. 1.1) an arbitrariness is involved. This is resolved in a local formulation, in which A_μ and the charged fields commute at equal times, since then the local field $A_\mu(k,t)$ in the charged sectors has the same representation as in the vacuum sector

$$< \psi(t)\Psi_o,\ A_\mu(t)\ \psi(t)\ \Psi_o > = < \psi^+(t)\psi(t)\Psi_o,\ A_\mu(t)\Psi_o > .$$

Such additional information is not available in the Coulomb gauge, since $[A_\mu(k,t),\ \psi(k',t)] \neq 0$, there.

Quite generally, by applying local charged fields to the vacuum one obtains representations of A^{as}_{μ} which exhibit the (covariant) BN infrared singularities.

2.3 *Construction of physical charged states*

A general problem of gauge theories is that, in the renormalizable gauges (the ones available by standard perturbative methods), the theory is formulated in terms of correlation functions $W(x_1,\ldots,x_n)$ of local fields and, to get physical matrix elements, one has to construct physical charged states starting from the local states, defined by the Wightman functions $W(x_1,\ldots,x_n)$.

In the QED case, given the set of local charged states one has to construct (a linear space of) physical charged states characterized by the validity of the Gauss law (see Sect. 1.2)

$$< \Phi_{phys}, (\partial F-j)\Psi_{phys} > = 0 \quad , \tag{2.15}$$

$$\lim_{R\to\infty} < \Psi_{phys}, Q^{el}_R \Psi_{phys} > = q < \Psi_{phys},\Psi_{phys} > \tag{2.16}$$

The strategy of getting the charged states from the local chargeless states (the vacuum sector), through a morphism (i.e. a mapping which preserves the algebraic relations) on the gauge invariant algebra, has been advocated by Doplicher, Haag and Roberts[15] (see also Ref. 16). The difficulty is that the Gauss' law constraint is a non-local one. In the GB formulation, the Gauss' law constraint involves a free field ($\partial A = \partial A^{in} = \partial A^{out}$) and one may try to get the physical states from the local charged states (so that eq. (2.16) is easily satisfied), through a morphism σ, which commutes with the charge. Eq. (2.15) then requires

$$0 = \sigma(\partial F-j) = \sigma(\partial\partial A) = \sigma(\partial\partial A^{as}) \ . \tag{2.17}$$

In terms of the asymptotic e.m. field algebra, the problem is equivalent to finding a morphism σ such that

$$\sigma(F^{as}_{\mu\nu}) = F^{as}_{\mu\nu} + f_{\mu\nu} \quad , \tag{2.18}$$

with

$$\partial^{\mu} f_{\mu\nu} = \partial_{\nu}\partial A \ .$$

In fact, eq. (2.18) implies eq. (2.17)

$$\sigma(-\partial\partial A^{as}) = \sigma(-\partial\partial A^{as} + \Box A^{as}) = \sigma(\partial F^{as}) = \partial F^{as} + \partial\partial A^{as} = 0$$

and conversely.

A natural technical question is whether the morphism σ can be described by an operator T. Formal solutions[32] involve exponentials of the interacting field A_{μ}, smeared with test functions which are singular on surfaces (or lines) in p- space. In the above approach T can be constructed in terms of the free field A_μ^{as}. A proper solution essentially involves the construction of limits of local states: since physical charged states cannot be local states, a topology is needed to reach them from the local states. In the GB formulation, the Wightman functions only define a vector space with an indefinite inner product. Different Hilbert topologies can be associated to the Wightman functions (Hilbert space structures (HSS)). They yield different representations of the field algebra, and are related to the infrared behaviour of the corresponding states[31,33]. In the standard positive metric QFT's, different Hilbert structures appear as inequivalent representations; in the GB formulation they can be obtained as different closures of the same local structure.

Also the strategy of constructing a set of non-local gauge invariant operators to produce physical states from the vacuum[32] implicitly involves a choice very similar to the choice of an HSS. The point is that the construction of non-local gauge invariant operators requires a choice of a set of sequences of local states, whose scalar products converge; the resulting space of states crucially depends on such a choice, since in general different sets give rise to inequivalent representations of the observable algebra (e.g. those corresponding to physical charged states with different Lienard-Wiechert potentials labelled by the superselected parameter $\vec{v}$).

We give a brief outline of the construction[31]. Given Ψ = local state, with $\langle\Psi,\Psi\rangle > 0$, let

$$H_\Psi \equiv \overline{\{\mathfrak{A}^{in}\ \Psi\}} ,$$

where the bar denotes the closure with respect to a suitable, as yet unspecified Hilbert space topology. The existence of a vector $\Phi \in H_\Psi$ satisfying the GB condition

$$k^- \cdot a^- \ \Phi = 0 ,$$

where $a_\mu^\pm(\vec{k})$ are the destruction and creation operators corresponding to A_μ^{as}, is equivalent to the existence of a Fock vacuum in H_Ψ for $\chi^- \equiv k^- \cdot a^-$. To this purpose one defines

$$d^-(\vec{k}) \equiv \chi^-(k)\ M(k)$$

and its Hilbert space adjoint $d^+(\vec{k})$, where $M(k)$ is a suitable normalization factor such that

$$[d^-(\vec{k}),\ d^+(\vec{k}')] = \delta(\vec{k}-\vec{k}') \quad .$$

Construct

$$N_d = \sum_i d^+(f_i)\ d^-(f_i) \quad ,$$

where $\{f_i\}$ denotes a complete orthonormal set of L^2 functions $f_i(\vec{k})$. By a well-known theorem[34] the Fock vacuum exists iff

$$\langle N_d \rangle_{\Psi'} < \infty , \quad \text{for one} \quad \Psi' \in H_\Psi \ .$$

One may consider metric operators η, $\eta^2 = 1$, of the form

$$\eta A_\mu^{in}(k)\eta = A_\mu^{in}(k) + 2c_\mu(k)\,\bar{c}^\nu(k)\,A_\nu^{in}(k) + B_\mu(k),\; [B_\mu(k), \mathcal{A}^{in}] = 0 ,$$

$$\bar{c}^\nu(k)\quad c_\nu(k) = -1 \quad .$$

Then, $M(k) = \sqrt{2}|k^-\cdot c(k^-)|$ and

$$\langle N_d\rangle_{\Psi = \psi(h)\Psi_o} = \sum_i ||[d^-(f_i),\ \psi(h)]\Psi_o||^2 = q\int d^3k\,\frac{1}{|\vec{k}|\ |k^-\cdot c(k^-)|^2}$$

$$\{\int d^4x d^4x'\ \bar{h}(x)h(x')\ (\psi(x)\Psi_o,\ \psi(x')\Psi_o)\ e^{-ik^-(x-x')}\} \;.$$

The expression in curly brackets is bounded and measurable in k provided the Hilbert space structure is such that the positive two-point function $(\psi(x)\Psi_o, \psi(x')\Psi_o)$ is a tempered distribution, so that all the Schwartz seminorms are k-continuous, i.e. $||e^{ikx} - e^{ik'x})\ h(x)||_{n,m} \to 0$, as $k \to k'$. The high k integration is govered by the Hamiltonian being well defined. The low k integration converges if the Hilbert structure, i.e. the metric η, is so chosen that

$$(\sqrt{|\vec{k}|}\ \ |k\cdot c(k)|)^{-1} \in L^2(|k| < \rho) \;.$$

A more explicit construction is possible[31] starting from local states which are coherent states for $k\cdot a^-$:

$$k\cdot a^-(k)\Psi = q(2\pi)^{-3/2}(2|\vec{k}|)^{1/2}\ G(k)\ \Psi \;,$$

with $G(0) = 1$, $G(k) = G(-k)$. Then the transformation T can be realized by a unitary operator U

$$U = \exp[d^+(f) - d^-(f)]\ \exp[\eta d^+(f)\eta - \eta d^-(f)\eta],$$

$$f(\vec{k}) = -q(2\pi)^{-3/2}\ (2\sqrt{|\vec{k}|}\ \ |k\cdot c|)^{-1}\ G(k) \;,$$

(U is well defined since $f \in L^2(|k| < \rho)$). The vector

$$\hat{\Psi} = U\Psi$$

then satisfies the Gauss' law and $\langle \hat{\Psi}, A_\mu^{in}(k)\hat{\Psi}\rangle$ has non-covariant BN infrared behaviour.

Remark 6 It is worthwhile to stress the general features which underlie the breaking of the Lorentz group in QED, namely that the symmetry of the local field correlations is lost in the physical space, because the transformations or morphism, needed to satisfy the Gauss' law constraint, maps the (local state) representation, in which the symmetry is unbroken, into one in which it is broken. In our opinion this mechanism may play a role in gauge QFT as a general symmetry breaking mechanism. In the abelian (QED) case the fulfilment of the Gauss' law constraint leads to the breaking of the Lorentz group; in the non-abelian case, semieuristic considerations indicate that the construction of coloured states, which satisfy the Gauss' law, leads to the breaking of the space-time translations, i.e. to states with infinite energy (confinement). Similar situations in QED_4 and QED_3 are discussed in the next section.

2.4 *Implications on confinement of massless charged particles in QED_4 and of charged particles in QED_3*

The experimental fact that all charged particles are massive raises the question of whether this can be understood on general grounds. An argument in this direction, for spin $S > 1$ has been given by Case and Gasiorowicz[35], but it crucially depends on the pathologies of higher spin relativistic equations with minimal coupling. The question has also been discussed by means of the renormalization group analysis (see e.g. Ref. 27, Ch. 9). One takes the limit of vanishing electron mass in QED_4. Under suitable approximations (leading order summation etc.), one derives the following transverse photon propagator

$$\frac{d(k^2)}{k^2}, \quad d(k^2) = (1 - \frac{\alpha(\lambda^2)}{3\pi} \log \frac{k^2}{\lambda^2})^{-1} , \tag{2.19}$$

where λ^2 is the renormalization point. Eq. (2.19) leads to the vanishing of the electric charge defined through the $k \to 0$ limit of the vertex function $<\bar{\psi}(p_1)\, A_\mu(k)\psi(p_2)>$. The result however raises some non-trivial questions: i) the $m \to 0$ limit is interchanged with the leading term summation of the perturbative expansion, ii) the infraparticle problem is neglected, in particular the infrared dressing transformation needed to get physical states from local states is not considered, iii) the above expression for $d(k^2)/k^2$ violates positivity of the two point function of $F_{\mu\nu}$ (also the $\delta(k^2)$ singularity is lacking in the $F_{\mu\nu}$ spectral measure), so that $F_{\mu\nu}$ would not be an observable operator. The same problem has recently been discussed by Gribov[36]. His analysis is based on the discussion of the amplitudes for pair creation in finite times and on a resummation of the Feynman diagrams equivalent to the renormalization group for the photon propagator. The above questions again apply. It seems therefore worthwhile to clarify whether basic properties of QFT conflict with the existence of massless charged particles.

We shall base our analysis[37] on the framework discussed previously. Under the assumption that the asymptotic photon fields exist (III). One characterizes the physical charged states as representations of $\mathcal{A}^{as}$, one decomposes $P = P^{ph} + P^{c}$ and identifies the charged particle mass through the spectrum of P^c.

<u>THEOREM 2.2</u> If charged massless particle states Ψ with infrared Bloch-Nordsieck mean values

$$<\Psi,\, F^{as}_{\mu\nu}(k)\Psi> \simeq iq \left[\frac{k_\mu p_\nu - k_\nu p_\mu}{k\cdot p} - \frac{k_\mu v_\nu - k_\nu v_\mu}{k\cdot v}\right] \delta(k^2)\, \varepsilon(k), \qquad k \to 0$$

and with finite energy exist, then the Hamiltonian cannot be bounded from below.

Proof Suppose that H_{ph} is bounded from below; then by Theorem 1.2, H_{ph} can be approximated by

$$H_\rho = \frac{1}{2}\int_{k>\rho} d^3k: E(\vec{k})^2 + B(\vec{k})^2: + C_\rho \equiv H^o_\rho + C_\rho , \quad [C_\rho, \mathfrak{A}^{as}] = 0.$$

The locally Fock property in momentum space implies

$$\text{Inf } \sigma(H^o_\rho) = 0 ,$$

so that

$$\text{Inf } \sigma(H_{ph}) = \lim_{\rho\to 0} \text{Inf } \sigma(H_\rho) = \lim_{\rho\to 0} C_\rho .$$

Let Ψ be a physical charged state with finite mean energy, then

$$C_\rho = \langle H_\rho \rangle_\Psi - \langle H^o_\rho \rangle_\Psi , \quad \lim_{\rho\to 0} \langle H_\rho \rangle = \langle H_{ph} \rangle \tag{2.20}$$

$$\langle H^o_\rho \rangle = \langle \int_{k>\rho} d^3k(\vec{E}^+\cdot\vec{E}^- + \vec{B}^+\cdot\vec{B}^-)\rangle \geq \int_{k\geq\rho} d^3k \, (|\langle\vec{E}(\vec{k})\rangle|^2 + |\langle\vec{B}(\vec{k})\rangle|^2)$$

$$|\langle E(\vec{k})\rangle|^2 = |\int dk_o \langle F_{oi}(\vec{k},k_o)\rangle|^2 \geq g(k)|\vec{p} - \vec{k}\cdot\vec{p}\,\vec{k}/|\vec{k}|\,|/(p\cdot k)^2 ,$$

with g(k) a suitable function which is strictly positive in the neighbourhood of k = 0. Then the integral on the r.h.s. of eq. (2.20) is logarithmically divergent

$$\lim_{\varepsilon\to 0} \int_0^{\pi-\varepsilon} (1+\cos\theta)^{-1} \, d\cos\theta \sim - \lim_{\varepsilon\to 0} \log \varepsilon$$

so that

$$\lim_{\rho\to 0} C_\rho = -\infty$$

and the Hamiltonian cannot be bounded from below*.

Remark The assumption about the infrared behaviour can be justified by using a local formulation, as previously. If the vacuum is the lowest energy state, then massless charged states have infinite energy, equivalently the space time translations are broken in the charged sectors. Thus, charged massless particle states cannot be physically realized (i.e. they are confined). In conclusion, the existence of asymptotic (massless) photons and of charged massless particles (taking care of the infraparticle problem) is incompatible with the spectral condition.

By a similar argument one can also prove the confinement of charged particles in QED_3. In fact, in low space dimensions, $s \leq 2$, the BN factors $v_\mu/v\cdot k$ have worse infrared singularity (with respect to $d^sk/|\vec{k}|$) than in three dimensions (s = 3). Within the same general framework discussed above we have

* For coherent states, with the same infrared singularity, the unboundedness of the Hamiltonian from below also follows from the analysis of Ref.8.

THEOREM 2.3[37] If charged particle states exist in QED_3 with BN infrared behaviour and finite mean energy, then H_{ph} is not bounded from below.

If the energy is (re)normalized so that the vacuum is the lowest energy state, then charged particle states have infinite energy (i.e. they are confined).

III. HIGGS PHENOMENON AND SYMMETRY BREAKING ORDER PARAMETER

3.1 *Symmetry breaking order parameter and particle spectrum*

As emphasized in the introductory talk by Wightman, one of the fundamental problems of gauge quantum field theories is the relation between the Higgs phenomenon, the existence of a symmetry breaking and the absence of Goldstone bosons. The conventional picture is that the Higgs phenomenon is characterized by a symmetry breaking order parameter, $< \phi > \neq 0$, and that the Goldstone theorem is evaded because the Goldstone mode is "eaten up" by the vector boson, which becomes massive[38]. The support for this picture comes from a perturbative expansion around a classical field configuration (which minimizes the potential) and clearly, since mean field approximations are known to often lead to misleading results, a non-perturbative control of the Higgs effect is needed.

The disentangling of symmetry breaking and existence of Goldstone bosons (whose strict relation is the conclusion of the Goldstone theorem) can be simply understood in the local gauges as a consequence of the local Gauss' law[39].

THEOREM 3.1[39] Let β^λ be a one parameter group of automorphisms of the local field algebra, generated by a local current j_μ, which satisfies a local Gauss' law, in the sense that there is a local antisymmetric field $F_{\mu\nu}$ such that

$$< j_\mu - \partial^\nu F_{\nu\mu} >_\Psi = 0 \quad , \tag{3.1}$$

for any physical state Ψ. Then, if β^λ is spontaneously broken, there is a $\delta(p^2)$ singularity in the Fourier transform of the correlation function $<j_\mu(x)A>_o$, A a local operator with $<\delta A>_o \neq 0$, but such singularity cannot contribute to physical matrix elements and cannot correspond to a physical massless particle.

Since gauge groups are characterized by generators (of global gauge transformations) which obey a local Gauss' law[39], the above Theorem provides a non-perturbative explanation of the absence of massless particles accompanying the gauge symmetry breaking. It remains to establish whether the Higgs effect is indeed characterized by a symmetry breaking order parameter. Recent results by Kennedy and King[40] indicate that this is the case in the Landau gauge (for which, however, the relation between cluster property and pure phases is not obvious since the space time translations are not described by unitary operators and peculiar phenomena might occur[33]). The situation is under control in the axial or temporal gauge on a lattice, where the following theorem has been proved[41].

THEOREM 3.2 In the axial gauge, the correlation functions of local fields are invariant under the group G of global gauge transformations (no gauge symmetry breaking).

Furthermore the gauge invariant (bilocal) two point function of the Higgs field has an exponential decay at large distances, i.e. it satisfies the strong cluster property.

Remark The absence of a local symmetry breaking order parameter is clearly in conflict with the conventional perturbative expansion and clearly, if $<\phi> = 0$ at a non-perturbative level, the perturbative expansion leading to $<\phi>_{pert} \neq 0$ cannot be asymptotic. Furthermore, if the symmetry is unbroken, one expects G-degenerate multiplets, which are not seen in elementary particle physics and it becomes a problem to explain the conventional mass splitting, (e.g. $m_{el} \neq m_{\nu}$, $M_Z \neq M_{\gamma}$ etc.), since the standard formulae for the mass generation depend on the order parameter $<\phi>$, (e.g. $m_{fermion} = f<\phi>$ etc.). As we shall see, the above non-perturbative result can be reconciled with the elementary particle mass spectrum by a correct analysis of the physical degrees of freedom.

3.2 *Higgs phenomenon, charge screening and symmetry breaking*

In contrast with the standard picture, we shall not base our discussion on the existence of a gauge dependent order parameter $<\phi> = \bar{\phi}$, but we shall only make reference to a distinguished orbit $\{\bar{\phi}\}$ (a gauge invariant concept). Such an orbit may be identified with the set of classical (constant) Higgs field configurations which minimize the Higgs potential. Let $\phi(x)$ and $\psi(x)$ denote fields transforming according to irreducible representations R_H and R respectively, of a compact Lie group G; let $G_{\bar{\phi}}$ denote the residual group at $\bar{\phi} \in \{\bar{\phi}\}$ and $G_{\{\bar{\phi}\}}$ the abstract group to which all the $G_{\bar{\phi}}$, $\bar{\phi} \in \{\bar{\phi}\}$, are isomorphic (residual group of $\{\bar{\phi}\}$).

THEOREM 3.3[41] Let $G_{\{\bar{\phi}\}}$ = identity, (total breaking), then there exists a linear correspondence between the fields ψ_α of R and the set of G-invariant (composite) fields $\Psi^i = P^i(\phi)\cdot\psi$, linear in ψ and polynomial in ϕ; the correspondence is one to one modulo fields of the same form vanishing on $\{\bar{\phi}\}$.

Lemma[41] Given $\{\bar{\phi}\}$, any function $F(\phi)$ defined on $\{\bar{\phi}\}$ and R-covariant under G

$$F(R_H(g)\phi) = R(g)\, F(\phi), \qquad \forall g \in G, \tag{3.2}$$

is the restriction of an R-covariant polynomial to $\{\bar{\phi}\}$.

Proof of Theorem 3.3 Given $\Psi^i(\phi) = P^i_\alpha(\phi)\psi_\alpha$, by fixing $\phi = \bar{\phi} \in \{\bar{\phi}\}$, one gets a component $P^i_\alpha(\bar{\phi})\psi_\alpha \equiv v^i\cdot\psi$ of R (depending on i). Conversely, let v_i be a vector of R. For any $\phi \in \{\bar{\phi}\}$, there exists a unique $g_\phi \in G$, such that

$$R_H(g_\phi)\,\bar{\phi} = \phi,$$

(since $G_{\{\bar{\phi}\}}$ = ident.). Now, define

$$F(\phi) \equiv R(g_\phi)\, v_i \quad ,$$

then $F(\phi)$ is R-covariant under G and by the Lemma is the restriction to $\{\bar{\phi}\}$ of a polynomial $P^i(\phi)$. Clearly $\Psi^i(\phi) \equiv P^i(\phi)_\alpha \psi_\alpha$ is G-invariant and $\Psi^i(\bar{\phi}) = v^i \cdot \psi$.

As a direct consequence of the above Theorem, if the representation R is n-dimensional, one has exactly n independent invariants $P^i(\phi)\cdot\psi$ and therefore one can use them instead of the original gauge dependent fields ψ_α to obtain one particle states from the vacuum. Actually, in the limiting case in which in the Euclidean approach one considers the orbit as fixed, $\phi \in \{\bar{\phi}\}$, one can show[41] that, in the prototypic case of total breaking, $G_{\{\bar{\phi}\}} = 1$, the two point functions of the invariant fields $\Psi^i(\phi)$ have the same masses as in the standard approach, in the lowest perturbative orders. As a matter of fact, the correlation functions of the gauge invariant fields $\Psi^i(\phi)$ coincide with the correlation functions of the corresponding gauge dependent fields ψ_i in the unitary gauge. This and the above theorem support the picture that, in a genuine Higgs phenomenon (i.e. in the case of total breaking), the particle states are described by gauge invariant (i.e. G-neutral) fields, i.e. the (gauge) charge is screened off.

This explains why, even if G is unbroken, the standard particle states are not degenerate, because they correspond to different singlets and there is no symmetry relation between them. A non-trivial mass pattern is possible as well as mass differences. Thus, the Higgs phenomenon does not require in general a symmetry breaking order parameter, but rather a "dominant" critical orbit $\{\bar{\phi}\}$, with $G_{\{\bar{\phi}\}}$ smaller than G. The above picture can be taken as a basis for a perturbative expansion around a critical orbit[41]. Non-perturbative corrections due to field configurations, which are far from the critical orbit, may be relevant especially for a widely separated mass spectrum and in particular may provide a possible mechanism for solving the gauge hierarchy problem[42].

By further exploiting the above discussion, one can agrue that the Hilbert space of states of the unitary gauge is generated by applying local gauge invariant fields to the vacuum. For simplicity we consider the prototypic case in which the Higgs field ϕ takes values on a single orbit $\{\bar{\phi}\}$. Then, the unitary gauge is described by a field algebra $\mathcal{F}_u$ generated by the gauge dependent fields ψ_α, A_μ, ϕ etc., which by definition can be obtained from the gauge invariant fields $\Psi^i(\phi)$ by freezing $\phi = \bar{\phi}$, through the gauge fixing[43], (see the proof of Theor. 3.3). Now the correlation functions of the gauge invariant fields Ψ^i are independent of the gauge fixing* and, by the standard reconstruction theorem, define the Hilbert space $\mathcal{H}_{inv}$ of the local gauge invariant states. On the other hand, the correlation functions

* To be concrete we consider the theory formulated on a lattice (see the lectures by Profs. Jaffe and Balaban in these proceedings).

of the gauge dependent fields ψ_α etc., in the unitary gauge, can be obtained by functional integration of the corresponding polynomial of the gauge invariant fields Ψ^i, with the gauge fixing $\phi = \bar{\phi}$. Thus

$$< \Psi^{i_1}(\phi)_1 \ldots \Psi^{i_n}(\phi)_n >_0 = < \Psi^{i_1}(\phi)_1 \ldots \Psi^{i_n}(\phi)_n >_{0,\phi=\bar{\phi}}$$

$$= < \psi_1 \ldots \psi_n >_0 \, . \qquad (3.3)$$

Since the fields ψ etc. generate F_u and the correlation functions of F_u define the Hilbert space H_{phys} of the unitary gauge,

$$H_{phys} = \overline{\{F_u \psi_o\}} \, , \qquad (3.4)$$

we get

$$H_{phys} = H_{inv} = \overline{\{F_o \psi_o\}} \, . \qquad (3.5)$$

Thus, if the unitary gauge provides a suitable description of the Higgs phenomenon, the latter can be characterized by the property that all the physical states can be obtained by applying local gauge invariant fields to the vacuum, eq. (3.5), (*charge screening*).

It is worthwhile to discuss the above picture in the various gauges. Each gauge g has its own field algebra F_g and its own Hilbert space $H_g = \overline{\{F_g \Psi_o\}}$; according to the above picture of the Higgs phenomenon all the gauges define the same distinguished physical space H_{phys}.

In the axial gauge, by the Gauss' law constraint only the gauge invariant subalgebra $F_o \subset F_{axial}$ is represented in H_{phys}[44]; there is no breaking of the gauge symmetry defined on F_{axial} and all the physical states have zero charge.

In the Coulomb gauge there is no distinguished subspace specified by the Gauss' law condition, since the latter is satisfied as an operator equation. Then

$$H_C = \overline{\{F_C \Psi_o\}} = H_{phys} = \overline{\{F_o \Psi_o\}} \, . \qquad (3.6)$$

Now, F_C properly contains F_o and the gauge symmetry in the Coulomb gauge has a non-trivial action on F_C. This, together with eq. (3.6), implies that the gauge symmetry is spontaneously broken in the Coulomb gauge as a consequence of the following

THEOREM 3.4 (Charge screening and symmetry breaking) Let β be a non-trivial group G of translation invariant automorphisms of a field algebra F of bounded operators, which is generated by fields transforming as a finite dimensional representation of G, and satisfies asymptotic abelianess

$$\text{w-}\lim_{|\vec{x}|\to\infty} [A_{\vec{x}}, B] = 0, \quad A, B \in F \, . \qquad (3.7)$$

Let F_o denote the subalgebra of F which is neutral under β and π a representation of F with unique (cyclic) translational invariant ground state Ψ_o. Then, β is broken down to the identity in π iff

$$\overline{\{F\Psi_o\}} = \overline{\{F_o\Psi_o\}} , \tag{3.8}$$

i.e. <u>all the states have the same "charge" of the vacuum.</u>

<u>Proof</u>. If β is broken down to a non-trivial group H ⊂ G, then H is implemented by a non-trivial unitary operator U_H and

$$\overline{\{F\Psi_o\}} \neq \overline{\{F_o\Psi_o\}} .$$

Conversely, let $\{A_i\}$ be a set of polynomials of the fields which break β down to the identity, in the sense that

$$<\beta_g(A_i)>_o = <A_i> , \quad \forall i \Rightarrow g = \text{identity}.$$

Then, as in Theorem 3.3, ∀ n-dimensional representation R, $\psi_\alpha \in R$, one can construct G-invariant fields

$$\Psi^i = P^i_\alpha(A)\,\psi_\alpha , \quad i = 1, \dots n. \tag{3.9}$$

Furthermore

$$A^i_V = \frac{1}{V}\int_V d^3x\, A^i_{\vec{x}}$$

converges strongly in π to the ergodic mean A_∞[45] and $\pi(A^i_\infty) = a^i$ are c-numbers. Then, the gauge invariant fields

$$\Psi^i_V \equiv P^i_\alpha(A_V)\,\psi_\alpha$$

converge strongly in π to the gauge dependent fields

$$\Psi^i_\infty = P^i_\alpha(a)\,\psi_\alpha ,$$

which generate F.

<u>Remark</u> The same Theorem holds for unbounded fields, with the cluster property replacing the asymptotic abelianess. The proof is the same, with the ergodic means replaced by translation of the operators A_i in eq. (3.9) to (spacelike) infinity.

It is worthwhile to mention that, since the field algebra F depends on the gauge, also the automorphisms defined on F depend on the gauge. In particular, one may construct non-local operators[40], such that the symmetry, as defined on the local operators is unbroken (local correlation functions are symmetric), whereas a suitable extension of the symmetry to non-local operators is broken (non-local correlation functions are not symmetric) (see the Coulomb and the temporal gauge and more generally Ref. 46). In conclusion, symmetry breaking and charge screening are equivalent in the physical gauges in which the gauge automorphism is well defined and non-trivial.

The occurence of gauge symmetry breaking associated to the Higgs phenomenon in the Coulomb gauge raises the question mentioned at the beginning of Sect. III, namely the explanation of mass/energy gap associated to symmetry breaking, in contrast with the Goldstone theorem.

IV LONG RANGE INTERACTIONS AND SYMMETRY BREAKING

4.1 *Symmetry breaking and energy spectrum*

The phenomenon of symmetry breaking in the case of local or short range interactions is well understood and it is not necessary to recall the power of Goldstone's theorem in providing non-perturbative exact information on the excitation spectrum associated to the spontaneous symmetry breaking (SSB)[47,48]. Its application to many body systems, statistical mechanics, current algebra etc. are well known. On the other side, most of the many body systems involve the long range Coulomb interaction and, with the advent of gauge theories, the definition itself of gauge symmetry breaking as well as the discussion of the equations of motion involve the introduction of non-local (charged) field variables. For those cases an extension of the standard Goldstone's theorem is needed in order to understand the relation between symmetry breaking and energy spectrum and possibly get exact informa tion on the low k behaviour of the excitation spectrum, especially in those cases in which it is non-trivial, $\omega(\vec{k} \to 0) \neq 0$. It also appears that many questions about the general common features exhibited by different physical phenomena, (like the plasmon gap generation, the phenomenon of superconductivity, the possible analogies between the BCS theory and the Higgs phenomenon beyond the semiclassical Ginzburg-Landau-Gorkov treatment, the field theoretical understanding of the U(1) problem and in general the breaking of global symmetries in grand unified theories etc.) are deeply rooted in the interplay between symmetry breaking and long range interactions. As we shall see, the clarification of this structure[49] will also provide an understanding of (and a rigorous basis for) the mechanism of "seizing of the vacuum" advocated by Kogut and Sussking on the basis of two dimensional models[50]. Moreover, we shall clarify the symmetry breaking aspects of the Coulomb gauge in the Stückebberg-Kibble model[51].

Quite generally[47,49] the spectrum of the Goldstone excitations is related to the time dependence of the charge commutator function

$$J(-t) = i \lim_{R\to\infty} \int_{|\vec{x}|<R} d^3x < [j_o(x,t),\ A] >_o \equiv i \lim_{R\to\infty} < [Q_R(t),A]>_o \quad (4.1)$$

and in fact it is given by (the support of) the Fourier transform $\tilde{J}(\omega)$. (For a more precise meaning a suitable space and time smearing[47] should be introduced.)

Then, the time independence of J(t) implies $\tilde{J}(\omega) \propto \delta(\omega)$ and the energy spectrum of the Goldstone excitations reduces to the single point $\omega = 0$ in the limit $\vec{k} \to 0$ (equivalently $R \to \infty$). Now, if the finite volume or infrared cutoffed Hamiltonian H_V is symmetric, it is not obvious how one can evade the vanishing of

$$<[[Q_R, H_V], A]>_o ,$$

in the infinite volume limit or in the removal of the infrared cutoff. Actually, if the interaction is local or (sufficiently) short range this conclusion is unavoidable. The reason is that in this case if A is an (observable) operator localized in a bounded region $\mathcal{O}$ at $t = 0$, then at a later time t, A_t is also essentially localized, since the commutator $[H_V,A]$ contains operators with localization (essentially) independent of V, plus small contributions from operators localized near the boundary of V. The short range of the potential implies a weak coupling with the boundary as far as the dynamics is concerned, i.e. the boundary contributions to the time evolution of a localized operator go to zero as $V \to \infty$. It then follows that, if H_V is symmetric, the equations of motion

$$i \frac{d}{dt} A = [H_V, A] ,$$

which are symmetric for V finite, remain so also in the limit $V \to \infty$. Then, roughly speaking,

$$i \frac{d}{dt} <[Q(t), A] > = < [[H,Q], A] > = 0 ,$$

i.e.

$$\lim_{R\to\infty} < [Q_R(t), A] >_o = \text{indep. of t.} \tag{4.2}$$

More precisely one has

<u>THEOREM 4.1</u> Let β^λ be a one parameter group of automorphisms of the local algebra $\mathcal{A}_o$

$$\mathcal{A}_o = \underset{V}{\cup} \mathcal{A}_V ,$$

$\mathcal{A}_V \equiv$ Von Neumann algebra associated to the finite volume V, and let β^λ be generated by a local charge Q_R (affiliated to some $\mathcal{A}_V$) in the sense that $\forall A \in \mathcal{A}_o$

$$\beta^\lambda(A) = || \, ||\text{-}\lim_{R\to\infty} e^{iQ_R\lambda} A e^{-iQ_R\lambda} . \tag{4.3}$$

If the finite volume (or infrared cutoffed) dynamics $\alpha_V^t(A)$ converges in norm $\forall A \in \mathcal{A}_o$, as $V \to \infty$, and $\beta^\lambda \alpha_V^t = \alpha_V^t \beta^\lambda$, then

i) β^λ has a unique extension to an automorphism of the norm closure $\mathcal{A}$ of $\mathcal{A}_o$;

ii) β^λ is still generated by Q_R, eq. (4.3), on $\mathcal{A}$, and it commutes with the infinite volume dynamics $\alpha^t \equiv \lim_V \alpha_V^t$.

Proof Let $A \in \mathfrak{A}$ and $A_n \to A$, $A_n \in \mathfrak{A}_o$; then

$$\beta^\lambda(A) = \|\ \| - \lim_n \beta^\lambda(A_n) = \|\ \| - \lim_n \|\ \| - \lim_R \beta^\lambda_R (A_n)$$

and, since

$$\|\beta^\lambda_R (A_n) - \beta^\lambda_R (A_m)\| = \|A_n - A_m\| ,$$

the limit $n \to \infty$ is uniform with respect to R, so that the two limits can be interchanged.

Under the (additional) technical assumption that the limit in eq. (4.3) is uniform in λ together with the derivative with respect to λ, one then gets that if $\alpha^t_V(A)$ converges in norm, eq. (4.2) holds, i.e. the Goldstone theorem applies. In conclusion, all the standard wisdom on the Goldstone theorem relies on the idea of spontaneous symmetry breaking characterized by symmetric equations of motion and non-symmetric correlation functions. In a constructive approach this can be understood as the fact that in the case of local or short range interactions, in the limit $V \to \infty$, the boundary conditions cannot affect the equations of motion but only the ground state.

4.2 *Long range interactions and variables at infinity*

The situation changes drastically with respect to the previous section if the interactions are long range, more precisely if the finite volume (or infrared cutoffed) dynamics α^t_V does not converge in norm as $V \to \infty$. This is the case of spin systems with interaction decaying slower than $|\vec{x}|^{-3}$, many body systems with 1/r interactions (Coulomb systems) and also gauge field theories in the physical (positive) gauges, since they involve non-local field variables and instantaneous long range interactions.

The point is that, when the interaction is long range, the commutator $[H_V, A]$, with A a localized variable, involves in a substantial way operators localized on or near the boundary of V, which, in the limit $V \to \infty$, converge to variables at infinity[55] i.e. to variables localized outside any bounded region. Typical examples are the ergodic means

$$\lim_{V\to\infty} \frac{1}{V} \int_V d^3x\, A(\vec{x}) = A_\infty ,$$

or the infinite volume limit of averages around the boundary

$$\lim_{R\to\infty} \int d^3x\, f_R(\vec{x}) \quad A(x) \quad , \tag{4.4}$$

with $f_R(x)$ a regular function different from zero only in the region $R < |\vec{x}| < R\,(1+a)$ and normalized in such a way that $\int f_R(x)\, d^3x = 1$. Such limits still converge to A_∞ (in a suitable topology related to the family of representations under consideration[49]). The variables at infinity commute with all local observables, if the algebra satisfies asymptotic abelianess, eq. (3.7)[45], a property which will be always assumed in the following (this condition is in fact necessary for a reasonable physical interpretation of measurements and cannot be dispensed with).

To be more concrete, we briefly discuss two prototypic examples of the mechanism by which variables at infinity occur in the time evolution of a local observable, due to the interplay of the kinetic term and of the long range 1/r interaction[49,46].

Example 1 (ELECTRON GAS WITH UNIFORM BACKGROUND[46])

The systems is formally described by the following Hamiltonian

$$H = \frac{1}{2m}\int|\nabla\psi|^2\, d^3x + \frac{1}{2}\int d^3x d^3y\ \psi^*(x)\psi^*(y)\ U(x-y)\psi(y)\psi(x)$$

$$- \int d^3x d^3y\ \psi^*(x)\psi(x)\ U(x-y)\rho_B\ ,$$

where ρ_B is the background density and

$$U(x) = e^2/|\vec{x}|$$

is the Coulomb potential. To properly define the dynamics an infrared regularization is necessary; the specific form of such regularization is not relevant and we shall choose

$$U(x) \to U_L(x) \equiv U(x)f(|\vec{x}|/L)\ , \tag{4.5}$$

with f a regular function which is one inside a sphere of radius one and vanishes outside a sphere of radius 1+a. The corresponding Hamiltonian H_L defines the infrared cutoffed dynamics α_L^t. Consider in particular the equation of motion for the local variable $\Delta\psi$ at t=0:

$$\Delta\dot{\psi} = (i/2m)\Delta^2\psi - 2i[\partial_i U_L * (\rho-\rho_B)]\partial_i\psi - i[U_L * (\rho-\rho_B)]\Delta\psi -$$

$$- i[\Delta U_L * (\rho-\rho_B)]\psi\ ,$$

($\rho \equiv \psi^*\psi$). We focus our attention on the last term; we have

$$\Delta U_L(x) = -4\pi e^2\delta(x) + \sigma_L(x)\ , \tag{4.6}$$

with $\sigma_L(x) \neq 0$ only in the region $L < |\vec{x}| < L(1+a)$ and

$$\int\sigma_L(x)\, d^3x = 4\pi e^2. \tag{4.7}$$

The σ_L term describes an average around the boundary of V of the same type of eq. (4.4); then, as $L \to \infty$,

$$\sigma_L * \rho \to \rho_\infty .$$

Example 2 (STUCKELBERG-KIBBLE MODEL[51])

The model corresponds to the abelian Higgs-Kibble model with frozen modulus of the Higgs field $\chi = \rho e^{i\phi}$. The Lagrangian is (ρ = 1)

$$L = -\frac{1}{4}F_{\mu\nu}^2 - \tfrac{1}{2}(\partial_\mu\phi + eA_\mu)^2 + \text{gauge fixing.}$$

In the Coulomb gauge the formal Hamiltonian is

$$H = \frac{1}{2}\int d^3x\,[|\nabla\phi|^2 + \pi^2] + \frac{1}{2}\int d^3x d^3y\,\pi(x)\,U(x-y)\pi(y) \tag{4.8}$$

and as in the previous example an infrared regularization is necessary to properly define the dynamics. Again we choose $U(x) \to U_L(x) = U(x)\, f(|\vec{x}|/L)$, as in eq. (4.5). The infrared regularized equations of motion are

$$\dot{\phi} = \pi + \int U_L(x-y)\,\pi(y)\,d^3y \quad , \qquad \dot{\pi} = \Delta\phi \tag{4.9}$$

and yield

$$\ddot{\phi} = \Delta\phi - 4\pi e^2\phi + \int \sigma_L(x-y)\phi(y)\,d^3y \quad .$$

As in the previous example, the last term on the r.h.s. converges to a variable at infinity, $4\pi e^2\phi_\infty$, and

$$\ddot{\phi} = \Delta\phi - 4\pi e^2\phi + 4\pi e^2\phi_\infty \quad . \tag{4.10}$$

Thus, the time evolution reads

$$\phi(\vec{k},t) = (1-\cos\omega(0)t)\phi_\infty + \cos(\omega(k)t)\,\phi(\vec{k},0) + \\ + (\vec{k}^2/\omega(k))^{-1}\sin(\omega(k)t)\,\pi(\vec{k},0) \quad , \tag{4.11}$$

where

$$\omega^2(k) = \lim_{L\to\infty} \omega_L^2(k) \equiv \lim_{L\to\infty} \vec{k}^2(1+\tilde{U}_L(k)) \tag{4.12}$$

In conclusion, the above discussion should make clear that in the case of long range interactions (in particular for Coulomb systems) the time evolution of a localized observable involves variables at infinity, in general, i.e.

$$A_t = F_t\,(B_\ell, B_\infty) \tag{4.13}$$

where B_ℓ are "essentially local" operators and B_∞ variables at infinity. Examples of this structure occur in Coulomb systems, Spin models, BCS model, gauge models of the Higgs phenomenon etc.[49,46].

The main point of the above discussion (and in particular of the two examples above) is that the occurrence of variables at infinity is a general phenomenon related to long range interactions, not peculiar to those simple models (like the BCS model[52], the Spin models in the molecular field approximation[49], the Kogut-Susskind model[50] etc.) in which the variables at infinity are put in by hand in the definition of the models themselves.

The above eq. (4.13) can be given a precise meaning. The infinite volume dynamics α^t is defined as the limit of the finite volume (or infrared cutoffed) dynamics α_V^t, with respect to the weak topology defined by a full folium F of "physically relevant" states on the quasi local algebra $\mathfrak{A}$[49]. <u>The quasi local algebra</u> $\mathfrak{A}$ <u>is not stable under time evolution</u> α^t, but the algebraic dynamics is defined in the weak closure $\mathfrak{M} = \bar{\mathfrak{A}}$, which contains variables at infinity and therefore has a non-trivial center. This makes clear that the treatment of systems with long range interactions requires substantial changes with respect to the standard Haag-Kastler algebraic framework[53]. This change, in particular the need to consider infinitely delocalized variables, is inevitable for most non-relativistic theories; it is also necessary for gauge theories in order to solve problems which are formulated in terms of non-local field variables (like gauge symmetry breaking, U(1) problem etc.).

In each irreducible representation π of $\mathfrak{A}$, the variables at infinity are represented by c-numbers and therefore the "effective dynamics" with reference to the representation π

$$F_t(B_\ell, \langle B_\infty \rangle_\pi) \equiv \alpha_\pi^t(A) \tag{4.14}$$

depends on the boundary conditions which fix the representation. (This is a crucial difference with respect to the local or short range case where the boundary conditions can only affect the correlation functions, not the equations of motion). Thus, in the long range case, even if the finite volume Hamiltonian H_V is symmetric, the equations of motion in irreducible representations are in general not symmetric. The point is that, if H_V is symmetric, F_t is a symmetric function of its variables, but the symmetry is lost when the variables at infinity are frozen to their expectation values in π. The above framework provides a generalization of Haag's treatment[52] of the BCS model*, gives a clear meaning to the so-called seizing of the vacuum[50], also in much more general situations than the simple two-dimensional models*, and, as we shall see in the next section, it provides a general mechanism for explaining and predicting mass/energy gap generation associated to symmetry breaking.

4.3 *Symmetry breaking and energy gap generated by variables at infinity*

We are now in the position of revisiting the Goldstone theorem for systems with long range interactions. If H_L denotes the infrared cutoffed Hamiltonian and

$$Q_R = \int_{|x|<R} d^3x \; j_o(\vec{x})$$

a local charge generating a symmetry of H_L, we have

$$\lim_R \, [Q_R \, , H_L] = 0. \tag{4.15}$$

But in each irreducible representation π, with ground state ϕ_o,

$$\lim_{R\to\infty} \lim_{L\to\infty} \langle [[Q_R, H_L], A] \rangle_{\phi_o} \tag{4.16}$$

* There, the occurrence of variables at infinity is not so surprising, since their introduction in the equations of motion is at the basis of the (mean field) approximation which defines the model.

is in general not vanishing, and the interchange of limits (eq. (4.15) and eq. (4.16)) is very delicate. On the other hand, the formal Hamiltonian has in general no clear meaning and therefore in order to even speak of symmetries of the Hamiltonian an infrared regularization is necessary. Then, in order to relate the charge commutator function J(t) to the excitation spectrum, the infrared cutoff has to be removed first and only then one may look at the long wavelength behaviour ($\vec{k} \to 0$, equivalently $R \to \infty$)

$$J(-t) = \lim_{\vec{k}\to 0} J(\vec{k},-t) = i \lim_{R\to\infty} \lim_{L\to\infty} \langle [\alpha_L^t(Q_R), A] \rangle \quad . \tag{4.17}$$

The above argument can be made precise by considering a one-parameter group β^λ, $\lambda \in \mathbb{R}$, of automorphims of $\mathcal{Q}$ which commute with the space translations α_x and are generated by a local charge Q_R on an "essentially local algebra" $\mathcal{Q}_\ell$, on which the effective dynamics α_π^{tR} acts as an automorphism (for more details see Ref. 49). Under these conditions, one can prove that the <u>energy spectrum</u> at $\vec{k} \to 0$ <u>associated to the breaking of</u> β^λ <u>is in general non-trivial</u> ($\omega(\vec{k} \to 0) \neq 0$) and <u>it is related to the classical motion of variables at infinity</u>, (*generalized Goldstone's theorem*)[49].

In fact, if ϕ_o is the translationally invariant ground state which defines the representation π, and $A \in \mathcal{Q}_\ell$

$$J(t) = i \lim_{R\to\infty} \langle [Q_R, A^t] \rangle_{\phi_o} = i \lim_{R\to\infty} \langle [Q_R, \alpha_\pi^t(A)] \rangle_{\phi_o} =$$

$$= \frac{d}{d\lambda} \langle \beta^\lambda \alpha_\pi^t(A) \rangle_{\phi_o} \Big|_{\lambda=0} = \frac{d}{d\lambda} \langle \alpha_\pi^t (A) \rangle_{\phi_o^\lambda} \Big|_{\lambda=0} \quad ,$$

where ϕ_o^λ is the ground state obtained from ϕ_o through β^λ. Since β^λ commutes with the space translations also ϕ_o is translationally invariant and

$$J(t) = \frac{1}{V} \frac{d}{d\lambda} \int_V d^3x \langle \alpha_{\vec{x}} \alpha_\pi^t(A) \rangle_{\phi_o^\lambda} \Big|_{\lambda=0} = \frac{d}{d\lambda} \langle \alpha_\pi^t(\frac{1}{V} \int_V d^3x \, A_x) \rangle_{\phi_o^\lambda} \Big|_{\lambda=0}$$

$$= \frac{d}{d\lambda} \langle \alpha_\pi^t(A_\infty) \rangle_{\phi_o^\lambda} \Big|_{\lambda=0} \quad . \tag{4.18}$$

Thus, the time dependence of J(t) is related to the evolution of the variable at infinity A_∞ under the effective dynamics α_π^t. The evolution of A_∞ under α_π^t takes values in an <u>abelian</u> algebra defined[49] as the algebra at infinity identified by a family of states stable under the action of α_π^t and β^λ. One then has a "classical" dynamical system. Clearly, since by definition ϕ_o is invariant under the effective time evolution α_π^t we have

$$\langle \alpha_\pi^t(A_\infty) \rangle_{\phi_o} = \langle A_\infty \rangle_{\phi_o} \quad ,$$

whereas

$$\langle \alpha_\pi^t(A_\infty) \rangle_{\phi_o^\lambda}$$

is in general not constant in time and the "classical motion" is non-trivial.

By eq. (4.18), the spectrum of J(t) (which gives the spectrum at $\vec{k} \to 0$ of the generalized Goldstone bosons) is given by the linearized motion around the stable point $< A_\infty >_{\phi_0}$ [49]. Moreover, if only a finite number of charges $Q_R(t)$, $t \in \mathbb{R}$, are independent, in the sense of commutators

$$\lim_{R\to\infty} <[Q_R(t), A^\tau]>_0 \quad ,$$

(or if the group generated by β^λ and α_π^t is a finite dimensional Lie group), then the linearized motion is quasi periodic and its frequencies ω_j give a discrete excitation spectrum at $\vec{k} \to 0$[49].

4.4 *The Stückelberg-Kibble model*

We can now discuss the symmetry breaking aspects and the associated mass generation in the Stückelberg-Kibble model discussed in Sect. 4.2. The infrared cutoffed Hamiltonian is symmetric under the gauge transformations β^λ:

$$\phi \to \phi + \lambda,$$

which are generated by the local charge

$$Q_R = \int_{|x|<R} d^3x \; \pi(x) \quad ,$$

where π is the momentum canonically conjugated to ϕ

$$[\phi(\vec{x}), \pi(\vec{y})] = i\,\delta(\vec{x}-\vec{y}) \quad .$$

If Ψ_0 is a translationally invariant ground state defining a (regular) representation of the Weyl algebra $\mathcal{Q}$ generated by ϕ and π (smeared with test functions in $S(R^3)$), then

$$\lim_{R\to\infty} <[Q_R, \phi]>_{\Psi_0} \neq 0 \quad ,$$

i.e. the gauge symmetry is spontaneously broken.

The general structures discussed in the previous sections can be proven[49] to be realized in this model and in particular the classical motion associated to the variable at infinity ϕ_∞ is given by

$$\phi_\infty(t) = (\cos \omega t)\phi_\infty + \omega^{-1}(\sin \omega t)\left(\frac{1}{4\pi r} * \pi\right)_\infty \quad ,$$

where

$$\omega^2 = \lim_{\vec{k}\to 0} \lim_{L\to\infty} \vec{k}^2(1+\tilde{U}_L(\vec{k})) = 4\pi e^2$$

and

$$\left(\frac{1}{4\pi r} * \pi\right)_\infty = \lim_{L\to\infty} \frac{1}{L^3} \int d^3x \;\; h(|x|/L)\left(\frac{1}{4\pi r} * \pi\right)(x) \;\; , \quad \int h(x)d^3x = 1.$$

The elementary excitations associated to the breaking of the gauge symmetry (generalized Goldstone bosons) have correspondingly a finite mass $4\pi e^2$.

The standard discussions of this model[51,48] have missed the following essential features, namely that the correct treatment of the dynamics requires an infrared regularization and a careful handling of the limit $L \to \infty$, that the long range interaction gives rise to variables at infinity in the time evolution of the field variables and that the finite mass of the generalized Goldstone boson, associated to the gauge symmetry breaking, is related to the non-trivial motion of the variables at infinity (under the "effective" dynamics).

4.5 *Breaking of the Galilei group and plasmon energy gap*

The electron gas with uniform background provides a physically non-trivial realization of the structures discussed in Sects. 4.1-4.3. It is also of interest for the interdisciplinary aspects exhibited by the phenomenon. First of all, the standard derivations of the plasma frequency (at the classical and at the quantum level) involve suitable approximations and eventually a linearization of the non-linear equations for the electron charge density; so that it is of some interest to improve the situation. Secondly, the plasmon gap generation and its relation with the effective screening of the Coulomb potential qualifies as a non-trivial dynamical mechanism having deep analogies in elementary particle physics, where one hopes to get the Higgs field dynamically, through fermion condensation. Another interesting question is the possible relation between the generation of the plasmon energy gap and of the energy gap in superconductivity. In the standard treatment they appear as prototypes of two different mechanisms: the energy of the longitudinal charge oscillations (plasma frequency) is related to the fermion condensation in the neutral channel, $< \psi^*\psi >$, whereas the superconductor energy gap is associated to the breaking of the charge symmetry and to the charged order parameter $< \psi\psi >$.

The starting point of our analysis is to consider the Galilei group which defines a symmetry of the electron gas in a uniform background (jellium model), with an infrared regularization of the Hamiltonian ($H \to H_L$, see Ex. 1 in Sect. 4.2). (For details about the local generators of the Galilei group for this system see Ref. 46). The generator of the Galilei boosts (in the i-th direction) is given by

$$G_R^i = m \int_{|x|<R} d^3x \, x^i \rho(x)$$

and the infinitesimal variation of the current is given by

$$\delta^i j_k(\vec{y}) = i \lim_{R\to\infty} [G_R^i, j_k(\vec{y})] = \delta_{ik}\rho(\vec{y}) .$$

Thus, if Ψ_o denotes a translationally (and rotationally) invariant ground state

$$< \delta^i j_k >_{\Psi_o} = \delta_{ik} < \rho >_{\Psi_o} \neq 0 ,$$

i.e. the Galilei boosts are spontaneously broken as a consequence of the finite electron density. We shall see that this symmetry breaking gives rise to generalized Goldstone bosons (the plasmons), with finite energy in the $\vec{k} \to 0$ limit (plasmon energy gap).

To this purpose we consider the infrared regularized equations of motion

$$\frac{d}{dt}\,\psi(\vec{x},t) = i(\Delta/2m)\psi(\vec{x},t) - i(U_L * (\rho-\rho_B))(\vec{x},t)\psi(\vec{x},t) .$$

As discussed in the previous sections, the removal of the infrared cut-off and the convergence of the infrared cutoffed dynamics α_L^t is a delicate issue. This convergence requires that reference be made to a family F of states[49] sufficiently regular at infinity (so that α_L^t (A) converges strongly, for A an element of the (canonical) algebra $\mathcal{A}$ generated by ψ, ψ^*). From an inspection of the above equations of motion, one sees that the existence of the correlation functions of the electric field

$$E_i(\vec{x},t) = - \int d^3y\ \partial_i V(\vec{x}-\vec{y})\ (\rho(\vec{y},t) - \rho_B),$$

where t denotes the time evolution corresponding to $\lim_{L\to\infty} \alpha_L^t$, requires that

A) The correlation functions of the density $\rho(\vec{x}) - \rho_B$ (in the states ϕ of the family F) must decay faster than $|\vec{x}|^{-1}$, in such a way that

$$|\vec{x}|^{-2}\ \phi(A(\rho(\vec{x},t) - \rho_B)B)$$

($\phi \in$ F, A,B polynomials of ψ^*,ψ) is absolutely integrable in $\vec{x}$.*

We then consider the unequal time commutator

$$\lim_{R\to\infty} [G_R^i(t),\ \psi(\vec{x})]$$

Proposition 4.2[46] Within the above framework

$$\lim_{R\to\infty} [\frac{d^2}{dt^2}\,\vec{G}_R(t) + \omega_p^2\,\vec{G}(t),\ \psi(\vec{z})]\Big|_{t=0} = 0, \quad \omega_p^2 \equiv 4\pi e^2\ \rho_\infty/m$$

The same equation holds if ψ is replaced by any local polynomial of ψ^*,ψ.

The proof[46] requires a very delicate handling of the limit $L \to \infty$, before the limit $R \to \infty$. To determine the spectrum of excitations associated to the Galilei breaking one has to consider the charge commutator function

$$J(-t) \equiv J_i^i(-t) = i \lim_{R\to\infty} <[G_R^i(t),\ j_i]>_o \quad \text{(no index summation)}$$

and Proposition 4.2 implies

---

* Considerably stronger conditions are at the basis of the analysis of Coulomb systems by Gruber, Martin, Lebowitz[54].

$$\ddot{J}(0) = -\omega_p^2 \, J(0).$$

By exploiting positivity and by careful estimates on the limit $L \to \infty$, one can further prove[46]

Proposition 4.3 The Galilei charge commutator function has the following time dependence

$$J(t) = \rho_\infty \cos \omega_p t$$

This shows that

i) the plasmon energy gap is generated by the spontaneous breaking of the Galilei boosts;

ii) the plasmon spectrum at $\vec{k} \to 0$ can be calculated exactly, by exploiting a generalized Goldstone theorem; the excitation spectrum associated to the Galilei breaking consists of a single point ω_p at $\vec{k} \to 0$ and the plasmon is the corresponding generalized Goldstone boson;

iii) the sharp point in the energy spectrum implies that the plasmons are excitations with a lifetime which goes to infinity when $\vec{k} \to 0$, i.e. the plasmons are quasi particles.

Finally, one can also show that the plasmon energy gap is related to a non-trivial classical motion of variables at infinity

$$\vec{j}_\infty(t) = (\cos \omega_p t)\vec{j}_\infty + (m\omega_p)^{-1} \sin \omega_p t(:\rho\vec{E}:)_\infty,$$

where E is the electric field and

$$(:\rho\vec{E}:)_\infty = (\psi^*\vec{E}\psi)_\infty .$$

REFERENCES

1. P. Blanchard and R. Sénéor, Ann. Inst. H. Poincaré XXIII 147 (1975); see also the contributions by H.Epstein, V. Glaser, J.H. Löwenstein, P. Breitenlohner and D. Maison in "Renormalization Theory", Proc. Int. School Math. Physics, Erice 1975 G. Velo and A.S. Wightman eds., D. Reidel (1976).
2. J.M. Jauch and F. Rohrlich, "The Theory of Photons and Electrons", 2nd exp. ed., 2nd corr. print, Springer Verlag (1980).
3. J. Schwinger, Phys Rev. 76:790 (1949); see also the review by L.C. Maximon, Rev. Mod. Phys. 41:193 (1969).
4. V. Chung, Phys. Rev. 140B:1110 (1965).
5. T.W. Kibble, Phys. Rev. 173:1527 (1968); 174:1882 (1968); 175:1624 (1968).
6. D.R. Yennie, S.C. Frautschi and H. Suura, Ann. Phys.(N.Y.) 13:379 (1961).
7. F. Bloch and A. Nordsieck, Phys. Rev. 52:54 (1937).
8. G. Roepstorff, Comm. Math. Phys. 19:301 (1970).
9. P.P. Kulish and L.D. Fadde'ev, Teor. Matem. Fiz 4:153 (1970)[Theor. Math. Phys. 4:745 (1971)]; P. Blanchard, Comm. Math. Phys. 15:156 (1969)

10. J.D. Dollard, Jour. Math. Phys. 5:729 (1964).
11. D. Zwanziger, Phys. Rev. D14:2570 (1976) and refs. therein.
12. R. Ferrari, L.E. Picasso and F. Strocchi, Comm. Math. Phys. 35:25 (1974); Nuovo Cim. A39:1 (1977).
13. F. Strocchi, Gauss' law in local quantum field theory in: "Field Theory, Quantization and Statistical Physics", ed. E. Tirapegui, D. Reidel (1981).
14. F. Strocchi and A.S. Wightman, Journ. Math. Phys. 15:2198 (1974).
15. S. Doplicher, R. Haag and J.E. Roberts, Comm. Math. Phys. 23:199 (1971); 35:49 (1974).
16. D. Buchholz and K. Fredenhagen, Comm. Math. Phys. 84:1 (1982).
17. For a detailed account see e.g. R. Jost, in: "The General Theory of Quantized Fields", Am. Math. Soc. (1965) and K. Hepp, in: "Brandeis Lectures 1965", Vol. I, M. Chrétien and S. Deser eds., Gordon and Breach (1966).
18. D. Buchholz, Comm. Math. Phys. 42:269 (1975); 52:147 (1977).
19. D. Buchholz, Comm. Math. Phys. 85:49 (1982).
20. J. Fröhlich, G. Morchio and F. Strocchi, Ann. Phys. (N.Y.) 119:241 (1979).
21. J. Fröhlich, G. Morchio and F. Strocchi, Phys. Lett. 89B:61 (1979).
22. J. Fröhlich, Comm. Math. Phys. 54:135 (1977); W. Driessler and J. Fröhlich, Ann. Inst. H. Poincaré 27A:167 (1977).
23. H.J. Borchers, R. Haag and B. Schroer, Nuovo Cim. 29:148 (1963).
24. J. Dixmier, "Les Algèbres d'operateurs dans l'espace Hilbertien", Paris Gauthier-Villars (1957).
25. H. Stapp, Phys. Rev. Lett. 50:467 (1983); Phys. Rev. D28:1386 (1983).
26. B. Schroer, Fortschr. Physik 11:1 (1963).
27. N.N. Bogoliubov and D.V. Shirkov, "Introduction to the Theory of Quantized Fields", 3rd ed. Interscience, Ch. 8 §46 (1980).
28. J. Tarski, Jour. Math. Phys. 7:560 (1966).
29. G. Morchio and F. Strocchi, Infrared problem in QED and electric charge renormalization, Ann. Phys. (in press).
30. G. Morchio and F. Strocchi, in preparation.
31. G. Morchio and F. Strocchi, Nucl. Phys. B211:471; B232:547 (1984).
32. P.A.M. Dirac, Can. J. Phys. 33:650 (1955); O. Steinmann, Perturbative QED in Terms of Gauge Invariant Fields, Bielefeld preprint BI-TP 83/19; E. D'Emilio and M. Mintchev, Fortschr. Physik 32:473 (1984).
33. G. Morchio and F. Strocchi, Ann. Inst. H. Poincaré XXIII, n.3 251 (1980).
34. G.F. Dell'Antonio and S. Doplicher, Jour. Math. Phys. 8:663 (1967); J.M. Chaiken, Ann. Phys. (N.Y.) 42:23 (1967).
35. K.M. Case and S.G. Gasiorowicz, Phys. Rev. 125:1055 (1962).
36. V. Gribov, Nucl. Phys. B206:103 (1982).
37. G. Morchio and F. Strocchi, submitted for publication.
38. S. Coleman, Secret Symmetries, in: Erice Summer School 1973, A. Zichichi ed., Academic Press (1975).
39. F. Strocchi, Comm. Math. Phys. 56:57 (1977); Gauss' law in local quantum field theory, in: "Field Theory, Quantization and Statistical Physics", E. Tirapegui ed., D. Reidel (1981).
40. T. Kennedy and C. King, Phys. Rev. Lett. 55:776 (1985).
41. J. Fröhlich, G. Morchio and F. Strocchi, Phys. Lett. 97B:249 (1980); Nucl. Phys. B190:553 [FS3] (1981).
42. G. Morchio and F. Strocchi, Phys. Lett. 104B:277 (1981).

43. S. Weinberg, Phys. Rev. D7:1068 (1973).
44. M. Creutz and T.N. Tudron, Phys. Rev. D17:2619 (1978).
45. See e.g. O. Brattelli and D.W. Robinson, "Operator Algebras and Quantum Statistical Mechanics," Vol. I, Springer-Verlag (1979), pp. 396-397.
46. G. Morchio and F. Strocchi, Spontaneous breaking of the Galilei group and the plasmon energy spectrum, to be published in Ann. Phys.
47. J. Swieca, Goldstone theorem and related topics, in: "Cargèse Lectures," Vol . 4, Gordon and Breach (1970).
48. R.F. Streater, Spontaneously Broken Symmetries, in "Many Degrees of Freedom in Field Theory," L. Streit ed., Plenum Press (1978).
49. G. Morchio and F. Strocchi, Comm. Math. Phys. 99:153 (1985).
50. J. Kogut and L. Susskind, Phys. Rev. D11:3594 (1975).
51. T.W. Kibble, Proc. Int. Conf. Elementary Particles, Oxford 1965, Oxford University Press (1965).
52. R. Haag, Nuovo Cim. 25:1078 (1962).
53. R. Haag and D. Kastler, Jour. Math. Phys. 5:848 (1964).
54. Ch. Gruber, J.L. Lebowitz and Ph. A. Martin, Jour. Chem. Phys. 75: 944 (1981); J.L. Lebowitz and Ph. A. Martin, Phys. Rev. Lett. 54:1506 (1985).
55. O.E. Lanford, II and D. Ruelle, Comm. Math. Phys. 13:194 (1969).

STOCHASTIC QUANTIZATION OF GAUGE FIELDS

Daniel Zwanziger*

Physics Department
New York University
New York, New York 10003

SUMMARY. Stochastic quantization of gauge fields is reviewed, with particular attention to the mechanism by which stochastic gauge-fixing resolves the Gribov ambiguity, and the light which this casts on the geometry of gauge orbits. Some geometric properties of gauge orbits are derived, and the difficulty which they present for the Faddeev-Popov method is reviewed. Stochastic quantization is introduced, and it is shown how the Euclidean probability distribution may be determined by a diffusion process with a drift force $\mathbf{K}=-\nabla S$, where $S=S(A)$ is the classical Yang-Mills action. The equivalence of the diffusion and Langevin equations is demonstrated. Stochastic gauge-fixing is explained, and it is demonstrated that an additional drift force, called the gauge-fixing force, may be introduced which does not affect expectation values of gauge-invariant quantities because it is everywhere tangent to the gauge orbits. The gauge-fixing force is non-conservative and thus cannot be accommodated in an action formalism. Properties of the gauge-fixing force are explained and it is shown how, by a mechanism of "instability stabilizes", it concentrates the probability near the interior Ω of the first Gribov horizon. As a new result, the diffusion equation is solved for large values of the gauge-fixing force, the solution being a

*Research supported in part by the National Science Foundation under grant no. PHY 84-13569

modified Faddeev-Popov formula, whereby the integration extends only over Ω. Another new result is the $\hbar$ or loop expansion of the effective action for stochastic gauge-fixing, and it is shown that all primitive divergences arize in the iterative calculation of an effective drift force $\mathbf{K}_{eff}$.

1. INTRODUCTION

The method of stochastic quantization was invented by Parisi and Wu.[1] It may be used to advantage in non-singular systems.[2] Its main attraction however would appear to be the new possibilites which it offers for quantizing systems with constraints arising from symmetries. This is at present a topic of active research in the quantization of the gravitational field[3] and it may prove to be of value for quantizing other systems such as strings. In the present lectures we shall deal only with its application to gauge fields. Current literature on this subject includes, to mention only some articles, a study of the equivalence of stochastic perturbation theory to conventional Feynman diagrams with Faddeev-Popov ghosts[4], proposals to use the fifth time as a gauge-invariant regulator[5,6], functional integral representations of the solution[7], studies of the dependence of gauge-dependent Green's functions on the initial conditions in the fifth time[8], and numerical simulation of the underlying stochastic process[9].

The basic idea of stochastic quantization is that the Euclidean probability distribution is determined by a diffusion process with a drift force, $\mathbf{K}$. In the original formulation of Parisi and Wu[1], the drift force is given by $\mathbf{K}=-\nabla S$, where S is the classical Yang-Mills action. With stochastic gauge-fixing[10], this force is modified by the addition of a force that does not affect the expectation values of gauge-invariant observables because it is tangent to the gauge orbits. This force is of necessity non-conservative, and thus cannot be accommodated in an action formalism, but it has the restoring property required to maintain an equilibrium, or time-independent, probability distribution. The analog in quantum gravity would be to introduce a force tangent to the orbits under the group of diffeomorphisms.

The approach of the present lectures is based upon the method of stochastic gauge-fixing which was reviewed previously by E. Seiler[11]. Our review also includes two new results: 1) We solve the equilibrium diffusion equation in the limit of a large gauge-fixing force, and obtain, as solution, a modified Faddeev-Popov formula. (See Section 8.) 2) We derive the $\hbar$ or loop expansion of the effective action and prove a theorem which states that all primitive divergences arise in the iterative calculation of an effective force, K_{eff}. (See Appendix.)

The motivation for the stochastic approach to the quantization of gauge fields is that it is free of the Gribov ambiguity[12] which afflicts the Faddeev-Popov method[13], that it is correct non-perturbatively, that it is ghost free, and that is it is defined by a local stochastic process. The same may be said of Wilson's lattice gauge theory[14] which has the further advantage of being regularized. However the relation between Wilson's lattice theory and continuum Euclidean quantum field theory (if it exists) remains obscure. It is even an open question as to whether Wilson's theory and the continuum theory have the same perturbative expansion[15]. (The answer depends on whether the weak coupling expansion and continuum limits are interchangable.) It must also be stated that the non-perturbative advantage of the stochastic method over the Faddeev-Popov method would be illusory without a non-perturbative regularization that preserves gauge invariance, at least in the cut-off or continuum limit. We briefly return to this issue in the concluding section. [There is little doubt that the two methods give the same perturbative expansion for gauge invariant quantities (see ref. 4, and Section 8 of the present lectures).]] The Gribov difficulty is by-passed in stochastic quantization because it is not necessary to make a section of the gauge orbits, which would require an *a priori* knowldedge of what they look like. On the contrary, because gauge orbits must ultimately receive equal weight, the method of stochastic gauge fixing (if it is correct) provides information about the gauge orbits, as we shall see.

In Section 2 we review some elementary geometrical properties of gauge orbits. In Section 3 we give a brief critique of the Faddeev-Popov method in relation to the issue of Gribov copies. In Section 4 we review the diffusion and Langevin equations in their traditional physical context as descriptions of diffusion under the influence of a drift force. In Section 5, stochastic quantization, and stochastic gauge-fixing by means of a non-conservative gauge fixing force are effected. In Section 6 some properties of the gauge-fixing force are derived. In Section 7 we see explicitly how the gauge-fixing force pushes the probability weight away from the Gribov copies. In Section 8 we derive a modified Faddeev-Popov formula by solving the diffusion equation in the limit of a large gauge-fixing force. In the Appendix we derive the $\hbar$ or loop expansion of the effective action and prove that all primitive divergences arise in the iterative calculation of an effective force K_{eff}. Our conclusions are presented in Section 9.

2. SOME ELEMENTARY PROPERTIES OF GAUGE ORBITS

In this section we present results from ref. 16. For Euclidean space-time we take a compactification of R^n, for example, a box with periodic boundary conditions which defines the torus T^n. To simplify the presentation, we shall initially assume that the vector potential $A_\mu(x)$, $\mu = 1, \ldots n$, is globally defined, which corresponds to zero instanton sector. More precisely, $A_\mu(x) = A_\mu{}^a(x)\lambda^a$ is a globally defined function with values in the Lie algebra of a semi-simple Lie group G. Here the λ^a are anti-hermitian matrices, normalized to $\mathrm{tr}\lambda^a\lambda^b = -\delta^{ab}$. Such a potential $A_\mu(x)$ corresponds to a point A in A-space.

Let A^g be the gauge transform of the point A by the local gauge transformation $g(x) \in G$,

$$A_\mu{}^g(x) = g^{-1} A_\mu g + g^{-1} \partial_\mu g. \tag{2.1}$$

Observe that the Hilbert distance between two points A_1 and A_2 is gauge invariant,

$$\| A_2 - A_1 \| = \| A_2{}^g - A_1{}^g \|,$$

where

$$\| A \|^2 \equiv \int d^n x \, \Sigma_{\mu,a} [A_\mu{}^a(x)]^2 = - \int d^n x \, \Sigma_\mu \, \mathrm{tr}[A_\mu(x)^2].$$

This motivates us to stipulate that A be an element of a Hilbert space H. We also stipulate that the local gauge transformations $g(x)$ be such that $A_1{}^g$ be an element of H for some $A_1 \in H$. We have, for any A_2,

$$\|A_2^g\| = \|A_2^g - A_1^g + A_1^g\| \le \|A_2^g - A_1^g\| + \|A_1^g\| = \|A_2 - A_1\| + \|A_1^g\|.$$

so $A_2^g \in H$. Thus our restriction on g(x) is independent of the choice of A_1 and implies that A^g exists for all $A \in H$. It follows that this restriction is compatible with the group property.

As g(x) varies over all local gauge transformations, A^g describes the gauge orbit of the point A. We consider the question: When, for a given A, is A^g as small as possible? More precisely, when is the Hilbert square norm of A^g,

$$S_A[g] \equiv \|A^g\|^2 = -\int \Sigma_\mu \operatorname{tr}[A_\mu{}^g(x)^2]\, d^n x \qquad (2.2)$$

a minimum? The answer is very simple[16].

Theorem1: For given A, the Hilbert norm $\|A^g\|$ is a minimum when 1)

$$\partial \cdot A^g(x) = 0 \qquad (2.3)$$

and 2), the Faddeev-Popov operator

$$L(A^g) \equiv -D(A^g)\cdot\partial = -\partial\cdot D(A^g) \qquad (2.4)$$

is non-negative. Here $D_\mu(A)$ is the covariant derivative. It is an anti-hermitian operator that acts on functions $\psi(x)$, that are elements of the Lie algebra according to $D_\mu(A)\psi = \partial_\mu\psi + [A_\mu,\psi]$. (Note that $[\partial_\mu, D_\mu(A)]\psi = [\partial\cdot A,\psi]$, so (2.4) is hermitian in virtue of (2.3).) To prove this we replace g by ge^w and obtain, to second order in w,

$$S_A[ge^w] = (A^g, A^g) - 2(\partial\cdot A^g, w) + (w, -\partial\cdot D(A^g)w). \qquad (2.5)$$

At a minimum, the first variation vanishes and the second variation is non-negative. QED. It would be valuable to prove that the action (2.2) actually does achieve its minimum. On the assumption that this is true, we conclude that it is always possible to choose a gauge such that the Landau gauge condition $\partial\cdot A = 0$ holds and that moreover the Fadeev-Popov operator $-D(A)\cdot\partial$ is non-negative. We call "optimal" the gauge choice which minimizes the Hilbert norm on each gauge orbit. The fundamental problem of Gribov copies is that, as Gribov has shown[12], there are, in general, more than one point on each gauge orbit such that $\partial\cdot A = 0$, so that the Landau gauge condition does not, in fact fix the gauge uniquely. An important open question which we shall address later is whether the additional condition that the Faddeev-Popov operator be non-negative suffices to fix the gauge.

The linear equation $\partial\cdot A=0$ defines a hyperplane in A-space. Our proof shows that every stationary point of the action $S_A[g] = \|A^g\|^2$ lies in this hyperplane and the every point of the hyperplane is a stationary point of this action (for some orbit). We call Ω the part of the hyperplane $\partial\cdot A=0$ where the Faddeev-Popov operator is non-negative and which coincides with minima of $S_A[g]$. Note that $-D(A)\cdot\partial$ vanishes trivially on all ψ in the linear subspace H_0 defined by $\partial_\mu\psi = 0$. If we consider $-D(a)\cdot\partial$ as an operator on H_0', the orthogonal complement to H_0, then the interior of Ω is the set of points (with $\partial\cdot A = 0$) where $-D(A)\cdot\partial$ is strictly positive or, in other words, where all its eigenvalues $\lambda_i(A)$ are strictly positive, $\lambda_i(A) > 0$. On the boundary of Ω the lowest eigenvalue $\lambda_0(A)$ vanishes, $\lambda_0(A) = 0$, and outside of Ω, at least one eigenvalue is negative, $\lambda_0(A) < 0$. The interior of Ω is non-empty because for $A = 0$, the Faddeev-Popov operator reduces to $-\partial^2$ which is strictly positive on H_0'. (Recall that our Euclidean space-time is a periodic box.) The boundary of Ω is known as the first Gribov horizon. The nth Gribov horizon occurs where the nth eigenvalue $\lambda_n(A)$ vanishes.

The shape of Ω is described by the following two theorems[16].

<u>Theorem 2</u>: Ω is convex.
We have to prove that if $A_1 \in \Omega$ and $A_2 \in \Omega$, then also $A\equiv\alpha A_1+(1-\alpha)A_2$ is also in Ω, where α is in the interval $0<\alpha<1$. To prove this, we note that $\partial\cdot A=\alpha\partial\cdot A_1+(1-\alpha)\partial\cdot A_2$, so $\partial\cdot A_1=\partial\cdot A_2=0$ implies $\partial\cdot A=0$. Similarly,

$$-D(A)\cdot\partial = \alpha[-D(A_1)\cdot\partial] + (1-\alpha)[-D(A_2)\cdot\partial]$$

and thus, for any ψ

$$(\psi, -D(A)\cdot\partial\psi) = \alpha\,(\psi, -D(A_1)\cdot\partial\psi) + (1-\alpha)\,(\psi, -D(A_2)\cdot\partial\psi).$$

Because, by hypothesis, the two terms on the right hand side are positive, so is the left hand side. QED.

<u>Theorem 3</u>: For a semi-simple group, the Gribov horizon is at a finite distance in any direction. In other words, if $A\neq 0$ is in Ω, then for some

positive λ, $A' \equiv \lambda A$ lies outside Ω. (This is a slightly modified version of a theorem of Singer[17].)
We have

$$(\psi, -D(\lambda A)\cdot\partial\psi) = (\psi, -\partial^2\psi) + \lambda\,(\psi, -A\cdot\partial\psi),$$

so, for sufficiently large $\lambda > 0$, the sign of the right hand side is the sign of $M(A) \equiv (\psi, -A\cdot\partial\psi)$. It is sufficient to show that for some ψ, the sign of $M(A)$ is negative. Pose $\psi^a = \exp(ick\cdot x)\varphi^a$, so

$$M(A) = -ick_\mu(\varphi^a, f^{abc}A_\mu{}^b\varphi^c) + (\varphi, -A\cdot\partial\varphi).$$

Hence, for sufficiently large $|c|$, $M(A)$ will have the sign of

$$N(A) \equiv ck_\mu u^{a*}(-if^{abc})(S, A_\mu{}^b S)u^c,$$

where we have taken φ to be of the form $\varphi^a(x) = S(x)u^a$, with $S(x)$ independent of the color index, and u^a independent of x. Provided $A_\mu{}^b(x)$ does not vanish identically, then one can choose $S(x)$ such that $v_\mu{}^b \equiv (S, A_\mu{}^b S)$ does not vanish identically. (For example, let $A_\mu{}^b(x)$ be non-zero and continuous at x_0, and let $S(x)$ have support in a sufficiently small neighborhood of x_0.) Choose k_μ such that the color vector $w^b \equiv k_\mu v_\mu{}^b$ does not vanish identically. Then we have $N(A) = cu^{a*}\lambda_{ac}u^c$, where $\lambda_{ac} \equiv -if^{abc}w^b$ is a hermitian generator in the adjoint representation of a semi-simple group. Consequently it has at least one non-vanishing eigenvalue ν. Let u be the corresponding normalized eigenvector, so $N(A) = c\nu$. As the sign of c is at our disposal, we may make $N(A)$ negative. QED.

Note that we used the fact that the gauge group was semi-simple to conclude that the Gribov horizon is at a finite distance in all directions. (For an Abelian group, the Gribov horizon is at infinity.) Because A-space is infinite dimensional, it does not follow that Ω is compact or even that it is uniformly bounded. However for the discretization of A-space by a finite lattice given in ref. 9 (SSZ, section 2), Ω is in fact compact.

The situation described by these theorems is depicted in Fig. 1. The Gribov copies outside of Ω are saddle points of $S_A[g]$. A Gribov copy inside Ω would be a local relative minimum of $S_A[g]$.

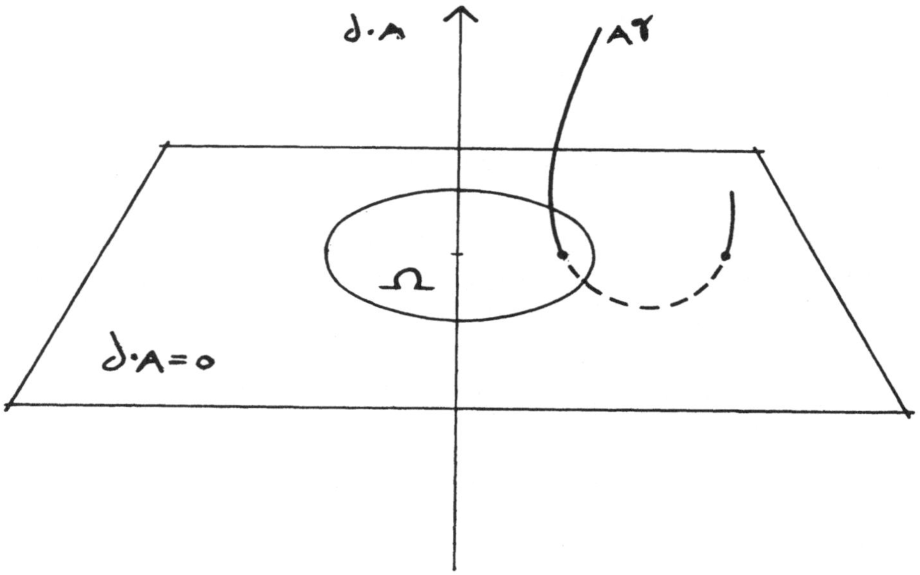

Fig. 1. In this three-dimensional perspective drawing of A-space, the horizontal plane represents the hyperplane defined by the gauge condition $\partial \cdot A=0$. It contains the shaded area which represents Ω, the convex interior of the first Gribov horizon, which is bounded in every direction. The curve labelled A^g represents the gauge orbit through a generic point A. Its intersections with the horizontal plane are Gribov copies.

Our considerations may be extended to non-zero instanton sector, i.e. a non-trivial bundle, as follows. The connection A is a function on various patches I, which cover the base space, $A=\{A_I\}$. On the overlap between patches, the A_I are related by the transition formula,

$$A_{\mu K}=U_{IK}^{-1}(A_{\mu I}+\partial_\mu)U_{IK},$$

where the transition functions U_{IK} are fixed gauge transformations. Let $A^*=\{A_I^*\}$ be a fixed connection, for example, the instanton solution itself. The quantity $a\equiv A-A^*$ transforms homogeneously under these transition formulae which are thus compatible with the linear gauge condition $D(A^*)\cdot a=0$. One considers the action

$$S_{A,A^*}(g) = \|A^g - A^*\|^2,$$

and similar conslusions follow as above[16].

<u>On the assumption</u> that the action $S_A[g]$ actually achieves its minimum, our first theorem implies that every gauge orbit passes through Ω. It would be valuable to prove this, and we shall adapt it as a working hypothesis.[18] If every gauge orbit intersects Ω only once, (apart from a set which would be, in some sense, "of measure zero") then Ω would be a convient image of the orbit space, and would be a natural configuration space for a gauge field theory. An interesting point is that Ω is convex, and thus topologically trivial, whereas, as beautifully explained in the lectures of Alvarez-Gaumé, the famous anomalies are precisely due to the topological non-triviality of the orbit space. This apparent contradiction is resolved if the surface of Ω is sewn together in some terrible way. If some gauge orbits intersect Ω more than once, presumably a subset of Ω would be the configuration space. Nobody should have any doubt however that a bound on the configuration space is significant for the dynamics. Cutkosky[18] has proposed that in a non-Abelian gauge theory, the bound provided by the Gribov horizon is responsible for a mass gap, which accords with Gribov's original suggestion[12].

{The phenomenon of the Gribov horizon may also be exhibited on the Wilson lattice. Let U_ℓ be a set of link variables with values in a Lie group, and let $U_\ell{}^g$ be their gauge orbit. More explicitly, the gauge transformation g is a set of site variables g(x), with values in the Lie group, where x labels a lattice site, and $U_\ell{}^g = g(x) U_\ell g^{-1}(x')$, where the link ℓ joins the sites x and x'. Consider the action

$$S_U[g] = -\sum_\ell \mathrm{Re}\,\mathrm{tr}\, U_\ell{}^g, \qquad (2.6)$$

considered as a function of the g(x), for fixed U_ℓ, on a finite periodic lattice. It is stationary when a lattice analog of the gauge condition $\partial \cdot A = 0$ holds, which defines a gauge manifold M in U-space. [U-space is of dimension nCV, and M is of dimension CV, where n is the dimension of space-time, C is the dimension of the Lie group, and V is the (finite) number of sites of the lattice.] The second variation is a square matrix $L(U^g)$, of dimension (CV)x(CV) which corresponds to the Faddeev-Popov operator. Because the g(x) are defined on a compact manifold, we know that the action $S_U[g]$ does achieve its minimum. At the minimum, barring accidental degeneracies, C(V-1) eigenvalues of $L(U^g)$ are positive, and the remaining C, which correspond to global gauge transformations, are zero. The action $S_U[g]$ is bounded above as well as below, so it also achieves its maximum where, again barring accidental degeneracies, $L(U^g)$ has $C(V-1)$ negative eigenvalues and C that are zero. The gauge manifold M is divided into $C(V-1)+1$ regions, according to the number of positive eigenvalues of $L(U^g)$. Consequently one expects Ω, the region where there are no negative eigenvalues (and whose boundary defines the lattice analog of the Gribov horizon), to be a rather restricted part of the gauge manifold. Nevertheless we know that every gauge orbit passes through Ω because the action $S_U[g]$ does achieve its minimum. The gauge choice corresponding to the absolute minimum of $S_U[g]$ is "optimal" in the sense that it picks out a point on each orbit such that the link variables U_ℓ are all as close to unity as possible.}

3. SOME ASPECTS OF THE FADDEEV-POPOV METHOD

Consider the Faddeev-Popov weight in the Landau gauge,

$$dA \exp[-S_{Cl}]\, \delta(\partial\cdot A)\, \det[-D(A)\cdot\partial]. \qquad (3.1)$$

(To be meaningful, this expression of course requires a renormalization prescription, but we restrict our discussion to its formal properties. The issue of Gribov copies is in fact a purely classical question about gauge orbits, and shows up to zeroth order in $\hbar$, where no renormalization is required. See the discussion at the end of the Appendix.) The weight is concentrated on the hyperplane $\partial\cdot A=0$, which corresponds to the Landau gauge. In consequence, the operator $-D(A)\cdot\partial$ is hermitian, and the determinant may be defined by

$$\Pi_{i=0}^{\infty} [\lambda_i(A)/\lambda_i(0)] \qquad (3.2)$$

where the $\lambda_i(A)$ are the real eigenvalues of $-D(A)\cdot\partial$. The problem of the Gribov copies is that a given gauge orbit in general intersects the hyperplane $\partial\cdot A=0$ more than once. (This also occurs in the Coulomb gauge in Minkowski space.) Thus if the absolute value of the determinant were understood in (3.1) some orbits would be counted too many times. It has been argued that Gribov copies occur in pairs with opposite sign of the determinant, so that their contribution cancels out of the expectation values of gauge invariant quantities[19]. In this sense, formula (3.1) may be correct non-perturbatively, but positivity in Euclidean A-space is lost, so powerful probabilistic methods are not directly applicable. The non-positivity of the determinant is clear from (3.2), because, as we have seen, outside the first Gribov horizon (the boundary of Ω) some of the eigenvalues $\lambda_i(A)$ of $-D(A)\cdot\partial$ are negative, and it is easy to construct examples where the number of negative eigenvalues is odd. If, as we have suggested, every gauge orbit intersects Ω, then one could, in principle, replace det by $\det_+$ in (3.1), where $\det_+ = \det$, when there are no negative eigenvalues, and is zero otherwise. Both det and $\det_+$ have the same perturbative expansion (because $\det(0)=1$). We may represent $\det_+$, for example, by

$$\det{}_+[-D(A)\cdot\partial] = [\int d\varphi \exp(\varphi, D(A)\cdot\partial\varphi)]^{-2}. \tag{3.3}$$

For if $-D(A)\cdot\partial$ has a negative eigenvalue, the Gaussian integral blows up, giving zero when raised to the negative power. Here one divides by the Gaussian integral over bose ghosts instead of multiplying by the integral over the usual Faddeev-Popov fermi ghosts, so there is no corresponding local action. We turn next to the method of stochastic quantization which determines a positive weight in Euclidean A-space by a local stochastic process.

4. DIFFUSION AND LANGEVIN EQUATIONS

In this section we abandon gauge fields to briefly review the apparently unrelated subject of diffusion theory. The diffusion equation,

$$\partial/\partial t\, P(\mathbf{x},t) = \nabla\cdot\{[D\nabla - \gamma\mathbf{K}(\mathbf{x})]\, P(\mathbf{x},t) \tag{4.1}$$

describes the evolution of the probability distribution $P(\mathbf{x},t)$ of an ensemble of Brownian particles subject to an external time-independent drift force $\mathbf{K}(\mathbf{x})$. This equation may be justified physically as follows. The drift force is assumed to cause a drift current $\gamma\mathbf{K}(\mathbf{x})P(\mathbf{x},t)$ which, as in Ohm's law, is proportional to the external force $\mathbf{K}(\mathbf{x})$ and to the density of carriers $P(\mathbf{x},t)$. (As we shall see shortly, γ is the inverse of the coefficient of friction.) Similarly, the concentration gradient $\nabla P(\mathbf{x},t)$ is assumed to cause a diffusion current, $-D\nabla P(\mathbf{x},t)$, proportional to it but oppositely directed. (Heat flows from hot to cold.) The proportionality constant D is known as the diffusion coefficient. The total current $\mathbf{j}$ is given by,

$$\mathbf{j}(x,t) = -D\nabla P(\mathbf{x},t) + \gamma \mathbf{K}(\mathbf{x})P(\mathbf{x},t). \tag{4.2}$$

The diffusion equation (4.1) is simply the statement of local conservation of numbers of particles,

$$(\partial/\partial t)P(\mathbf{x},t) = -\nabla\cdot\mathbf{j}(\mathbf{x},t). \tag{4.3}$$

It is obvious that total probability is conserved, $(\partial/\partial t)\int P(\mathbf{x},t)d^n x=0$. One may show that the diffusion equation has the property that $P(\mathbf{x},t)$ stays positive, $P(\mathbf{x},t)>0$ for $t>0$, if it was so initially, $P(\mathbf{x},0)\geq 0$, as is required for physical interpretation. An important theorem, related to the Frobenius theorem for a discrete Markov process, states that if the diffusion equation possesses a positive, normalized, time-independent solution $P_E(\mathbf{x})$,

$$\nabla\cdot(D\nabla P_E - \gamma \mathbf{K} P_E) = 0, \tag{4.4}$$

then this solution, called the "equilibrium distribution", is unique, and moreover, every positive normalized solution relaxes to it,

$$\lim_{t\to\infty} P(\mathbf{x},t) = P_E(\mathbf{x}). \tag{4.5}$$

In particular, if the drift force is conservative, $\mathbf{K}=-\nabla V$, then an equilibrium distribution exists and is given by $P_E(\mathbf{x})=c\exp[-\gamma V(\mathbf{x})/D]$, provided that this expression is normalizable. To be consistent with the Boltzmann distribution, $c\exp[-\beta V(x)]$, where $\beta=(kT)^{-1}$, one must have

$$\gamma = D\beta, \tag{4.6}$$

a relation first discovered by Einstein, and an instance of the fluctuation-dissipation theorem. For it relates the fluctuations, characterized by the diffusion constant D to the dissipation characterized by the frictional coefficient $\alpha=\gamma^{-1}$. (See (4.9) and (4.10), below.)

If the force is non-conservative, $dK \neq 0$, then it is not possible, in general, to write down the exact equilibrium distribution $P_E(\mathbf{x})$. Moreover in this case there is a steady-state non-zero equilibrium current

$$\mathbf{j}_E(\mathbf{x}) = -D\nabla P_E(\mathbf{x}) + \gamma \mathbf{K}(\mathbf{x}) P_E(\mathbf{x}). \qquad (4.7)$$

For with $\mathbf{v} = \mathbf{j}_E/P_E$, we have $\mathbf{v} = -D\,\nabla(\ln P_E) + \gamma\mathbf{K}$, so $dv = \gamma dK \neq 0$, and consequently $\mathbf{v}$ and thus $\mathbf{j}_E$ is non-zero.

It is possible to describe the same diffusion process in terms of the trajectory $\mathbf{x}(t)$ of an individual particle in Brownian motion. Let it have mass m, and be subject to an external force $\mathbf{K}(\mathbf{x})$, a frictional force, $-\alpha\dot{\mathbf{x}}$, where α is the frictional coefficient. It is also subject to a random fluctuating force $\alpha\boldsymbol{\eta}(t)$ which describes the force due to random collisions with the particles in the surrounding medium. (The frictional coefficient has been factored out of $\boldsymbol{\eta}$ for later convenience.) It is represented mathematically as white noise, or in other words, as a free field, or Gaussian random variable, with mean and 2-point function given by

$$\langle \eta_i(t) \rangle = 0, \qquad \langle \eta_i(t)\, \eta_j(t') \rangle = 2D\delta(t-t'). \qquad (4.8)$$

The frictional force arises because the particle suffers more random collisions on the side toward which it as moving, as anyone who has ridden a bicycle in the rain knows. The equation of motion of the particle is given by Newton's second law,

$$m d^2\mathbf{x}/dt^2 = -\alpha d\mathbf{x}/dt + \mathbf{K}(\mathbf{x}) + \alpha\boldsymbol{\eta}(t). \qquad (4.9)$$

An important limiting case of this equation occurs when the inertial force $md^2\mathbf{x}/dt^2$ is negligible compared to the other terms. (This is typical of a small particle, for example a living cell, in a fluid medium.) This gives the Langevin equation

$$d\mathbf{x}/dt = \gamma\mathbf{K}(\mathbf{x}) + \boldsymbol{\eta}(t), \qquad (4.10)$$

where we have written $\gamma = \alpha^{-1}$. For a given initial condition, $\mathbf{x}(0) = \mathbf{x}_0$, this equation is equivalent to the integral equation

$$\mathbf{x}(t) = \mathbf{x}_0 + \int_0^t [\gamma \mathbf{K}(\mathbf{x}(t')) + \boldsymbol{\eta}(t')] \, dt'. \tag{4.11}$$

Given any $\boldsymbol{\eta}(t)$, this equation has a unique solution for which $\mathbf{x}(t)$ is a functional of $\boldsymbol{\eta}$,

$$\mathbf{x}(t) = \mathbf{x}_{[\eta]}(t). \tag{4.12}$$

Because $\boldsymbol{\eta}(t)$ is a random variable, so is $\mathbf{x}_{[\eta]}(t)$. An important property which follows from the integral equation is that $\mathbf{x}_{[\eta]}(t)$ only depends on $\boldsymbol{\eta}(t')$ for $0<t'<t$.

The probability distribution for finding the particle at $\mathbf{x}$ at time t is given by

$$P(\mathbf{x},t) = \langle \delta(\mathbf{x} - \mathbf{x}_{[\eta]}(t) \rangle_\eta \tag{4.13}$$

where $\langle ... \rangle_\eta$ means average over the free field η. We shall show that the probability distribution defined in this way satisfies the diffusion equation (4.1). From

$$\mathbf{x}_{[\eta]}(t+\epsilon) = \mathbf{x}_{[\eta]}(t) + \int_t^{t+\epsilon} [\gamma \mathbf{K}(\mathbf{x}(t')) + \boldsymbol{\eta}(t')] \, dt',$$

we obtain

$$\begin{aligned} P(\mathbf{x},t+\epsilon) = {} & \langle \delta(\mathbf{x} - \mathbf{x}_{[\eta]}(t) \rangle_\eta \\ & - (\partial/\partial x_i) \langle \delta(\mathbf{x} - \mathbf{x}_{[\eta]}(t)) \int_t^{t+\epsilon} [\gamma K_i(\mathbf{x}(t')) + \eta_i(t')] \, dt' \rangle_\eta \\ & + (1/2) (\partial^2/\partial x_i \partial x_j) \langle \delta(\mathbf{x} - \mathbf{x}_{[\eta]}(t)) \int_t^{t+\epsilon} \eta_i(t') dt' \int_t^{t+\epsilon} \eta_j(t'') dt'' \rangle_\eta \end{aligned}$$

plus higher order terms. The white noise property of $\eta(t)$ gives

$$\langle f[\eta] \, g[\eta] \rangle_\eta = \langle f[\eta] \rangle_\eta \langle g[\eta] \rangle_\eta$$

provided $f[\eta]$ depends on $\eta(t)$ and $g[\eta]$ on $\eta(t')$ only for $t<t'$. In particular we have

$$\langle f[\eta] \, \eta_i(t') \rangle_\eta = \langle f[\eta] \rangle_\eta \langle \eta_i(t') \rangle_\eta = 0$$

and

$$\langle f[\eta]\, \eta_i(t')\, \eta_j(t'') \rangle_\eta = \langle f[\eta]\rangle_\eta \langle \eta_i(t')\, \eta_j(t'') \rangle_\eta$$

$$= \langle f[\eta]\rangle_\eta\, 2D\delta_{ij}\, \delta(t'-t''),$$

provided $f[\eta]$ depends on $\eta(t)$ only for $t<t'$, and $t<t''$. This gives, to order ϵ,

$$P(\mathbf{x},t+\epsilon) = P(\mathbf{x},t) - (\partial/\partial x_i)[P(\mathbf{x},t)\ \gamma K_i(\mathbf{x})]\epsilon + (\partial/\partial x_i)D(\partial/\partial x_i)P(\mathbf{x},t)\epsilon,$$

where we have used

$$\int_t^{t+\epsilon} \int_t^{t+\epsilon} 2D\delta_{ij}\, \delta(t-t')dt'dt'' = 2D\delta_{ij}\,\epsilon.$$

The diffusion equation follows by letting $\epsilon \to 0$. This shows the equivalence of the Langevin and diffusion equations, and we verified, incidentally that $\gamma=\alpha^{-1}$, where α is the frictional coefficient.

An ergodic theorem holds which states that the time average over the motion of a *single* Brownian particle approaches, for sufficiently long times, the ensemble average over the equilibrium distribution, $P_E(\mathbf{x})$, if it exists. This is the basis for calculation of ensemble averages by Monte Carlo *simulation*, that is, by following the evolution of a single particle as it evolves under the stochastic process described by the Langevin equation.

5. STOCHASTIC QUANTIZATION OF GAUGE FIELDS

Consider a Euclidean quantum field theory with probability distribution $Z^{-1}\exp(-S/\hbar)$, where $S=S[\varphi]$ is a local action, and φ is a Bose field. By what has been said, this is the unique distribution to which any solution of the diffusion equation

$$(\partial/\partial t)P = \sum_i \partial/\partial\varphi_i[(\hbar\partial/\partial\varphi_i + \partial S/\partial\varphi_i)P] \qquad (5.1)$$

relaxes, provided that $\exp(-S/\hbar)$ is normalizable. Note that $\hbar$ has been identified with the classical diffusion constant D, and so, by the magic of Euclidean quantum field theory, the quantum fluctuations are assimilated to classical fluctuations. We will henceforth set $\hbar=1$, except in the Appendix. (We have also taken $\gamma=1$.) Here $P=P(\varphi,t)$, the discrete index i represents the continuum of Euclidean space-time points x, the partial derivative $\partial/\partial\varphi_i$ symbolically represents the functional derivative $\delta/\delta\varphi(x)$, and $\sum_i$ means $\int d^n x$. We will generally omit the summation sign.

A gauge theory presents the difficulty that $\exp\{-S_{cl}[A]\}$, where $S_{cl}[A]$ is a gauge-invariant functional of A, is not normalizable because of the infinite volume of the gauge orbits. Parisi and Wu[1] pointed out that the time-dependent diffusion equation for $P=P(A,t)$,

$$\partial P/\partial t = \partial/\partial A_i[(\partial/\partial A_i + \partial S_{cl}/\partial A_i)P] \qquad (5.2)$$

possesses (modulo renormalization) a normalized solution for finite times t, provided $P(A,0)$ is normalized, even though the limit as $t \to \infty$ of $P(A,t)$ does not exist as a normalized distribution. (Here the index i represents Lorentz and color index as well as Euclidean space-time position x.) This is entirely analogous to the fact that the solution to the free diffusion equation

$$P(x,t) = (2\pi Dt)^{-1/2} \exp(-x^2/2Dt) \qquad (5.3)$$

has no equilibrium limit as $t \to \infty$. Pointwise in x, it converges to 0, hardly a normalizable distribution! From this point of view, the problem of a gauge theory is the trivial absence of a restoring force along the direction of the gauge orbits. Parisi and Wu observed that since one is only interested in the expectation value $\langle I \rangle$ of gauge invariant objects, $I(A)$, one does not really care how the probability is distributed along the gauge orbits. They made the intriguing suggestion that although the solution $P(A,t)$ of Eq. (5.2) has no equilibrium limit as $t \to \infty$, the expectation of gauge-invariant objects

$$\langle I \rangle_t = \int I(A)P(A,t)\, dA \qquad (5.4)$$

should have a finite limit

$$\langle I \rangle = \lim_{t \to \infty} \langle I \rangle_t, \qquad (5.5)$$

which is moreover independent of the initial normalized distribution $P(A,0)$. The analogy presented by $P(x,t)$, Eq. (5.3), is that the (trivial) expectation value of $I(x)=1$, namely $\langle I\rangle_t=\int P(x,t)dx=1$ has a perfectly well defined limit as $t \to \infty$, namely 1, even though $P(x,t)$ does not.

This approach runs into the practical difficulty that for purposes of renormalization, one wants individual diagrams to be finite or renormalizable, not just their formally gauge invariant sum. For example, in lowest order, the gluon propagator is of the form

$$(\delta_{\mu\nu} - k_\mu k_\nu/k^2)/k^2 + 2tk_\mu k_\nu/k^2$$

if the initial distribution is $P(A,0)=\delta(A)$. Here the longitudinal part (which is time dependent and thus should cancel out when different diagrams are combined) is of order k^0 instead of k^{-2}, so the ultraviolet divergences in individual diagrams are worse than usual.

This difficulty may be overcome by the method of stochastic gauge fixing. Consider the Langevin equation which is equivalent to (5.2). With the substitution

$$\partial/\partial A_i \to \delta/\delta A_\mu{}^a(x),$$

the drift force in (5.2) is given explicitly by

$$K_{cl\mu}{}^a(x) \equiv - \delta S_{cl}/\delta A_\mu{}^a(x) = (D_\lambda F_{\lambda\mu})^a(x), \tag{5.6}$$

corresponding to the classical Yang-Mills action,

$$S_{cl} = (1/4)\int\sum (F_{\mu\nu}{}^a)^2\, d^nx. \tag{5.7}$$

Thus the Langevin equation (4.10) corresponding to (5.2) is

$$(d/dt)A_\mu{}^a(x,t) = (D_\lambda F_{\lambda\mu})^a(x) + \eta_\mu{}^a(x,t), \tag{5.8}$$

which is supplemented by any convenient initial condition, $A_\mu{}^a(x,0)=A_\mu{}^a(x)_0$. Here η is the Gaussian random variable with mean and 2-point function

$$\langle \eta_\mu{}^a(x,t)\rangle = 0 \qquad (5.9a)$$

$$\langle \eta_\mu{}^a(x,t)\eta_\nu{}^b(y,t')\rangle = 2\delta_{\mu\nu}\delta^{ab}\delta(x-y)\delta(t-t'). \qquad (5.9b)$$

Observe now that we may add to the right hand side of (5.8) the generator of a gauge transformation, namely a force of the form $(D_\mu v)^a$, without changing the value of any gauge invariant function of A(t). Here $v^a(x)$ may in principle be anything. It is convenient to choose v^a to be a Lorentz scalar, which is a local function of A, with dimension such that the "gauge-fixing force" $(D_\mu v)^a$ has the same dimension as the classical force (5.6), and which is a global color vector. This gives $v=\partial\cdot A$, and we have

$$K_{gf\mu}{}^a(x) = (D_\mu \partial\cdot A)^a(x). \qquad (5.10)$$

With

$$K_\mu{}^a(x) \equiv K_{cl\mu}{}^a(x) + \alpha K_{gf\mu}{}^a(x), \qquad (5.11a)$$

where α is a free parameter, we have

$$K_\mu{}^a(x) = (D_\lambda F_{\lambda\mu})^a(x) + \alpha(D_\mu \partial\cdot A)^a(x). \qquad (5.11b)$$

Thus the Langevin equation (5.8) may be replaced by

$$(d/dt)A_\mu{}^a(x,t) = K_\mu{}^a(x) + \eta_\mu{}^a(x,t). \qquad (5.12)$$

The corresponding diffusion equation follows from the equivalence of the Langevin and diffusion equations, and we may replace Eq. (5.2) by

$$\partial P/\partial t = \partial/\partial A_i[(\partial/\partial A_i - K_i)P] \qquad (5.13a)$$

or more explicitly by

$$\partial P/\partial t = \int d^n x\, \delta/\delta A_\mu{}^a(x)[(\delta/\delta A_\mu{}^a(x) - K_\mu{}^a(x))P]. \qquad (5.13b)$$

Here $K_\mu{}^a$ is given in Eq. (5.11) and is the sum of both K_{cl} and K_{gf}. The purpose of the gauge-fixing force is to give a diffusion equation which has a time-independent equilibrium solution. It is determined by an equation from which all reference to the fictitious fifth time is eliminated,

$$\int d^n x\, \delta/\delta A_\mu^a(x) [(\delta/\delta A_\mu^a(x) - K_\mu^a(x))P] = 0. \qquad (5.14)$$

The remaining sections will be devoted to a discussion of the solution to this equation. In ref. (10), an alternate derivation of the gauge-fixing force is presented which is based on the diffusion equation, without reference to the Langevin equation.

Note that the gauge-fixing force is non-conservative,

$$[\delta/\delta A_\mu^a(x)]\, K_{gf\nu}^b(y) - [\delta/\delta A_\nu^b(y)]\, K_{gf\mu}^a(x) \neq 0 \qquad (5.15)$$

so it cannot be expressed as the gradient of an action,

$$K_{gf} \neq -\delta S_{gf}/\delta A. \qquad (5.16)$$

The possibility of a non-conservative but local gauge fixing force is the new feature which is available in stochastic quantization that is not present in an action formalism.

6. PROPERTIES OF THE GAUGE-FIXING FORCE

1. <u>The gauge-fixing force is a restoring force</u>. Consider the classical motion

$$\partial A/\partial t = \alpha K_{gf}. \qquad (6.1)$$

We have

$$(d/dt)\|A\|^2 = 2(A, \partial A/\partial t) = 2\alpha(A, K_{gf}) = 2\alpha(A, D\partial\cdot A) = -2\alpha(D\cdot A, \partial\cdot A)$$

$$(d/dt)\|A\|^2 = -2\alpha(\partial\cdot A, \partial\cdot A) = -2\alpha\|\partial\cdot A\|^2 \leq 0, \qquad (6.2)$$

where we have used Hilbert space notation. Thus the gauge-fixing force reduces the Hilbert norm of A, and in this sense it is a restoring force, provided we choose $\alpha > 0$.

2. The locus of points of equilibium under the gauge fixing force consists of the hyperplane $\partial \cdot A = 0$. From its definition, (5.10) the gauge-fixing force vanishes whenever $\partial \cdot A = 0$. Moreover the inequality in (6.2) holds whenever $\partial \cdot A \neq 0$, so the gauge-fixing force vanishes only on this hyperplane.

3. The equilibrium is stable inside the Gribov horizon and unstable outside it. We have

$$(d/dt)\|\partial \cdot A\|^2 = 2(\partial \cdot A, \partial \cdot (\partial/\partial t)A) = -2\alpha(\partial \cdot A, [-\partial \cdot D(A)]\partial \cdot A). \qquad (6.3)$$

The Faddeev-Popov operator has made its appearance inside the expectation value. Consider now a point A in the neighborhood of the hyperplane $\partial \cdot A = 0$, and write $A = A^0 + \epsilon A^1$, where A^0 is in the hyperplane ($\partial \cdot A^0 = 0$), and A^1 is in the orthogonal subspace. To leading order in ϵ, we have

$$(d/dt)\|\partial \cdot A\|^2 = -2\alpha(\partial \cdot A^1, [-\partial \cdot D(A^0)]\partial \cdot A^1)\epsilon^2. \qquad (6.4)$$

As discussed in Section (2), the Faddeev-Popov operator is strictly positive in Ω, the region inside the Gribov horizon, Thus, if A^0 (with $\partial \cdot A^0 = 0$) lies inside the Gribov horizon, then, for all sufficiently close neighboring points, the left-hand side is negative, and $\|\partial \cdot A\|^2$ decreases under the action of K_{gf}, and so Ω is an attractor. On the other hand, if A^0 (with $\partial \cdot A^0 = 0$) lies outside the Gribov horizon, then the Faddeev-Popov operator has at least one negative eigenvalue, and so, for some neighboring points $\|\partial \cdot A\|^2$ is increasing, and the equilibrium is unstable. This situation is described in Fig. 2.

7. INSTABILITY STABILIZES

It is instructive to consider the Langevin equation

$$A_\mu{}^a(x,t) = K_{gf\mu}{}^a(x) + \eta_\mu{}^a(x,t), \qquad (7.1)$$

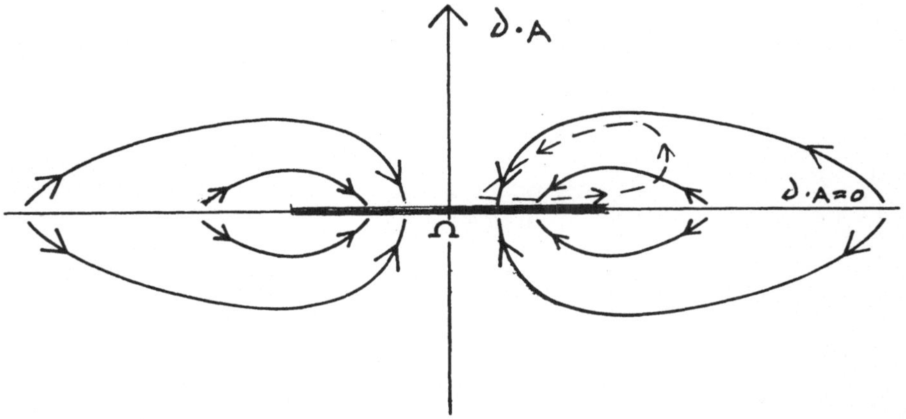

Fig. 2. Instability stabilizes. In this two-dimensional picture of A-space, the horizontal line represents the hyperplane $\partial \cdot A=0$, of which the thickened part represents the interior Ω of the first Gribov horizon. The solid curves represent field lines of the gauge-fixing force $D_\mu \partial \cdot A$. The dotted line is a typical Brownian path in A-space of a "particle" under the influence of this drift force and random thermal forces.

where the only force that acts is the gauge-fixing force. Let the corresponding Brownian particle start out near Ω. Its motion is represented by the dotted line in Fig. 2. For sufficiently large α, the gauge-fixing force initially keeps the particle near Ω. However under the influence of the fluctuating force η, it may drift horizontally along Ω (where K_{gf} vanishes), and out through the Gribov horizon to the region where the hyperplane $\partial \cdot A=0$ is no longer an attractor. Because of the fluctuating force η, it inevitably picks up a component in the unstable subspace. The gauge-fixing force then drives it away from the hyperplane $\partial \cdot A=0$, and it develops an exponentially growing longitudinal componant. At this point the restoring nature of the gauge-fixing force makes itself felt, $(d/dt)\|A\|^2=-2\alpha\|\partial \cdot A\|^2$, and it is driven back toward the origin, instead of escaping to infinity along the hyperplane of equilibrium.

This scenario, which we call "instability stabilizes" is supported[22] by numerical studies[9], of which Fig. 3 is a sample illustration. It may be understood by considering, as an analog of the the gauge-fixing force, a simple drift force in 2 dimensions of the form

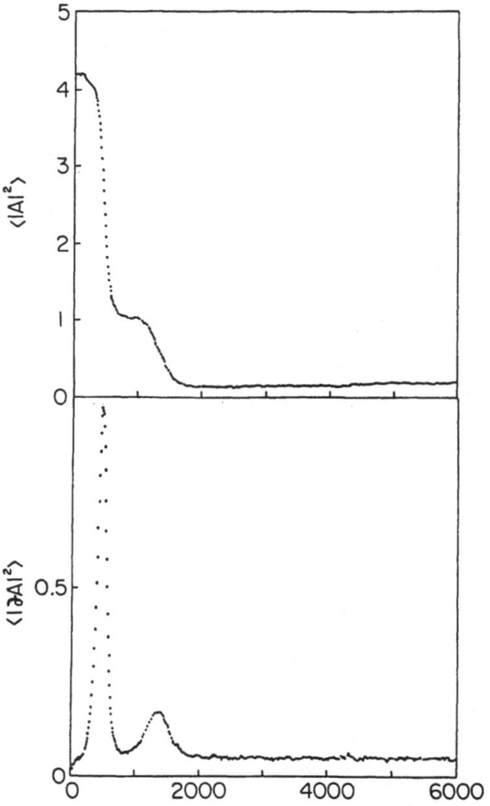

Figs. 3a, b. Data, provided by the courtesy of D. Nesic[9], from a numerical simulation of a 4^3 lattice discretization of the Langevin equation (5.12). The initial point is a non-equilibrium point beyond the second Gribov horizon. The initial, non-equilibrium motion shows $\langle \| A \|^2 \rangle$ decreasing rapidly where $\langle \| \partial \cdot A \|^2 \rangle$ is large, in accordance with Eq. (6.2). The two maxima in $\langle \| \partial \cdot A \|^2 \rangle$ are interpreted as passage through two Gribov horizons. The x-axis counts sweeps.

$$\mathbf{K} = (K_x, K_y) = (-\gamma x y^2, (1-x^2)y) \tag{7.2}$$

where $\alpha>0$ and $\gamma\geq 1$ are constants. With $(d/dt)(x,y)=\alpha(K_x,K_y)$, we find easily

$$(d/dt)(x^2 + y^2) = -2\alpha[1 + (\gamma-1)x^2]y^2 \tag{7.3a}$$

$$(d/dt)(y^2) = -2\alpha(1-x^2)y^2 \tag{7.3b}$$

which are the analogs of (6.2) and (6.3). We see that for $y\neq 0$ the length of the radius vector is decreasing and thus that the force is restoring. Moreover the line $y=0$ is attractive for $x^2<1$, and repulsive fo $x^2>1$. Here y plays the role of $\partial\cdot A$, the region Ω is the interval $-1\leq x\leq 1$, and the Gribov horizon consists of the points $x=\pm 1$. From (7.2) the lines of flow satisfy

$$dy/dx = (1-x^2)/(\gamma x y). \tag{7.4a}$$

with solution

$$\gamma y^2 = \ln x^2 - x^2 + c, \tag{7.4b}$$

and are shown in figure 4. The correctness of the "instability stabilizes" scenario of this two-dimensional model rests on whether or not the diffusion equation with drift force (7.2) possesses a normalized time-independent equilibrium solution, $P_E(x,y)$. This was first answered in the affirmative by computer simulations of I. O. Stamatescu[20]. More recently, this was proven analytically by G. Iona-Lasinio and G. Dell'Antonio[21,22], who constructed the appropriate Liapunov function. For the case $\gamma=2$ and $\alpha=9$, the exact solution was found by E. Seiler

$$P(x, y) = (6\pi\sqrt{2})^{-1}\exp[-(3/2)x^2 \ -3y^2].$$

As a warm-up for the following section, it is clarifying to find the equilibrium probability distribution for this model in the limit $\alpha\to\infty$. For finite α it satisfies the time-independent diffusion equation

$$\nabla\cdot(\nabla P - \alpha\mathbf{K}P) \tag{7.5}$$

with $\mathbf{K}$ given in (7.2) This force is non-conservative, so we cannot write

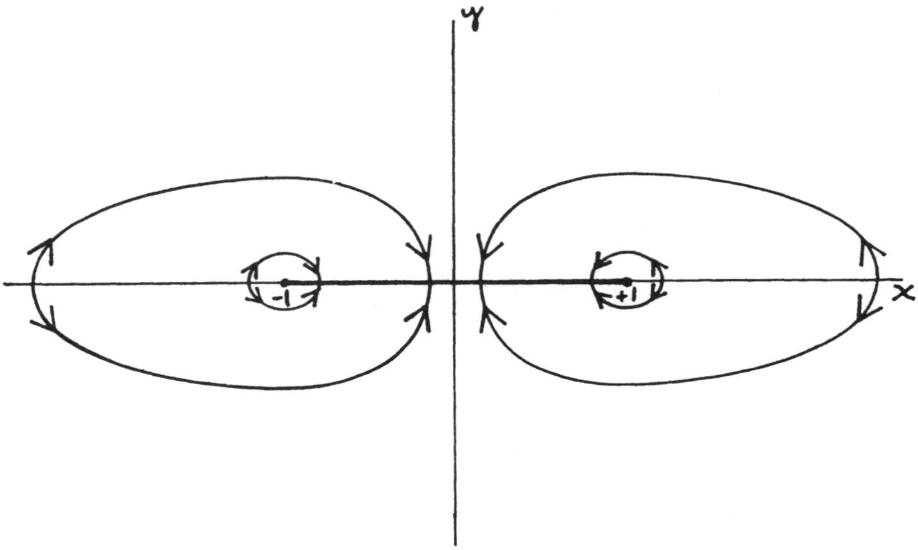

Fig. 4. Lines of force for the 2-dimensional force field, Eq. (7.2) or (7.4). The points of equilibrium form the line $y=0$, which is the analog of the gauge condition $\partial\cdot A=0$. The points of stable equilibrium form the line segment $y=0$, $|x|<1$, which is the analog of the interior of the Gribov horizon.

the solution in closed from. Instead we pose

$$P = \exp[-\alpha S_0 + S_1 + O(\alpha^{-1})]. \tag{7.6}$$

To order α^2 we obtain

$$\nabla S_0 \cdot (\nabla S_0 + \mathbf{K}) = 0. \tag{7.7}$$

It follows that $\nabla S_0=0$ when $\mathbf{K} = 0$. This happens when $y=0$, so we have,

$$(\partial/\partial x)S_0(x, y)\big|_{y=0} \qquad (\partial/\partial y)S_0(x, y)\big|_{y=0}. \tag{7.8}$$

Thus, if we expand S_0 in powers of y we have, apart from an irrelevant additive constant,

$$S_0(x, y) = (1/2)M(x)y^2 + O(y^3). \tag{7.9}$$

Note that as α gets large, we obtain a Gaussian dependence on y. To determine $M(x)$, we return to (7.8), and obtain, to order y^2,

$$M(x)\ [-M(x) + (1-x^2)]$$

and so,

$$M(x) = (1-x^2). \tag{7.10}$$

Thus, for large α the leading terms in the expansion of S in powers of α^{-1} give

$$P(x,\ y) = \sqrt{[\alpha(1-x^2)/(2\pi)]}\ \exp[-(1/2)\alpha(1-x^2)y^2]\ F(x,\ y) \tag{7.11}$$

where an overall normalization has been factored out, and F is independent of α. There is an exploding Gaussian dependence in y for $|x|>1$, so this is not a consistent solution, unless $F(x, y)$ vanishes there. (In fact it is a general feature of diffusion equations that, as the strength of the drift force (i.e. α) becomes infinite, the equilibrium solution, if it exists, concentrates at the points of <u>stable equilibrium</u> of the drift force.) Thus in the present case we are concerned only with a small neighborhood of the interval $y=0$, $|x|<1$. Let $Q(x)$ be the probability distribution in x, if y is not observed,

$$Q(x) = \int dy\ P(x,\ y). \tag{7.12}$$

A diffusion equation for $Q(x)$ is obtained by integrating (7.5) with respect to y, namely,

$$(\partial/\partial x)\ [\partial Q/\partial x\ \ - \alpha\langle K_x\rangle Q] = 0, \tag{7.13}$$

where

$$\langle K_x\rangle = \int dy\ K_x(x,\ y)\ \sqrt{[\alpha(1-x^2)/(2\pi)]}\ \exp[-(1/2)\alpha(1-x^2)y^2]$$

$$= -\gamma x\langle y^2\rangle = -\gamma x/[\alpha(1-x^2)].$$

Note that this force becomes infinite as the "Gribov horizon", $x=\pm 1$, is approached from inside. We have

$$\partial/\partial x\ [\partial Q/\partial x\ \ + \gamma x(1-x^2)Q] = 0,$$

with solution

$$Q(x) = c(1-x^2)^{\gamma/2}.$$

In conclusion, we find that as $\alpha\to\infty$, the equilibrium distribution approaches

$$P(x, y) = c(1-x^2)^{\gamma/2}\ \theta(1-x^2)\ \delta(y). \tag{7.14}$$

Note that as the Gribov horizon, $x=\pm1$, is approached, $Q(x)$ approaches zero smoothly. The reason is that, as a Brownian particle approaches the Gribov horizon, it diffuses further from $y=0$, and the restoring nature of the drift force, Eq. (7.3) comes into effect, as illustrated in Fig. 4.

8. DERIVATION OF MODIFIED FADDEEV-POPOV FORMULA FROM STOCHASTIC QUANTIZAION

As we noted in Section 2, under a gauge transformation $A\to A^g=g^{-1}(Ag+\partial g)$, the Hilbert distance between two connections A^1 and A^2 remains invariant,

$$\|A^1 - A^2\| = \|(A^1)^g - (A^2)^g\|. \tag{8.1}$$

Let the corresponding infinitesimal distance squared between two connections with inifinitesimal difference $\delta A_\mu{}^a(x)$ be designated by

$$ds^2 = \int d^nx \sum [\delta A_\mu{}^a(x)]^2. \tag{8.2}$$

Here δA is one-form in A space. (We do not write it as dA to avoid confusion with $dA=A_\mu dx^\mu$ which is a one-form in x-space.) It is the gauge invariance of this line element which assures that the diffusion equation (5.2) is gauge invariant. In order to exploit the gauge properties of Eq. 5.14), introduce curvilinear coordinates in A-space according to

$$A_\mu = g^{-1}(B_\mu + \partial_\mu)g \tag{8.3a}$$

$$\partial\cdot B = 0, \tag{8.3b}$$

whereby the variable A is replaced by the pair (B, g). (I am grateful to E. Seiler for suggesting the use of these coordinates.) This change of coordinates is well defined globally, with B parametrizing gauge orbits and g parametrizing local gauge transformations, only if we stipulate that B runs over the orbit space C. At the moment, this is not explicitly known. On the other hand, this coordinate transformation is well defined in a sufficiently small neighborhood of $\partial{\cdot}A=0$, that does not include a Gribov horizon. The difficulty with the Faddeev-Popov method is that one needs to know the orbit space C. In this section we will solve the diffusion equation with stochastic gauge-fixing and thereby derive the Faddeev-Popov formula, modified by a specification of the orbit space.

In order to write the diffusion equation in the curvilinear coordinates, we require the metric tensor in this basis. An elementary calculation gives

$$\delta A_\mu = g^{-1}[\delta B_\mu + D_\mu(B)(\delta g g^{-1})]g, \qquad (8.4)$$

where,

$$D_\mu(B)(\delta g g^{-1}) = \partial_\mu(\delta g g^{-1}) + [B_\mu,(\delta g g^{-1})]. \qquad (8.5)$$

Let w(x) be a set of independent coordinates for the elements g(x), of the Lie group G, for example, $g(x)=\exp[\lambda_a w^a(x)]$ where the λ_a are an anti-hermitian basis of the Lie albegra. We write

$$\delta g g^{-1} = R(w)\delta w = R_a(w)\delta w^a, \qquad (8.6)$$

where $R_a(w(x))=\lambda_i R^i{}_a(w(x))$ is an anti-hermitian element of the Lie algebra. In terms of the new variables, the line element (8.2) becomes

$$ds^2 = -\int d^n x \; \mathrm{tr}\{[\delta B + D_\mu(B)(R\delta w)]^2\},$$

$$ds^2 = -\int d^n x \; \mathrm{tr}[\delta B_\mu P \delta B_\mu + 2\delta B_\mu P D_\mu R\delta w + \delta w R^\dagger D_\mu{}^\dagger D_\mu R \delta w], \qquad (8.7)$$

where it is understood that $D=-D^\dagger=D(B)$. Here P is the projector on tranverse vector fields which we are free to insert, since B is transverse, $(PB)_\mu \equiv (1/\partial^2)(\partial^2 B_\mu + \partial_\mu \partial{\cdot}B) = B_\mu$. We identify the metric tensor G in this coordinate system from

$$ds^2 = [\delta B, \delta w]\, G \begin{bmatrix} \delta w \\ \delta B \end{bmatrix}$$

as the block matrix

$$G = \begin{pmatrix} P & PDR \\ R'D'P & R'D'DR \end{pmatrix} \qquad (8.8)$$

Let us calculate the determinant of the metric tensor. We have, to within a multiplicative constant,

$$\det G = \det \begin{pmatrix} P & 0 \\ -R'D' & 1 \end{pmatrix} \begin{pmatrix} P & PDR \\ R'D'P & R'D'DR \end{pmatrix}$$

$$= \det \begin{pmatrix} P & PDR \\ 0 & R'D'(1-P)DR \end{pmatrix}$$

$$= \det \begin{pmatrix} P & PDR \\ 0 & R'D'(1-P)DR \end{pmatrix} \begin{pmatrix} 1 & -PDR \\ 0 & 1 \end{pmatrix}$$

$$= \det \begin{pmatrix} P & 0 \\ 0 & R'D'(1-P)DR \end{pmatrix} ,$$

which gives, to within a multiplicative constant,

$$\det G = \det [R'D'(1-P)DR].$$

Next we note that

$$D^\dagger(1-P)D = -D_\mu \partial_\mu (\partial^2)^{-1} \partial_\nu D_\nu = L(-\partial^2)^{-1} L,$$

where the Faddeev-Popov operator $L=L(B)=D\cdot\partial = D(B)\cdot\partial=L^\dagger$, has made its appearance. Thus we find

$$\det G = \det [R^\dagger L(-\partial^2)^{-1} LR], \qquad (8.9)$$

and so, to within a multiplicative constant,

$$(\det G)^{1/2} = \det^{1/2}(R^\dagger R) \det^{1/2}(L^2). \qquad (8.10)$$

With $L=L(B)$, and $R=R(w)$, we see that the determinant of the metric tensor and its square root factorizes into a factor which depends only on the orbit, and one which depends only on the local gauge transformation. The former is none other than the Faddeev-Popov determinant det L, and the latter, given by,

$$\det^{1/2}(R^\dagger R) = \Pi_x \det^{1/2}(R^\dagger R(x)), \qquad (8.11)$$

is, for each x, the Haar measure of the gauge group.

In order to write the diffusion equation in the new coordinates we also require the inverse of the metric tensor which may be verified to be, (again in block form),

$$G^{-1} = \begin{pmatrix} P + PDL^{-1}(-\partial^2)L^{-1}D^\dagger P & -PDL^{-1}(-\partial^2)L^{-1}R^{\dagger -1} \\ -R^{-1}L^{-1}(-\partial^2)L^{-1}D^\dagger P & R^{-1}L^{-1}(-\partial^2)L^{-1}R^{\dagger -1} \end{pmatrix} \qquad (8.12)$$

The diffusion equation in $u=(B,w)$ coordinates is given by

$$\surd(1/\det G)\ \partial/\partial u^i[\surd(\det G)\ J^i] = 0 \qquad (8.13a)$$

$$J^i = -G^{ij}\ \partial P/\partial u^j + K^j P. \qquad (8.13b)$$

We must find the force K^j as a tangent vector in the new coordinate system. For this purpose, we note

$$\int dy\, K_\nu{}^b(y)\, \delta/\delta A_\nu{}^b(y) = \int dx\, [K_{B\mu}{}^a(x)\, \delta/\delta B_\mu{}^a(x) + K_w{}^a(x)\, \delta/\delta w^a(x)]. \qquad (8.14)$$

A simple way to evaluate this formula for the gauge-fixing force, K_{gf}, is to make use of the fact that K_{gf} is tangent to the gauge orbits. Consider the infinitesimal gauge transformation (1+J)F, where F=F(A) is an arbitrary function of A, and J is the generator of the local gauge group given by

$$J = \epsilon K_{gf\mu}{}^{a}\, \delta/\delta A_{\mu}{}^{a} = [D_{\mu}(A)v]^{a}\delta/\delta A_{\mu}{}^{a} \qquad (8.15a)$$

where

$$v^{b} = \epsilon\partial\cdot A. \qquad (8.15b)$$

We have, to first order in ϵ,

$$(1 + J)F(A) = F(A + D(A)v) = F(A^{(1+v)}) = F(B^{g(1+v)}), \qquad (8.16)$$

where $A^{(1+v)}$ is the gauge transform of A by the infinitisimal gauge transformation 1+v, and we have used the multiplicative property of gauge transformations. Next we note that, with g parametrized by w, under the coordinate transformation from A to (B,w), F transforms according to

$$F(A) = F(B^{g}) = F'(B, w) \qquad (8.17a)$$

and

$$F(B^{g(1+v)}) = F'(B, w+\delta w), \qquad (8.17b)$$

where, by (8.6),

$$R(w)\delta w = \delta g g^{-1} = gvg^{-1},$$

which gives

$$(1 + J)F'(B, w) = F'(B, w+R^{-1}(gvg^{-1})),$$

and so $J = [R^{-1}(gvg^{-1})]\delta/\delta w$, or, by (8.15),

$$K_{gf\mu}{}^{a}\delta/\delta A_{\mu}{}^{a} = [R^{-1}\ (g\partial\cdot Ag^{-1})]^{a}\delta/\delta w^{a}. \qquad (8.21)$$

This gives, by (8.14)

$$K_{gfB\mu}{}^{a} = 0 \qquad (8.22a)$$

$$K_{gfw}{}^{a} = [R^{-1}\,(g\partial\cdot Ag^{-1})]^{a}\,. \qquad (8.22b)$$

Our calculation will be much simplified by the fact that K_{gfB} vanishes. With $A = B^{g} = g^{-1}(B+\partial)g$, one finds easily

$$g\partial\cdot Ag^{-1} = \partial_{\mu}(\partial_{\mu}gg^{-1}) + [B_{\mu}, \partial_{\mu}gg^{-1}] \equiv D(B)_{\mu}(\partial_{\mu}gg^{-1}) = D(B)_{\mu}(R\partial_{\mu}w)$$

and so

$$K_{gfw}{}^{a} = \{R^{-1}\,[D(B)_{\mu}(R\partial_{\mu}w)]\}^{a}. \qquad (8.22c)$$

To calculate the components of $K_{cl}=-\delta S_{cl}/\delta A$ is very simple, because it is the gradient of a gauge-invariant functional. We have

$$S_{cl}(A) = S_{cl}(B^{g}) = S_{cl}(B),$$

so the covariant components of K_{cl} in the (B,w) basis are

$$-(\delta S_{cl}/\delta B,\ \delta S_{cl}/\delta w) = -(\delta S_{cl}/\delta B,\ 0) \qquad (8.23)$$

Thus the contravariant componants of K_{cl} are

$$(K_{clB}, K_{clw}) = -[(G^{-1})_{BB}\,\delta S_{cl}/\delta B,\ (G^{-1})_{wB}\,\delta S_{cl}/\delta B], \qquad (8.24)$$

where G^{-1} is given in block form in (8.12).

We will derive the modified Faddeev-Popov formula from stochastic gauge fixing, initially under the assumption that we know what the orbit space C is. Let $\rho(w)$ designate the Haar measure of the local gauge group,

$$\rho(w) \equiv \det{}^{1/2}(R^{\dagger}R), \qquad (8.25)$$

so

$$\sqrt{\det G} = \rho(w)\det L(B) \qquad (8.26)$$

We pose

$$Q(B) \equiv \int dw\ \rho(w)\ P(B, w). \qquad (8.27)$$

For gauge invariant observables O(A) this is the only relevant probability distribution, for we have $O(A)=O(B^g)=O(B)$ and

$$\langle O\rangle = \int_C dB\ dw\ \rho(w)\ \det L(B)\ O(B)\ P(B, w) \qquad (8.28a)$$

$$\langle O\rangle = \int_C dB\ \det L(B)\ O(B)\ Q(B), \qquad (8.28b)$$

where we integrate B over the as yet undetermined orbit space C.

To obtain an equation for Q, we return to the diffusion equation (8.13) which we write as

$$[1/(\rho\det L)]\ \{\partial/\partial B^i\ [(\rho\det L)\ J_B{}^i] + \partial/\partial w^i\ [(\rho\det L)\ J_w{}^i]\} = 0$$

or

$$(1/\det L)\ \partial/\partial B^i\ (\det L\ J_B{}^i) + (1/\rho)\partial/\partial w^i\ (\rho\ J_w{}^i)\} = 0, \quad (8.29a)$$

where

$$J_B{}^i = G_{BB}{}^{ij}\ \partial P/\partial B^j + G_{Bw}{}^{ij}\ \partial P/\partial w^j - K_B{}^i P \qquad (8.29b)$$

$$J_w{}^i = G_{wB}{}^{ij}\ \partial P/\partial B^j + G_{ww}{}^{ij}\ \partial P/\partial w^j - K_w{}^i P. \qquad (8.29c)$$

We apply $\int dw\rho$ to (8.29a) and obtain

$$(1/\det L)\ \partial/\partial B^i\ (\det L\ J^*{}_B{}^i) = 0, \qquad (8.30a)$$

where

$$J^*{}_B{}^i = \int dw\rho\ J_B{}^i$$

$$J^*{}_B{}^i = G_{BB}{}^{ij}\ (\partial Q/\partial B^j - \partial S_{cl}/\partial B^j Q). \qquad (8.30b)$$

Here we have used the properties of $G^{ij}=(G^{-1})_{ij}$, displayed in (8.12), namely that $G_{BB}{}^{ij}$ is independent of w, and that $\int dw\rho\, G_{Bw}{}^{ij}\,\partial P/\partial w^j$, contains the factor $\int dw\rho R'^{-1}\partial P/\partial w$ which vanishes because of the properties of the Haar measure and the Lie group generator. Note that this system is completely independent of the gauge parameter α. The obvious solution to (8.30) is

$$Q(B) = \exp(-S_{cl}). \tag{8.31}$$

This, together with (8.28b) give us the Faddeev-Popov formula,

$$\langle O\rangle = \int_C dB\, \det L(B)\, O(B)\, \exp(-S_{cl}), \tag{8.32}$$

modified by the specification that B is to be integrated only over the orbit space C.

Among other things, we have verified that the expectation values of gauge invariant objects are independent of the gauge parameter α which appears in

$$K = K_{cl} + \alpha K_{gf}. \tag{8.33}$$

We now use stochastic gauge-fixing to learn something about the orbit space C by solving the equilibrium diffusion equation (8.13) in the limit as α get very large. (At this point, the reader may wish to review the preceding section from (7.5) on.) We pose

$$P = \exp[-\alpha\Gamma_0 - \Gamma_1 + O(\alpha^{-1})], \tag{8.34}$$

and find, to leading order in α,

$$\partial\Gamma_0/\partial u^i(G^{ij}\,\partial\Gamma_0/\partial u^j + K_{gf}{}^i) = 0, \tag{8.35}$$

where $G^{ij}=(G^{-1})_{ij}$. Note that only the gauge-fixing force appears here. With u=(B,w), this reads,

$$\partial\Gamma_0/\partial B^i(G_{BB}{}^{ij}\,\partial\Gamma_0/\partial B^j + G_{Bw}{}^{ij}\,\partial\Gamma_0/\partial w^j)$$
$$+\,\partial\Gamma_0/\partial w^i(G_{wB}{}^{ij}\,\partial\Gamma_0/\partial B^j + G_{ww}{}^{ij}\,\partial\Gamma_0/\partial w^j + K_{gfw}{}^i) = 0, \tag{8.36}$$

where we have used the fact that $K_{gfB}=0$. Because the metric G is a positive matrix, it follows that $\partial\Gamma_0/\partial u=(\partial\Gamma_0/\partial B,\partial\Gamma_0/\partial w)=0$ whenever $K_{gfw}=0$. By (8.22c), this occurs when $w=0$. This gives

$$\partial\Gamma_0(B, 0)/\partial B = \partial\Gamma_0(B, 0)/\partial w = 0. \tag{8.37}$$

Upon expanding Γ_0 in a functional power series in w, we have, apart from an irrelevant additive constant,

$$\Gamma_0(B, w) = (1/2)w^t M(B)w + O(w^3), \tag{8.38}$$

where we have used matrix notation, and M(B) is an operator to be determined. To order w^2, the relevant terms in (8.36) are

$$\partial\Gamma_0/\partial w^i(G_{ww}{}^{ij}\ \partial\Gamma_0/\partial w^j + K_{gfw}{}^i) = 0, \tag{8.39}$$

To lowest order in w, we have, by (8.6), $R=1$, and, by (8.22c), $K_{gfw}=D(B)\cdot\partial w=-Lw$, and, by (8.12),

$$H \equiv (G^{-1})_{ww}|_{(w=0)} = L^{-1}(-\partial^2)L^{-1} \tag{8.40}$$

[recall that $G^{ij} = (G^{-1})_{ij}$], so that, to order w^2, (8.39) reads,

$$w^t M(HMw - Lw) = 0. \tag{8.41}$$

Because w is a generic vector, it follows that the symmetric part of the sandwiched operator vanishes,

$$2MHM - ML - LM = 0. \tag{8.42}$$

With,

$$N(B) \equiv M^{-1}(B), \tag{8.43}$$

this gives

$$LN + NL = 2H. \tag{8.44}$$

This is a linear equation for N. Suppose that B lies outside the first Gribov horizon, or, in other words, that the Faddeev-Popov operator

$L=L(B)$ has a negative eigenvalue λ, with corresponding eigenvector φ, $L(B)\varphi=\lambda\varphi$. Then we have, upon taking the expectation value of the last equation, $\lambda(\varphi,N\varphi)=(\varphi,H\varphi)$. Because H is a metric tensor, the right-hand side is positive, from which it follows that $(\varphi,N\varphi)$ is negative. In this case N, and hence also $M=N^{-1}$, is not a positive operator, and so (8.38) gives a solution which is an exploding Gaussian in some direction (as $\alpha \to \infty$) for the probability distribution $P \sim c\exp[-\alpha w^t M(B) w]$. As this must join smoothly onto a normalized solution, we conclude that c and thus P vanishes outside the first Gribov horizon.

Let B lie inside Ω, i. e. inside the first Gribov horizon, so that $L=L(B)$ is a positive operator. Then (8.44) has the unique solution

$$N(B) = M^{-1}(B) = 2\int_0^\infty dt\ e^{-Lt} H e^{-Lt}, \qquad (8.45)$$

which is a positive operator because L and H are. The equilibrium solution for large α is

$$P(B, w) = \chi_\Omega(B)\exp[-\alpha(1/2)w^t M(B)w - \Gamma_1(B, w)], \qquad (8.46)$$

where $\chi_\Omega(B)=1$ for B in Ω and $\chi_\Omega(B)=0$ otherwise. As $\alpha \to \infty$, the dependence on w approaches $\delta(w)$, because Γ_1 is independent of α. Hence we may rewrite this as

$$P(B, w) = \chi_\Omega(B)\ \ [1/\sqrt{\det M(B)}]\ \ \exp[-\alpha(1/2)w^t M(B)w]\ Q(B). \qquad (8.47)$$

We know by our previous reasoning (apply $\int dw\rho(w)$ to the diffusion equation for P and use $\rho(0)=1$) that Q(B) satisfies (8.30) throughout Ω because the change of coordinates from A to (B,w) is regular in a sufficiently small neighborhood of Ω. The obvious solution to this equation is $Q(B)=c\exp[-S_{cl}(B)]$. It would indeed be strange if there were another normalizable positive solution which is regular throughout Ω. We thus conclude that for large α, the solution to the equilibrium diffusion equation with stochastic gauge fixing is

$$P(B, w) = c\chi_\Omega(B)\ \ [1/\sqrt{\det M(B)}]\ \ \exp[-\alpha(1/2)w^t M(B)w]\ \exp[-S_{cl}(B)], \quad (8.48)$$

or

$$P(B, w) = c\ \delta(w)\ \chi_\Omega(B)\ \exp[-S_{cl}(B)]. \qquad (8.49)$$

By our previous reasoning leading to Eq. (8.28), this leads to the modified Faddeev-Popov formula, for the expectation value of gauge invariant observables O(A),

$$\langle O \rangle = \int_{\Omega} dB \, \det L(B) \, O(B) \exp[-S_{cl}(B)]. \qquad (8.50)$$

The modification consists in the specification that the orbit space is Ω. This is the interior of the first Gribov horizon which, as we recall from Section 2, is convex and bounded in every direction.

9. CONCLUSION

We have not discussed numerical simulation of the underlying stochastic process, for which we refer the reader to the original literature[9] and the lectures of E. Seiler[11]. (See however Fig. 3.) It is done by replacing the continuum of Euclidean space-time by a 4-dimensional hypercubic lattice, the variable A is reduced to a link variable, and the force K, also a link variable, is expressed as a cubic polynomial which is a lattice analog or our continuum force.

Another topic which has not been discussed is a suitable non-perturbative regularization of the continuum theory and its gauge properties, without which the non-perturbative superiority of stochastic quantization over the perturbatively correct Faddeev-Popov method is rather empty. The lattice discretization introduced in ref. 9 (SSZ, Section2) for the purposes of numerical simulation is, in fact, such a regularization, but its gauge properties have not been explored. Whether gauge invariance is assured in the continuum limit is a topic which is left for another occasion. An interesting feature of this discretization is that under it Ω remains convex and bounded in every direction (ref. 9, SSZ). Moreover, if the lattice has a finite volume, so A-space becomes finite dimensional, then Ω is, in fact, compact. This "self-compactifying" property of the A-variables which are *a priori* unbounded, allows a strong-coupling expansion, as in Wilson's theory. G. Dell'Antonio has recently kindly informed me[22] that he has proven the existence of the equilibrium probability distribution for this lattice discretization, which is a very encrouraging result for this approach.

Our general conclusion is that stochastic quantization with stochastic gauge-fixing appears to provide a consistent approach to quantization of gauge fields that overcomes the Gribov ambiguity. In particular, in Section 8, we derived a modified Faddeev-Popov formula that is consistent with what we learned about gauge orbits in Section 2. In the Appendix we found a unique expansion of the effective action, in powers of $\hbar$ and A and seen how Gribov copies are suppressed at order $\hbar^0$.

ACKNOWLEDGMENTS

I am grateful to the organizers of the Summer School, Professor Velo and Professor Wightman, for the opportunity to review the subject in a more extended form than is possible in a research article. I am also grateful to many of the participants in the Summer School for stimulating discussions which led me to correct some, if not all, of my errors. I would like to express my appreciation for the hospitality of the Max-Planck Institut für Physik und Astrophysik, Munich, where these lectures were prepared, and for valuable discussions there with D. Maison and E. Seiler.

APPENDIX: LOOP OR $\hbar$ EXPANSION OF THE EFFECTIVE ACTION

In this Appendix, we shall derive the $\hbar$ expansion and show that it is a loop expansion. We will prove a theorem which states that all primitive divergences arise in the iterative calculation of K_{eff}, the effective force. This suggests that the primitive divergences can be absorbed in a redifinition of the drift force K, a local quantity. At the end of this Appendix we show how Gribov copies are suppressed at order $\hbar^0$.

The generating functional of Green's functions is defined by

$$G(J) \equiv \int dA\, e^{J \cdot A/\hbar}\, P(A). \qquad \text{(A.1)}$$

Here, $\hbar$ (which corresponds to the diffusion constant in physical diffusion problems) is a parameter which, as we shall see, serves to count closed loops. In a theory with Euclidean action S(A), the probability distribution is given by

$$P(A) = e^{-S(A)/\hbar} \qquad (A.2a)$$

However, in the case of stochastic gauge-fixing of gauge fields, the drift force is non-conservative and the explicit form of P is not known. Instead the equilbrium probability distribution P is determined by the time-independent diffusion equation (5.14),

$$\{\partial/\partial A_i[\hbar\partial/\partial A_i - K_i(A)]\} P(A) = 0. \qquad (A.2b)$$

In the first case, G(J) satisfies the Schwinger-Dyson equation

$$[J - \nabla S(\hbar\partial/\partial J)]G(J) = 0. \qquad (A.3a)$$

In the second case we obtain a substitute for the Schwinger-Dyson equation by applying $\int dA e^{J\cdot A/\hbar}$ to (A.2b) which gives

$$J\cdot[J + K(\hbar\partial/\partial J)]G(J) = 0. \qquad (A.3b)$$

These equations are supplemented by the normalization condition G(0)=1. The generating functional of connected Green's functions W(J) is defined by

$$G(J) = \exp[W(J)/\hbar]. \qquad (A.4)$$

It satisfies[23]

$$J - \nabla S(\partial W/\partial J + \hbar\partial/\partial J)1 = 0 \qquad (A.5a)$$

or

$$J\cdot[J + K(\partial W/\partial J + \hbar\partial/\partial J)1] = 0, \qquad (A.5b)$$

with normalization condition W(0) = 0.

Let $\langle F\rangle_J$ designate expectation values with (unnormalized) probability distribution $\exp(J\cdot A/\hbar)P$, so that

$$\langle F\rangle_J=[F(\hbar\partial/\partial J)G(J)]/G(J). \qquad (A.6)$$

Observe that

$$\partial^2 W(J)/\partial J_i \partial J_k = \hbar[\partial^2 G(J)/\partial J_i \partial J_k]/G(J) - \hbar[\partial G(J)/\partial J_i\ \partial G(J)/\partial J_k]/G(J)^2$$

so

$$\hbar \partial^2 W(J)/\partial J_i \partial J_k = \langle A_i A_k \rangle_J - \langle A_i \rangle_J \langle A_k \rangle_J$$

is the connected two-point function. This quantity is a positive matrix because, upon contraction with $f_i f_k$, it equals the fluctuation $\langle f^2 \rangle_J - \langle f \rangle_J^2$ of $f \equiv f_i A_i$, which is a positive quantity. This is equivalent to the statement that $W(J)$ is a convex function. Let A_{cl} be defined by

$$A_{cli} = \partial W(J)/\partial J_i \tag{A.7}$$

It follows from the positivity of the matrix of second derivatives and the implicit function theorem that this equation is invertible, and we have,

$$J_i = J_i(A_{cl}). \tag{A.8}$$

The effective action Γ is defined by the Legendre transform

$$\Gamma(A_{cl}) = J_i A_{cli} - W(J), \tag{A.9}$$

where J is given in (A.8). As usual, we have

$$J_i(A_{cl}) = \partial \Gamma(A_{cl})/\partial A_{cli}. \tag{A.10}$$

In the remainder of this Appendix we omit the subscript cl *on* A. If these expressions are substituted into (A.5) one obtains corresponding equations for Γ,

$$\nabla\Gamma(A) - \nabla S(A + \hbar \mathcal{D}(A)\partial/\partial A)1 = 0 \tag{A.11a}$$

or

$$\nabla\Gamma(A) \cdot [\nabla\Gamma(A) + K(A + \hbar \mathcal{D}(A)\partial/\partial A)1] = 0, \tag{A.11b}$$

with $\Gamma(0)=0$. Here $\mathcal{D}(A)\partial/\partial A$ has components $\mathcal{D}(A)_{ik}\partial/\partial A_k$, where

$$\mathcal{D}(A)_{ik} = \partial A_k/\partial J_i = \partial^2 W/\partial J_i \partial J_k \quad \text{(A.12)}$$

so $\hbar \mathcal{D}$ is the connected gluon propagator. It may be expressed in terms of Γ by means of

$$\mathcal{D}(A)_{ik} = (\partial J/\partial A)^{-1}{}_{ik} = (\partial^2\Gamma/\partial A \partial A)^{-1}{}_{ik}. \quad \text{(A.13)}$$

The vacuum corresponding to $J=0$ is found from $\partial\Gamma/\partial A=0$.

Equations (A.11) and (A.13) provide the basis for an expansion of the effective action in powers of $\hbar$. The explict form of K is given in (5.11b), namely

$$K_\mu{}^a(x) = (D_\lambda F_{\lambda\mu})^a(x) + \alpha(D_\mu \partial\cdot A)^a(x). \quad \text{(A.14a)}$$

This is a cubic expresssion in A, and we write it in component notation as

$$K_i(A) = -Q_{ij}A_j + (1/2!)R_{ijk}A_jA_k + (1/3!)T_{ijkl}A_jA_kA_l \quad \text{(A.14b)}$$

The explicit minus sign in front of Q has been inserted so that Q_{ij} is a positive matrix. (It is the matrix corresponding to the operator $Q_{\lambda\mu}=-[(\partial^2\delta_{\lambda\mu}-\partial_\lambda\partial_\mu)+\alpha\partial_\lambda\partial_\mu].$) The important point here is that the coefficients Q, R, T are local, so that K(x) only depends on A(x) and its first and second derivatives, at the same x. We define

$$K_{eff}(A) \equiv K(A + \hbar\mathcal{D}(A)\partial/\partial A)1, \quad \text{(A.15)}$$

which gives

$$K_{eff\,i}(A) = K_i(A) + \hbar(1/2!)[R_{ijk}\mathcal{D}(A)_{jk} + T_{ijk\ell}A_j\mathcal{D}(A)_{k\ell}] - \hbar^2(1/3!)T_{ijk\ell}\mathcal{D}_{jr}\mathcal{D}_{ks}\mathcal{D}_{\ell t}\,\partial^3\Gamma/\partial A_r\partial A_s\partial A_t, \quad \text{(A.16)}$$

where we have used (A.13) and

$$(\partial/\partial A_r)\mathcal{D}_{k\ell} = -\mathcal{D}_{ks}\,\partial^3\Gamma/\partial A_r\partial A_s\partial A_t\,\mathcal{D}_{t\ell}.$$

The important point is that when expression (A.16) is substituted into Eq. (A.11) for Γ, the only explicit powers of $\hbar$ which appear are in K_{eff}. A

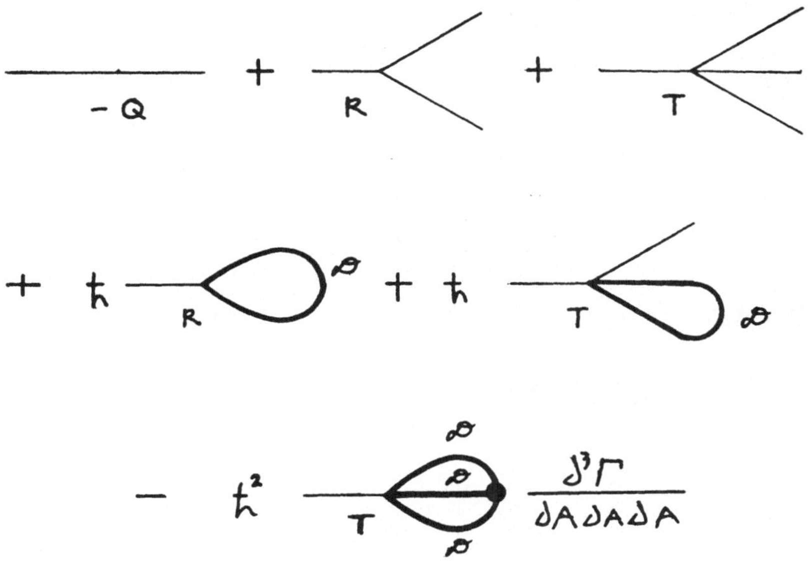

Fig. 5. Graphical representation of the effective force K_{eff}, Eq. (A.16). Here D represents the full gluon propagator.

graphical representation of K_{eff} is shown in Fig.5. Note that each closed loop which appears is multiplied by a power of $\hbar$. We shall see that the only closed loops which appear when Γ is calculated as a power series in $\hbar$ are generated by these insertions of closed loops into K_{eff}, so that the $\hbar$ expansion is a loop expansion, as in an action formalism. Simple power counting shows that primitive divergences are associated with these closed loops. We shall see that no other divergences arise when Γ is expressed as a power series in $\hbar$. We remark also that the expresssion for Green's functions in terms of the derivatives of Γ depends only on the form of the Legendre transformation, so it is the same in the present approach and in an action formalism, namely, by (A.6),

$$\langle F \rangle = F(\partial W/\partial J + \hbar\partial/\partial J)1,$$

evaluated at J=0, or

$$\langle F \rangle = F(A + \hbar D\partial/\partial A)1, \qquad \text{(A.17)}$$

evaluated at $\partial\Gamma/\partial A_i = 0$.

We pose

$$\Gamma = \Gamma_0 + \hbar\Gamma_1 + \hbar^2\Gamma_2 + \dots \tag{A.18}$$

This induces a corresponding $\hbar$ expansion for $\mathcal{D}$ by (A.13), namely,

$$\mathcal{D} = \mathcal{D}_0 + \hbar\mathcal{D}_1 + \hbar^2\mathcal{D}_2 + \dots \tag{A.19}$$

and hence also for K_{eff}, by (A.16),

$$K_{eff} = K_0 + \hbar K_1 + \hbar^2 K_2 + \dots \tag{A.20}$$

where

$$K_0 = K. \tag{A.21}$$

The equations for Γ_n follow from (A.11), namely,

$$\partial\Gamma_0/\partial A_i[\partial\Gamma_0/\partial A_i + K_i] = 0. \tag{A.22a}$$

$$(2\partial\Gamma_0/\partial A_i + K_i)\,\partial\Gamma_1/\partial A_i = -K_{1i}\,\partial\Gamma_0/\partial A_i \tag{A.22b}$$

and, in general, for $n>1$,

$$(2\partial\Gamma_0/\partial A_i + K_i)\partial\Gamma_n/\partial A_i = -K_{ni}\,\partial\Gamma_0/\partial A_i - \Sigma_{r=1}{}^{n-1}\,\partial\Gamma_{n-r}/\partial A_i[\partial\Gamma_r/\partial A_i + K_{ri}]. \tag{A.22c}$$

Note that by the $\hbar$ expansion the second order functional differential equation for Γ has been converted into a series of first order functional differential equations for the Γ_n. Only the equation for Γ_0 is non-linear. It is the time-independent Hamilton-Jacobi equation corresponding to the Hamiltonian

$$H(J, A) = J_i[J_i + K_i(A)],$$

where J_i and A_i are canonically conjugate. The remaining equations are linear.

[Suppose the Hamilton-Jacobi eqution (A22.a) were solved by the method of characteristics, so that one had in hand the solutions to the canonical system

$$dJ_i/dt = -\partial H/\partial A_i$$

$$dA_i/dt = \partial H/\partial J_i = 2J_i + K_i,$$

with boundary conditions $A_i(-\infty)=0$, and $A_i(0)=A_i$. Then Γ_0 would be given by

$$\Gamma_0(A) = \int_{-\infty}^{0} L\, dt,$$

where L is the lagrangian corresponding to H, and the integral extends along the trajectory. Moreover with $J_i=\partial\Gamma_0/\partial A_i$ along the trajectory, we see that $dA_i/dt=2\partial\Gamma_0/\partial A_i+K_i$ is satisfied along the trajectory. This reduces the remaining equations (A.22) to quadratures for they are seen to be of the form, for n>0,

$$dA_i/dt\ \partial\Gamma_n/\partial A_i = S_n,$$

with solution

$$\Gamma_n = \int_{-\infty}^{0} S_n(A(t))\, dt,$$

where S_n is known in terms of the Γ_m for m<n.]

We shall find the unique solution to the system of equations (A.22), under the assumption that the Γ_n may be expanded in a power series in A. Consider first the equation for Γ_0. Note that $K_i(0)=0$, by (A.14b), so that $\partial\Gamma_0/\partial A_i=0$ at A=0, by (A.22a). Recall also that $\Gamma(0)=0$, for all $\hbar$ so $\Gamma_n(0)=0$ for all n., Thus the leading term in the expansion of Γ_0 in powers of A is quadratic in A. We write

$$\Gamma_0 = \Gamma_0^2 + \Gamma_0^3 + \ldots, \qquad \text{(A.23a)}$$

$$K_i = K_i^1 + K_i^2 + \ldots, \qquad \text{(A.23b)}$$

where Γ_0^n and K_i^n are homogeneous in A of degree n. Then (A.22a) gives

$$\partial\Gamma_0^2/\partial A_i[\partial\Gamma_0^2/\partial A_i + K^1_i] = 0. \qquad \text{(A.24a)}$$

$$(2\partial\Gamma_0^2/\partial A_i + K^1_i)\,\partial\Gamma_0^3/\partial A_i = -K^2_i\,\partial\Gamma_0^2/\partial A_i \qquad \text{(A.24b)}$$

and for n>3,

$$(2\partial\Gamma_0^2/\partial A_i + K^1_i)\,\partial\Gamma_0^n/\partial A_i = -K^{n-1}{}_i\,\partial\Gamma_0^2/\partial A_i$$
$$- \sum_{r=2}^{n-2} \partial\Gamma_0^{n-r+1}/\partial A_i[\partial\Gamma^{r+1}/\partial A_i + K^r{}_i]. \qquad \text{(A.24c)}$$

Only equation (A.24a) for Γ_0^2 is non-linear, and the remaining equations constitute a set of linear recursion relations for the Γ_0^n, n>2. From (A.14b) we have $K^1{}_i = -Q_{ij}A_j$, where Q is a positive symmetric matrix. We also write

$$\Gamma_0^2 = (1/2)A_i M_{ij} A_j, \qquad \text{(A.25)}$$

where M is a symmetric matrix which is also positive because of the convexity of Γ. Then (A.24a) reads

$$M_{ij}A_j(M_{ik}A_k - Q_{ik}A_k).$$

This gives for the symmetrized coefficients of A_jA_k the matrix equation for M

$$QM + MQ = 2M^2. \qquad \text{(A.26)}$$

To solve this, choose a basis where M is diagonal, $M_{ij} = \mu_i\delta_{ij}$, which gives,

$$(\mu_i + \mu_j)Q_{ij} = 2\mu_i^2\delta_{ij}.$$

Thus Q is diagonal in this basis, $Q_{ij} = \lambda_i\delta_{ij}$, and we have

$$\mu_i \lambda_i = \mu_i^2, \qquad \text{(no summation)}$$

with solution $\mu_i=0$ or $\mu_i=\lambda_i$. In the former case the matrix $M_{ij}=\partial^2\Gamma/\partial A_i \partial A_j$ is not invertible. This solution must be rejected because the inverse is precisely the gluon propagator at $A=0$ and $\hbar=0$. Thus we must choose the solution $\mu_i=\lambda_i$, that is, $M=Q$, and we have

$$\Gamma_0^2 = (1/2)A_i Q_{ij} A_j. \qquad \text{(A.27)}$$

The remaining equations (A.24) are of the form

$$A_j Q_{ji}\, \partial\Gamma_0^n/\partial A_i = R^n, \qquad \text{(A.28)}$$

where R^n is expressed in term of Γ_0^m for $m<n$. With

$$\Gamma_0^n = (1/n!)\, \Gamma_{0 i_1 i_2 \ldots}\, A_{i_1} A_{i_2} \ldots$$

$$R^n = (1/n!)\, R_{i_1 i_2 \ldots}\, A_{i_1} A_{i_2} \ldots,$$

(A.28) reads, in the basis where Q is diagonal

$$[1/(n-1)!]\;\; \lambda_{i_1} \Gamma_{0 i_1 i_2 \ldots}\, A_{i_1} A_{i_2} \ldots = (1/n!)\, R_{i_1 i_2 \ldots}\, A_{i_1} A_{i_2} \ldots$$

This gives, after symmetrization of the coefficient on the left, the unique solution

$$\Gamma_{0 i_1 i_2 \ldots} = R_{i_1 i_2 \ldots} / (\lambda_{i_1} + \lambda_{i_2} \ldots). \qquad \text{(A.29)}$$

The denominator is a finite sum of positive terms and cannot vanish. Thus Γ_0 has been determined uniquely as a power series in A. [As a parenthetic remark, we note that if K_i is conservative, $K_i=-\partial S/\partial A_i$, then a solution is $\Gamma_0=S$. However, as our iterative solution is unique (with $K_i(0)=0$), it necessarily coincides with S in this case.]

To find the remaining Γ_n, observe that the remaining equations (A.22) are of the form

$$(2\partial\Gamma_0/\partial A_i+K_i)\partial\Gamma_n/\partial A_i = S_n, \qquad (A.30)$$

where Γ_0 has just been found, and S_n is expressed in terms of the Γ_m for $m<n$, and is assumed to be known. Expand Γ_0, K_i, Γ_n, and S_n in powers of A, as before. A simple induction argument shows that the leading terms in the expansion of Γ_n and S_n are quadratic in A. We find

$$A_j Q_{ji}\, \partial\Gamma_n{}^2/\partial A_i = S_n{}^2, \qquad (A.31a)$$

and for $r>2$

$$A_j Q_{ji}\, \partial\Gamma_n{}^r/\partial A_i = -\Sigma_{s=1}{}^{t-2}(2\partial\Gamma_0{}^{r-s+1}/\partial A_i+K^{r-s}{}_i)\partial\Gamma_n/\partial A_i + S_n{}^r. \qquad (A.31b)$$

The solution to these equations is given by an expression similar to (A.29). Thus all the Γ_n have been uniquely determined as power series in A.

Let us now consider where divergences may arise when Γ is calculated in this way. We have observed that K_{eff} contains closed loops in its graphical representation, Fig. (5), so that divergences occur in the calculation of K_{eff}. Suppose that they have been removed by a renormalization procedure, to some order. No new divergences appear in the expressions $\Sigma_i\partial\Gamma_n{}^m/\partial A_i\ K_r{}^s{}_i$ or in $\Sigma_i\partial\Gamma_n{}^m/\partial A_i\ \partial\Gamma_r{}^s/\partial A_i$, which appear in the recusion relations for the $\Gamma_n{}^m$, as may be verified by writing them out in momentum space. Moreover we have solved all the equations for the $\Gamma_n{}^m$, and in all cases found a simple expression of the type (A.29). It is not divergent. We have established the

Theorem: The only divergences which arise in calculating Green's functions by the above iterative procedure appear in the closed loops associated with calculating K_{eff} in each new order.

Although we have given a simple way of calculating the Γ_n as power series in A, the reader may wonder why these expressions are relatively more complicated than in the Faddeev-Popov method. For example, in the latter method one has simply $(\Gamma_0)_{FP}=S_{cl}+(1/2)\int d^n x\, (\partial\cdot A)^2$, which is the integral of a local polynomial in A. (There is an additional ghost term, but if one is only interested in Green's functions of the A the ghost term is irrelevant.) In our case, Γ_0 is an infinite series in A, with non-local coefficients. However, consider the matter of Gribov copies. This is a question of purely classical gauge fields, having to do with the classical gauge orbits, and which should be settled to zeroth order in $\hbar$. Let A_1 be a Gribov copy of $A_0=0$, so $\partial \cdot A_1=0$, and $S_{cl}(A_1)=0$, and we have $(\Gamma_0(A_1))_{FP}=0$. Thus, to zeroth order in $\hbar$, the Faddeev-Popov effective action has two absolute minima at A_0 and A_1. In our method, this will not be the case. One may see the mechanism which insures that $\Gamma_0(A_1)>\Gamma(A_0)$. Let A_1 lie outside the Gribov horizon. From our general discussion in Section (6), we know that under the flow,

$$(d/dt)A = \alpha K_{gf}, \tag{A.32}$$

there is an unstable direction leading away from A_1, and that if we follow the flow line in this direction it will end at a Gribov copy of A_1, which we suppose to be $A_0=0$. As this flow line lies along the gauge orbit passing through zero, we know that $K_{cl}=D\cdot F$ vanishes along the entire flow line C from A_1 to A_0. Thus along C, the equation

$$(d/dt)A = K(A) = K_{cl} + K_{gf}, \tag{A.33}$$

is satisfied. Consider now the variation of Γ_0 along this flow line. We have

$$(d/dt)\Gamma_0(A(t)) = K_i\, \partial\Gamma_0(A(t))/\partial A_i = -[\partial\Gamma_0(A(t))/\partial A_i]^2 < 0,$$

by (A.22a). Thus, Γ_0 decreases monotonically along the flow line from A_1 to A_0.

REFERENCES

1. G. Parisi and Y.-S. Wu, Sci. Sinica **24** (1981) 483.
2. G. Jona-Lasinio and P. K. Mitter, "On the Stochastic Quantization of Field Theory", Comm. Math. Phys. (to be published);
   C. R. Doering, U. of Texas preprints;

J. R. Klauder, "Stochastic Quantization" in Proceedings of the 1983 Schladming School, Acta Physica Auistrica XXV, Suppl. Springer Verlag (1983) 251.

3. J. R. Klauder, "Stochastic Quantization: Application to Field Theory and General Relativity", preprint, Bell Labs., Murray Hill, N. J.
4. H. Hüffel and P. V. Landshoff, "Stochastic Diagrams and Feynman Diagrams", preprint, CERN-TH. 4120/85
5. J. D. Breit, S. Gupta, and A. Zaks, Nucl. Phys. **B233** (1984) 61;
   J. Alfaro, "Stochastic Analytic Regularization", preprint, Trieste, IC/84/92.
6. The approach of Ref. 5 has been criticized, Z. Bern and M. B. Halpern, preprint, LBL-19714, UCB-PTH-85/24 and an alternative stochastic regularization has been proposed, Z. Bern, M. B. Halpern, L. Sadun, and C. Taubes, LBL-19900, UCB-PTH-85/29.
7. C. M. Bender and F. Cooper, Nucl. Phys **B219** (1983) 61;
   E. Gozzi, Phys. Rev. D **28** (1983) 1922.
8. M. Namiki et al, Prog. Theor. Phys **69** (1983) 1580;
   E. Gozzi, Phys. Rev. D **31** (1985) 1349.
9. E. Seiler, I. O. Stamatescu and D. Zwanziger, Nucl. Phys. **B239** (1984) 177 and 204;
   D. Nesic, "Numerical Study of three-dimensional QCD with Stochastic Gauge Fixing", thesis, New York University (1985).
10. D. Zwanziger, Nucl. Phys. **B192** (1981) 259.
11. E. Seiler, "Stochastic quantization and Gauge Fixing in Gauge Theories", in Proceedings of the 1984 Schladming School, Acta Physica Austriaca, Suppl. XXVI, (1984) 259. We refer the reader to this review for a discussion of numerical simulation of the underlying stochastic process.
12. V. N. Gribov, Nucl. Phys. **B139** (1978) 1.
13. L. D. Faddeev and V. N. Popov, Phys. Lett. **25B** (1967) 29.
14. K. G. Wilson, Phys. Rev. **D10** (1974) 2445.
15. I am grateful to D. Maison and E. Seiler for discussions of this point.
16. D. Zwanziger, Nucl. Phys. **B209** (1982) 336. The non-trivial bundle is also discussed here.
17. I. M. Singer, Comm. Math. Phys. **60** (1978) 7.
18. R. E. Cutkosky "The Gribov Horizon", preprint, Carnegie Mellon University CMU-HEP84-1, has argued that every gauge orbit passes through Ω only once.
19. P. Hirschfeld, Nucl. Phys. **B157** (1979) 37.
20. As reported in Ref. 11.
21. Private communication.

22. An important result, obtained by G. Dell'Antonio (private communication) after the completion of these lectures, is the contruction of the Liapunov function for the case of a lattice discretization with finite volume, thereby proving the existence of the equilibrium distribution for this case, and the correctness of the "instability stabilizes" scenario. He has also proved the existence of the equilibrium distribution in the limit $\alpha \to \infty$.

23. Equations for the connected Green's function W are given in E. Floratos, J. Iliopoulos, D. Zwanziger, Nucl. Phys. **B241** (1984) 221, and A. Thomas, "Stochastic Quantization -- the Classical Limit", preprint, Plymouth Polytechnic.

SEMINARS WITH REFERENCES FOR FURTHER INFORMATION

L. Alvarez-Gaume, Two-Dimensional Field Theories and String Compactification
Proceedings of the Argonne Meeting on Geometry, Anomalies and Topology, to appear.

L. Baulien, Cohomological Nature of Gauge Symmetries
L. Baulien in preparation and IPTENS 85/7 Gauge Symmetries in Curved Space, submitted to Nucl. Phys. B.

C. Hagen, A Gauge Theory in 2 + 1 Space
Ann. of Phys. (NY) 157 (1984) 342
Phys. Rev. D31 (1985) 848.

J. Hoppe, Quantum Theory of a Relativistic Surface
Ph.D. Thesis MIT (1982) (Adviser: Jeffrey Goldstone).

G. Morchio, Symmetries of Systems with Long Range Interactions
G. Morchio and F. Strocchi, Commun. in Math. Phys. 99 (1985) 153-175.

A. Soffer, Scattering Theory for N Particles
I.M. Sigal and A. Soffer, N-Particle Scattering Problem: Asymptotic Completeness for Short-Range Systems, preprint 1985.

G. Velo, The Cauchy Problem for Classical Yang-Mills Equations
D. Eardley and V. Moncrief, Commun. Math. Phys. 83 (1982) 193-212.
J. Ginibre and G. Velo, Ann. Inst. H. Poincare 36 (1982) 59-78.

G. Morchio and F. Strocchi, Spontaneous Symmetry Breaking and Energy Gap Generated by Variables at Infinity
Commun. Math. Phys. 99 (1985) 153-175.

INDEX